ArcGIS 9

ArcGIS® Server Administrator and Developer Guide

Published by
ESRI
380 New York Street
Redlands, California 92373-8100

Contributing Writers

Eric Bader, Euan Cameron, Chris Davies, Shelly Gill, Sean Jones, Andy MacDonald, Glenn Meister, Mike Minami, Dan O'Neill, Anne Reuland, Rohit Singh, Steve Van Esch, Zhiqian Yu

Contents

Introducing
ArcGIS Server

ArcGIS® Server is a platform for building enterprise geographic information system (GIS) applications that are centrally managed, support multiple users, include advanced GIS functionality, and are built using industry standards. ArcGIS Server manages geographic resources, such as maps, locators, and software objects, for use by applications.

This chapter will introduce you to ArcGIS Server, how you use it, and its different components. Topics covered in this chapter include:

• an overview of ArcGIS 9 development • the ArcGIS Server product • what you can do with ArcGIS Server • the ArcGIS Server developer kits • a description of this book

WHO SHOULD READ THIS BOOK?

This book will be of greatest use to programmers who want to use ESRI® ArcGIS Server to build server applications, such as Web services and Web applications, that do simple mapping or that include advanced GIS functionality. However, this book provides a general explanation of the use of ArcGIS Server and the possibilities when building and deploying custom applications and solutions. Several scenarios will illustrate, with code examples, some of the different types of applications that can be developed with the ArcGIS Server and the Application Developer Framework (ADF) developer kits.

This book will also be of use to ArcGIS Server administrators who need to administrate aspects of an ArcGIS Server, such as its set of server objects, its output directories, and so on.

ArcGIS 9 OVERVIEW

ArcGIS 9 is an integrated family of GIS software products for building a complete GIS. It is based on a common library of shared GIS software components called ArcObjects™. ArcGIS 9 consists of four key parts:

- ArcGIS Desktop—an integrated suite of advanced GIS applications.

- ArcGIS Engine—embeddable GIS component libraries for building custom applications using multiple application programming interfaces (APIs).

- ArcGIS Server—a platform for building server-side GIS applications in enterprise and Web computing frameworks. Used for building both Web services and Web applications.

- ArcIMS®—GIS Web server to publish maps, data, and metadata through open Internet protocols.

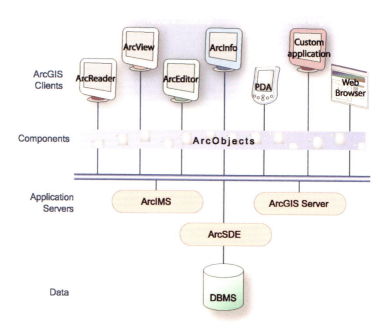

Each of the GIS frameworks also includes the ArcSDE® gateway, an interface for managing geodatabases in numerous relational database management systems (RDBMSs).

ArcGIS 9 extends the system with major new capabilities in the areas of geoprocessing, 3D visualization, and developer tools. Two new products, ArcGIS Engine and ArcGIS Server, are introduced at this release, making ArcGIS a complete system for application and server development.

There is wide range of possibilities when developing with ArcGIS. Developers can:

* Configure/Customize ArcGIS applications such as ArcMap™ and ArcCatalog™
* Extend the ArcGIS architecture and data model
* Embed maps and GIS functionality in other applications with ArcGIS Engine
* Build and deploy custom desktop applications with ArcGIS Engine
* Build Web services and applications with ArcGIS Server

ArcGIS 9 has a common developer experience across all ArcGIS products (Engine, Server, and Desktop). This book focuses on building and deploying server applications using the ArcGIS Server. Developers wishing to customize the ArcGIS Desktop applications or work with the ArcGIS Engine should refer to the *ArcGIS Desktop Developer Guide* and *ArcGIS Engine Developer Guide*.

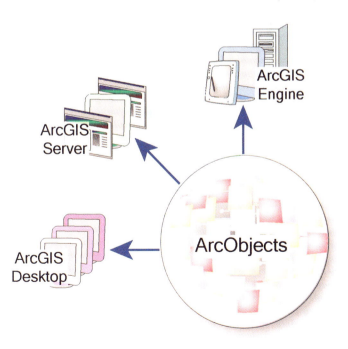

The ArcGIS system is built and extended using software components called ArcObjects. ArcObjects components are software objects that have multiple developer APIs. These include Component Object Model (COM), .NET, Java™, and C++. Developers can use these APIs to build applications that make use of ArcObjects functionality. ArcObjects is at the core of all the ArcGIS products: ArcGIS Desktop, ArcGIS Engine, and ArcGIS Server.

The ArcGIS Server provides a new set of deployment options and resources for developers as well as new and improved tools for developers to work with. ArcGIS Server is a set of the core ArcObjects and a framework for running ArcObjects in a server. The ArcGIS Server ADF is a set of components and Web controls that allow developers to build and deploy Web services and Web applications that make use of ArcObjects running within a server.

ArcGIS Server is a platform for building enterprise GIS applications that are centrally managed, support multiple users, include advanced GIS functionality and are built using industry standards. ArcGIS Server manages geographic resources, such as maps, locators, and software objects, for use by applications.

Developers can use ArcGIS Server to build Web applications, Web services, and other enterprise applications, such as those based on Enterprise JavaBeans™ (EJBs). Developers can use ArcGIS Server to build desktop applications that interact with the server in client/server mode. ArcGIS Server also supports out-of-the-box use by ArcGIS Desktop applications for server administration, simple mapping, and geocoding over a local area network (LAN) or over the Internet.

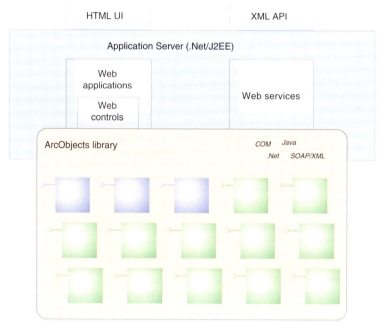

The ArcGIS Server consists of two components: a GIS server and an ADF for .NET and Java. The GIS server hosts ArcObjects for use by Web and desktop applications. It includes the core ArcObjects library and provides a scalable environment for running ArcObjects in the server. The ADF allows you to build and deploy .NET or Java desktop and Web applications that use ArcObjects running within the GIS server.

The ADF includes a software developer kit (SDK) with software objects, Web controls, Web application templates, developer help, and code samples. It also includes a Web application runtime, which allows you to deploy Web applications without having to install ArcObjects on your Web server.

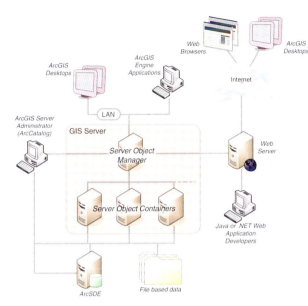

The ArcGIS Server is a distributed system that provides ArcObjects functionality to both Web applications and client/server desktop applications.

KEY FEATURES OF ArcGIS SERVER

Standard GIS framework

ArcGIS Server provides a standard framework for developing GIS server applications. The world's most popular GIS software (ArcView®, ArcEditor™, and ArcInfo™) is built from this same set of software objects. ArcGIS Server is both robust and extensible, and its rich functionality allows developers to concentrate on solving organizational problems, not building GIS functionality from scratch.

Cost-effective deployment

ArcGIS Server supports enterprise applications, such as Web applications, running on servers, and supporting many users. The ADF runtime is not licensed. This allows multiple server applications to run on multiple Web servers, incurring the cost of licensing the GIS server to support the number of users of those applications.

Web controls

ArcGIS Server provides a set of Web controls. These Web controls simplify the programming model for including mapping functionality in your Web application, and allow developers to focus on more advanced GIS functionality aspects of their applications.

Web application templates

ArcGIS Server provides a set of Web application templates as a starting point for developers who want to build Web applications using the Web controls, and as an example of how to use the Web controls to build Web applications.

Cross-platform functionality

The ArcGIS Server ADF for Java runs on a variety of UNIX® platforms and supports numerous Web servers. Your Java Web applications and Web services will fit within your standard Web server environments.

The GIS server itself and the ADF for .NET are available on a number of Windows® platforms.

Cross developer languages

ArcGIS Server supports a variety of developer languages for its use including .NET and Java for building Web applications and Web services; COM and .NET for extending the GIS server with custom components; and COM, .NET, Java, and C++ for building desktop client applications. This allows the objects to be programmed using a wide range of tools and should not require your programming staff learn a new or proprietary language.

ArcGIS Server options

The ArcGIS Server developer kits include the extended functionality of ArcGIS 3D Analyst™, ArcGIS Spatial Analyst, and ArcGIS StreetMap™.

Developer resources

The ArcGIS Server developer kits provide a help system along with object model diagrams (OMDs), Web application templates, and sample code to help developers get started.

WHY ArcGIS SERVER?

ArcGIS Server allows developers and system designers to implement a centrally managed GIS. This gives the advantage of lower cost of ownership through single GIS applications (such as a Web application) that can scale to support multiple users and saves the cost of installing desktop applications on each user's machine. This, along with the ability of ArcGIS Server to leverage Web services, makes it ideal for integration with other critical information technology (IT) systems, such as relational databases, Web servers, and enterprise application servers.

ArcGIS Server complements the ESRI family of server products: ArcIMS, ArcSDE, and ArcGIS Server. ArcIMS provides high-performance Web geopublishing of maps and metadata, ArcGIS Server is a centrally managed GIS for advanced GIS applications, and ArcSDE manages data access for ArcGIS Server and ArcIMS.

There are a number of different roles that a user of ArcGIS Server may take:

WEB APPLICATION USERS

This is the largest group of users of ArcGIS Server. A Web application user uses an Internet browser to connect to a Web application written and deployed by a Web application developer. That user interacts with the Web application to make use of the GIS and other functionality it presents. Web application users themselves may have little or no knowledge that they are using GIS functionality provided by a GIS server.

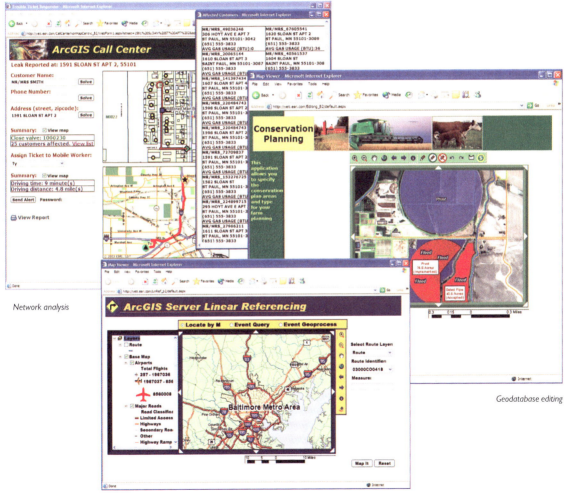

Network analysis

Geodatabase editing

Linear referencing and map composition

Web application users will use their browsers to connect to GIS Web applications built and deployed by a developer.

WEB APPLICATION AND WEB SERVICE DEVELOPERS

Web application and Web services developers will use the ADF to build and deploy .NET and Java Web applications and Web services. These Web applications and Web services include advanced functionality by connecting to a GIS server and make use of ArcObjects running within the server. Developers can build Web services to expose maps and address locators for use by ArcGIS Desktop users over the Internet. Developers can also build application Web services that encapsulate GIS functionality and are consumable by other programs.

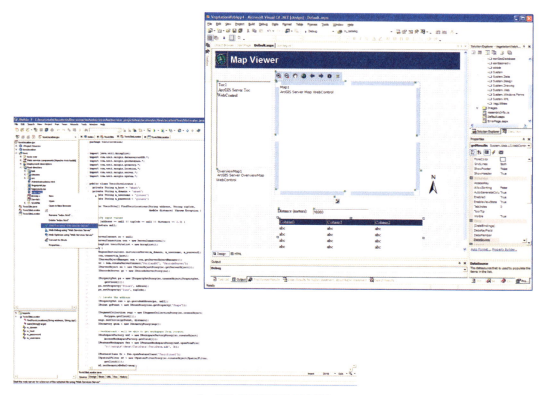

Web application and Web service developers will use standard development environments, such as Microsoft Visual Studio .NET and JBuilder, to build ArcGIS Server applications.

ArcGIS DESKTOP USERS

ArcGIS Desktop users can use ArcMap and ArcCatalog to connect to a GIS server on their local network over the LAN or to a Web service catalog over the Internet. In both cases, users can use the maps and address locators published as map and geocode server objects to do basic mapping and geocoding.

ArcGIS Desktop is also the software that you use to create the data that is used by ArcGIS Server applications. ArcGIS Desktop provides the tools to build the databases, map documents, and address locators that are served by ArcGIS Server.

ArcGIS Desktop users can connect to a GIS server directly over a LAN or over the Internet and make use of the map and geocode server objects running in the GIS server.

ArcGIS Desktop, ArcGIS Engine developers

ArcGIS Engine and ArcGIS Desktop developers can also develop applications that connect to a GIS server and make use of ArcObjects running within the server. This allows the integration of desktop functionality with server functionality.

Developers can also build desktop applications that work with the GIS server in a client/server mode. These desktop applications can be built using the ArcGIS Server ADFs or with the ArcGIS Engine developer kit.

GIS SERVER ADMINISTRATORS

The GIS server administrator uses ArcCatalog to connect to a GIS server on the local network and administer aspects of the server itself and the set of server objects running in the server. The GIS server administrator will add machines to and remove them from the system to perform the server's GIS processing, manage the server's output directories, and monitor statistics and output logs to troubleshoot any errors or performance problems.

The GIS administrator will also use ArcCatalog to manage and configure the server objects running in the server that are used by desktop and Web applications. In each case, the administrator will work with the application developer to understand the nature of the application and the number of users it needs to support to make decisions about its configuration.

The GIS administrator must make use of operating system tools to provide appropriate privileges to the ArcGIS Server accounts on data and output directories needed to run server objects and support any applications. The GIS administrator must also make use of operating system tools to control user access to the GIS server.

The GIS server administrator uses ArcCatalog to connect to a GIS server on the local network and administer aspects of the server itself and the set of server objects running in the server.

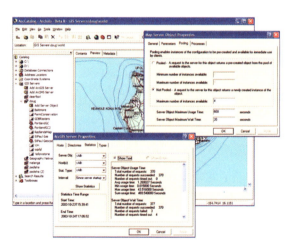

As an ArcGIS Desktop user, you can perform the following functions using ArcGIS Server:

- Connect to a GIS server on your local network over the LAN
- Connect to an ArcGIS Server Web services catalog

Through your LAN or Internet connection you can:

- Use ArcCatalog to preview, pan, zoom throughout a map server, and identify features on a map server
- Use ArcCatalog to batch geocode addresses using a geocode server
- Use ArcMap to add a map server as a layer to your local map document
- Use ArcMap to pan and zoom throughout a map server
- Use ArcMap to identify, search for, and find features on a map server
- Use ArcMap to interactively toggle layers on and off
- Use ArcMap to find addresses with a geocode server

As a developer, you can implement these and many other functions in Web applications and Web services built with the ADF developer kit:

- Display a map with multiple map layers such as roads, streams, and boundaries
- Pan and zoom throughout a map
- Identify features on a map by pointing at them
- Search for and find features on a map
- Display labels with text from field values
- Draw images from aerial photography or satellite imagery
- Draw graphic features such as points, lines, circles, and polygons
- Draw descriptive text
- Select features along lines and inside boxes, areas, polygons, and circles
- Select features within a specified distance of other features
- Find and select features with a Structured Query Language (SQL) expression
- Render features with thematic methods such as value map, class breaks, and dot density
- Dynamically display real-time or time series data
- Find locations on a map from a street address or intersection you provide
- Transform the coordinate system of your map data
- Perform geometric operations on shapes to create buffers; calculate differences; or find intersections, unions, or inverse intersections of shapes
- Perform advanced spatial and attribute queries
- Perform network analysis
- Manipulate the shape or rotation of a map

- Create and update geographic features and their attributes

- Perform geodatabase management tasks such as reconciling versions and validating topology

You'll find ArcGIS Server suitable for building basic mapping to advanced GIS applications. In addition to the above core functionality, ArcGIS server can be enhanced to include support for specialized options. The available options for ArcGIS Server are:

SPATIAL OPTION

The ArcGIS Server Spatial option provides a powerful set of functions that allows you to create, query, and analyze cell-based raster data. Using the spatial option to the GIS server, your applications can derive information about your data, identify spatial relationships, find suitable locations, and calculate cost of traveling from one point to another.

The Spatial option for ArcGIS Server provides a powerful set of tools that allows you to create, query, and analyze cell-based raster data.

3D OPTION

The ArcGIS Server 3D option provides a powerful set of functions that allows your applications to create and analyze surfaces.

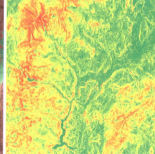
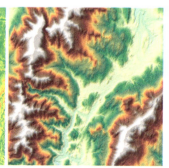

The 3D option for ArcGIS Server allows you to create and analyze surfaces. These functions include viewshed, slope, aspect, hillshade analysis, and more.

StreetMap

The ArcGIS Server StreetMap option provides street-level mapping and address matching. StreetMap layers automatically manage, label, and draw features such as local landmarks, streets, parks, water bodies, and other features. All data comes compressed on CD–ROM.

The StreetMap option for ArcGIS Server provides street-level mapping and address matching.

Each component of ArcGIS Server includes a developer kit: the GIS server developer kit for developers who want to extend the GIS server, the Java ADF developer kit for developers who want to build Java applications, and the .NET ADF developer kit for developers who want to build .NET applications.

Each developer kit contains common developer resources to support your development task. An integrated help system is provided for several APIs (COM, Java, .NET, and C++) along with object model diagrams and samples for each part of the core ArcObjects components. Each developer kit provides access to a large collection of ArcObjects for you to exploit to include any range of GIS functionality in your application.

The GIS server developer kit lets you, the programmer, develop custom COM components to extend the GIS server using COM development languages, such as Visual Basic® (VB), C++, and .NET. To help facilitate this, the GIS server developer kit also contains a collection of integrated development environment (IDE) plug-ins and utilities to make developing COM objects with ArcObjects easier.

The .NET and Java ADF developer kits are a set of Web controls and helper objects that let you add dynamic mapping and GIS capabilities into new or exist-

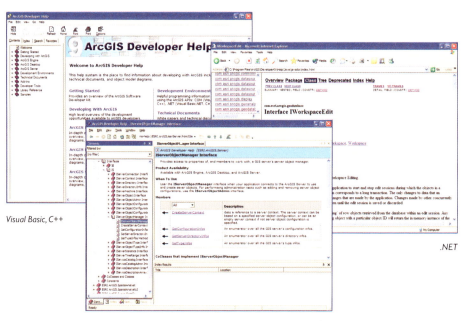

Visual Basic, C++

.NET

Java

Each developer kit contains an integrated help system for several APIs.

ing Web applications. The ADFs include the following Web controls to assist with Web application development:

- Map
- Overview map (.NET) or Overview (Java)
- Table of contents (Toc)
- Page layout
- North arrow
- Toolbar
- Scale bar
- GeocodeConnection (.NET) or Geocode (Java)
- Impersonation

These controls are available as .NET Web controls and Java Web controls exposed as JavaServer Pages™ (JSP) tags. These controls can be combined with other Web controls and components to create customized Web applications.

In addition, the ADF developer kits include a collection of Web application templates that serve as both a starting point for your Web application and an example of how to use the Web controls and the ArcGIS Server API to build Web applications. The Web application templates include:

- Map Viewer template, which provides basic map display capabilities.

- Search template, which provides a search-centric interface for finding features on a map.
- Page Layout template, which displays the entire page layout for a map.
- Thematic template, which adds thematic mapping capabilities on top of the Map Viewer template.
- Geocode template, which provides an interface for finding map locations using an address.
- Buffer selection template, which allows you to find features in one layer of the map based on their location relative to features in another layer.

Both the Java and .NET ADFs include a collection of Web application templates that developers can extend to include their own functionality.

Developers building Web services or desktop applications, rather than Web applications, will also benefit from the ADF developer kits' documentation and code samples.

The ArcGIS Server developer kits are not for end users. They are for people who are developing server applications or extending the server to support server applications. As a developer, you can build applications based on ArcGIS Server and deploy those applications either to end users or on application servers. An important feature of ArcGIS Server is that the application you create can treat a map as a central or incidental element of the application. In the case of some applications (especially Web services) there may be no mapping component at all.

ADF RUNTIMES

In order to deploy a Web application or Web service on an application server, or deploy a server desktop application written with the ADF, you need the ADF runtime. Because server applications use ArcObjects that are running within the server, the application server or desktop machine running a server application does not need ArcObjects installed on it. The ADF runtime includes only those components necessary for applications to connect to the GIS server and make use of ArcObjects running in the server.

The .NET ADF runtime installs a collection of .NET assemblies and corresponding COM object libraries. The Java ADF runtime installs a set of Java archive (JAR) files for working with ArcObjects. These runtimes do not require a license; however, the applications deployed with the runtimes require a GIS server to connect to, which must be licensed to run the ArcObjects components required by the application.

The ArcGIS Server is a distributed system that supports both desktop and server (e.g., Web) applications. The ADF runtimes provide the necessary components for applications to connect to the GIS server and make use of ArcObjects running within the server.

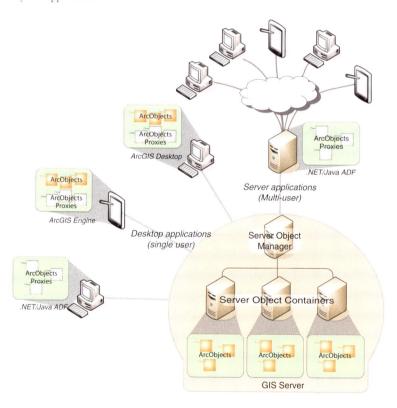

Once you have installed the various components of ArcGIS Server and the server is up and running, the next steps are to configure the GIS server, add server objects and start developing applications.

For more information on configuring and administrating your GIS server, see Chapter 3.

CONFIGURING THE GIS SERVER

Configuring the GIS server is the job of the GIS server administrator. The following administration tasks must be completed before you can start adding server objects to your GIS server:

- Use operating system tools to grant access to the GIS server to users and administrators.

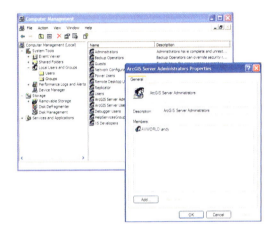

The ArcGIS Server administrator will use operating system tools to manage access to the GIS server and manage privileges on both data and output directories.

- Use operating system tools to grant privileges on the server's output directories to the GIS server account.
- Use operating system tools to grant privileges on the server's data to the GIS server account.
- Use ArcCatalog to add server object container machines to the GIS server.
- Use ArcCatalog to add server directories to the GIS server.

ADDING SERVER OBJECTS

Once the GIS server is configured, you can use ArcCatalog to add, configure, and start map and geocode server objects. Once you have added server objects to your GIS server, you can start using them with the ArcGIS Desktop applications. You can use ArcMap and ArcCatalog to help you determine whether your server object is configured properly by drawing your map servers and finding addresses with your geocode servers.

The ArcGIS Server administrator will use ArcCatalog to configure the server and manage its server objects.

BUILDING APPLICATIONS

The real power of ArcGIS server can be exploited by developers building server-based applications. Use the server objects running in the GIS server to build applications with mapping and advanced GIS functionality. Developers can take advantage of the software objects, Web controls, and template applications in the ADF to get started building applications that make use of ArcObjects running in the GIS server.

The best way to get started once your server is configured is to implement one of the applications outlined in the developer scenarios in Chapter 7. These scenarios are step-by-step walk-throughs of the entire process from creating the necessary server objects to programming and deploying the application.

The *ArcGIS Server Administrator and Developer Guide* is an introduction for anyone who wants to configure ArcGIS Server and create desktop and Web applications using COM, .NET, or Java.

This guide will help you become a server developer by stepping through numerous code samples and developer scenarios. Although the samples documented in this guide may not solve your immediate problem, they will serve as a framework or template on which you can build a more specific or complex solution.

To serve the widest base of developers, most of the code samples in this guide are written in VB. As necessary, some code samples are written in Microsoft® C#, VB.NET, or Java.

The first three chapters of this book provide an overview of the ArcGIS Server product and its capabilities and an overview of the ArcGIS Server architecture and its administration. The remaining chapters focus on developing applications that make use of objects running in the GIS server, including using ArcObjects in the server, using the ADFs to build applications, developer scenarios, and methods for deploying those applications.

CHAPTER GUIDE

Chapter 1, 'Introducing ArcGIS Server', gives you an overview of the ArcGIS Server product, its developer kits, and additional resources.

Chapter 2, 'The ArcGIS Server architecture', describes the various aspects of ArcGIS Server and how they interact, including how the GIS server manages ArcObjects, how applications interact with the objects in the GIS server, and GIS server security.

Chapter 3, 'Administering an ArcGIS Server', describes the various responsibilities of the GIS server administrator. It provides descriptions of how to use ArcCatalog and various operating system tools to administer, monitor, and troubleshoot your GIS server.

After reading this chapter you should have enough information to configure your GIS server, add server objects, and troubleshoot any problems. Once this is complete, developers can use the server objects in your GIS server to build GIS functionality into their applications.

Chapter 4, 'Developing ArcGIS Server applications', describes how to use the server API to access and work with ArcObjects in the GIS server. This chapter includes important programming rules and best practices for developing applications with ArcGIS Server.

Chapter 5, 'Developing Web applications with .NET', describes how to use the Web controls, application templates, and other objects in the .NET ADF to build Web applications that use functionality available through ArcGIS Server.

Chapter 6, 'Developing Web applications with Java', describes how to use the Web controls, application templates, and other objects in the Java ADF to build Web applications that use functionality available through ArcGIS Server.

Chapter 7, 'Developer scenarios', guides you through the creation and deployment of the several types of Web applications and Web services you can build with both .NET and Java.

This book also contains a number of appendixes that provide detailed information about the GIS server API's object model and information on how to use and understand the GIS server's log and configuration files. Additional appendixes provide background information for all ArcObjects developers, including language-specific notes and a how-to guide to reading OMDs.

In addition, Appendix C, 'Developing ArcGIS Server applications with EJBs', describes how a developer would build EJBs that consume services from the ArcGIS Server. EJBs can call and work with ArcGIS Server objects through a connection to a Java Connector Architecture (JCA)-compliant resource adapter, which is provided in the Java ADF. Recommended usage and programming models for JCA and ArcGIS Server are given in this section.

The following topics describe some of the additional resources available to you as a developer. These include books, guides, and various help systems.

ArcGIS SOFTWARE DEVELOPER KIT

The ArcGIS SDK is the collection of diagrams, utilities, add-ins, samples, and documentation geared to help developers implement custom ArcGIS functionality.

C:\Program Files\ArcGIS\DeveloperKit
- Addins
- Diagrams
- Documentation
- Help
- samples
- Templates
- tools

A typical SDK installation

ArcGIS Developer Help system

The ArcGIS Developer Help system is the gateway to all the SDK documentation including help for the add-ins, developer tools, and samples; in addition, it serves as the complete syntactical reference for all object libraries.

Each supported API has a version of the help system that works in concert with it. Regardless of the API you choose to use, you will see the appropriate library reference syntax and have a help system that is integrated with your development environment. For example, if you are a Visual Basic 6 developer you will use ArcGISDevHelp.chm, which has the VB6 syntax and integrates with the VB6 IDE, thereby providing F1 help support in the code window.

The help systems reside in the DeveloperKit\Help folder but are typically launched from the start menu or F1 help in Visual Basic 6 and Visual Studio® .NET 2003. The graphic below shows the start menu options for opening the help systems.

Samples

The ArcGIS developer kit contains more than 600 samples, many of which are written in several languages. The samples are described in the help system and the source code and project files are installed in the DeveloperKit\samples folder. The help system's table of contents for the samples section mirrors the samples directory structure.

The help system organizes samples by functionality. For example, all the Geodatabase samples are grouped under Samples\Geodatabase. Most first-tier groupings are further subdivided. You can also find samples in the SDK using the 'Query the Samples' topic in the help system, which lists all the samples alphabetically; in addition, you can sort the list by language. For example, you can elect to only list the available Java samples.

You can use the 'Query the Samples' topic in the help system to find specific samples you are interested in.

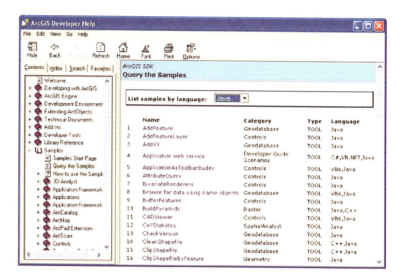

Installing the samples source code and project files is an option in the install. The samples are installed under the ArcGIS\DeveloperKit\samples folder. If you don't have this folder on your computer, you can rerun the install program and check Samples under Developer Kit.

Developer tools

The ArcGIS developer tools are executables that ESRI has provided to facilitate your ArcObjects development. You may find some of these tools essential. For example, if you are a Visual Basic 6 desktop developer you will likely use the Categories.exe tool to register components in component categories.

The following developer tools are available with each product. Please refer to the help system for more developer tool details and instructions.

- *Component Categories Manager*—Registers components within a specific component category.

- *Fix Registry Utility*—Fixes corruptions in the Component Categories section of the registry.

- *GUID Tool*—Generates globally unique identifiers (GUIDs), in registry format for use within source code.

- *Library Locator*—Identifies object library containing a specified interface, coclass, enumeration, or structure.

The developer tools are installed under the DeveloperKit\tools folder. There is one exception—the Component Category Manager is located in the ArcGIS\bin folder.

Add-ins

The ESRI add-ins automate some of the tasks performed by the software engineer when developing with ArcObjects, as well as provide tools that make debugging code easier. ESRI provides add-ins for the Visual Basic 6 IDE and the Visual Studio .NET IDE. The table below lists the add-ins available for these development environments.

Visual Basic 6

- *ESRI Align Controls with Tab Index*—Ensures control creation order matches tab index.

- *ESRI Automatic References*—Automatically adds ArcGIS library references.

- *ESRI Code Converter*—Converts projects from ArcGIS 8.x to ArcGIS 9.x.

- *ESRI Command Creation Wizard*—Facilitates the creation of commands and tools.

- *ESRI Compile and Register*—Aids in compiling components and registering these in desired component categories.

- *ESRI ErrorHandler Generator*—Automates the generation of error handling code.

- *ESRI ErrorHandler Remover*—Removes the error handlers from the source files.

- *ESRI Interface Implementer*—Automatically stubs out implemented interfaces.

- *ESRI Line Number Generator*—Adds line numbers to the appropriate lines within source files.

- *ESRI Line Number Remover*—Removes the line numbers from source files.

Visual Studio .NET

- *ESRI Component Category Registrar*—Stubs out registration functions to enable self component category registration.

- *ESRI GUID Generator*—Inserts a GUID attribute.

The .NET add-ins are automatically installed during setup if a version of Visual Studio .NET 2003 is detected; the Visual Basic 6 add-ins are only installed if you select them on the install.

THE ArcGIS DEVELOPER SERIES

This book is one in a series of books for ArcGIS developers.

The *ArcGIS Engine Developer Guide* provides information for developers who wish to create applications based on ArcGIS Engine. ArcGIS Engine allows you to embed GIS functionality within other applications and create desktop-like applications using the supplied ArcGIS controls, such as the MapControl, Toolbar, and PageLayout controls. ArcGIS Engine is also based on ArcObjects components and may be programmed through a number of APIs.

Visual Basic 6 add-ins are only installed if you select them on the install.

The *ArcGIS Desktop Developer Guide* is for developers who want to customize or extend the ArcGIS Desktop applications, such as ArcMap and ArcCatalog, using Visual Basic for Applications (VBA) and extend the applications using Visual Basic, .NET, or C++.

ArcGIS DEVELOPER ONLINE WEB SITE

ArcGIS Developer Online is the place to find the most up-to-date ArcGIS 9 developer information including sample code, technical documents, object model diagrams, and the complete object library reference.

The Web site is a reflection of the ArcGIS Developer Help system except it is online and, therefore, more current. The Web site has some additional features including an advanced search utility that enables you to control the scope of your searches. For example, you can create a search that only scans the library reference portion of the help system.

Visit the site today (*http://arcgisdeveloperonline.esri.com*).

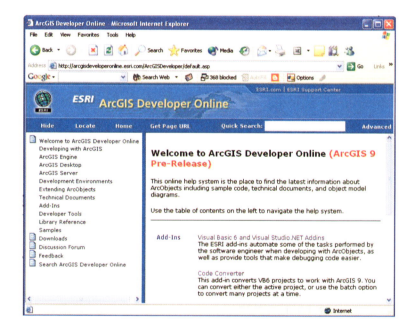

The ESRI Support Center at http://support.esri.com

ESRI SUPPORT CENTER

The ESRI Support Center at *http://support.esri.com* contains software information, technical documents, samples, forums, and a knowledge base for all ArcGIS products.

ArcGIS developers can take advantage of the forums, knowledge base, and samples sections to aid in development of their ArcGIS applications.

TRAINING

ESRI offers a number of instructor-led and Web-based training courses for the ArcGIS Desktop developer. These courses range from the introductory level for VBA to the more advanced courses in component development with specific APIs.

For more information, visit *http://www.esri.com* and select the Training and Events tab.

The ESRI Virtual Campus can also be found directly at *http://campus.esri.com*.

2 The ArcGIS Server architecture

ArcGIS Server is an object server for ArcObjects. The ArcGIS Server software system is distributed across multiple machines. Each aspect of ArcGIS Server plays a role in managing GIS functionality and data and making that functionality useful to end users.

This chapter provides an overview of the ArcGIS software architecture and details of the ArcGIS Server architecture, specifically the various aspects of the server and how they interact, including:

• the role of the GIS server • the server object manager • server object containers • GIS server objects • the Web application server

Before discussing the details of the ArcGIS Server architecture, it's important to discuss the ArcGIS system architecture as a whole. The ArcGIS architecture has evolved over several releases of the technology to be a modular, scalable, cross-platform architecture implemented by a set of software components called ArcObjects. This section focuses on the main themes of this evolution at ArcGIS 9 and introduces the reader to the libraries that compose the ArcGIS system.

The ArcGIS software architecture supports a number of products, each with its unique set of requirements. ArcObjects, the components that make up ArcGIS—ArcObjects—are designed and built to support this. This chapter introduces ArcObjects.

ArcObjects is a set of platform-independent software components, written in C++, that provides services to support GIS applications on the desktop in the form of thick and thin clients and on the server.

For a detailed explanation of COM see the Microsoft COM section of Appendix D, 'Developer environments'.

As stated, the language chosen to develop ArcObjects was C++; in addition to this language, ArcObjects makes use of the Microsoft Component Object Model. COM is often thought of as simply specifying how objects are implemented and built in memory and how these objects communicate with one another. While this is true, COM also provides a solid infrastructure at the operating system level to support any system built using COM. On Microsoft Windows operating systems, the COM infrastructure is built directly into the operating system. For operating systems other than Microsoft Windows, this infrastructure must be provided for the ArcObjects system to function.

Not all ArcObjects components are created equally. The requirements of a particular object, in addition to its basic functionality, vary depending on the final end use of the object. This end use broadly falls into one of the three ArcGIS product families:

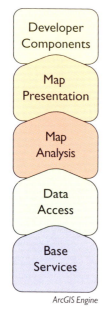

ArcGIS Engine

- ArcGIS Engine—Use of the object is within a custom application. Objects within the Engine must support a variety of uses; simple map dialog boxes, multithreaded servers, and complex Windows desktop applications are all possible uses of Engine objects. The dependencies of the objects within the Engine must be well understood. The impact of adding dependencies external to ArcObjects must be carefully reviewed, since new dependencies may introduce undesirable complexity to the installation of the application built on the Engine.

- ArcGIS Server—The object is used within the server framework, where clients of the object are most often remote. The remoteness of the client can vary from local, possibly on the same machine or network, to distant, where clients can be on the Internet. Objects running within the server must be scalable and thread safe to allow execution in a multithreaded environment.

- ArcGIS Desktop—Use of the object is within one of the ArcGIS Desktop applications. ArcGIS Desktop applications have a rich user experience, with applications containing many dialog boxes and property pages that allow end users to work effectively with the functionality of the object. Objects that contain properties that are to be modified by users of these applications should have property pages created for these properties. Not all objects require property pages.

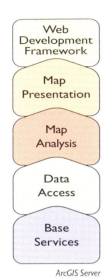

ArcGIS Server

ArcGIS Desktop

Many of the ArcObjects components that make up ArcGIS are used within all three of the ArcGIS products. The product diagrams on these pages show that the objects within the broad categories of base services, data access, map analysis, and map presentation are contained in all three products. These four categories contain the majority of the GIS functionality exposed to developers and users in ArcGIS.

This commonality of function among all the products is important for developers to understand, since it means that when working in a particular category, much of the development effort can be transferred between the ArcGIS products with little change to the software. After all, this is exactly how the ArcGIS architecture is developed. Code reuse is a major benefit of building a modular architecture, but code reuse does not simply come from creating components in a modular fashion.

The ArcGIS architecture provides rich functionality to the developer, but it is not a closed system. The ArcGIS architecture is extendable by developers external to ESRI. Developers have been extending the architecture for a number of years, and the ArcGIS 9 architecture is no different; it, too, can be extended. However, ArcGIS 9 introduces many new possibilities for the use of objects created by ESRI and you. To realize these possibilities, components must meet additional requirements to ensure that they will operate successfully within this new and significantly enhanced ArcGIS system. Some of the changes from ArcGIS 8 to ArcGIS 9 appear superficial, an example being the breakup of the type libraries into smaller libraries. That, along with the fact that the objects with their methods and properties that were present at 8.3 are still available at 9, masks the fact that internally ArcObjects has undergone some significant work.

The main focus of the changes made to the ArcGIS architecture at 9 revolves around four key concepts:

- Modularity—A modular system where the dependencies between components are well-defined in a flexible system.

- Scalability—ArcObjects must perform well in all intended operating environments, from single user desktop applications to multiuser/multithreaded server applications.

- Multiple Platform Support—ArcObjects for the Engine and Server should be capable of running on multiple computing platforms.

- Compatibility—ArcObjects 9 should remain equivalent, both functionally and programmatically, to ArcObjects 8.3.

MODULARITY

The esriCore object library, shipped as part of ArcGIS 8.3, effectively packaged all of ArcObjects into one large block of GIS functionality; there was no distinction between components. The ArcObjects components were divided into smaller groups of components, these groups being packaged in dynamic link libraries (DLLs). The one large library, while simplifying the task of development for external developers, prevented the software from being modular. Adding the type information to all the DLLs, while possible, would have greatly increased the burden on external developers and, hence, was not an option. In addition, the DLL structure did not always reflect the best modular breakup of software components based on functionality and dependency.

ESRI has developed a modular architecture for ArcGIS 9 by a process of analyzing features and functions and matching those with end user requirements and deployment options based on the three ArcGIS product families. Developers who have extended the ArcGIS 8 architecture with custom components are encouraged to go through the same process to restructure their source code into similar modular structures.

There is always a trade-off in performance and manageability when considering architecture modularity. For each criterion, thought is given to the end use and the modularity required for support. For example, the system could be divided into many small DLLs with only a few objects in each. Although this provides a flexible system for deployment options, at minimum memory requirements, it would affect performance due to the large number of DLLs being loaded and unloaded. Conversely, one large DLL containing all objects is not a suitable solution either. Knowing the requirements of the components allows them to be effectively packaged into DLLs.

The ArcGIS 9 architecture is divided into a number of libraries. It is possible for a library to have any number of DLLs and executables (EXEs) within it. The requirements that components must meet to be within a library are well-defined. For instance, a library, such as Geometry (from the base services set of modules), has the requirements of being thread safe, scalable, without user interface (UI) components, and deployable on a number of computing platforms. These requirements are different from libraries, such as ArcMap (from the applications category), which does have user interface components and is a Windows-only library.

An obvious functionality split to make is user interface and nonuser interface code. UI libraries tend to be included only with the ArcGIS Desktop products.

All the components in the library will share the same set of requirements placed on the library. It is not possible to subdivide a library into smaller pieces for distribution. The library defines the namespace for all components within it and is seen in a form suitable for your chosen ArcObjects API.

- Type Library—COM
- .NET Interop Assembly—.NET
- Java Package—Java
- Header File—C++

SCALABILITY

The ArcObjects components within ArcGIS Engine and ArcGIS Server must be scalable. Engine objects are scalable because they can be used in many different types of applications; some require scalability, while others do not. Server objects are required to be scalable to ensure that the server can handle many users connecting to it, and as the configuration of the server grows, so does the performance of the ArcObjects components running on the server.

The scalability of a system is achieved using a number of variables involving the hardware and software of the system. In this regard, ArcObjects supports scalability with the effective use of memory within the objects and the ability to execute the objects within multithreaded processes.

There are two considerations when multithreaded applications are discussed: thread safety and scalability. It is important for all objects to be thread safe, but simply having thread-safe objects does not automatically mean that creating multithreaded applications is straightforward or that the resulting application will provide vastly improved performance.

Thread safety refers to concurrent object access from multiple threads.

The ArcObjects components contained in the base services, data access, map analysis, and map presentation categories are all thread safe. This means that application developers can use them in multithreaded applications; however, programmers must still write multithreaded code in such a way as to avoid application failures due to deadlock situations and so forth.

In addition to the ArcObjects components being thread safe for ArcGIS 9, the apartment threading model used by ArcObjects was analyzed to ensure that ArcObjects could be run efficiently in a multithreaded process. A model referred to as "Threads in Isolation" was used to ensure that the ArcObjects architecture is used efficiently.

The classic singleton per process model means that all threads of an application will still access the main thread hosting the singleton objects. This effectively reduces the application to a single-threaded application.

This model works by reducing cross-thread communication to an absolute minimum or, better still, removing it entirely. For this to work, the singleton objects at ArcGIS 9 were changed to be singletons per thread and not singletons per process. The resource overhead of hosting multiple singletons in a process was outweighed by the performance gain of stopping cross-thread communication where the singleton object is created in one thread, normally the Main single-threaded apartment (STA), and the accessing object is in another thread.

ArcGIS is an extensible system, and for the Threads in Isolation model to work, all singleton objects must adhere to this rule. If you are creating singleton objects as part of your development, you must ensure that these objects adhere to the rule.

MULTIPLE PLATFORM SUPPORT

As stated earlier, ArcObjects components are C++ objects, meaning that any computing platform with a C++ compiler can potentially be a platform for ArcObjects. In addition to the C++ compiler, the platform must also support some basic services required by ArcObjects.

Microsoft Windows is a little endian platform, while Sun Solaris is a big endian platform.

Although many of the platform differences do not affect the way in which ArcObjects components are developed, there are areas where differences do affect the way code is developed. The byte order of different computing architectures varies between little endian and big endian. This is most readily seen when objects read and write data to disk. Data written using one computing platform will not be compatible if read using another platform, unless some decoding work is performed. All the ArcGIS Engine and ArcGIS Server objects support this multiple platform persistence model. ArcObjects components always persist themselves using the little endian model; when the objects read persisted data, it is converted to the appropriate native byte order. In addition to the byte order differences, there are other areas of functionality that differ between platforms; the directory structure, for example, uses different separators for Windows and UNIX—'\' and '/', respectively. Another example is the platform-specific areas of functionality, such as Object Linking and Embedding database (OLE DB).

COMPATIBILITY

Maintaining compatibility of the ArcGIS system between releases is important to ensure that external developers are not burdened with changing their code to work with the latest release of the technology. Maintaining compatibility at the object level was a primary goal of the ArcGIS 9 development effort. Although this object-level compatibility has been maintained, there are some changes between the ArcGIS 8 and ArcGIS 9 architectures that will affect developers, mainly related to the compilation of the software.

While the aim of ArcGIS releases is to limit the change in the APIs, developers should still test their software thoroughly with later releases.

Although the changes required for software created for use with ArcGIS 8 to work with ArcGIS 9 are minimal, it is important to understand that to realize any existing investment in the ArcObjects architecture at ArcGIS 9, you must review your developments with respect to ArcGIS Engine, ArcGIS Server, and ArcGIS Desktop.

ESRI understands the importance of a unified software architecture and has made numerous changes for ArcGIS 9 so the investment in ArcObjects can be realized on multiple products. If you have been involved in creating extensions to the ArcGIS architecture for ArcGIS 8, you should think about how the new ArcGIS 9 architecture affects the way your components are implemented.

The remainder of this chapter focuses on the components of ArcGIS Server that make it possible to run ArcObjects in a server environment. For more information about developing with ArcGIS Engine and ArcGIS Desktop, refer to *ArcGIS Engine Developer Guide* and *ArcGIS Desktop Developer Guide*, respectively.

ArcGIS Server is fundamentally an object server that manages a set of GIS server objects. These server objects are software objects that serve a GIS resource such as a map or a locator. Developers make use of server objects in their custom applications. Server objects are ArcObjects. ArcObjects is a collection of software objects that make up the foundation of ArcGIS.

ArcObjects components have multiple developer application programming interfaces. These include COM, .NET, Java, and C++. Developers can use these APIs to build applications that make use of ArcObjects functionality.

ArcObjects is at the core of all the ArcGIS products: ArcGIS Desktop, ArcGIS Engine, and ArcGIS Server. ArcGIS Server adds the framework for running ArcObjects in a server. ArcGIS Server also provides a framework for developers to build advanced GIS Web services and Web applications using ArcObjects in standard application server frameworks such as .NET and J2EE.

As described above, at the core of ArcGIS Server is a rich ArcObjects library that can be exploited in Web applications and Web services to deliver advanced GIS functionality to a wide range of users who interact with the server through Web browsers and other thin client applications. Web applications that run in the Web server use distributed object technology to communicate with ArcObjects components, which are themselves COM objects, running in the GIS server over a local area network or wide area network (WAN).

Distributed object technology allows applications running in one process to use COM objects that are running in another process seamlessly using local interprocess communication. Distributed object technology also allows applications to use COM objects running on other machines by using network protocols to provide the communication between the client and COM server. This allows the Web application or Web service developer to make use of remote ArcObjects running in the server using the same programming interfaces a developer writing a desktop application with ArcObjects would use.

A proxy object is a local representation of a remote object. The proxy object controls access to the remote object by forcing all interaction with the remote object to be via the proxy object. The supported interfaces and methods on a proxy object are the same as those supported by the remote object. You can make methods calls on and get and set properties of a proxy object as if you were working directly with the remote object.

ArcObjects include APIs for both .NET and Java. These APIs allow developers to write Web applications using both .NET and Java and to host those Web applications in standard server frameworks such as .NET and J2EE. To facilitate the development of such applications, the ArcGIS Server Application Developer Framework (ADF) includes a set of Web controls and ArcObjects proxies for both .NET and Java around which to build GIS Web applications. These Web applications can deliver advanced GIS functionality to the end user through browsers.

ArcGIS Server also includes a Simple Object Access Protocol (SOAP) toolkit that contains ArcObjects components that support SOAP request handling via an eXtensible Markup Language (XML) API. Once an ArcObjects component is exposed as a Web service, this XML API allows applications to remotely use that object running in the server over standard Internet protocols.

The remainder of this chapter explores how the different pieces of the ArcGIS Server work together to make the use of ArcObjects in server applications possible. Detail is given to the ArcGIS Server programming model in Chapter 4, and the application developer frameworks are discussed in detail in Chapters 5 and 6.

ArcGIS Server is a distributed system consisting of several components that can be distributed across multiple machines. Each component in the ArcGIS Server system plays a specific role in the process of managing, activating, deactivating, and load balancing the resources that are allocated to a given server object or set of server objects.

The components of ArcGIS Server can be summarized as:

- GIS server—Hosts and runs server objects. The GIS server consists of a server object manager (SOM) and one or more server object containers (SOCs).

- Web server—Hosts Web applications and Web services that use the objects running in the GIS server.

- Web browsers—Used to connect to Web applications running in the Web server.

- Desktop applications—Connect over HyperText Transfer Protocol (HTTP) to ArcGIS Web services running in the Web server or connect directly to GIS servers over a LAN or WAN.

The ArcGIS Server is a distributed system that consists of a server object manager, server containers, and clients to the server, such as desktop and Web applications.

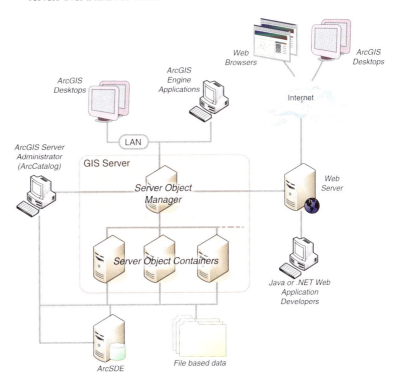

THE GIS SERVER

The GIS server, responsible for hosting and managing server objects, is the set of objects, applications, and services that make it possible to run ArcObjects components on a server. Before describing the various aspects of the GIS server, server object and the role of ArcObjects in ArcGIS Server will be defined.

A server object is a software object that manages and serves a GIS resource such as a map or a locator. For example, a server object named RedlandsMap may serve a map document of data for the city of Redlands, while the server object RedlandsGeocode may serve an address locator for geocoding addresses. ArcGIS Server objects are themselves ArcObject components.

Server objects are managed and run within the GIS server. A server object may be preconfigured and preloaded in the server and can be shared between applications. Server applications make use of server objects and may also use other ArcObjects that are installed on the GIS server.

The server object manager

The GIS server is composed of a SOM, which is a Windows service running on a single machine, and SOCs, which run on one or more machines (container machines). The SOM manages the set of server objects that are distributed across one or more container machines. When an application makes a direct connection to a GIS server over a LAN or WAN, it is are making a connection to the SOM, so the parameter that is provided for the connection to be made is the name or Internet Protocol (IP) address of the SOM machine.

The server object containers

The container machine or machines actually host the server objects that are managed by the SOM. Each container machine is capable of hosting multiple container processes. A container process is a process in which one or more server objects is running. Container processes are started and shut down by the SOM. The objects hosted within the container processes are ArcObjects components that are installed on the container machine as part of the installation of ArcGIS Server.

All server objects run on all container machines and are balanced equally across all container machines. So, it's important that all container machines have access to the resources and data necessary to run each server object. It's also important to note that the GIS server assumes that all container machines are configured equally, such that they are all capable of hosting the same number of server objects. Server object resources and data are discussed in more detail in the next section.

The server directories

A server directory is a location on a file system. The GIS server is configured to clean up any files it writes to a server directory. By definition, a server directory can be written to by all container machines.

The GIS server hosts and manages server objects and other ArcObjects components for use in applications. In many cases, the use of those objects requires writing output to files. For example, when a map server object draws a map, it writes images to disk on the server machine. Other applications may write their own data; for example, an application that checks out data from a geodatabase may write the check out personal geodatabase to disk on the server.

Typically, these files are transient and need only be available to the application for a short time—for example, the time for the application to draw the map or the time required to download the check out database. As applications do their work and write out data, these files can accumulate quickly. The GIS server will automatically clean up its output if that output is written to a server directory.

A server directory can be configured such that files created by the GIS server in it are cleaned based on either file age or time since they were last accessed. The maximum file age is a property of a server directory. All files created by the GIS server that are older than or have not been accessed for the time defined by the maximum age are automatically cleaned up by the GIS server.

THE WEB SERVER

The Web server hosts server applications and Web services written using the ArcGIS Server API. These server applications use the ArcGIS Server API to connect to a SOM, make use of server objects, and create other ArcObjects for use in their applications.

These Web services and Web applications can be written using the ArcGIS Server Application Developer Framework, which is available for both .NET and Java developers. Examples of Web applications include mapping applications, disconnected editing applications, and any other application that makes use of ArcObjects and is appropriate for Web browsers.

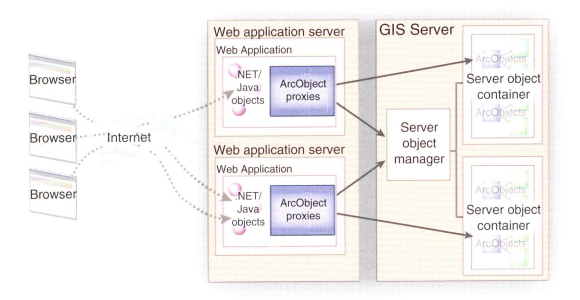

Web applications running in the Web server connect to the SOM and work with proxies to objects running within containers in the GIS server.

Examples of Web services include Web services for exposing map and geocode server objects that desktop GIS users can connect to and consume over the Internet. It is possible to create your own native .NET or Java Web services whose parameters are not ArcObjects types, but do perform a specific GIS function. For example, it is possible to write a Web service called FindNearestHospital that accepts x,y coordinates as input and returns an application-defined Hospital object that has properties such as the address, name, number of beds, and so on.

A more detailed description of the application developer's framework and how it is used to create Web services and Web applications is given later in this book.

Web applications connect to GIS servers within their organization over the LAN. In this sense, the Web application or Web service is a client of the GIS server. Users connect to Web applications and Web services over the Internet or Intranet, but all of the Web application's logic runs in the Web server and sends HyperText Markup Language (HTML) to the browser client. The Web application itself makes use of objects and functionality running within the GIS server. This allows the development of Web applications to make use of ArcObjects in the server as would a desktop application connecting to the GIS server in client–server mode over the LAN or WAN.

As users interact with their browser, it makes requests to the Web application, which in turn makes requests on the SOM. The SOM hands back a proxy to a server object or server objects that are running within the GIS server. The Web application uses the proxy to work with the object as if it existed in the Web application's process, but all execution happens on the GIS server.

Web and desktop server application development will be discussed in greater detail in later chapters of this book.

ArcGIS DESKTOP APPLICATIONS

Users can connect to ArcGIS Server using ArcGIS Desktop applications to make use of map and geocode server objects running in the server. Users can use ArcCatalog to connect to a GIS server directly on the LAN or WAN. They can also specify the URL of a Web service catalog to indirectly connect to a GIS server over the Internet to make use of map and geocode server objects exposed by that Web service catalog.

Additionally, the set of server objects and their properties are managed by the GIS server administrator using ArcCatalog. Administrators can connect to the GIS server over the LAN/WAN and use ArcCatalog to add and remove map and geocode server objects as well as configure how server objects should be run, including the set of container machines that are available for the server and the directories on the server they can use to write any output.

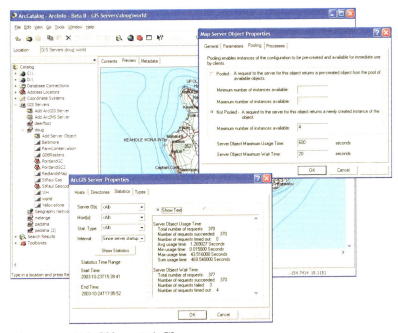

Users can connect to ArcGIS Server using ArcGIS Desktop applications to consume and administer server objects.

As described earlier in this chapter, a server object is a software object that manages and serves a GIS resource such as a map or a locator. Server objects are managed by the GIS server and run within processes on container machines.

A server object is simply a coarse-grained ArcObjects component, that is, a high-level object that simplifies the programming model for doing certain operations and hides the fine-grained ArcObjects that do the work. These coarse-grained objects allow clients to perform large units of work, such as drawing a map or geocoding a table of addresses, using a single method call. These coarse-grained objects use the finer-grained ArcObjects components on the server to draw the map and geocode the addresses.

Server objects also have SOAP interfaces for handling SOAP requests to execute methods and returning results as SOAP responses. This support for SOAP request handling makes it possible to expose server objects as Web services that can be consumed by clients across the Internet.

ArcGIS Server 9.0 includes two coarse-grained server objects: the GeocodeServer and the MapServer.

Note that a server object also has other associated objects that a developer can get to and make use of; for example, a developer working with a MapServer object can get to the Map and Layer objects associated with that map. These are the same Map and Layer objects that a desktop or engine developer would work with, except they live in the server. This is discussed in much greater detail in Chapter 4. For now, the concentration will be on how the MapServer and GeocodeServer objects are managed.

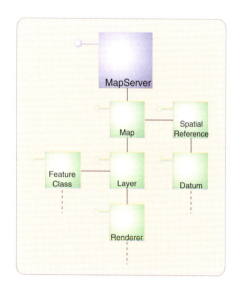

A server object is a coarse-grained ArcObjects component that has other ArcObjects components associated with it.

A server object, unlike other ArcObjects components, can be preconfigured by a GIS server administrator. Once these server objects are preconfigured, they can be used by developers and end users who can connect to the server through ArcGIS Desktop applications.

When a server object is configured by the server administrator to run in a GIS server, the following configuration aspects of the server object need to be specified:

• Name of the server object

• Type of server object

• The initialization data and parameters for the server object

• Whether or not the object is pooled and the minimum and maximum number of instances that can be running

• How long a client can wait for and use a server object

• The isolation level of the server object

• Whether the object is recycled

Each of these aspects of server objects will be described in more detail.

The GIS server includes the MapServer and GeocodeServer server object types.

Server object type

All server objects have a type, which dictates what its initialization parameters are and what methods and properties it exposes to developers. At ArcGIS 9.0, there are two server object types: the MapServer, which is in the Carto library, and the GeocodeServer, which is in the Location library.

Initialization data and parameters

As already stated, a GIS server object manages and serves a GIS resource. When a server object is configured, this resource and other required parameters associated with the server object must be specified such that when a server object is initialized, it knows what resource to bind to.

The MapServer server object's initialization data is the map document (.mxd) or published map document (.pmf) that it's going to serve. When instances of a particular MapServer object are initialized on the server, the map document is loaded.

The GeocodeServer server object's initialization data is the address locator that it will use to perform address matching against. The address locator may be a locator file (.loc), an ArcView 3 address locator (.mxs), or an ArcSDE address locator. In addition to the address locator, the GeocodeServer also has a batch size parameter that indicates the number of records it will process at a time when doing batch geocoding.

Note that when configuring the initialization data for a server object, both the resource (that is, map document or locator) and the data that the resource references are accessible by the GIS server's container machines. This becomes an especially important consideration when the GIS server has multiple container machines. In these cases, the location of both the resource and the data it uses must be on a shared file system (in the case of file-based information) or on an ArcSDE server that all the container machines can connect to.

For example, a map document that acts as a resource to a MapServer object hosted on multiple container machines must be located on a shared network drive. All of the data for the various layers in the map that reference file-based data must also be on a shared network drive, and any layers that reference data in an ArcSDE geodatabase must be able to connect to that database.

Note that server objects can run on any of the container machines configured in the GIS server. All container machines must have access to the data and resources needed for a particular server object.

Server object pooling

A server object that is running on the GIS server is available for use by users who connect to the server through ArcGIS Desktop or by developers who create applications that connect to and make use of the server. Any server object may be used by a number of different users. For example, a MapServer object containing features from a land records database may be used in an application that users query for information about their land parcel. The same MapServer may also be used by editors who update the database in a disconnected editing application.

A server object may be configured to be pooled or non-pooled. Non-pooled server objects are created new for each application use and destroyed when released by the application to the server. The creation of the object includes creating the object and loading up any initialization data, such as the map document associated with a MapServer server object. Each user of a server application that makes use of a non-pooled server object requires an instance of that object dedicated to the user's application.

The number of users on the system at any one time has a 1:1 correlation with the number of running server object instances, so the number of concurrent users the GIS server can support is equal to the number of server objects that it can support effectively at any one time. When configuring a server object to be non-pooled, the maximum number of instances can be limited by specifying it as a property of the configuration. Once the maximum number of instances has been reached (that is, the maximum number of concurrent users of that server object), additional users will be queued until the number of users drops below the maximum.

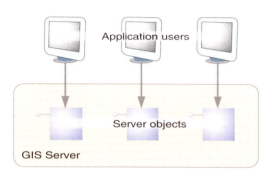

The number of non-pooled server objects running in the GIS server is 1:1 with the number of application users making use of that server object.

ArcGIS Server allows you to pool instances of server objects so they can be shared between multiple application sessions at the per request level. This allows you to support more users with fewer resources allocated to a particular server object. When a server object is configured to be pooled, the minimum and maximum number of instances must be specified as properties of the server object configuration. When the server object is started, the GIS server will precreate and initialize the minimum number of server objects. When an application asks the server object manager for an instance of that server object, it will get a reference to one of the preloaded server objects in the pool.

The advantages of pooling server objects include:

• Separate the potentially expensive initialization of a server object from the actual work that the object performs for each client.

• Share the cost of expensive initialization and acquisition of resources, such as database connections and so on, across all clients.

• Precreate objects at server object manager startup, before any client requests come in.

• Administratively configure pooling to take best advantage of available hardware resources.

Applications that use pooled server objects keep their reference on that object for the duration of their request (for example, draw map, identify feature, geocode address), then return the object to the server. When the object is returned to the server, it's available for use by another user's request. Users of such an application may be working with a number of different instances of a server object in the pool as they interact with the application, all of which is transparent to the users.

If there are more simultaneous requests on a pooled server object than the minimum, new instances of the server object will be created until the maximum number of instances is reached, at which point the user is queued until objects in the pool become free.

Pooling server objects allows the GIS server to support more users. Because applications can share a pool of server objects, the number of concurrent users on the system is greater than the number of server objects that the GIS server can effectively support at one time.

Non-pooled and pooled server objects support different types of server applications. Applications are expected to make stateless use of pooled server objects, meaning they do not make changes to the server object when they are using it, and they are released back to the pool in a timely manner when the request has been processed.

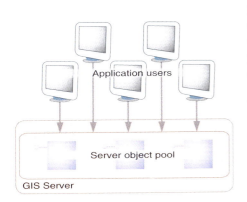

Pooled server objects can support more users because application sessions share a collection of objects in the pool.

For example, a stateless mapping application that wants to draw a certain extent of a MapServer's map will get a reference to an instance of a MapServer server object from the pool, execute a method on the MapServer to draw the map, then release it back to the pool. The next time the application needs to draw the map, this is repeated. Each draw of the map may use a different instance of the pooled server object; therefore, each pooled object must be the same (have the same set of layers, the same renderer for each layer, and so on). If the state of one of the pooled objects is changed (for example, a new layer is added, a layer's renderer is changed), then as a user pans and zooms around the map, inconsistent results will be seen.

Non-pooled server objects are created for each application session that uses it. Since server objects can take time to initialize, the application that makes use of a non-pooled object typically holds a reference to the server object for the duration of the application's session. Since the server object is destroyed when it's returned to the server, the application is free to change any aspect of the server object's state.

Stateful versus stateless use of server objects, as well as the use of application session state in ArcGIS Server applications, will be discussed in greater detail in Chapter 4.

Creation time, wait time, and usage time

When server objects are created in the GIS server, either as a result of the server starting or in response to a request for a server by a client, the time it takes to initialize the server object is referred to as its creation time. The GIS server maintains a maximum creation time-out that dictates the amount of time a server object has to start before the GIS server will assume its startup is hanging and cancel the creation of the server object.

When the maximum number of instances of a pooled or non-pooled server object is in use, a client requesting a server object will be queued until another client releases one of the server objects. The amount of time it takes between a client requesting a server object and getting a server object is called the wait time. A server object can be configured to have a maximum wait time. If a client's wait time exceeds the maximum wait time for a server object, then the request will time out.

Once a client gets a reference to a server object, it can hold onto that server object for as long as it wants before releasing it. The amount of time between when a client gets a reference to a server object and when it releases it is called the usage time. To ensure that clients don't hold references to server objects for too long (that is, they don't correctly release server objects), each server object can also be configured with a maximum usage time. If a client holds on to a server object for longer than the maximum usage time, then the server object is automatically released and the client will lose its reference to the server object.

Maximum usage time also protects server objects from being used to do larger volumes of work than the administrator intended. For example, a server object that is used by an application to perform geodatabase check outs may have a maximum usage time of 10 minutes. In contrast, a server object that is used by applications that only draw maps may have a maximum usage time of one minute.

The GIS server maintains statistics both in memory and in its log files about wait time, usage time, and other events that occur within the server. The server administrator can use these statistics to determine if, for example, the wait time for a server object is high, which may indicate a need to increase the maximum number of instances for that server object.

For more information about how to view these statistics, see Chapter 3 and Appendix B.

Server object isolation

Server objects run within processes on the container machines. Server objects can be configured such that they run in a dedicated process on the server, or they can be configured to run in processes they share with other server objects. How they share processes is referred to as their isolation level.

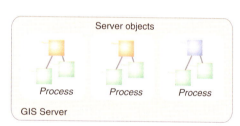

Server objects with high isolation run in dedicated processes on the GIS server.

Server objects with high isolation do not share a process with other server objects. Each instance of a server object with high isolation has its own dedicated process on the server. Server objects with low isolation can share processes with other server objects of the same type.

Up to four server objects can share the same process. When more than four server objects of a particular type (for example, four RedlandsMap server objects) are created, an additional process is started for the next four server objects, and so on. As server objects are created and destroyed, they will vacate and fill spaces in these running processes.

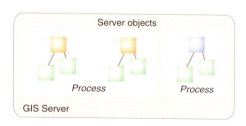

Server objects with low isolation can share processes with other server objects of the same type.

Instances of server objects whose isolation level is high require more resources on the server to run, as they require dedicated processes. Since instances of server objects with low isolation can share processes, they make more efficient use of server resources. However, isolation does have its benefits: since server objects with high isolation do not share processes, if an error occurs on the object, causing its process to shut down or crash, it will not affect other server objects. If a server object is sharing its process with other server objects, however, and the process is shut down or crashes, all the server objects in that process will be destroyed.

Server object recycling

Recycling allows server objects that have become unusable to be destroyed and replaced with fresh server objects; recycling also reclaims resources taken up by stale server objects. This process allows you to keep the pool of server objects fresh and cycle out stale or unusable server objects.

Pooled server objects are typically shared between multiple applications and users of those applications. Through reuse, a number of things can happen to a server object to make it unavailable for use by applications. For example, an application may incorrectly modify a server object's state, or an application may incorrectly hold a reference to a server object, making it unavailable to other applications or sessions. In some cases, server objects may become corrupted and unusable.

Non-pooled server objects whose isolation level is low can also be recycled. This recycling will shut down and restart the processes in which non-pooled objects are started up and run.

In each case, recycling occurs as a background process on the server. The time between recycling events is called the recycling interval. A server object's recycling interval can be configured by the administrator. During recycling, instances of server objects in use by clients are not recycled until released, so recycling occurs without interrupting the user of a server object.

The ArcGIS Server is a secure server and grants connections only to those users who are authorized to connect by the GIS server administrator. There are two levels at which security can be configured for an ArcGIS Server: at the GIS server level itself and at the level of a Web application or Web service that runs in the Web server. Each of these will be discussed separately.

GIS SERVER SECURITY

ArcGIS Server security is based on authenticating operating system user accounts. Connections will be granted to the server for those users who are members of the ArcGIS Server users group (agsusers). The agsusers group is an operating system group, created by the ArcGIS Server install on the SOM machine and all container machines. When a user runs an application that connects to the server, that user's login is authenticated against the users group. If the user is a member of that group, then access is granted to the server; if not, the connection is rejected.

Once connected to the server, the user or application can make use of the server objects running in the server and create new objects in the server for the application's use.

Members of the users group have consumer-level privileges on the GIS server. Consumers may not perform administrative tasks, such as add, remove, or modify preconfigured server objects, or modify properties of the server itself, such as adding and removing machines and so on.

Users may also connect to the server if they are a member of the ArcGIS Server administrators group (agsadmin). These users are granted administrator privileges on the GIS server. Once connected to the server as an administrator, the user or application can administer aspects of the server, such as:

• Add or remove container machines.

• Add, remove, or modify server directories.

• Add, delete, or modify server objects.

• Start, stop, or pause server objects.

• View statistical information.

When applications make connections to the GIS server, they are authenticated against the agsusers and agsadmin users groups on the GIS server.

The users (agsusers) and administrators (agsadmin) groups are created as local user groups by the ArcGIS Server install on both the SOM machine and container machines. When adding a user account to one of these groups, it's important to make sure that the account is added to the group on each machine (the SOM machine and all container machines). One strategy for doing this is to create a domain user group for server users and another for server administrators, then add those domain groups to the users and administrators groups, respectively, on all machines. Then your task of adding and removing users privileges to the ArcGIS Server can be managed by adding and removing them from those domain user groups.

If your login user is not a member of the users or administrators groups, you can use the "runas" command to run ArcCatalog as another user who is in the appropriate group.

A desktop client application to ArcGIS Server will run as the user account that started the application. For example, if you are logged into a desktop computer as the domain user ANDY on the domain AVWORLD, and you are running an application such as ArcCatalog, the identity of the application is AVWORLD\ANDY. When you connect to an ArcGIS Server in that ArcCatalog session, you are connecting as AVWORLD\ANDY. As long as AVWORLD\ANDY is a member of the users group (agsusers) on the SOM, you will be able to connect. If AVWORLD\ANDY is a member of the administrators group (agsadmin), you will have administrator privileges on the server through that connection.

When you connect to ArcGIS Server with ArcCatalog, you will have more commands when connected as an administrator than you would as a consumer—that is, the commands necessary for administering the GIS server. For more details on how to use ArcCatalog to administer ArcGIS Server, see Chapter 3.

IMPERSONATION

Web applications running in the Web server must connect to the ArcGIS Server as a valid GIS server user (i.e., a member of the users group). The Web application must use impersonation to connect to the server as a user account in the users group. For more information on impersonation in Java and .NET, see Chapters 5 and 6.

At the application level, Web applications and Web services define their own security model based on ASP.NET, the .NET implementation of Active Server Pages, and J2EE. Based on this standard security infrastructure, you can build anonymous applications and Web services that are open to all who know the URL. You can also build secure applications with their own users, authentication, and authorization independent of the GIS server. Using these security infrastructures, you can limit the accessibility of your Web application or Web service and deny unauthorized users from having access to your GIS server through the Internet.

These technologies are beyond the scope of this book. For more information on securing ASP.NET and J2EE Web applications, consult the literature on those technologies.

IDENTITY

The GIS server itself runs as two distinct operating system accounts: the server account and the container account. These accounts can have any name and can be assigned to any account already existing in your organization. However, it is important to understand why these accounts are necessary before deciding whether existing accounts in your domain should be used or if you should let the ArcGIS Server installation create these accounts for you.

The server account is the account that runs the Server Object Manager Windows service and process. This process manages the container processes on the container machines as well as the GIS server's configuration information and log files. So, the server account has privileges to write to the locations where the server configuration information and log files are stored. It also has privileges to start container processes on the container machines.

The GIS server postinstallation application allows you to specify the server and container accounts.

The container process actually hosts the server objects and does the work. Container processes are started by the server object manager but run as the container account. Therefore, the container account must have read access to any GIS resources (maps, locators, data) that preconfigured and application-specific server objects require to do their work. In addition, the container account must have write access to the server directories of the GIS server so that server objects running in container processes can write their output. These aspects of the container account are important for administering your site, especially when considering privileges on shared network drives and so on.

One important aspect of the container account is that, since the container processes runs as that account, a user who connects to the GIS server can do anything that the container account can do. Because developers are free to create their own objects on the server, they have access to a wide range of functionality, including the ability to read data that the container account has read privileges on. More important, developers can edit, delete, and otherwise affect files that the container account has write privileges to.

It can be dangerous to use a domain account with many privileges as the container account for your GIS server. The container account should only have enough privileges to access necessary data and perform the task of running server objects. The ArcGIS Server installation can create the server container account with the following minimum privileges on each container machine:

• Ability to launch container processes

• Write access to the system temp directory

It is up to the GIS server administrator to grant this account access to any necessary data and write privileges to the server's output directories.

3 Administering an ArcGIS Server

Administering a GIS server means managing the server itself and the server objects running on it. These server objects are, for example, the maps and locators you need to support client applications that require GIS functionality embedded in them. Just create the map or locator you want to host on your GIS server, then use the administration tools in ArcCatalog to add the server object to your server. Once added to the GIS server, your client applications can remotely access the server objects.

Administration is not a one-time event; it's an ongoing process during which you'll create, delete, and monitor the server objects that run on your server. You'll also manage the machines and server directories that comprise your server, making sure the system is performing optimally.

This chapter describes how to manage the server using ArcCatalog. Topics covered in this chapter include:

• connecting to a GIS server • adding and removing GIS server objects
• managing the server and its server objects

You can think of ArcCatalog as the user interface to a GIS server. ArcCatalog lets you view and, if you're an administrator, manage the set of server objects running on the server. ArcCatalog provides two distinct views of your GIS server, one for administrators and one for those with consumer privileges.

When you connect to a GIS server without administrative privileges, ArcCatalog simply displays the list of server objects available to you. You can use these server objects—for example, display a map hosted on the server in ArcMap—but you can't manage them in any way—for example, delete them. When you connect to a GIS server as an administrator, you'll see some extra tools that allow you to manage the server objects as well.

As an administrator, ArcCatalog lets you manage the set of server objects running on your server and the set of machines that comprise the server. You can monitor how client applications consume individual server objects and whether there are enough resources to satisfy demand. At times, you may need to increase the amount of computer resources allocated to a particular server object; other times you may need to add new computers to handle the load.

If you've just installed ArcGIS Server, there are a few things you need to do before you can start creating server objects and allowing client applications to access these objects. The steps below provide a summary of the things you need to think about and do to get started. The remaining sections of the chapter provide the details for these steps.

- **Identify who can access your GIS server.**

  This doesn't have to be an exhaustive process at this point. At a minimum, you will need to grant yourself administrative access to your GIS server. This way, you'll be able to add server objects to the server and configure other aspects of the server itself—such as add container machines. Optionally, you can identify those people who should have administrative access and others who should have consumer-level access to the server objects.

- **Set up access to the GIS server.**

  The ArcGIS Server security model utilizes the operating system's security model to determine who can connect to and administer the server. You will need to add the list of people you identified above to the ArcGIS Server administrators group (agsadmin) and ArcGIS Server users group (agsusers) on the GIS server. You'll add members to these groups using the operating system's computer management tools.

- **Connect to your GIS server.**

  To administer your server, you connect to it as a Local Server in ArcCatalog.

- **Add container machines to host server objects.**

  During installation, you should have installed software on one or more machines to function as server object containers (SOCs), or container machines. The container machines host the server objects and are the workhorses of your GIS server. The first time you connect to your GIS server in ArcCatalog, you'll need to link these container machines to the server object manager (SOM).

- **Organize your GIS data and set the appropriate directory permissions.**

  The server objects you run on your GIS server are created from the same resources that you work with in ArcGIS Desktop. These are resources such as ArcMap maps. To host them on your server, the data that a map references, for example, must be accessible to the container machines and the operating system account the container machines run as.

- **Add server objects to your GIS server.**

  You'll add to the server the server objects that are required by your client applications and configure them as needed—specifying whether they are pooled or non-pooled and how many instances to create.

- **Create client applications or use ArcGIS Desktop to access your server objects.**

  ArcGIS Desktop clients can directly access the server objects. Connect to a GIS server in ArcCatalog, and you'll be able to use, for example, the map server objects in ArcMap. Alternatively, you can create Web applications that use the server objects. See Chapter 4, 'Developing ArcGIS Server applications', Chapter 5, 'Developing Web applications with .NET', and Chapter 6, 'Developing Web applications with Java', for more information on developing Web applications.

Access to a GIS server—and the server objects running on it—is managed by the operating systems of the server machines that comprise the GIS server. In much the same way the operating system allows you to create and delete files on your own computer, yet prevents you from doing so on your colleague's computer, the operating systems on the GIS server machines grant some users access to the server objects running on the server machines, while denying access to others. When you log in to your computer, the username and password you specify identifies you as a valid user on your network. Based on your operating system account, you are allowed to perform a certain set of actions—one of which might be to access a GIS server.

Before you can begin to use your GIS server, you need to establish who can access it. Once you've done that, you'll be able to connect to your GIS server and add server objects to it.

IDENTIFYING WHO CAN ACCESS THE SERVER

To whom should you grant access to your GIS server? The answer to this question will depend on what kind of server objects you run on your server and how you plan to use them. In some cases, the server objects you place on your server should be made available to everyone. In other cases, you might want to restrict access because a server object contains sensitive information that only certain individuals should see.

As the administrator of the GIS server, you allow people to access the GIS server by adding their operating system account to the list of users who should have access to the server. In reality, there are actually two lists: a list of users who can use the server objects running on the server and a list of users who can administer the server itself (that is, add, delete, and modify server objects). Because it is the operating system that ultimately controls access to the server, you use two operating system user groups—ArcGIS Server users and ArcGIS Server administrators—to manage the two lists of users. A user group simply defines the set of users who have access to a particular resource, in this case, a GIS server.

In general, the list of accounts you add to the ArcGIS Server users and administrators groups will depend on what clients you anticipate will connect to the server. Each operating system account from which ArcGIS Desktop is run will need to be added to the ArcGIS Server users group if you want that client to access the GIS server. Additionally, each Web application you create can connect to the GIS server through a particular operating system account. Each account your Web applications utilize should be added to the ArcGIS Server users group as well.

CONTROLLING ACCESS TO A GIS SERVER

When you install ArcGIS Server, the install program automatically creates the ArcGIS Server users and administrators groups for you. Specifically, the users group is named *agsusers* and the administrators group is named *agsadmin*. These groups are created as local operating system groups on the SOM machine and on each container machine. The ArcGIS Server install program doesn't automatically add any users to these groups for you. Thus, you will need to add users to these groups, depending on the type of access each user should have. Of course, the first account you'll need to add to the administrators group is your own.

How you choose to add user accounts to these groups depends on how your organization manages users in general. If your organization has a number of user groups already established, you may choose to add particular groups as members of the ArcGIS Server users or administrators groups. By allocating users based on other groups, you can minimize the amount of work you need to do to change access to your GIS server. For example, if a new employee is hired and is added to one of your organization's existing groups, access to your GIS server will automatically be granted because the employee is a member of a group that is a member of the ArcGIS Server users group.

Alternatively, you can add individual users to the ArcGIS Server users or administrators groups. This approach requires a little more work because the SOM machine and each container machine have their own local group for ArcGIS Server users and ArcGIS Server administrators. Thus, to add a user account, you need to add that account to the SOM machine's ArcGIS Server users group and to each container machine's ArcGIS Server users group. Similarly, to remove an account, you'll need to remove it from the group on each machine.

Because access to the GIS server is managed by the operating system, users can have either full access or no access to the server. Thus, someone who has access to the GIS server can work with all the server objects running on the server and someone who doesn't have access can't work with any server objects. This level of access control does not allow you to restrict access to individual server objects running on the server. Instead, you'll need to restrict access to individual server objects in other ways. For example, instead of allowing your users to directly connect to the server, you can provide more restricted access through a Web service catalog. A Web service catalog is a Web service that only exposes the subset of server objects that you choose to expose.

The ArcGIS Server Application Developer Framework for .NET and Java contains a template that you can use to build a Web service catalog that contains the server objects you specify. See Chapter 5 or 6 for information on creating a Web service catalog.

Adding administrative users to the ArcGIS Server administrators group, agsadmin, in Windows

The first thing you need to do before you attempt to connect to your GIS server is grant yourself administrative access to it. You do this by adding your operating system account to the list of users who can administer the GIS server. If there are others in your organization that need to administer the GIS server, you can add their accounts as well. Don't worry if you don't have the complete list of administrators; you can always add and remove accounts later. It's important to note that the accounts you add to this group will be able to add, delete, and modify server objects running on the GIS server.

1. On the SOM machine, start the Computer Management application. Computer Management can be found in the Control Panel under Administrative Tools.

2. Expand System Tools, then Local Users and Groups, then Groups.

3. Right-click the ArcGIS Server administrators group, named agsadmin, and click Properties.

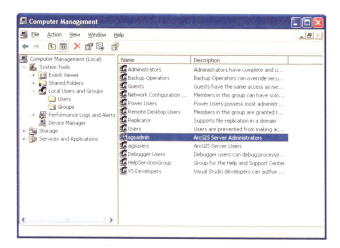

4. On the Property page, click Add and, in the dialog box that appears, add your operating system account and any other accounts to which you want to grant administrative access to the GIS server. The account you add will be the account that you typically run ArcCatalog through.

5. Repeat steps 1 through 4 on each server object container machine.

The users of the accounts you add to the ArcGIS Server administrators group may need to log off before the new settings take affect.

Adding users to the ArcGIS Server users group, agsusers, in Windows

The accounts that you add to the ArcGIS Server users group, agsusers, are those users that you expect to access the objects with ArcGIS Desktop or that your custom Web applications will connect to the server as. These users will not be able to administer the GIS server. Don't worry if you don't have the complete list of accounts. You can always add and remove accounts later.

1. On the SOM machine, start the Computer Management application. Computer Management can be found in the Control Panel under Administrative Tools.

2. Expand System Tools, then Local Users and Groups, then Groups.

3. Right-click the ArcGIS Server users group, named agsusers, and click Properties.

4. On the Property page, click Add and, in the dialog box that appears, add those operating system accounts to which you want to grant access to the GIS server. These are typically the accounts that people use to log in to the network from their own computer. Note: You don't need to add administrative users to this group. Administrative users already have user-level privileges to the server objects.

5. Repeat steps 1 through 4 on each server object container machine.

The users of the accounts you add to the ArcGIS Server users group may need to log off before the new settings take affect.

Connections to ArcGIS Servers appear in the Catalog Tree under the GIS Servers node.

CONNECTING TO A GIS SERVER

Once you've set up access to the server, you can connect to it. ArcCatalog provides two options for connecting to a GIS server. You can connect to a local server running at your site over the LAN, or you can connect to a remote server over the Internet. When you connect to a local server, you can see and utilize all the server objects running on the server. When you connect over the Internet, you're actually connecting to the GIS server through a Web service catalog that runs on a Web server. It's important to note, however, that you can only administer the GIS server when connected to it through the LAN and not over the Internet.

You can create as many Web service catalogs as you like and organize the server objects within them to suit your needs. Why would you want to create a Web service catalog? Perhaps the most compelling reason is that Web service catalogs allow client applications—such as ArcGIS Desktop—to access the server objects running on the server over the Internet. For example, if the people who need to access the server objects running on your GIS server are not within your organization, you can create a Web service catalog and allow them access over the Internet.

Another reason you might want to create a Web service catalog—for people within your organization—is to better organize the server objects for general consumption. By creating one or more Web service catalogs, you can organize server objects for specific groups of people. For example, you might create a Web service catalog for each department in your organization; each Web service catalog would only contain the specific server objects used by the people in that department. Organizing server objects in Web service catalogs also allows you to restrict access to particular server objects that contain sensitive information. The ArcGIS Server Application Developer Frameworks for .NET and Java both contain a template that will help you build a Web service catalog. For more information on creating a Web service catalog, see Chapter 5 or 6, depending on your development environment.

Connecting to a local server

When you connect to a GIS server through the LAN, you can administer the server objects running on the server and aspects of the GIS server's configuration—assuming you have administrative access to the server. Follow the steps below to connect to a GIS server in ArcCatalog.

1. Double-click GIS Servers in the Catalog tree.

2. Double-click Add ArcGIS Server.

3. Click Local Server and type the name of the server you want to connect to. While your GIS server may be configured with several machines, the server machine you should specify is the one that's running the server object manager.

   The first time you connect to the GIS server after installation, you'll need to add one or more container machines to it. This is described in the next section titled 'Adding container machines to host server objects'.

4. Optionally, check the specific server objects you want to display and click OK.

If you are unable to connect to the server or you can connect but don't have administrative access, it means your operating system account does not have the right permissions for accessing your GIS server. See the section, 'Controlling access to a GIS server' earlier in this chapter for more information.

Connecting to an Internet server

If your organization connects to the Internet through a proxy server, see the section titled 'Using a proxy server' later in this chapter for more information.

To connect to a GIS server over the Internet, you specify the URL address of the particular Web service catalog you want to connect to.

1. Double-click GIS Servers in the Catalog tree.

2. Double-click Add ArcGIS Server.

3. Click Internet Server and type the URL of the Web service catalog you want to connect to. Typically, the URL will be something like this: http://www.server.com/MyWebCatalog/default.aspx

4. Optionally, type the username and password if connecting to the Web service catalog requires this and click OK.

Running ArcCatalog under a different operating system account

As you read earlier, to administer a GIS server, you must be logged in with an operating system account that has access to the GIS server, where that account is a member of the agsadmin users group. In general, it's good practice to use an account with more restrictive permissions when performing your daily tasks and to use the administrative account only when you need to administer your server. To run ArcCatalog under the administrative account without logging off and back on again, you can use the Windows runas command.

The runas command allows you to run ArcCatalog under an operating system account other than your current logon. You can run the runas command at the command prompt and, in some operating system versions, as a right-click menu option on the ArcCatalog executable, located in the bin folder of your install location. For example, at the command prompt type:

```
runas /user:username "C:\Program Files\ArcGIS\bin\ArcCatalog.exe"
```

Alternatively, navigate to the ArcCatalog.exe file in the bin folder, or locate your shortcut to it, right-click the executable, and click Run as. Consult the Windows documentation for more information on using the runas command.

ADDING CONTAINER MACHINES TO HOST SERVER OBJECTS

The first time you connect to your GIS server, you will need to add one or more server object containers, or container machines, to it. During installation, you should have installed software on one or more machines to function as container machines. The container machines host the server objects and are the workhorses of your GIS server. For your GIS server to utilize these container machines, you need to link the SOM machine to the container machines. Follow the steps below to add container machines to your GIS server.

1. In the Catalog tree, right-click the name of your GIS server and click Server Properties.

2. Click the Hosts tab and click Add.

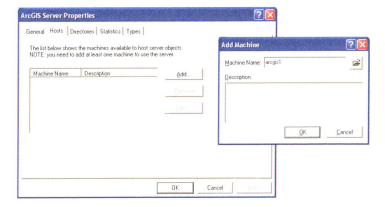

3. Type the machine name or click the Browse button to locate it on the network. This is the machine on which you installed the SOC software. For smaller system configurations, this may be the same machine on which you installed the server object manager.

4. Click OK.

5. Add any other container machines you want to connect to the GIS server.

Now you're ready to add server objects to your server. This is covered in the next section.

The server objects you run on your GIS server are derived from the same resources that you work with in ArcGIS Desktop. These resources are, for example, the maps and locators you use in your everyday work. By hosting them on the server, you make them available to any client application that has access to the server. For example, if you want to quickly share a map among several departments, you could add it to your GIS server. Then, the people in each department could connect to the server and view the map in ArcGIS Desktop.

ORGANIZING YOUR GIS DATA

Before you add a server object to your GIS server, you need to consider what that server object needs to work properly. For example, a map document contains layers that reference shapefiles, geodatabases, coverages, images, and so on. When you open the map in ArcMap, the layers won't display if the data referenced by the layers are not accessible to you. The same holds true when you add a server object, such as a map, to your GIS server.

All the data that is required for the server object must be accessible from the container machines that comprise your GIS server, and the server object container account must have the appropriate access permissions. Follow these guidelines when organizing the GIS data that is needed for your server objects.

- All data required for the server object must be accessible from all container machines that comprise your GIS server through shared network directories.

- The ArcGIS Server Container account you established during the postinstall should have read access to any shared network directories.

- Any file-based references to the data in map documents should be specified using universal naming convention (UNC) pathnames—for instance, \\server\data\layer1.

- ArcSDE connections must be saved in map documents.

ADDING A SERVER OBJECT TO THE SERVER

At the initial release, ArcGIS Server supports two types of server objects: map server objects and geocode server objects. You need to create the GIS resource (that is, map document or locator) before you can add it to the GIS server. Once created, follow the steps below to add the resource as a server object to the GIS server.

1. Connect to the GIS server to which you want to add a server object.

2. In the Catalog tree, expand the server folder and double-click Add Server Object. The Add Server Object Wizard appears.

   Note: If this is the first time you're accessing the server, you'll need to add one or more container machines to it before you can add a server object. See the section called 'Adding container machines to host server objects' earlier in this chapter for information on adding container machines.

3. Type the name of the server object. This is the name people will see and use to identify the server object. Be descriptive. The name can only contain alphanumeric characters and underscores. No spaces or special characters are allowed. The name cannot be more than 120 characters in length.

4. Click the Type dropdown arrow, click the server object type, and click Next.

5a. If you're creating a map server object, browse to the map document and choose the data frame to display. Optionally, specify an output directory. If you don't specify an output directory, the GIS server will only return images as MIME data. By specifying an output directory that has an associated virtual directory, the returned images will also be accessible via a Uniform Resource Locator (URL) address. See the section titled 'Configuring server directories' later in this chapter for more information.

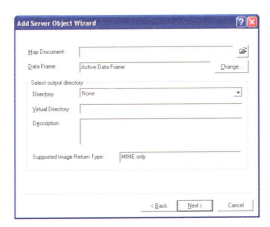

5b. If you're creating a geocode server object, browse to the locator and set the batch size. Click Next.

6. Click Pooled or Not Pooled and optionally change the maximum usage and wait times. Click Next. See Chapter 2, 'The ArcGIS Server architecture', for more information on server object pooling, usage time, and wait time.

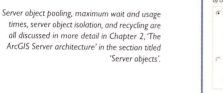

Server object pooling, maximum wait and usage times, server object isolation, and recycling are all discussed in more detail in Chapter 2, 'The ArcGIS Server architecture' in the section titled 'Server objects'.

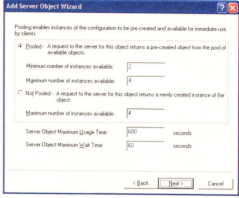

7. Set the process isolation level and the recycling parameters. Click Next. See Chapter 2, 'The ArcGIS Server architecture', for more information.

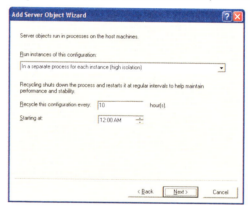

8. Click Yes to start the server object now and click Finish.

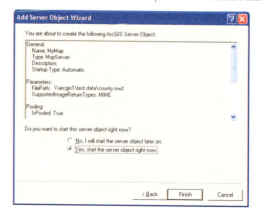

For information on using server objects in ArcGIS Desktop, consult Using ArcCatalog *and* Using ArcMap.

9. Verify that your server object is working properly. You can preview a map server object or display the properties of a geocode server object in ArcCatalog to ensure that the server object is correctly configured. If, for some reason, your server object is not working as expected, you can review the log files for errors. For more information on accessing and reading log files, consult Appendix B, 'Configuration and log files'.

ADDING MULTIPLE SERVER OBJECTS TO THE SERVER

ArcCatalog also allows you to add more than one server object to the server at one time. The default configuration will create pooled objects, with a minimum of two and a maximum of four instances, and a high process isolation that is recycled every 10 hours starting at midnight. You can change the configuration of a server object through its property pages once you've added it to the server.

To add more than one server object to the server, follow the steps below.

1. Connect to the ArcGIS server you want to add server objects to.

2. Navigate the Catalog tree and locate the resources you want to add, for example, all the maps you want to add to the server.

3. Select the resources and drag and drop them onto the server.

4. To start the server objects, from the ArcGIS Server Object Administration toolbar, click Start.

REMOVING A SERVER OBJECT FROM THE SERVER

When you remove a server object from your server, any client application accessing the server object will no longer be able to access it, which may result in an error being returned to the client. In general, you may want to pause a server object before deleting it. This will prevent any subsequent client's access and allow existing clients to finish using the object. The next section discusses starting, stopping, and pausing server objects in more detail.

1. Connect to the GIS server containing the server object you want to delete.

2. Right-click the server object and click Delete.

Once you've added server objects to your server, the server will respond to clients that need to utilize your server objects. Using ArcCatalog, you can monitor how well the server responds to clients and troubleshoot any problems that may arise. Over time, demand for your server objects may require that you make a change to your server. For example, you might allocate more instances for a particular server object in high demand, or you might add more machines to your server to handle the increased load.

STARTING, STOPPING, AND PAUSING SERVER OBJECTS

You can start, stop, and pause a server object at any time. In ArcCatalog, right-click a server object to reveal the context menu for starting, stopping, and pausing the server object. Alternatively, you can display the ArcGIS Server Object Administration toolbar that contains these same operations.

The ArcGIS Server Object Administration toolbar contains start, stop, and pause buttons.

Starting a server object makes it available for client access. When you start a pooled server object, the server instantiates the minimum number of objects you specified. As clients request the object, the server returns an available object from the pool of objects. If none are currently available, the server instantiates more server objects to meet demand until the maximum number of objects is reached. At this point, any new clients that make a request for the object will be placed in a queue for the next available instance. The server object will continue to run with the maximum number of instances until you stop the server object. When you start a non-pooled object, the server creates instances as clients request them, up to the maximum number of objects that you specify.

When you stop a server object, the server immediately removes all instances of that object from the server. This frees up any machine resources that were dedicated to the server object. Clients that were using the object may fail to work properly because the object is no longer available.

When you pause a server object, the server refuses any new client requests for the object. However, existing clients can complete their use of the object. Pausing a server object doesn't remove instances from the server. You might pause a server object if a data source required by the server object is not available. For example, if your map server objects reference an ArcSDE server containing a geodatabase, and that ArcSDE server is down for maintenance, you might pause the map server object until the database is available again. Because the instances of the server object are not removed when you pause it, it's much quicker to start a paused server object than one that is stopped.

MONITORING PERFORMANCE

As you've read in previous chapters, you control the performance of server objects by setting options such as whether the object is pooled or not, whether it's running in its own process (high or low isolation), and how many instances you've allocated to it. ArcCatalog allows you to set and manage all these properties. But how can you tell whether or not you need to make adjustments to these settings?

Before you put a server object on the server, you should have some idea as to how it will be used. Will the server object be accessed from, for example, a Web application? How many people do you expect to use the application? Knowing the answers to these and other similar questions will help you determine how to initially configure the server object.

When you first add a server object to the server, you'll set some initial values for its configuration. As clients begin accessing the server object, you can monitor its performance by examining its statistics. You can review statistics for your entire GIS server as well as each individual server object. You can examine how many requests get processed per unit of time, what the average wait time is for a client, and how many requests timed out and didn't get a response back from the server.

Use your common sense when evaluating the statistics. If you notice a high number of clients requesting the object and many time-outs, you can try increasing the number of instances for the server object to handle the load. At some point, increasing the number of instances won't improve performance because you've reached the capacity of your server machines. To alleviate this issue, you can try reducing the number of instances allocated to other server objects or, if that's not possible, you can add new server machines to your ArcGIS Server system.

The server log files can also provide useful information on the status of a particular server object. For more information on log files, see Appendix B, 'Configuration and log files'.

DISPLAYING STATISTICS FOR A SERVER OBJECT

Statistics for a server object can be found on the property pages of the server itself.

1. Right-click the name of the GIS server that contains the server object for which you want to obtain statistics and click Server Properties.

2. Click the Statistics tab.

3. Click the Server Obj dropdown arrow and click the particular server object you want statistics for.

4. Click Show Statistics.

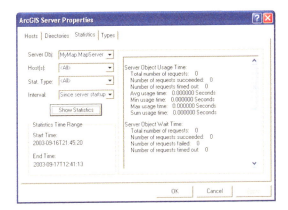

LIMITING THE SIZE OF QUERY RESULTS

When working with map and geocode server objects, it's up to you, as the developer, to ensure that you use these objects appropriately. For example, you don't want your application to send a query to the server that can potentially return all the records in a particular table. Doing so will undoubtedly have a negative impact on the performance of the server and of the database.

Both the *MapServer* and *GeocodeServer* have built-in limits on the number of results returned from a query. These limits are exposed as properties of the configuration itself. ArcCatalog does not expose any of these properties in its user interface. To change the default limits, you need to directly edit the *MapServer* and *GeocodeServer* configuration files and change the value associated with the appropriate XML tag. For more information about the GIS server's configuration files and how to modify them, see Appendix B.

The properties that limit queries against the *MapServer* and *GeocodeServer* are described below.

MapServer query limits

MaxRecordCount—Limits the number of records returned from a query. Applies specifically to the following methods on *IMapServer*: *QueryFeatureData*, *Find*, and *Identify*. The default value is 500.

For additional information on these limits, see the section titled 'Limiting the size of query results and output' in Chapter 4.

MaxBufferFeatures—Limits the number of buffers that can be drawn around the selected set of features of an *ILayerDescription*. The default value is 100.

GeocodeServer query limits

MaxResultsSize—Limits the number of records returned by the *FindAddressCandidates* and *GeocodeAddresses* methods. The default value is 500.

MaxBatchSize—Limits the number of input records that can be passed into the *GeocodeAddresses* method.

LIMITING THE SIZE OF OUTPUT

Another situation where you need to be careful is when your application writes output to a server directory. If your application generates large files, it may consume a large amount of disk space and also require a large amount of resources to produce. MapServer has built-in limits for the size of images that the *ExportMapImage* method will produce. You set this maximum through two properties, *MaxImageWidth* and *MaxImageHeight*.

MaxImageWidth—The maximum width of the exported image in pixels. The default value is 2048.

MaxImageHeight—The maximum height of the exported image in pixels. The default value is 2048.

You can modify these properties in the *MapServer*'s configuration file. For more information about the GIS server's configuration files and how to modify them, see Appendix B.

Your GIS server consists of one or more server machines that together host the GIS server objects you want to make available. You manage these server machines with ArcCatalog in much the same way you manage the individual server objects running on your system. At times, you may need to add new machines to your server to increase performance or handle increased traffic. Other times, you may need to remove a particular machine for maintenance. The server itself also has a set of properties that governs how it runs—for example, where output files are generated and who has access to the server.

ADDING AND REMOVING CONTAINER MACHINES

One of the first things you'll need to do when you first install ArcGIS Server is add one or more machines (server object containers, or container machines) to host your server objects. As time passes, you'll periodically need to add or remove machines for various reasons. When you add a machine to your server, you'll be able to take advantage of the additional computing power provided by the new machine, which the GIS server will start using immediately.

When you remove a server machine from your system, the GIS server will more heavily utilize the machine resources of the remaining container machines in your system, which may affect performance of the GIS server as a whole. The server objects that had been running on the machine you remove will be reallocated to other machines.

Adding a container machine to your server

1. Connect to your GIS server in the Catalog tree.

2. Right-click the name of the server and click Server Properties.

3. Click the Hosts tab.

4. Click Add.

5. Type the machine name or click the Browse button to locate it on the network.

6. Click OK.

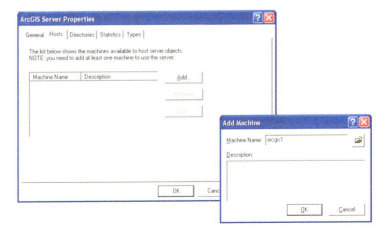

Removing a container machine from your server

1. Connect to your ArcGIS Server in the Catalog tree.

2. Right-click the name of the server and click Server Properties.

3. Click the Hosts tab.

4. Click the name of the server machine you want to remove and click Remove.

CONFIGURING SERVER DIRECTORIES

A server directory represents a physical directory on your network that is accessible to all the container machines of your GIS server. The GIS server manages one or more server directories and periodically deletes files contained within them—at an interval you specify. Typically, the files written to a server directory are transient and need only be available to a client application for a short period of time. A server directory may also have an associated virtual directory. By specifying a virtual directory, you allow the contents of the server directory to be available through a Web server via a URL address. You must create the virtual directory beforehand in your Web server and link it to the physical directory on disk. For more information on creating a virtual directory, consult the documentation for your Web server.

If you don't specify a server directory when configuring a map server object, all images generated by the GIS server will be returned as MIME data. By specifying a server directory with a map server object, output images will also be written to the specified server directory and available through the Web server via the virtual directory. Thus, you should specify a server directory when your client application requires it. ArcGIS Desktop applications and the Web templates distributed with the Application Developer Frameworks for .NET and Java can all work with MIME data.

Any custom applications you create can also write files to the server directories. If you want the GIS server to clean up your custom files, prefix them with "_ags_" (one underscore on each side of ags). Thus, a file named _ags_myfile will get deleted by the GIS server. For additional information on server directories, see Chapter 2, 'The ArcGIS Server architecture'. Follow the steps below to create a server directory.

1. Right-click the server in the Catalog tree and click Server Properties.

2. Click the Directories tab and click Add.

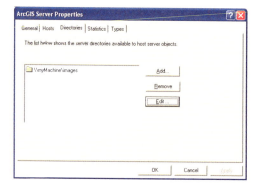

3. Set the Output directory, for example, \\mymachine\output. The ArcGIS Server Container user account that you established during the postinstall phase must have write access to this location. You will need to use the operating system tools to set the appropriate access to this location.

4. Optionally, set the Virtual directory, for example, http://server_name/output. This virtual directory should point to the same disk location as the output directory. You must create the virtual directory in your Web server and link it to the physical directory on disk. The virtual directory provides Web applications access to files created by the GIS server via a URL address.

5. Set the cleaning mode and maximum file age. You can have the server delete files at regular intervals based either on the age of the file or when the file was last accessed by a client.

SPECIFYING THE LOG FILE LOCATION

ArcGIS Server records system messages in log files. When you suspect something isn't working right, you can examine the log files to see exactly what is happening with your GIS server. You can specify where the server writes its log files, or simply accept the default location in the ArcGIS Server installation directory, <ArcGIS Server Install directory>\log on the SOM machine. Follow the steps below to view or change the current log directory.

1. Right-click the name of the GIS server for which you want to view or modify the log file location.

2. Click the General tab.

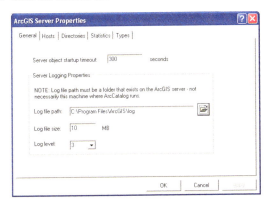

3. Type a location or click the Browse button to set the location for the log files.

4. Set the maximum file size for an individual log file. When a log file reaches this size, the server will create a new one.

5. Set the logging level.

0—No logging
1—Errors only
2—Errors and warnings
3—Errors, warnings, and brief administrative messages
4—Errors, warnings, and detailed administrative messages
5—Errors, warnings, and verbose messages used for debugging purposes

For more information on reading the log files and deciphering messages, see Appendix B, 'Configuration and log files'.

DISPLAYING STATISTICS FOR THE SERVER

Statistics for the server can be found on the property pages of the server.

1. Right-click the name of the GIS server for which you want to obtain statistics and click Server Properties.

2. Click the Statistics tab.

3. Choose the statistics you wish to view and click Show Statistics.

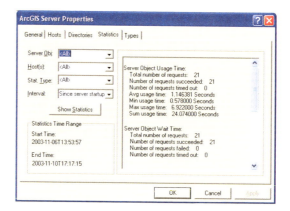

USING A PROXY SERVER TO CONNECT TO THE INTERNET

A proxy server is a computer on your LAN that connects to the Internet without compromising the security of your internal network. If your organization uses a proxy server to connect to the Internet, you need to configure ArcCatalog so that it can connect to GIS servers on the Internet through the proxy server.

When configuring a proxy server, all Internet connections—not Local Server connections—to GIS servers (both ArcGIS and ArcIMS servers) will utilize this proxy server. Follow the steps below to configure ArcCatalog to use a proxy server.

1. In ArcCatalog, click the Tools menu and click Options.

2. Click the Proxy Server tab.

3. Check Use a proxy server for your ArcGIS/ArcIMS server connections.

4. Type the address and port number of the proxy server you want to use to connect to the Internet.

5. Optionally, type a username and password, if required.

4

Developing ArcGIS Server applications

Programming ArcGIS Server applications is about programming with ArcObjects that are running remotely on a server. Developers can become effective ArcGIS Server application developers if they understand the ArcObjects programming model and some key rules for programming with the server.

This chapter contains important information including rules and best practices for developing ArcGIS Server applications. If you are building Web applications and Web services with .NET or Java, this chapter is a must read and complements later chapters that focus on .NET and Java development.

This chapter covers topics such as:

• the ArcGIS Server API • working with ArcObjects in the server • working with server contexts • stateful versus stateless use of the ArcGIS Server • application performance and scalability

Programming ArcGIS Server applications is all about programming ArcObjects. The key to programming ArcGIS Server applications is that these applications work with ArcObjects that actually run remotely on container machines managed by your GIS server.

An application developer who can use ArcObjects to build an application that runs on the desktop can also build ArcGIS Server applications by learning some rules and programming patterns for working with remote ArcObjects. ArcGIS Server is a platform for building Web applications and Web services. While the ArcObjects programming model for ArcGIS Server will be familiar to ArcObjects developers, developing Web applications does require knowledge of Internet programming using frameworks such as ASP.NET and J2EE.

This chapter will focus on the ArcObjects programming rules and patterns to program ArcGIS Server applications. The information in this chapter is complementary to information about using ArcGIS Server that is more specific to developing Web applications using the .NET and Java ADFs in Chapters 5 and 6, respectively.

It is recommended that developers first become familiar with the ArcObjects programming model, which is discussed in great detail in the ArcGIS Developer Help system and in the appendixes of this book. This chapter assumes knowledge of ArcObjects, except where the programming APIs and patterns are specific to ArcGIS Server development.

ArcGIS SERVER DEFINITIONS AND CONCEPTS

Before diving into the details, it's important to review some key terms and concepts from Chapters 1 and 2 that will be used throughout this chapter:

GIS server: The GIS server is responsible for hosting and managing server objects. The GIS server is the set of objects, applications, and services that make it possible to run ArcObjects components on a server. The GIS server consists of a server object manager and one or more server object containers.

Server object manager: The SOM is a Windows service that manages the set of server objects that are distributed across one or more container machines. When an application makes a connection to an ArcGIS Server over a LAN, it is making a connection to the SOM.

Server object container: A SOC is a process in which one or more server objects is running. SOC processes are started and shut down by the SOM. The SOC processes run on the GIS server's container machines. Each container machine is capable of hosting multiple SOC processes.

Server object: A server object is a coarse-grained ArcObjects component, that is, a high-level object that simplifies the programming model for doing certain operations and hides the fine-grained ArcObjects that do the work. Server objects support coarse-grained interfaces that have methods that do large units of work, such as "draw a map" or "geocode a set of addresses". Server objects also have SOAP interfaces, which make it possible to expose server objects as Web services that can be consumed by clients across the Internet.

Web server: The Web server hosts Web applications and Web services written using the ArcGIS Server API. These Web applications use the ArcGIS Server API to connect to a SOM to make use of server objects and to create ArcObjects for use in their applications.

Pooled server object: ArcGIS Server allows you to pool instances of server objects such that they can be shared between multiple application sessions at the per-request level. This allows you to support more users with fewer resources. Pooled server objects should be used by applications that make stateless use of the GIS server.

Non-pooled server object: Non-pooled server objects are created for exclusive use by an application session and destroyed when returned by the application to the server. The creation of the object includes creating the object and loading up any initialization data, for example, the map document associated with a map server object. Users of a server application that makes use of a non-pooled server object require an instance of that object dedicated to their session. Non-pooled server objects are for applications that make stateful use of the GIS server.

The ArcGIS Server is a distributed system that consists of a server object manager, server containers, and clients to the server, such as desktop applications and Web applications.

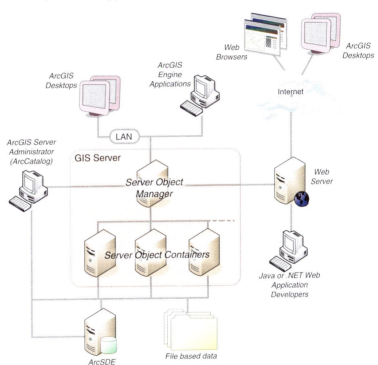

Application developer framework: A collection of Web controls, convenience classes and data objects, and application templates that make it easy to build and deploy .NET or Java Web applications that use ArcObjects running within the GIS server. The ADF includes a software developer kit with Web controls, Web application templates, developer help, and code samples. It also includes a Web application runtime that allows you to deploy applications without having to install ArcObjects on your Web server.

The ArcGIS Server has three application programming interfaces:

- The Server API
- The .NET Web Controls
- The Java Web Controls

This chapter focuses on the server API, and Chapters 5 and 6 focus on the .NET Web controls and Java Web controls, respectively. It's important to understand the relationship between the Web control APIs, server API, and server programming model if you want to build applications that go beyond simple map viewing and query.

The server API is a collection of object libraries that contains the ArcObjects necessary to write an application that connects to the GIS server and makes use of server objects. These object libraries are available to any developer who installs the ArcGIS Desktop, ArcGIS Engine, or ArcGIS Server products. The ArcGIS Server product also includes a set of Web controls and Web application templates as part of the ADF. Programming with the ADF and the Web controls is the subject of Chapters 5 and 6.

It is possible to write a number of different types of applications using ArcGIS Server. The developer who works with the ADF can build server applications,

Developers can build different types of ArcGIS Server applications. These include multiuser server applications, such as Web applications and Web services written using the ADF. ArcGIS Server also supports single-user desktop applications using the ADF, ArcGIS Engine, or ArcGIS Desktop products.

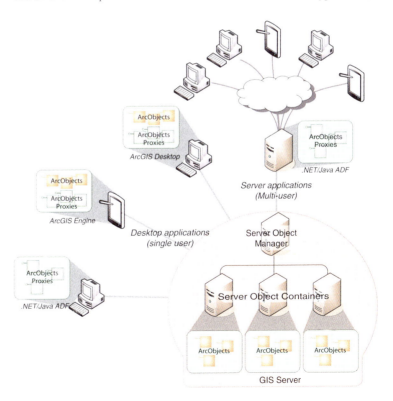

such as Web applications, Web services, and enterprise applications—for example, EJBs. Developers can also build desktop applications using .NET or Java. These applications are deployed using the ADF runtime.

ArcGIS Engine developers can build desktop applications that make use of ArcObjects running in the GIS server. ArcGIS Desktop developers can extend the ArcGIS Desktop applications to include functionality that makes use of the GIS server. In both of these cases, the deployment of the application itself requires an ArcGIS Engine runtime deployment license or an ArcGIS Desktop license.

In the case of ArcGIS Desktop and ArcGIS Engine, developers who write desktop applications that use the server will have ArcObjects installed on the machine on which the application is both developed and deployed on. When using these products to write applications that use ArcGIS Server, you must follow the same programming guidelines as a developer using the ADF to build a server application, such as a Web application that is deployed within a Web server. The only difference is that in the case of the desktop application, each instance of your application is bound to a single user session (though there might be multiple instances of your application running at any time), while Web applications or Web services are multiuser/multisession applications.

A proxy object is a local representation of a remote object. The proxy object controls access to the remote object by forcing all interaction with the remote object to be via the proxy object. The supported interfaces and methods on a proxy object are the same as those supported by the remote object. You can make method calls on, and get and set properties of, a proxy object as if you were working directly with the remote object.

The .NET ADF runtime and Java ADF runtime installation of ArcGIS Server do not include ArcObjects components, but they do include .NET assemblies and OLBs (in the case of the .NET runtime), and JAR files (in the case of the Java runtime) that provide proxies for working with ArcObjects running within the server (in addition to Web control APIs). Applications that are built and deployed using one of the ArcGIS Server runtime installs must follow the coding guidelines of the server, or they won't work. Those aspects of the coding guidelines will become more apparent later in this chapter.

THE SERVER API

Programming with the server API is all about remotely programming ArcObjects. Programming ArcObjects remotely is the same as programming ArcObjects for use in desktop applications, but there are some additional details and programming guidelines you need to follow. You need to understand:

- How to connect to the server.

- How to get objects that are running within the server.

- How to create new objects within the server.

- The best ways (dos and don'ts) for working with remote ArcObjects.

The rest of programming the server is just programming ArcObjects. Each aspect of programming the server listed above will be described in more detail in the following sections.

ArcGIS Server developers have access to a set of visual Web controls that permit the use of many properties, events, and methods. The server has no ArcGIS Desktop applications, such as ArcMap, or any user interface components except for the Web controls. Although a simple application can be built with just the Web controls, practical applications of the server require knowledge of the

For a comprehensive discussion on each library, refer to the library overview topics, a part of the library reference section of the ArcGIS Developer Help system.

object libraries that compose the ArcGIS Server. The libraries contained within the ArcGIS Server are summarized below. The diagrams that accompany this section indicate the library architecture of the ArcGIS Server. Understanding the library structure, their dependencies, and basic functionality will help you as a developer navigate through the components of ArcGIS Server. The libraries are discussed in dependency order. The diagrams show this with a number in the upper right corner of the library block.

Object libraries are logical collections of the programmable ArcObjects components, ranging from fine-grained objects (for example, individual geometry objects) to coarse-grained objects, which aggregate logical collections of functionality (for example, an ArcMap object to work with map documents). Programmers use a number of standards-based APIs (COM, .NET, Java, and C++). These same libraries are used to program the ArcGIS Desktop and the ArcGIS Engine.

SYSTEM

The System library is the lowest level library in the ArcGIS architecture. The library contains components that expose services used by the other libraries composing ArcGIS. There are a number of interfaces defined within System that can be implemented by the developer. The developer does not extend this library but can extend the ArcGIS system by implementing interfaces contained within this library.

SYSTEMUI

The SystemUI library contains the interface definitions for user interface components that can be extended within the ArcGIS system. These include the *ICommand*, *ITool*, and *IToolControl* interfaces. The objects contained within this library are utility objects available to the developer to simplify some user interface developments. The developer does not extend this library but can extend the ArcGIS system by implementing interfaces contained within this library.

GEOMETRY

Knowing the library dependency order is important since it affects the way in which developers interact with the libraries as they develop software. As an example, C++ developers must include the type libraries in the library dependency order to ensure correct compilation. Understanding the dependencies also helps when deploying your developments.

The Geometry library handles the geometry, or shape, of features stored in feature classes or other graphical elements. The fundamental geometry objects that most users will interact with are *Point*, *MultiPoint*, *Polyline*, and *Polygon*. Besides those top-level entities are geometries that serve as building blocks for polylines and polygons. Those are the primitives that compose the geometries. They are *Segments*, *Paths*, and *Rings*. Polylines and polygons are composed of a sequence of connected segments that form a path. A segment consists of two distinguished points, the start and the end point, and an element type that defines the curve from start to end. The types of segments are *CircularArc*, *Line*, *EllipticArc*, and *BezierCurve*. All geometry objects can have Z, M, and IDs associated with their vertices. The fundamental geometry objects all support geometric operations such as *Buffer* and *Clip*. The geometry primitives are not meant to be extended by developers.

Entities within a GIS refer to real-world features; the location of these real-world features is defined by a geometry along with a spatial reference. Spatial reference objects, for both projected and geographic coordinate systems, are included in the Geometry library. Developers can extend the spatial reference system by adding new spatial references and projections between spatial references.

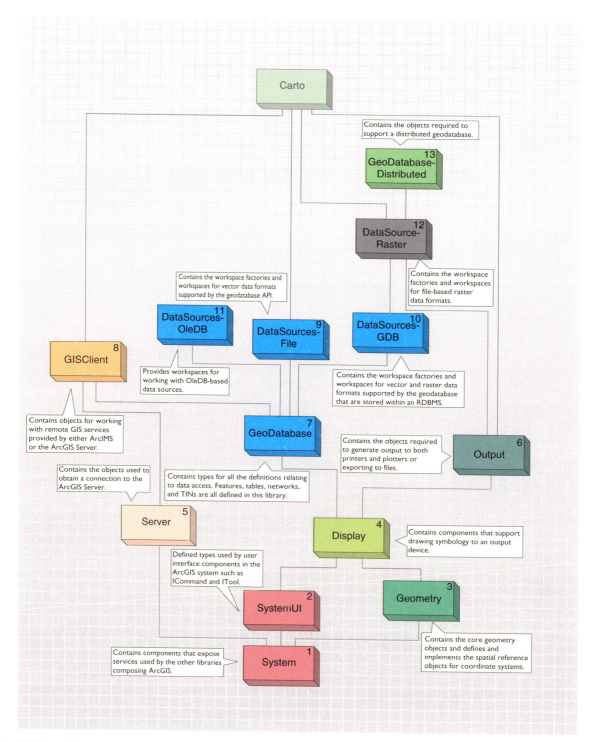

DISPLAY

The Display library contains objects used for the display of GIS data. In addition to the main display objects responsible for the actual output of the image, the library contains objects that represent symbols and colors used to control the properties of entities drawn on the display. The library also contains objects that provide the user with visual feedback when interacting with the display. Developers most often interact with Display through a view similar to the ones provided by the *Map* or *PageLayout* objects. All parts of the library can be extended; commonly extended areas are symbols, colors, and display feedbacks.

SERVER

The Server library contains objects that allow you to connect and work with ArcGIS Servers. Developers gain access to an ArcGIS Server using the *GISServerConnection* object. The *GISServerConnection* object gives access to the *ServerObjectManager*. Using this object, a developer works with *ServerContext* objects to manipulate ArcObjects running on the server. The Server library is not extended by developers. Developers can also use the GISClient library when interacting with the ArcGIS Server.

OUTPUT

The Output library is used to create graphical output to devices, such as printers and plotters, and hardcopy formats such as enhanced metafiles and raster image formats (JPG, BMP, and so on). The developer uses the objects in the library with other parts of the ArcGIS system to create graphical output. Usually these would be objects in the Display and Carto libraries. Developers can extend the Output library for custom devices and export formats.

GEODATABASE

The GeoDatabase library provides the programming API for the geodatabase. The geodatabase is a repository of geographic data built on standard industry and object relational database technology. The objects within the library provide a unified programming model for all supported data sources within ArcGIS. The GeoDatabase library defines many of the interfaces that are implemented by data source providers higher in the architecture. The geodatabase can be extended by developers to support specialized types of data objects (features, classes, and so forth); in addition, it can have custom vector data sources added using the *PlugInDataSource* objects. The native data types supported by the geodatabase cannot be extended.

GISCLIENT

The GISClient library allows developers to consume Web services; these Web services can be provided by ArcIMS and ArcGIS Server. The library includes objects for connecting to GIS servers to make use of Web services. There is support for ArcIMS Image and Feature Services. The library provides a common programming model for working with ArcGIS Server objects in a stateless manner either directly or through a Web service catalog. The ArcObjects components running on the ArcGIS Server are not accessible through the GISClient interface. To gain direct access to ArcObjects components running on the server, you should use functionality in the Server library.

DataSourcesFile

The DataSourcesFile library contains the implementation of the GeoDatabase API for file-based data sources. These file-based data sources include shapefile, coverage, triangulated irregular network (TIN), computer-aided drafting (CAD), smart data compression (SDC), StreetMap, and vector product format (VPF). The DataSourcesFile library is not extended by developers.

DataSourcesGDB

The DataSourcesGDB library contains the implementation of the GeoDatabase API for the database data sources. These data sources include Microsoft Access and relational database management systems supported by ArcSDE—IBM DB2, Informix, Microsoft SQL Server, and Oracle. The DataSourcesGDB library is not extended by developers.

DataSourcesOleDB

The DataSourcesOleDB library contains the implementation of the GeoDatabase API for the Microsoft OLE DB data sources. This library is only available on the Microsoft Windows operating system. These data sources include any OLE DB-supported data provider and text file workspaces. The DataSourcesOleDB library is not extended by developers.

DataSourcesRaster

Raster Data Objects is a COM API that provides display and analysis support for file-based raster data.

The DataSourcesRaster library contains the implementation of the GeoDatabase API for the raster data sources. These data sources include relational database management systems supported by ArcSDE—IBM DB2, Informix, Microsoft SQL Server, and Oracle—along with supported Raster Data Objects (RDO) raster file formats. Developers do not extend this library when support for new raster formats is required; rather, they extend RDO. The DataSourcesRaster library is not extended by developers.

GeoDatabaseDistributed

The GeoDatabaseDistributed library supports distributed access to an enterprise geodatabase by providing tools for importing data into and exporting data out of a geodatabase. The GeoDatabaseDistributed library is not extended by developers.

Carto

The Carto library supports the creation and display of maps; these maps can consist of data in one map or a page with many maps and associated marginalia. The *PageLayout* object is a container for hosting one or more maps and their associated marginalia: North arrows, legends, scale bars, and so on. The *Map* object is a container of layers. The *Map* object has properties that operate on all layers within the map—spatial reference, map scale, and so on—along with methods that manipulate the map's layers. There are many different types of layers that can be added to a map. Different data sources often have an associated layer responsible for displaying the data on the map: vector features are handled by the *FeatureLayer* object, raster data by the *RasterLayer*, TIN data by the *TinLayer*, and so on. Layers can, if required, handle all the drawing operations for

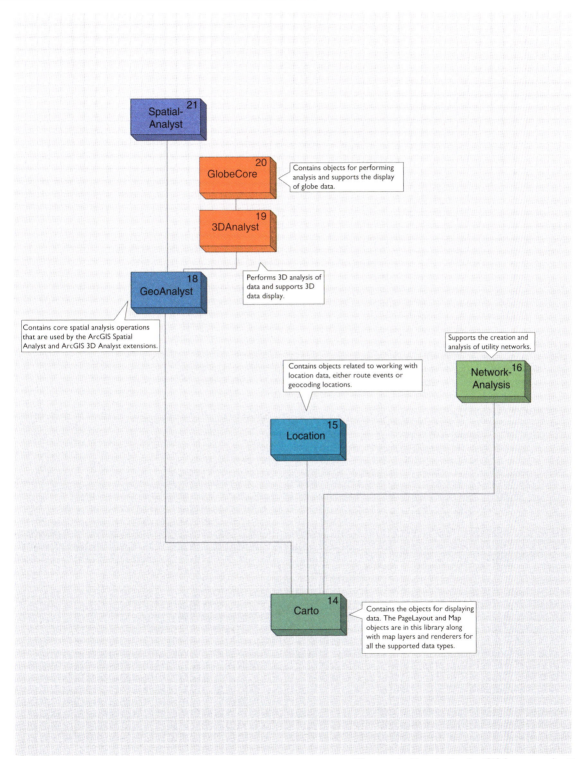

Contains objects for performing analysis and supports the display of globe data.

Performs 3D analysis of data and supports 3D data display.

Contains core spatial analysis operations that are used by the ArcGIS Spatial Analyst and ArcGIS 3D Analyst extensions.

Supports the creation and analysis of utility networks.

Contains objects related to working with location data, either route events or geocoding locations.

Contains the objects for displaying data. The PageLayout and Map objects are in this library along with map layers and renderers for all the supported data types.

their associated data, but it is more common for layers to have an associated *Renderer* object. The properties of the *Renderer* object control how the data is displayed in the map. Renderers commonly use symbols from the Display library for the actual drawing; the renderer simply matches a particular symbol with the properties of the entity that is to be drawn. A *Map* object, along with a *PageLayout* object, can contain elements. An element has geometry to define its location on the map or page, along with behavior that controls the display of the element. There are elements for basic shapes, text labels, complex marginalia, and so on. The Carto library also contains support for map annotation and dynamic labeling.

Although developers can directly make use of the *Map* or *PageLayout* objects in their applications, it is more common for developers to use a higher level object such as the *MapControl*, *PageLayoutControl*, or an ArcGIS application. These higher level objects simplify some tasks, although they always provide access to the lower level *Map* and *PageLayout* objects, allowing the developer fine control of the objects.

The ArcGIS Server uses the MapServer object for its MapService.

The *Map* and *PageLayout* objects are not the only objects in Carto that expose the behavior of map and page drawing. The *MxdServer* and *MapServer* objects both support the rendering of maps and pages, but instead of rendering to a window, these objects render directly to a file.

Using the *MapDocument* object, developers can persist the state of the map and page layout within a map document (.mxd), which can be used in ArcMap or one of the ArcGIS controls.

The Carto library is commonly extended in a number of areas. Custom renderers, layers, and so forth, are common. A custom layer is often the easiest method of adding custom data support to a mapping application.

LOCATION

The Location library contains objects that support geocoding and working with route events. The geocoding functionality can be accessed through fine-grained objects for full control, or the *GeocodeServer* objects offers a simplified API. Developers can create their own geocoding objects. The linear referencing functionality provides objects for adding events to linear features and rendering these events using a variety of drawing options. The developer can extend the linear reference functionality.

NETWORKANALYSIS

The NetworkAnalysis library provides objects for populating a geodatabase with network data and objects to analyze the network when it is loaded in the geodatabase. Developers can extend this library to support custom network tracing. The library is meant to work with utility networks: gas lines, electricity supply lines, and so on.

GEOANALYST

The GeoAnalyst library contains objects that support core spatial analysis functions. These functions are used within both the SpatialAnalyst and 3DAnalyst libraries. Developers can extend the library by creating a new type of raster operation. A license for either the ArcGIS Spatial Analyst or 3D Analyst extension or the ArcGIS Engine Runtime Spatial or 3D option is required to make use of the objects in this library.

3DAnalyst

The 3DAnalyst library contains objects for working with 3D scenes in a similar way that the Carto library contains objects for working with 2D maps. The *Scene* object is one of the main objects of the library since it is the container for data similar to the *Map* object. The *Camera* and *Target* objects specify how the scene is viewed regarding the positioning of the features relative to the observer. A scene consists of one or more layers; these layers specify the data in the scene and how the data is drawn.

It is not common for developers to extend this library. A license for either the ArcGIS 3D Analyst extension or the ArcGIS Engine Runtime 3D option is required to work with objects in this library.

GlobeCore

The GlobeCore library contains objects for working with globe data in a similar way that the Carto library contains objects for working with 2D maps. The *Globe* object is one of the main objects of the library since it is the container for data similar to the *Map* object. The *GlobeCamera* object specifies how the globe is viewed regarding the positioning of the globe relative to the observer. The globe can have one or more layers; these layers specify the data on the globe and how the data is drawn.

It is not common for developers to extend this library. A license for either the ArcGIS 3D Analyst extension or the ArcGIS Engine Runtime 3D option is required to work with objects in this library.

SpatialAnalyst

The SpatialAnalyst library contains objects for performing spatial analysis on raster and vector data. Developers most commonly consume the objects within this library and do not extend it. A license for either the ArcGIS Spatial Analyst extension or the ArcGIS Engine Runtime Spatial option is required to work with objects in this library.

.NETWebControls

The .NET WebControls assembly contains Web controls and convenience classes that make it easy to build and deploy .NET Web applications and Web services that use ArcObjects running within the GIS server.

JavaWebControls

The Java WebControls JAR contains Web controls and data objects that make it easy to build and deploy Java Web applications and Web services that use ArcObjects running within the GIS server.

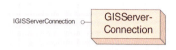

IGISServerConnection ○— GISServer-Connection

The GISServerConnection object provides connections to the GIS server and access to the ServerObjectManager and ServerObjectAdmin objects.

To make use of the GIS server to host ArcObjects by your application, the first thing that application must do is connect to the GIS server, that is, connect to the SOM. Connections to the GIS server are made through the *GISServerConnection* object. The *GISServerConnection* object supports a single interface (in addition to *IUnknown*): *IGISServerConnection*. *IGISServerConnection* has a *Connect* method that connects the application to the GIS server.

THE SERVERCONNECTION OBJECTS

You can work with the GIS server using COM, .NET, or Java. All of these runtimes have native server connection objects that you create to connect to the server. When developing ArcGIS Server applications using COM (for example, with VB or C++), the COM server connection object can be found in the Server object library. The following Visual Basic (VB6) code shows how to connect to a GIS server running on the machine "melange":

```
Dim pGISServerConnection As IGISServerConnection
Set pGISServerConnection = New GISServerConnection
pGISServerConnection.Connect "melange"
```

IGISServerConnection : IUnknown	Provides access to members that connect to a GIS server.
■— ServerObjectAdmin: IServerObjectAdmin	Gets the server object admin for the connected GIS server.
■— ServerObjectManager: IServerObjectManager	Gets the server object manager for the connected GIS server.
◄— Connect (in machine:Name: String)	Connects to the GIS server specified by the machineName.

If your application is written using .NET or Java and is deployed using the .NET ADF or Java ADF runtime (as is the case with a Web application or Web service), use the native .NET or Java server connection objects, called *ServerConnection*, to connect to the GIS server. The native .NET and Java server connection objects are in the Web controls assembly (.NET) and ArcObjects JAR (Java). If you use the COM connection object in the Server object library, you will get an error because the *GISServerConnection* COM object is not installed with the ADF runtime.

The following code demonstrates how to connect to the GIS server "melange" using the native .NET connection object:

```
C#:
ESRI.ArcGIS.Server.WebControls.ServerConnection connection = new
ESRI.ArcGIS.Server.WebControls.ServerConnection();
connection.Host = "melange";
connection.Connect();

VB.NET:
Dim connection As ESRI.ArcGIS.Server.WebControls.ServerConnection
connection = New ESRI.ArcGIS.Server.WebControls.ServerConnection
connection.Host = "melange"
connection.Connect()
```

The following code demonstrates how to connect to the GIS server "melange"

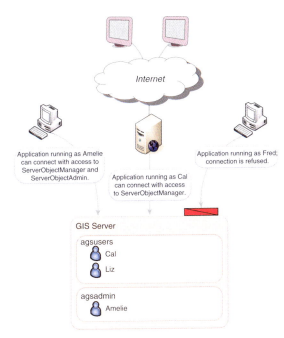

When applications make connections to the GIS server, they are authenticated against the agsusers and agsadmin users groups on the GIS server.

using the native Java connection object:

```
IServerConnection con = new ServerConnection();
con.connect("melange");
```

If your application is a Web application that uses the ArcGIS Server Web controls, the Web controls will connect to the GIS server for you, based on the properties you set for the control. For example, the *Map* control has properties for the server name it uses to make the connection.

CONNECTING TO THE GIS SERVER: SECURITY

For a client application to connect to the GIS server, the application must be running as an operating system user that is a member of one of the following two operating system user groups defined on the GIS server machines: the ArcGIS Server users group (*agsusers*) or ArcGIS Server administrators group (*agsadmin*). If the user the application is running as is not a member of either of those groups, then *Connect* will return an error.

In addition to the *Connect* method, the *IGISServerConnection* interface has two properties: *ServerObjectManager* and *ServerObjectAdmin*. If the application is running as a user in the users or administrators group, the application can access the *ServerObjectManager* property, which returns an *IServerObjectManager* interface. The *IServerObjectManager* interface provides methods for accessing and creating objects within the server for use by applications.

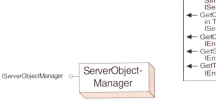

IServerObjectManager

The ServerObjectManager object provides methods for getting information about the GIS server and for creating server contexts for use by an application.

IServerObjectManager : IUnknown	Provides access to properties of, and members to work with, a GIS server's server object manager.
← CreateServerContext (in configName: String, in TypeName: String) : IServerContext	Gets a reference to a server context. The server context can be based on a specified server object configuration or can be an empty server context if no server object configuration is specified.
← GetConfigurationInfo (in Name: String, in TypeName: String) : IServerObjectConfigurationInfo	Gets the information for server object configuration with the specified Name and TypeName.
← GetConfigurationInfos: IEnumServerObjectConfigurationInfo	An enumerator over all the GIS server's configuration infos.
← GetServerDirectoryInfos: IEnumServerDirectoryInfo	An enumerator over all the GIS server's directory infos.
← GetTypeInfos: IEnumServerObjectTypeInfo	An enumerator over all the GIS server's type infos.

To access the *ServerObjectAdmin* property, the application must be running as a user who is a member of the administrators group. If the connected user is not a member of this group, attempts to access the *ServerObjectAdmin* property will fail. The *ServerObjectAdmin* property returns the *IServerObjectAdmin* interface, which provides methods for administering the various aspects of the server, such as server object configurations and server machines. Unless you are writing a GIS server administration application, your application does not need to make use of the *IServerObjectAdmin* interface.

ISeverObjectAdmin
IServerStatistics

ServerObject-Admin

The ServerObjectAdmin object provides methods for administrating the GIS server and its server objects.

ISeverObjectAdmin : IUnknown	Provide access to members that administrate the GIS server.
Properties: IPropertySet	The logging properties for the GIS server.
AddConfiguration (in config: IServerObjectConfiguration)	Adds a server object configuration (created with CreateConfiguration) to the GIS server.
AddMachine (in machine: IServerMachine)	Adds a host machine (created with CreateMachine) to the GIS server.
AddServerDirectory (in pSD: IServerDirectory)	Adds a server directory (created with CreateServerDirectory) to the GIS server.
CreateConfiguration: IServerObjectConfiguration	Creates a new server object configuration.
CreateMachine: IServerMachine	Creates a new host machine.
CreateServerDirectory: IServerDirectory	Creates a new server directory.
DeleteConfiguration (in Name: String, in TypeName: String)	Deletes a server object configuration from the GIS server.
DeleteMachine (in machineName: String)	Deletes a host machine from the GIS server, making it unavailable to host server objects.
DeleteServerDirectory (in Path: String)	Deletes a server directory such that its cleanup is no longer managed by the GIS server. It does not delete the physical directory from disk.
GetConfiguration (in Name: String, in TypeName: String) : IServerObjectConfiguration	Get the server object configuration with the specified Name and TypeName.
GetConfigurations: IEnumServerObjectConfiguration	An enumerator over all the server object configurations.
GetConfigurationStatus (in Name: String, in TypeName: String) : IServerObjectConfigurationStatus	Get the configuration status for a server object configuration with the specified Name and TypeName.
GetMachine (in Name: String) : IServerMachine	Get the host machine with the specified Name.
GetMachines: IEnumServerMachine	An enumerator over all the GIS server's host machines.
GetServerDirectories: IEnumServerDirectory	An enumerator over the GIS server's output directories.
GetServerDirectory (in Path: String) : IServerDirectory	Get the server directory with the specified Path.
GetTypes: IEnumServerObjectType	An enumerator over all the server object types.
PauseConfiguration (in Name: String, in TypeName: String)	Makes the configuration unavailable to clients for processing requests, but does not shut down running instances of server objects, or interrupt requests in progress.
StartConfiguration (in Name: String, in TypeName: String)	Starts a server object configuration and makes it available to clients for processing requests.
StopConfiguration (in Name: String, in TypeName: String)	Stops a server object configuration and shuts down any running instances of server objects defined by the configuration.
UpdateConfiguration (in config: IServerObjectConfiguration)	Updates the properties of a server object configuration.
UpdateMachine (in machine: IServerMachine)	Updates the properties of a host machine.
UpdateServerDirectory (in pSD: IServerDirectory)	Updates the properties of a server directory.

When connecting to the server using the native .NET or Java connection objects from a Web application or Web service, you must also think about Web application impersonation. As discussed in Chapter 2, your Web application must connect to the GIS server as a user in the users group. To do this, your Web application must impersonate such a user. If you do not use impersonation, then your Web application will attempt to connect to the GIS server with the identity of the Web server's worker process, for example, as the ASP.NET user for .NET applications. Impersonation strategies for both .NET and Java are discussed in more detail in Chapters 5 and 6.

IMapServer
IMapServerData
IMapServerInit
IMapServerLayout
IMapServerObjects
IMessageHandler
IObjectConstruct
IRequestHandler

MapServer

The MapServer object is a coarse-grained server object that provides access to the contents of a map document and methods for querying and drawing the map.

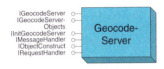

IGeocodeServer
IGeocodeServer-
    Objects
IInitGeocodeServer
IMessageHandler
IObjectConstruct
IRequestHandler

Geocode-Server

The GeocodeServer object is a coarse-grained server object that provides access to an address locator and methods for single address and batch geocoding.

Once connected to the GIS server your application can make use of server objects running within the server and create objects within the server for use by the application.

SERVER OBJECTS

A server object is a coarse-grained ArcObjects component that runs in a process on the SOC machine. ArcGIS Server comes with two out-of-the-box server objects:

- *esriCarto.MapServer*

- *esriLocation.GeocodeServer*

The *MapServer* object provides access to the contents of a map document and methods for querying and drawing the map. The *GeocodeServer* object provides access to an address locator and methods for performing single address and batch geocoding. These coarse-grained *MapServer* and *GeocodeServer* objects use the finer-grained ArcObjects on the server to perform mapping and geocoding operations, respectively. Application objects can use the high-level coarse-grained methods on the *MapServer* and *GeocodeServer* object and can also drill down and work with the fine-grained ArcObjects associated with them (feature layers, feature classes, renderers, and so on).

A server object, unlike other ArcObjects components, can be preconfigured by a GIS server administrator. To preconfigure a server object, the GIS server administrator has to set the object's properties using ArcCatalog before client applications can connect and make use of their mapping and geocoding functionality. When a server object is preconfigured, the administrator must specify a number of configuration properties (see Chapters 2 and 3), including the server object's pooling model. The pooling model will dictate the type of usage that an application will make of the server object. This aspect of server object usage will be discussed in more detail later when the concept of stateful versus stateless use of the server is discussed.

GETTING SERVER OBJECTS FROM THE SERVER

You get a server object by asking the server for a server context containing the object. You can think of a server context as a process, managed by the server, within which a server object runs. The details of server contexts are discussed later. Your application keeps a server object alive by holding on to its context, and must release the server object by releasing its context when it is done with it. The following VB6 code is an example of getting a *GeocodeServer* object from the GIS server and using it to locate an address:

```
Dim pServerContext As IServerContext
Set pServerContext = pSOM.CreateServerContext("RedlandsGeocode",
"GeocodeServer")

Dim pGCServer As IGeocodeServer
Set pGCServer = pServerContext.ServerObject

Dim pPropertySet As IPropertySet
```

```
Set pPropertySet = pServerContext.CreateObject("esriSystem.PropertySet")
pPropertySet.SetProperty "Street", "380 New York St"

Dim pResults As IPropertySet
Set pResults = pGCServer.GeocodeAddress(pPropertySet, Nothing)

Dim pPoint As IPoint
Set pPoint = pResults.GetProperty("Shape")

Debug.Print pPoint.X & ", " & pPoint.Y

pServerContext.ReleaseContext
```

As discussed earlier, a server object also has other associated objects that a developer can get to and make use of; for example, a developer working with a *MapServer* object can get to the *Map* and *Layer* objects associated with that map. These are the same *Map* and *Layer* objects that an ArcGIS Desktop or ArcGIS Engine developer would work with, except they reside in the server. The following VB6 code is an example of using the finer-grained ArcObjects associated with a *MapServer* object to work with a feature class associated with a particular layer:

```
Dim pServerContext As IServerContext
Set pServerContext = pSOM.CreateServerContext("RedlandsMap", "MapServer")

Dim pMapServer As IMapServer
Set pMapServer = pServerContext.ServerObject

Dim pMapServerObjs As IMapServerObjects
Set pMapServerObjs = pMapServer

Dim pMap As IMap
Set pMap = pMapServerObjs.Map(pMapServer.DefaultMapName)

Dim pFLayer As IFeatureLayer
Set pFLayer = pMap.Layer(0)

Dim pFeatureClass As IFeatureClass
Set pFeatureClass = pFLayer.FeatureClass

Debug.Print pFeatureClass.FeatureCount(Nothing)

pServerContext.ReleaseContext
```

In the above examples, *pSOM* is an *IServerObjectManager* interface that was previously retrieved from *IGISServerConnection.ServerObjectManager*. You will notice that when the server object's work is done and the application is finished using objects associated with the server object (in the first case, *pPoint*, in the second case, *pFeatureClass*), the server context (*pServerContext*) was explicitly released. It's important that the server context is released so other application sessions and other applications can make use of that server object. If your code does not explicitly release the context, it will be released once it goes out of scope and garbage collection kicks in. However, there may be a considerable lag between

Garbage collection is the process by which .NET and Java reclaim memory from objects that are created by applications. Garbage collection will happen based on memory allocations being made. When garbage collection occurs is when objects that are not referenced are actually cleaned up, which may be some time after they go out of the scope of your application.

when the server context variable goes out of scope and when garbage collection kicks in, and if you rely on this mechanism, then your server is at the mercy of .NET or Java garbage collection in terms of when objects are released.

MANAGING SERVER OBJECT LIFETIME

A key aspect of server object usage is managing the server object's lifetime. As described previously, server objects live in server contexts, and you get a server object by calling *CreateServerContext* containing a specified server object (see examples above).

The server object (for example, *RedlandsGeocode*, *RedlandsMap*) and all its associated objects are alive and may be used as long as you hold on to the context. Once you release the server context (by calling *ReleaseContext* or allowing the context to go out of scope), you may no longer make use of the server object or any other objects you obtained from the context. Once a server object's context is released, what happens to the context depends on whether the server object is pooled or non-pooled.

In the non-pooled case, when the context is released, it is shut down by the server. The next call to *CreateServerContext* for that server object will create a new instance of the server object in a new server context. If the server object is pooled, *ReleaseContext* returns the server object and its context to the pool, and the server will be free to give the server object to a request from another application session. This aspect of server object and server context behavior is critical when designing your application and how it manages state. This is discussed in the following section.

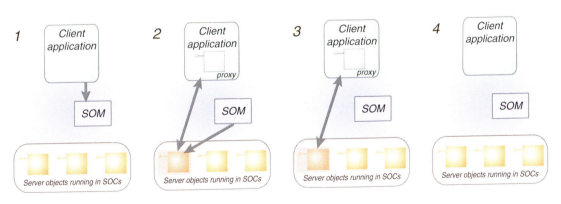

This diagram illustrates the server object lifetime for a pooled server object:

1. The client application makes a connection to the SOM and requests a server object.

2. The SOM returns to the client a proxy to one of the server objects available in the pool.

3. The client application works with the server object by making calls on its proxy.

4. When the client is finished with the server object, it releases it.

When the object is released, it is returned to the pool and is available to handle requests from other clients.

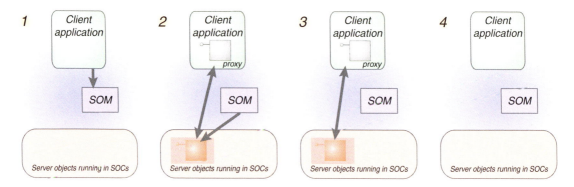

This diagram illustrates the server object lifetime for a non-pooled server object:

1. The client application makes a connection to the SOM and requests a server object.

2. The SOM creates a new instance of the server object and returns to the client a proxy to the server object.

3. The client application works with the server object by making calls on its proxy.

4. When the client is finished with the server object, it releases it.

When the object is released, it is destroyed. The SOM will create new instances of the server object to handle subsequent requests.

One key aspect of designing your application will be whether it is stateful or stateless. You can make either stateful or stateless use of a server object running within the GIS server. Stateless here refers to making read-only use of a server object, meaning your application does not make changes to the server object or any of its associated objects. Stateful refers to read–write use of a server object where your application does make changes to the server object or its related objects.

The question of state is important in server object usage because it dictates whether server objects can be shared across application sessions. If you make stateless use of server objects, then they can be shared across application sessions; if you make stateful use of server objects, then they cannot be shared.

GIS SERVER STATE AND OBJECT POOLING

This aspect of stateful versus stateless use and server object sharing relates directly to the pooling model for the server object. The following programming rules apply to using server objects:

- Client applications cannot change the properties of a pooled server object.

- Client applications can change the properties of a non-pooled server object.

Pooled server objects are expected to be used in a stateless manner. As a developer, you are responsible for making sure that the state of the server object, or its associated objects, has not changed when you return the object to the pool (by releasing its context via *ReleaseServerContext*). Each time a user or application session makes a request to create a pooled server object, it's indeterminate which running instance it will get out of the pool; therefore, all instances must have the same state or applications will experience inconsistent behavior.

Non-pooled server objects can be used in a stateful manner. Since non-pooled server objects and their contexts are destroyed when you release them, you need to hold onto them for as long as the state is important to you. When you call *ReleaseServerContext*, or you allow the server context to go out of scope, the server object and its context are destroyed, purging any state changes you made.

A server object is a coarse-grained ArcObjects component that has other associated ArcObjects.

STATEFUL VERSUS STATELESS USE OF SERVER OBJECTS

Methods and properties that are exposed by server object interfaces, such as *IMapServer* and *IGeocodeServer*, are by their nature stateless methods, such as *IMapServer.ExportMapImage* and *IGeocodeServer.GeocodeAddress*. These methods do not change any of the properties of the server object when they are called and are, therefore, safe to call on both pooled and non-pooled server objects. Changing the state of a server object typically involves making calls to get the finer-grained ArcObjects associated with a server object and making changes at that level.

Most GIS Web applications are not stateless. Typically, each user or session may have a current extent, each user or session may have a set of visible layers (that can be toggled on and off through the application), and each user or session may have different graphics visible on the map as a result of query operations such as network tracing. It is possible to write a stateful Web application that makes stateless use of server objects in the GIS server by maintaining aspects of application state, such as the extent of the map, layer visibility, and application-added graphics, using the Web application server's session state management capabilities. Such applications are called "shallowly stateful".

The GIS server also supports "deeply stateful" applications that use the GIS server to maintain application state. Examples of deeply stateful Web applications include:

- An application that starts a geodatabase edit session on behalf of a user and works with it across multiple requests in a session to support operations such as undo or redo.

- An application that allows a user to interactively compose a map across multiple requests within a session.

The following code is an example of a stateless use of a *MapServer* object. In this example, a request is made to the *MapServer* to draw itself at its default extent:

```
Dim pServerContext As IServerContext
Set pServerContext = pSOM.CreateServerContext("RedlandsMap", "MapServer")


Dim pMapServer As IMapServer
Set pMapServer = pServerContext.ServerObject


Dim it As IImageType
Dim idisp As IImageDisplay
Dim pID As IImageDescription


Set it = pServerContext.CreateObject("esriCarto.ImageType")
it.Format = esriImageFormat.esriImageJPG
it.ReturnType = esriImageReturnType.esriImageReturnMimeData


Set idisp = pServerContext.CreateObject("esriCarto.ImageDisplay")
idisp.Height = 400
idisp.Width = 500
idisp.DeviceResolution = 150


Set pID = pServerContext.CreateObject("esriCarto.ImageDescription")
pID.Display = idisp
pID.Type = it


Dim pMD As IMapDescription
Dim pMapServerInfo As IMapServerInfo
Set pMapServerInfo = pMapServer.GetServerInfo(pMapServer.DefaultMapName)
Set pMD = pMapServerInfo.DefaultMapDescription


Dim pMI As IImageResult
Set pMI = pMapServer.ExportMapImage(pMD, pID)
' do something with the image


pServerContext.ReleaseContext
```

The following is an example of a stateful use of a *MapServer* object. In this example, the first layer is removed from the map, then a request is made to the *MapServer* to draw itself at its default extent:

```
Dim pServerContext As IServerContext
Set pServerContext = pSOM.CreateServerContext("RedlandsMap", "MapServer")

Dim pMapServer As IMapServer
Set pMapServer = pServerContext.ServerObject

Dim pMapServerObjs As IMapServerObjects
Set pMapServerObjs = pMapServer

Dim pMap As IMap
Set pMap = pMapServerObjs.Map(pMapServer.DefaultMapName)

pMap.DeleteLayer pMap.Layer(0)
pMapServerObjs.RefreshServerObjects

Dim it As IImageType
Dim idisp As IImageDisplay
Dim pID As IImageDescription

Set it = pServerContext.CreateObject("esriCarto.ImageType")
it.Format = esriImageFormat.esriImageJPG
it.ReturnType = esriImageReturnType.esriImageReturnMimeData

Set idisp = pServerContext.CreateObject("esriCarto.ImageDisplay")
idisp.Height = 400
idisp.Width = 500
idisp.DeviceResolution = 150

Set pID = pServerContext.CreateObject("esriCarto.ImageDescription")
pID.Display = idisp
pID.Type = it

Dim pMD As IMapDescription
Dim pMapServerInfo As IMapServerInfo
Set pMapServerInfo = pMapServer.GetServerInfo(pMapServer.DefaultMapName)
Set pMD = pMapServerInfo.DefaultMapDescription

Dim pMI As IImageResult
Set pMI = pMapServer.ExportMapImage(pMD, pID)
' do something with the image

pServerContext.ReleaseContext
```

In the first example, no changes were made to the server object or any of its associated objects. Once the code finishes executing and the context is released, the server object is in the same state as when the application got it. In the second example, a layer was explicitly removed from the map using the *DeleteLayer* method on the *IMap* interface. This is an example of using a fine-grained ArcObjects component call to change the state of a server object.

Typically, if you are making state changes to server objects, you would hang on to a reference to its context for the duration of your application session. The above example releases the server context immediately after processing the request. You would not do this type of operation with a pooled server object, as subsequent use of this instance of the *MapServer* object will reflect the fact that the layer has been removed.

The state of a server object can be changed in a number of different ways. The example above demonstrates making direct changes to the properties of a server object—for example, removing a layer from a map. It's also possible to change the state of a server object indirectly through other objects in the server object's context. The following table summarizes the ways that you can change the state of a server object:

Stateful operation	Example
Call a stateful method on a server object	· Adding or removing a layer from a map server object · Changing the renderer for a layer in a map server object · Changing the locator properties for a geocode server object
Call a stateful method on an environment	· Changing the auto densify tolerance in the geometry environment · Changing the cell size in the raster analysis environment

Use a method that is stateful on a server object

A stateful method is one that modifies or changes the instance of the server object. There are many examples of stateful methods; some common examples include methods that add or remove layers from a map server object or methods that change a layer's renderer. These methods should never be called on a pooled server object unless the client application can return the object to its original state before releasing it back to the server.

Use a method that is stateful on an environment

Server objects run in contexts that have a number of environment settings associated with them. Some of these environments can be modified by developers. For example, the geometry environment can be manipulated through the *IGeometryEnvironment* interface. While changes to the geometry environment do not directly affect a server object, those changes may affect other operations that a client application may perform using a server object's context.

The following code is an example of how you can change the state of a server object's environment (in this case, the geometry environment) without directly changing the server object itself:

```
Dim pServerContext As IServerContext
Set pServerContext = pSOM.CreateServerContext("RedlandsMap", "MapServer")

Dim pGeomEnv As IGeometryEnvironment4
Set pGeomEnv =
pServerContext.CreateObject("esriGeometry.GeometryEnvironment")

pGeomEnv.DeviationAutoDensifyTolerance = 5.7
```

```
pGeomEnv.DicingEnabled = True

' perform a geometry operation

pServerContext.ReleaseContext
```

Changing the state of an environment is valid for both pooled and non-pooled server object use. To ensure that such changes to environments do not negatively impact operations made by client applications, applications should not rely on the environment being in a particular state before performing that operation. When performing operations that rely on a particular state of the environment, applications should set the required environment state before performing that operation, especially when using pooled server objects.

Working with cursors

Some objects that you can create in a server context may lock or use resources that the object frees only in its destructor. For example, a geodatabase cursor may acquire a shared schema lock on a file-based feature class or table on which it is based or may hold on to an ArcSDE stream.

Garbage collection is the process by which .NET and Java reclaim memory from objects that are created by applications. Garbage collection will happen based on memory allocations being made. When garbage collection occurs is when objects that are not referenced are actually cleaned up, which may be some time after they go out of the scope of your application.

While the shared schema lock is in place, other applications can continue to query or update the rows in the table, but they cannot delete the feature class or modify its schema. In the case of file-based data sources, such as shapefiles, update cursors acquire an exclusive write lock on the file, which will prevent other applications from accessing the file for read or write. The effect of these locks is that the data may be unavailable to other applications until all of the references on the cursor object are released.

In the case of ArcSDE data sources, the cursor holds on to an ArcSDE stream, and if the application has multiple clients, each may get and hold on to an ArcSDE stream, eventually exhausting the maximum allowable streams. The effect of the number of ArcSDE streams exceeding the maximum is that other clients will fail to open their own cursors to query the database.

Because of the above reasons, it's important to ensure that your reference to any cursor your application opens is released in a timely manner. If you are developing your application using Java, when the cursor (or any other COM object) goes out of scope, your reference will be removed immediately for you. If you are developing with .NET, your reference on the cursor (or any other COM object) will not be released until garbage collection kicks in. In a Web application or Web service that services multiple concurrent sessions and requests, relying on garbage collection to release references on objects will result in cursors and their resources not being released in a timely manner.

To ensure a COM object is released when it goes out of scope, the *WebControls* assembly contains a helper object called *WebObject*. Use the *ManageLifetime* method to add your COM object to the set of objects that will be explicitly released when the *WebObject* is disposed. You must scope the use of *WebObject* within a *Using* block. When you scope the use of *WebObject* within a using block, any object (including your cursor) that you have added to the *WebObject* using the *ManageLifetime* method will be explicitly released at the end of the using block.

The following C# example demonstrates this coding pattern:

```csharp
private void doSomething_Click(object sender, System.EventArgs e)
{
  using (WebObject webobj = new WebObject())
  {
    ServerConnection serverConn = new ServerConnection("doug",true);
    IServerObjectManager som = serverConn.ServerObjectManager;

    IServerContext ctx =
som.CreateServerContext("Yellowstone","MapServer");
    IMapServer mapsrv = ctx.ServerObject as IMapServer;
    IMapServerObjects mapo = mapsrv as IMapServerObjects;
    IMap map = mapo.get_Map(mapsrv.DefaultMapName);

    IFeatureLayer flayer = map.get_Layer(0) as IFeatureLayer;
    IFeatureClass fclass = flayer.FeatureClass;

    IFeatureCursor fcursor = fclass.Search(null, true);
    webobj.ManageLifetime(fcursor);

    IFeature f = null;
    while ((f = fcursor.NextFeature()) != null)
     {
      // do something with the feature
     }

    ctx.ReleaseContext();
  }
}
```

VB.NET does not have a *Using* clause. The following example demonstrates the coding pattern for VB.NET:

```vbnet
Private Sub doSomething_Click(ByVal sender As System.Object, ByVal e As
System.EventArgs) Handles doSomething.Click
  Dim webobj As WebObject = New WebObject
  Dim ctx As IServerContext = Nothing
  Try
    Dim serverConn As ServerConnection = New ServerConnection("doug",
True)
    Dim som As IServerObjectManager = serverConn.ServerObjectManager

    ctx = som.CreateServerContext("Yellowstone", "MapServer")
    Dim mapsrv As IMapServer = ctx.ServerObject
    Dim mapo As IMapServerObjects = mapsrv
    Dim map As IMap = mapo.Map(mapsrv.DefaultMapName)

    Dim flayer As IFeatureLayer = map.Layer(0)
    Dim fClass As IFeatureClass = flayer.FeatureClass

    Dim fcursor As IFeatureCursor = fClass.Search(Nothing, True)
    webobj.ManageLifetime(fcursor)
```

```
      Dim f As IFeature = fcursor.NextFeature()
      Do Until f Is Nothing
        ' do something with the feature
        f = fcursor.NextFeature()
      Loop


      Finally
        ctx.ReleaseContext()
        webobj.Dispose()
      End Try
  End Sub
```

The *WebMap*, *WebGeocode*, and *WebPageLayout* objects also have a *ManageLifetime* method. If you are using, for example, a *WebMap*, and scope your code in a using block, you can rely on these objects to explicitly release objects you add with *ManageLifetime* at the end of the using block.

MANAGING STATE IN THE WEB APPLICATION'S SESSION STATE— SHALLOWLY STATEFUL APPLICATIONS

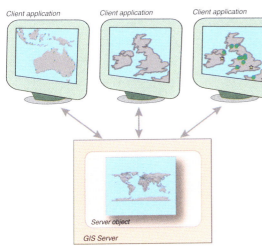

Client application *Client application* *Client application*

Server object

GIS Server

It is possible to write a stateful Web application that makes stateless use of server objects in the GIS server by maintaining aspects of application state, such as the extent of the map, layer visibility, and application-added graphics, using the Web application server's session state management capabilities.

This is not the end of the story when it comes to managing state of an ArcGIS Server application. As described above, it is possible to write a stateful Web application that makes stateless use of server objects in the GIS server by maintaining aspects of application state, such as the extent of the map, layer visibility, and application-added graphics, using the Web application server's session state management capabilities.

The *IServerContext* interface has methods that allow you to save GIS objects in session state by exporting them to strings. The server context also has methods to rehydrate the objects from strings as you need them. Objects that support the *IPersistStream* interface can be exported and rehydrated in this manner. A very common example illustrates this capability, which is managing the user's or session's current extent as it pans and zooms around the map.

This programming pattern is demonstrated using an ASP.NET example. The following C# code shows how on session startup you can save references to both the *ServerObjectManager* and a serialized copy of the *MapServer's* map description:

```
private void Page_Load(object sender, System.EventArgs e)
{
  // Put user code to initialize the page here
  if ( !Page.IsPostBack )
  {
    // Is this a new session?
    if ( Session.IsNewSession )
    {
      // connect to the server
      string m_host = "padisha";
```

```
ESRI.ArcGIS.Server.WebControls.ServerConnection connection = new
ESRI.ArcGIS.Server.WebControls.ServerConnection();
    connection.Host = m_host;
    connection.Connect();

    // save a reference to the SOM as an application variable
    // called "som" so the connection can be used again later
    // in the application
    IServerObjectManager som = connection.ServerObjectManager;
    Application.Set("som", som);

    IServerContext ctx =
som.CreateServerContext("RedlandsMap","MapServer");
    IMapServer map = ctx.ServerObject as IMapServer;

    IMapServerInfo mapinfo = map.GetServerInfo(map.DefaultMapName);
    IMapDescription md = mapinfo.DefaultMapDescription;

    // save the map description as a session variable called "md"
    string sMapDesc = ctx.SaveObject(md);

    Session["md"] = sMapDesc;

    ctx.ReleaseContext();
  }
 }
}
```

The following Java example shows the same programming pattern:

The objects and interfaces used for managing image displays can be found in the Carto object library. To learn more about these objects, refer to the online developer documentation.

```
public void connect() throws Exception
{
  // do only if it is a new session
  if (session.isNew())
  {
    // connect to the server
    IServerConnection connection = new ServerConnection();
    connection.connect("padisha");

    // save a reference to the SOM as an application variable
    // called "som" so the connection can be used again later
    // in the application
    IServerObjectManager som = connection.getServerObjectManager();
    session.setAttribute("som",som );
    IServerContext ctx =
som.createServerContext("RedlandsMap","MapServer");
    IMapServer mapServer = new IMapServerProxy(ctx.getServerObject());

    IMapServerInfo mapInfo =
mapServer.getServerInfo(mapServer.getDefaultMapName());
    IMapDescription mapDesc = mapInfo.getDefaultMapDescription();
```

```
// save the map description as a session variable called "mxd"
String sMapDesc;
sMapDesc = ctx.saveObject(mapDesc);
session.setAttribute("mapDesc",sMapDesc);

ctx.releaseContext();
  }
}
```

In the code examples above, the *MapServer's MapDescription* is being serialized to a string and saved in session state. Assume now that the user for this session wishes to zoom in by a fixed amount. The following code shows how this is done. The steps are:

1. Load the serialized map description to get the current map description for the session.

2. Shrink the extent for the map description.

3. Draw the map using the modified map description.

4. Export the modified map description to a string so that the session state is updated.

The ASP.NET example of this is illustrated below:

```
private void btnFixedZoomIn_Click(object sender, System.EventArgs e)
{
  IServerObjectManager som = (IServerObjectManager)
Application.Get("som");
  IServerContext ctx = som.CreateServerContext("RedlandsMap","MapServer");
  IMapServer map = ctx.ServerObject as IMapServer;

  // rehydrate the map description
  string smd = (string) Session["md"];
  IMapDescription md = ctx.LoadObject(smd) as IMapDescription;

  // get the extend, shrink it
  IMapArea ma = md.MapArea;
  IEnvelope env = ma.Extent;
  env.Expand(0.9,0.9,true);

  // set the extent into the MapDescription
  IMapExtent mx = ma as IMapExtent;
  mx.Extent = env;
  md.MapArea = ma;

  // create ImageDescription and export the map image
  IImageType it = ctx.CreateObject("esriCarto.ImageType") as ImageType;
  it.Format = esriImageFormat.esriImageJPG;
  it.ReturnType = esriImageReturnType.esriImageReturnURL;
  IImageDisplay idisp = ctx.CreateObject("esriCarto.ImageDisplay") as
IImageDisplay;
```

```
    idisp.Height = 400;
    idisp.Width = 500;
    idisp.DeviceResolution = 150;
    IImageDescription id = ctx.CreateObject("esriCarto.ImageDescription") as
    IImageDescription;
    id.Display = idisp;
    id.Type = it;

    IImageResult ir = map.ExportMapImage(md,id);
    Image1.ImageUrl = ir.URL;

    // export map description with the new extent and save it into session state
    string sMapDesc = ctx.SaveObject(md);
    Session["md"] = sMapDesc;

    ctx.ReleaseContext();
}
```

The following code shows the Java equivalent:

```
public void fixedZoom() throws Exception
{
    IServerObjectManager som = (IServerObjectManager)
session.getAttribute("mgr");
    IServerContext ctx =
som.createServerContext("RedlandsMap","MapServer");
    IMapServer mapServer = new IMapServerProxy(ctx.getServerObject());

    // rehydrate object from xml
    String smd = (String) session.getAttribute("mapDesc");
    IMapDescription mapDesc = new
IMapDescriptionProxy(ctx.loadObject(smd));

    // set the new extent
    IMapArea ma = mapDesc.getMapArea();
    IEnvelope env = ma.getExtent();
    env.expand(0.9, 0.9, true);

    // apply new extent to the MapDescription
    IMapExtent mx = (IMapExtent) ma;
    mx.setExtent(env);
    mapDesc.setMapArea(ma);

    // create ImageDescription and export the map image
    IImageType imgType = new
IImageTypeProxy(ctx.createObject(ImageType.getClsid()));

    imgType.setFormat(esriImageFormat.esriImageJPG);
    imgType.setReturnType(esriImageReturnType.esriImageReturnURL);

    IImageDisplay imgDisp = new
IImageDisplayProxy(ctx.createObject(ImageDisplay.getClsid()));
```

```
    imgDisp.setHeight(400);
    imgDisp.setWidth(500);
    imgDisp.setDeviceResolution(150);
    IImageDescription imgDesc = new
IImageDescriptionProxy(ctx.createObject(ImageDescription.getClsid()));
    imgDesc.setDisplay(imgDisp);
    imgDesc.setType(imgType);

    IImageResult imgResult = mapServer.exportMapImage(mapDesc,imgDesc);
    imgResult.getURL();

    // export map description with the new extent and save it into
    // session state
    String sMapDesc = ctx.saveObject(mapDesc);
    session.setAttribute("mapDesc",sMapDesc);
    ctx.releaseContext();
  }
```

MANAGING STATE IN THE GIS SERVER—DEEPLY STATEFUL APPLICATIONS

The example above is a shallowly stateful application, meaning it is stateful but its state is managed within the Web application server's session state. The GIS server also supports deeply stateful applications that use the GIS server to maintain application state.

Supporting such applications requires a server object instance dedicated to each application session. You can configure this by making your server object non-pooled. The fact that server objects necessary for such applications are non-pooled limits the number of concurrent sessions by the processing resources of the server.

When programming a deeply stateful Web application, you want to use the same server context and server object throughout the session. So, you want to get a server context at the beginning of the session and hold on to it until the session has ended. The following C# code is an example of how you would obtain a non-pooled server context and add it to your session state in an ASP.NET application:

```
private void Page_Load(object sender, System.EventArgs e)
{
  Session.Timeout = 5;

  // Put user code to initialize the page here
  if ( !Page.IsPostBack )
  {
   // Is this a new session?
   if ( Session.IsNewSession )
    {
     // connect to the server
     string m_host = "padisha";
```

```
        ESRI.ArcGIS.Server.WebControls.ServerConnection connection = new
ESRI.ArcGIS.Server.WebControls.ServerConnection();
      connection.Host = m_host;
      connection.Connect();

      IServerObjectManager som = connection.ServerObjectManager;
      IServerContext ctx = som.CreateServerContext("Farms","MapServer");

      // save the server context as a session variable called "context"
      Session["context"] = ctx;
    }
  }
}
```

The following C# code shows how you can make use of the server context in the following code, which alters the MapServer by removing the first layer:

```
private void btnDoSomthing_Click(object sender, System.EventArgs e)
{
  IServerContext ctx = Session["context"] as IServerContext;
  IMapServer map = ctx.ServerObject as IMapServer;

  IMapServerObjects mapObj = map as IMapServerObjects;
  IMap fgmap = mapObj.get_Map(map.DefaultMapName);

  fgmap.DeleteLayer(fgmap.get_Layer(0));
  mapObj.RefreshServerObjects();
}
```

The server context is held on to for the duration of the session and needs to be released at the end of the session. The following code shows how to release the context when the session ends:

```
protected void Session_End(Object sender, EventArgs e)
{
  IServerContext ctx = Session["context"] as IServerContext;
  ctx.ReleaseContext();
}
```

Note that the session ends based on a time-out that is set in the application. In this example, the session time-out was set to five minutes, meaning if the user does not interact with the running Web application session for five minutes, then the session will time out, and the server context will be released by the code above. It also means that once a user has ended the session by closing the Web browser, the server context will not actually be released for five minutes until the session time-out is triggered and the code above is executed.

These code examples are illustrations of how you work with server objects when building applications. It's important to note that if you are using the Web controls to build a Web application, the Web controls take care of many of these details, specifically, the *Map* control takes care of releasing the server context and takes care of saving the *MapDescription* in session state for you. The relationship between Web controls and these aspects of the server API are described in more detail later.

APPLICATION STATE AND SCALABILITY

The question of stateful versus stateless use of the GIS server is central to the scalability of your application. An application is more scalable than another application if it can support a larger number of users with the same amount of computer resources. The keys to scalability are:

- Make stateless use of the GIS server.

- Use pooled server objects.

- Minimize the time your application holds on to a server object. Release server objects as soon as possible and do not rely on .NET or Java garbage collection to do it for you.

Using the above criteria, it's clear that stateless or shallowly stateful applications can make use of object pooling and, therefore, are more scalable than deeply stateful applications. The question of stateful versus stateless use of the GIS server will be critical in designing your application.

This is not the end of performance and scalability when it comes to designing ArcGIS Server applications. More discussion on performance tuning of ArcGIS Server applications can be found in the 'ArcGIS Server application performance tuning' section later in this chapter.

The ServerContext object provides access to a context in the GIS server and methods for creating and managing objects within that context.

The ServerObjectManager object provides methods for getting information about the GIS server and for creating server contexts for use by an application.

To this point, this chapter has focused on the use of server objects to perform the functionality that they expose through their coarse-grained interfaces. The ADF's Web controls' functionality is based on the functionality exposed by these server objects. If you want your application to go beyond simple mapping and geocoding using server objects and Web controls, you must become familiar with working with server contexts as well as the ArcObjects programming model.

GETTING AND RELEASING SERVER CONTEXTS

You get a server context using the *CreateServerContext* method on *IServerObjectManager*, which hands back an *IServerContext* interface on the server context. The *IServerContext* interface has a number of methods for helping you manage the objects you create within server contexts. So far in the code examples in this chapter the use of three of these methods, specifically *CreateObject*, *SaveObject*, and *LoadObject*, have been shown. The use of these and other methods will be described in more detail in this section.

IServerObjectManager : IUnknown	Provides access to properties of, and members to work with, a GIS server's server object manager.
← CreateServerContext (in configName: String, in TypeName: String) : IServerContext	Gets a reference to a server context. The server context can be based on a specified server object configuration or can be an empty server context if no server object configuration is specified.
← GetConfigurationInfo (in Name: String, in TypeName: String) : IServerObjectConfigurationInfo	Gets the information for server object configuration with the specified Name and TypeName.
← GetConfigurationInfos: IEnumServerObjectConfigurationInfo	An enumerator over all the GIS server's configuration infos.
← GetServerDirectoryInfos: IEnumServerDirectoryInfo	An enumerator over all the GIS server's directory infos.
← GetTypeInfos: IEnumServerObjectTypeInfo	An enumerator over all the GIS server's type infos.

IServerContext : IUnknown	Provides access to members for managing a server context, and the objects running within that server context.
■— ServerObject: IServerObject	The map or geocode server object running in the server context.
← CreateObject (in CLSID: String) : IUnknown Pointer	Create an object in the server context whose type is specified by the CLSID.
← GetObject (in Name: String) : IUnknown Pointer	Get a reference to an object in the server context's object dictionary by its Name.
← LoadObject (in str: String) : IUnknown Pointer	Create an object in the server context from a string that was created by saving an object using SaveObject.
← ReleaseContext	Release the server context back to the server so it can be used by another client (if pooled), or so it can be destroyed (if non-pooled).
← Remove (in Name: String)	Remove an object from the server context's object dictionary.
← RemoveAll	Remove all objects from the server context's object dictionary.
← SaveObject (in obj: IUnknown Pointer) : String	Save an object in the server context to a string.
← SetObject (in Name: String, in obj: IUnknown Pointer)	Add an object running in the server context to the context's object dictionary.

When developing applications with ArcGIS Server, all ArcObjects that your application creates and uses live within a server context. A server context is a reserved space within the server dedicated to a set of running objects. Server objects also reside in a server context. To get a server object, you actually get a reference to its context, then get the server object from the context:

```
Dim pServerContext as IServerContext
Set pServerContext =
pServerObjectManager.CreateServerContext("RedlandsMap","MapServer")
```

```
Dim pMapServer as IMapServer
Set pMapServer = pServerContext.ServerObject
```

You can also create empty server contexts. You can use an empty context to create ArcObjects on the fly within the server to do ad hoc GIS processing:

```
Dim IServerContext as IServerContext
Set IServerContext = pServerObjectManager.CreateServerContext("","")
Dim pWorkspaceFactory as IWorkspaceFactory
Set pWorkspaceFactory =
pServerContext.CreateObject("esriDataSourcesGDB.SdeWorkspaceFactory")
```

An empty server context is useful when you want to create objects for your application's use in the server but do not require the use of a preconfigured server object. Empty contexts can be used to create any type of object, such as a connection to a workspace as shown above. Since server objects are ArcObjects, you can also use empty contexts to create server objects (*MapServer, GeocodeServer*) on-the-fly. Empty server contexts are non-pooled and have high isolation.

When your application is finished working with a server context, it must release it back to the server by calling the *ReleaseContext* method. If you allow the context to go out of scope without explicitly releasing it, it will remain in use and be unavailable to other applications until it is garbage collected. Once a context is released, the application can no longer make use of any objects in that context. This includes both objects that you may have obtained from or created in the context.

```
Dim pServerContext as IServerContext
Set pServerContext =
pServerObjectManager.CreateServerContext("RedlandsMap","MapServer")
Dim pMapServer as IMapServer
Set pMapServer = pServerContext.ServerObject
' Do something with the object
pServerContext.ReleaseContext
```

CREATING OBJECTS IN THE SERVER

Client machines (for example, the Web server machine) require only the ADF runtime be installed to run ArcGIS Server applications. The ADF runtime does not install ArcObjects, so these applications do not have the ability to create local ArcObjects. All ArcObjects that your application uses should be created within a server context using the *CreateObject* method on *IServerContext*. In previous examples, you have seen *CreateObject* used to create an *ImageDescription* object for use in *ExportMapImage*.

ArcGIS Server applications should not use New to create local ArcObjects but should always create objects within the server by calling *CreateObject* on *IServerContext*:

Incorrect:

```
Dim pPoint as IPoint
Set pPoint = New Point
```

Correct:

```
Dim pPoint as IPoint
Set pPoint = pServerContext.CreateObject("esriGeometry.Point")
```

Use *CreateObject* when you need to create an object for use in your application.

```
Dim pPointCollection as IPointCollection
Set pPointCollection = pServerContext.CreateObject("esriGeometry.Polygon")
```

A proxy object is a local representation of a remote object. The proxy object controls access to the remote object by forcing all interaction with the remote object to be via the proxy object. The supported interfaces and methods on a proxy object are the same as those supported by the remote object. You can make method calls on, and get and set properties of, a proxy object as if you were working directly with the remote object.

CreateObject will return a proxy to the object that is in the server context. Your application can make use of the proxy as if the object was created locally within its process. If you call a method on the proxy that hands back another object, that object will actually be in the server context, and your application will be handed back a proxy to that object. In the above example, if you get a point from the point collection using *IPointCollection::Point()*, the point returned will be in the same context as the point collection.

If you add a point to the point collection using *IPointCollection::AddPoint()*, you should create that point in the same context as the point collection.

```
Dim pPointCollection as IPointCollection
Set pPointCollection = pServerContext.CreateObject("esriGeometry.Polygon")

Dim pPoint as IPoint
Set pPoint = pServerContext.CreateObject("esriGeometry.Point")
pPoint.X = 1
pPoint.Y = 1


pPointCollection.AddPoint pPoint
```

It's important to understand how your application is making use of server contexts as it does its work because objects that are used together should be in the same context. For example, if you create a point object to use in a spatial selection to query features in a feature class, the point should be in the same context as the feature class. This becomes important if your application makes use of more than one server context. It may be necessary to copy objects from one context to another.

Also, you should not directly use objects in a server context with local objects in your application and vice versa. You can indirectly use objects or make copies of them. For example, if you have a *Point* object in a server context, you can get its X, Y properties and use them with local objects or use them to create a new local point. Don't directly use the point in the server context as, for example, the geometry of a local graphic element object.

Consider the following examples. In each example, assume that objects with Remote in their names are objects in a server context as in:

```
Dim pRemotePoint as IPoint
Set pRemotePoint = pServerContext.CreateObject("esriGeometry.Point")
```

while objects with Local in their name are objects created locally as in:

```
Dim pLocalPoint as IPoint
Set pLocalPoint = New Point
```

You can't set a local object to a remote object:

```
' this is incorrect
Set pLocalPoint = pRemotePoint
```

```
' this is also incorrect
Set pLocalElement.Geometry = pRemotePoint
```

Do not set a local object, or a property of a local object, to be an object obtained from a remote object:

```
' this is incorrect
Set pLocalPoint = pRemotePointCollection.Point(0)
```

When calling a method on a remote object, don't pass in local objects as parameters:

```
' this is incorrect
Set pRemoteWorkspace = pRemoteWorkspaceFactory.Open(pLocalPropertySet,0)
```

You can get simple data types (double, long, string, and so forth) that are passed by value from a remote object and use them as properties of a local object as in:

```
' this is OK
pLocalPoint.X = pRemotePoint.X
pLocalPoint.Y = pRemotePoint.Y
```

The *SaveObject* and *LoadObject* methods allow you to serialize objects in the server context to a string, then deserialize them back into objects. You have already seen how these methods can be used to manage state in your application while making stateless use of a pooled server object (see previous section). These methods also allow you to copy objects between contexts. Any object that supports *IPersistStream* can be saved and loaded using these methods. For example, in an application that makes use of a *MapServer* object for mapping and a *GeocodeServer*

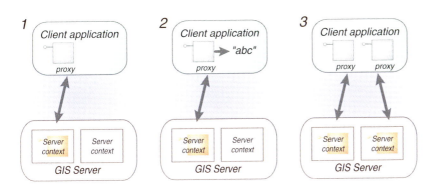

This diagram illustrates the use of SaveObject and LoadObject to copy objects between server contexts:

1. The client application gets or creates an object within a server context.

2. The application uses the SaveObject method on the object's context to serialize the object as a string that is held in the application's session state.

3. The client application gets a reference to another server context and calls the LoadObject method, passing in the string created by SaveObject. LoadObject creates a new instance of the object in the new server context.

object for geocoding, the *GeocodeServer* and *MapServer* will be running in different contexts. If you use the *GeocodeServer* object to locate an address and then you want to draw the resulting point that *GeocodeAddress* returns on your map, you need to copy the point into your *MapServer's* context:

```
Dim pServerContext as IServerContext
Set pServerContext = pSOM.CreateServerContext("RedlandsMap", "MapServer")

Dim pServerContext2 As IServerContext
Set pServerContext2 = pSOM.CreateServerContext("RedlandsGeocode",
"GeocodeServer")

Dim pGCServer As IGeocodeServer
Set pGCServer = pServerContext2.ServerObject

Dim pPropertySet As IPropertySet
Set pPropertySet = pServerContext2.CreateObject("esriSystem.PropertySet")
pPropertySet.SetProperty "Street", "380 New York St"

Dim pResults As IPropertySet
Set pResults = pGCServer.GeocodeAddress(pPropertySet, Nothing)

Dim pPoint As IPoint
Set pPoint = pResults.GetProperty("Shape")

' copy the Point to the Map's server context
Dim sPoint As String
sPoint = pServerContext2.SaveObject(pPoint)

Dim pPointCopy As IPoint
Set pPointCopy = pServerContext.LoadObject(sPoint)

pServerContext2.ReleaseContext

' add the point as a graphic to the map description and redraw the map

pServerContext.ReleaseContext
```

A PropertySet is a generic class that is used to hold any set of properties. A PropertySet's properties are stored as name/value pairs. Examples for the use of a property set are to hold the properties required for opening an SDE workspace or geocoding an address. To learn more about PropertySet objects, see the online developer documentation.

MANAGING OBJECTS IN A SERVER CONTEXT

As your application creates and uses various objects within a particular context, you may want a convenient place to store references to commonly used objects within the context. You can do this by using the server context's object dictionary to keep track of these objects during the lifetime of the context. You can use the context's dictionary as a convenient place to store objects that you create within the context. Note that this dictionary is itself valid only as long as you hold on to the server context and it's emptied when you release the context. You can use this dictionary to share objects created within a context between different parts of your application that have access to the context.

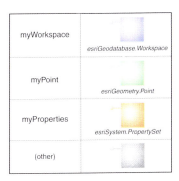

myWorkspace	esriGeodatabase.Workspace
myPoint	esriGeometry.Point
myProperties	esriSystem.PropertySet
(other)	

A server context contains an object dictionary that serves as a convenient place for you to store references to commonly used objects. Use the SetObject and GetObject methods on IServerContext to work with the object dictionary.

For example, your application may require a connection to a geodatabase workspace. Since making a geodatabase workspace connection can be expensive, you would want your application to make the connection once, then store the *Workspace* object in the context's dictionary so you can use it multiple times without having to re-create it and, therefore, reconnect to the workspace each time.

You add objects to and retrieve objects from the dictionary using the *SetObject* and *GetObject* methods, respectively. An object that is set in the context will be available until it is removed (by calling *Remove* or *RemoveAll*) or until the context is released.

```
Dim pPointCollection as IPointCollection
Set pPointCollection = pServerContext.CreateObject("esriGeometry.Polygon")

pServerContext.SetObject "myPoly", pPointCollection
Dim pPoly as IPolygon
set pPoly = pServerContext.GetObject("myPoly")
```

Use the *Remove* and *RemoveAll* methods to remove an object from a context that has been set using *SetObject*. Once an object is removed, a reference to it can no longer be obtained using *GetObject*. Note that if you do not explicitly call *Remove* or *RemoveAll*, you can still not get references to objects set in the context after the context has been released.

```
pServerContext.Remove "myPoly"
```

WRITING OUTPUT

When your application performs operations using ArcObjects running in server contexts, those operations may need to write out data to disk. For example, the *ExportMapImage* method on a map server object writes images to disk. You may have other applications that need to write data; for example, an application that creates geodatabase check outs needs to write personal geodatabases to disk where they can be downloaded from.

Typically, you will want these files to be cleaned up by the server after some period of time. To ensure this happens, your applications should write their output to a server directory.

The set of a GIS server's server directories is available by calling *GetServerDirectoryInfos* on the *IServerObjectManager* interface on the server object manager. For your files to be cleaned up when written into that server directory, they must follow a file naming convention. The GIS server will delete all files in a server directory that are prefixed with "_ags_". Any files written to an output directory that are not prefixed with "_ags_" will not be cleaned by the GIS server.

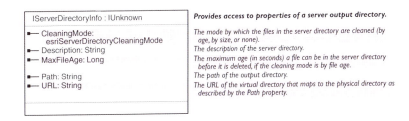

IServerDirectoryInfo : IUnknown

- CleaningMode:
  esriServerDirectoryCleaningMode
- Description: String
- MaxFileAge: Long

- Path: String
- URL: String

Provides access to properties of a server output directory.

The mode by which the files in the server directory are cleaned (by age, by size, or none).
The description of the server directory.
The maximum age (in seconds) a file can be in the server directory before it is deleted, if the cleaning mode is by file age.
The path of the output directory.
The URL of the virtual directory that maps to the physical directory as described by the Path property.

The following code shows how you can use the *GetServerDirectoryInfos* method on *IServerObjectManager* to get a server directory and create a new personal geodatabase. Note: The code creates two geodatabases, one named such that it will be cleaned by the GIS server, one named such that it will not be cleaned by the GIS server:

```
Dim pServerContext As IServerContext
Set pServerContext = pSOM.CreateServerContext("", "")

Dim pWSF As IWorkspaceFactory
Set pWSF =
pServerContext.CreateObject("esriDataSourcesGDB.AccessWorkspaceFactory")

Dim pEnumSDI As IEnumServerDirectoryInfo
Set pEnumSDI = pSOM.GetServerDirectoryInfos

Dim pSDI As IServerDirectoryInfo
Set pSDI = pEnumSDI.Next

Dim pProps As IPropertySet
Set pProps = pServerContext.CreateObject("esriSystem.PropertySet")

' this database will be cleaned by the GIS server
pProps.SetProperty "DATABASE", pSDI.Path & "\_ags_db1.mdb"
pWSF.Create pSDI.Path, "_ags_db1", pProps, 0

' this database will not be cleaned by the GIS server
pProps.SetProperty "DATABASE", pSDI.Path & "\db2.mdb"
pWSF.Create pSDI.Path, "db2", pProps, 0
```

A PropertySet is a generic class that is used to hold any set of properties. A PropertySet's properties are stored as name/value pairs. Examples for the use of a property set are to hold the properties required for opening an SDE workspace or geocoding an address. To learn more about PropertySet objects, see the online developer documentation.

WORKING WITH ArcGIS SERVER OPTIONS

If your server object container machines have licenses for the Spatial, 3D, or StreetMap options for ArcGIS Server, your applications can make use of the functionality that is unlocked by that license.

You do not need to do any explicit calls to get a license when you want to use an object that requires a license. Just create the object on the server and use it. If the server object container machine is not licensed, then any method calls you make on the object will fail.

In many cases, you will be building Web applications that make use of the GIS server through the Web controls. To build sophisticated GIS applications that go beyond the functionality exposed by the Web controls, it's important to understand both the server programming model as discussed in this chapter and the relationship between the server API and the Web controls API.

There are two key points that are important to discuss with respect to Web controls and the server API:

• Management of session state.

• Management of server contexts for pooled and non-pooled server objects.

The Web controls are described in much more detail in Chapters 5 and 6. The information presented here does not preclude the need to understand the material presented in those chapters.

WEB CONTROLS AND SESSION STATE

The *WebMap* (.NET) or *AGSWebContext* (Java) object of the map and overview map controls manages the *MapDescription* of the *MapServer* object in session state for you. When you use methods on *WebMap* or *AGSWebContext*, such as *Zoom*, *Pan*, and *CenterAt*, to navigate the map, *WebMap* or *AGSWebContext* internally uses *SaveObject* and *LoadObject* as described previously to maintain stateful aspects of the application, such as current extent, within session state. The page layout control's *WebPageLayout* (.NET) or *AGSWebContext* (Java) object does the same for the *PageDescription* and collection of *MapDescriptions* associated with the *MapServer*'s page layout. The TOC control has methods for turning the visibility of layers displayed in a map control on and off. Like the *WebMap* and *WebPageLayout* (.NET) and *AGSWebContext* (Java), using the methods on the TOC control to turn layers on and off will update the session's copy of the *MapDescription*.

If you want to manipulate the *MapDescription* yourself, for example, to add graphics to the map, you can ask the *WebMap* or *AGSWebContext* for its *MapDescription*, in which case you will be manipulating the copy of the map description that the *WebMap* or *AGSWebContext* is managing in session state for you.

WEB CONTROLS AND SERVER CONTEXT MANAGEMENT

The *WebMap* (.NET) or *AGSWebContext* (Java) object also manages the acquiring and releasing of server contexts. The remainder of this section will illustrate this concept using the .NET case (*WebMap* within an ASP.NET application). For Java specific examples of how the *AGSWebContext* does this, see Chapter 6.

If you scope the use of the *WebMap* within a Using block, the *WebMap* will get a server context from the server at the beginning of the Using block, then explicitly release it.

The following code is an example of using the *WebMap* in a Using block to use the *MapServer*'s context to count the number of features in the first layer of the map that intersect the current extent:

```
using (WebMap webMap = Map1.CreateWebMap())
{
  IMapServer map = webMap.MapServer;
```

The objects and interfaces used for performing spatial queries and for working with the results of those queries can be found in the GeoDatabase object library. To learn more about geodatabase objects, see the online developer documentation.

```
IMapServerObjects mapso = map as IMapServerObjects;

IMap fgmap = mapso.get_Map(map.DefaultMapName);
IFeatureLayer fl = fgmap.get_Layer(0) as IFeatureLayer;
IFeatureClass fc = fl.FeatureClass;

IServerContext sctx = webMap.ServerContext;
ISpatialFilter sf = sctx.CreateObject("esriGeoDatabase.SpatialFilter")
as ISpatialFilter;

IMapDescription md = webMap.MapDescription;
IMapArea ma = md.MapArea;

sf.SpatialRel = esriSpatialRelEnum.esriSpatialRelIntersects;
sf.Geometry = ma.Extent as IGeometry;
sf.GeometryField = fc.ShapeFieldName;

long lcount = fc.FeatureCount(sf);
}
```

The objects and interfaces used for creating and working with geometries can be found in the Geometry object library. To learn more about geometry objects, see the online developer documentation.

This example is valid to use for either a pooled or non-pooled server object since it does not change the state of any aspects of the server object and does not save any objects for later use in the server's context.

When the Map control's *ServerObject* property is set to be a pooled server object, when you get the *WebMap* from the Map control using the *CreateWebMap* method, the Map control is internally using *CreateServerContext* to create a new server context using the *ServerObject* property of the Web control to create a server context for the appropriate *MapServer* object. When the *WebMap* goes out of scope, the Map control internally calls *ReleaseContext* to release the context. The methods exposed by the Web controls are stateless, so a server object's state is never altered by calling one of the methods on a Web control.

When the Map control's *ServerObject* property is set to be a non-pooled server object, the Map control will create a server context at the beginning of the session and hold on to the same server context for the entire session. *CreateWebMap* uses the context already in session, and when the *WebMap* goes out of scope, the context is not released.

The next example uses the *WebMap* with a non-pooled server object to get a reference to the *Workspace* for a layer in the map and store that workspace object in the context's object dictionary. Since it is a non-pooled server object, the next time the *WebMap* is obtained from the Map control, it will be the same context. The following code would go in the *Page_Load* event of the application:

```
using (WebMap webMap = Map1.CreateWebMap() )
{
  // get the workspace from the first layer and add it to the context's
object dictionary
  IMapServerObjects mapo = (IMapServerObjects) webMap.MapServer;
  IMap map = mapo.get_Map(webMap.DataFrame);

  // get workspace from first layer and add it to the object dictionary
```

```
IFeatureLayer fl = (IFeatureLayer) map.get_Layer(0);
IDataset ds = (IDataset) fl.FeatureClass;
IWorkspaceEdit wse = (IWorkspaceEdit) ds.Workspace;
IServerContext sc = webMap.ServerContext;
sc.SetObject("theWS",wse);
}
```

Once the session ends, you must make sure that server context is released by looping through all of the Web controls and explicitly releasing their contexts:

```
protected void Session_End(Object sender, EventArgs e)
{
  IServerContext context;
  for ( int i = 0; i < Session.Count; i++)
  {
    context = Session[i] as IServerContext;
    if ( context != null )
      context.ReleaseContext();
  }
  Session.RemoveAll();
}
```

It's important to note that if you create any additional contexts within the Using block, it is still your responsibility to call ReleaseContext on them. The *WebMap* will only release the context that it creates.

There is more complete documentation on using the Web controls in Chapter 5 (.NET) and Chapter 6 (Java).

One aspect of programming ArcGIS Server is to create GIS Web applications that run in a browser and that people interact with. Developers can also use ArcGIS Server to create GIS Web services. Unlike a Web application, a Web service is a program that is used by other programs and not users directly. The types of Web services that ArcGIS Server supports can be divided into two categories: application Web services and ArcGIS Server Web services.

APPLICATION WEB SERVICES

A Web service is a set of related application functions that can be programmatically invoked over the Internet. An application Web service solves a particular problem, for example, a Web service that finds all of the hospitals within a certain distance of an address or performs some other type of GIS analysis. Application Web services can be implemented using the native Web service framework of your Web server, for example, ASP.NET Web service (WebMethod), or Java Web service (Axis), and so on.

When using native frameworks such as ASP.NET and J2EE to create and consume your application Web services, you need to use native or application-defined types as both arguments and return values from your Web methods. Clients of the Web service will not be ArcObjects applications, and as such, your Web service should not expect ArcObjects types as arguments and should not directly return ArcObjects types.

To learn more about WSDL, refer to http://www.w3.org.

Any development language that can use standard HTTP to invoke methods can consume a Web service. The Web service consumer can get the methods and types exposed by the Web service through its Web Service Description Language (WSDL).

The following is a simple example of a geocoding Web service written using ASP.NET, which connects to the ArcGIS Server and makes use of a *GeocodeServer* object to locate the address. Notice that its arguments are strings (an address and a ZIP Code), and it does not return an ArcObjects point object but returns the x and y coordinates of the point as a string:

```
[WebMethod]
public string LocateAddr(string address, string zipcode)
{
  string wsresult = "";

  if (address == null || zipcode == null)
    return wsresult;

  // connect to the GIS server
  string m_host = "padisha";

      ESRI.ArcGIS.Server.WebControls.ServerConnection connection = new
  ESRI.ArcGIS.Server.WebControls.ServerConnection();
      connection.Host = m_host;
      connection.Connect();

  // get a server context
  IServerObjectManager som = connection.ServerObjectManager;
```

```
  IServerContext sc =
som.CreateServerContext("RedlandsGC","GeocodeServer");

 try
 {
  IServerObject so = sc.ServerObject;
  IGeocodeServer gc = (IGeocodeServer)so;

  IPropertySet ps =
(IPropertySet)sc.CreateObject("esriSystem.PropertySet");
   ps.SetProperty("street",address);
   ps.SetProperty("Zone",zipcode);

   IPropertySet res = gc.GeocodeAddress(ps,null);
   ESRI.ArcGIS.Geometry.IPoint tempPt = (ESRI.ArcGIS.Geometry.IPoint)
res.GetProperty("Shape");

  wsresult = tempPt.X + "," + tempPt.Y;

  // release the server context
  sc.ReleaseContext();
 }
  catch
 {
  // release the server context
  sc.ReleaseContext();
 }

 return wsresult;
 }
```

Notice in the code above, all the standard aspects of programming with the ArcGIS Server API apply, including connecting to the server, working with server objects and server contexts, and releasing the server context at the end of the method. The following C# code is an example of a client to this Web service. In this example, *wsref* is the name of the Web reference made to this Web service:

```
wsref.FindAddress w = new wsref.FindAddress();
string slocation = w.LocateAddr("380 New York St","92373");
```

An application Web service is built by application developers, using native .NET or J2EE Web service frameworks. The Web service is processed and executed within the Web server and makes calls into the GIS server for GIS functionality.

ARCGIS SERVER WEB SERVICES

Administrators can expose *MapServer* and *GeocodeServer* objects as generic ArcGIS Server Web services for access across the Internet. These Web services are processed and executed from within the GIS server. SOAP requests are forwarded to the GIS server by a simple Web page. As described earlier, these server objects have SOAP interfaces for handling SOAP requests to execute methods and return results as SOAP responses. It is this support for SOAP request handling that makes it possible to expose server objects as Web services. ArcGIS Web services use all ArcObjects types, for example, ArcObjects geometry types.

The Web service templates organize published ArcGIS Server Web services into Web service catalogs. The ADF for both .NET and Java includes template applications for both ArcGIS Server Web services and Web service catalogs. To learn how to use these templates to create Web service catalogs, see Chapters 5 and 6.

ArcGIS Server Web services can be consumed by both application developers using .NET and J2EE and can be consumed directly over the Internet by ArcGIS Desktop applications and ArcGIS Engine applications. When making an Internet connection to an ArcGIS Server using ArcCatalog, you are actually connecting to a Web service catalog.

Accessing ArcGIS Web services with ArcObjects using GISClient

ArcGIS Desktop and ArcGIS Engine developers can consume ArcGIS Server Web services using the ArcGIS Server Client API (AGSClient), which is in the GISClient object library. The AGSClient includes objects and methods for connecting to GIS servers either directly or through Web service catalogs to make use of *MapServer* and *GeocodeServer* objects. The AGSClient differs from the server API in that it restricts clients to calling only the coarse-grained methods on the server object and does not provide access to the finer-grained ArcObjects associated with a server object.

The AGSServerConnection object provides connections to the GIS server and ArcGIS Server Web service catalogs.

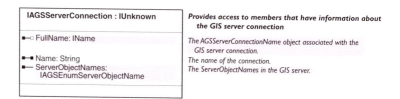

IAGSServerConnection : IUnknown	Provides access to members that have information about the GIS server connection
■○ FullName: IName	*The AGSServerConnectionName object associated with the GIS server connection.*
■■ Name: String ■— ServerObjectNames: IAGSEnumServerObjectName	*The name of the connection.* *The ServerObjectNames in the GIS server.*

The following VB6 code shows how to use the AGSClient to connect to a Web service catalog and list the set of server objects and their type that are exposed as Web services in the Web service catalog:

```
Dim pAGSConnectionFactory As IAGSServerConnectionFactory
Set pAGSConnectionFactory = New AGSServerConnectionFactory

Dim pAGSConnection As IAGSServerConnection
Dim pConnectionProps As IPropertySet
Set pConnectionProps = New PropertySet
pConnectionProps.SetProperty "URL", "http://padisha/redlandscatalog/
default.aspx"
```

```
Set pAGSConnection = pAGSConnectionFactory.Open(pConnectionProps, 0)

Dim pEnumSOName As IAGSEnumServerObjectName
Set pEnumSOName = pAGSConnection.ServerObjectNames

Dim pSOName As IAGSServerObjectName
Set pSOName = pEnumSOName.Next

Do Until pSOName Is Nothing
  Debug.Print pSOName.Name & ": " & pSOName.Type
  Set pSOName = pEnumSOName.Next
Loop
```

The GISClient objects can be used to connect to a GIS server either via a LAN connection (by specifying the machine property in the connection) or via a Web service catalog (by providing the URL property in the connection). The GISClient object provides a uniform programming interface for working with server objects through both LAN connections and Web service connections.

The following VB6 code shows how to use the AGSClient to connect to GIS server directly running on the machine "melange" and list the set of server objects and their type that are running in the GIS server:

```
Dim pAGSConnectionFactory As IAGSServerConnectionFactory
Set pAGSConnectionFactory = New AGSServerConnectionFactory

Dim pAGSConnection As IAGSServerConnection
Dim pConnectionProps As IPropertySet
Set pConnectionProps = New PropertySet
pConnectionProps.SetProperty "machine", "melange"

Set pAGSConnection = pAGSConnectionFactory.Open(pConnectionProps, 0)

Dim pEnumSOName As IAGSEnumServerObjectName
Set pEnumSOName = pAGSConnection.ServerObjectNames

Dim pSOName As IAGSServerObjectName
Set pSOName = pEnumSOName.Next

Do Until pSOName Is Nothing
  Debug.Print pSOName.Name & ": " & pSOName.Type
  Set pSOName = pEnumSOName.Next
Loop
```

You can make stateless use of server objects that you get from a GISClient connection and call methods exposed by the server object's stateless interfaces; however, you cannot use a server object obtained via GISClient to drill down to the finer-grained ArcObjects associated with the server object. You can get a reference to the *ServerObjectManager* through the GISClient connection, provided that connection is a LAN connection. You can then use the methods exposed by the *ServerObjectManager* to obtain references to server contexts and server objects to work with the server API.

Accessing ArcGIS Server Web services directly

ArcGIS Server Web services, similar to application Web services, can be accessed by any development language that can submit SOAP-based requests to a Web service and process SOAP-based responses. The Web service consumer can get the methods and types exposed by the Web service through its WSDL. ArcGIS Web services are based on the SOAP doc/literal format and are interoperable across Java and .NET.

On the Web server, the Web service catalog templates can be used by developers to publish any server object as a Web service over HTTP. For any published server object, the template creates an HTTP endpoint (URL) to which SOAP requests can be submitted using the standard HTTP POST method. The endpoint also supports returning the WSDL for the Web service using a standard HTTP GET method with "wsdl" as the query string. The implementation of the HTTP endpoint is thin, while the actual processing of the SOAP request and the generation of the SOAP response take place within the GIS server. The WSDLs that describe the definitions of the SOAP requests and responses are also part of the GIS server and are installed as part of the ArcGIS Server install under <install directory>\XMLSchema.

The Web service templates organize published ArcGIS Server Web services into Web service catalogs. Each Web service is a distinct HTTP endpoint/URL. The Web service catalog is itself a Web service with a distinct endpoint and can be queried to obtain the list of Web services in the catalog and the URL for each Web service in the catalog.

The types of Web services supported at version 9.0 are:

• Web service catalog

• MapServer

• GeocodeServer

Each Web service has its own WSDL.

The Web service templates are available for use on both .NET and Java Web servers and create identical endpoints. To a consumer of the Web services, the two different server implementations are identical in terms of WSDLs and how one obtains the WSDLs and interacts with the Web service over HTTP.

For example, assume that the ArcGIS Server administrator or developer at an organization named ACME, whose .NET Web server has the URL http://www.myserver.com, has used the ArcGIS Server Web service catalog templates to create a Web service catalog called "MyCatalog", which contains a MapServer Web service named "PortlandMap" and a GeocodeServer Web service "PortlandGeocode". The URLs for the Web services are:

Web service catalog:

```
http://www.myserver.com/MyCatalog/Default.aspx
```

Portland map server:

```
http://www.myserver.com/MyCatalog/PortlandMap.aspx
```

Portland geocode server:

```
http://www.myserver.com/MyCatalog/PortlandGeocode.aspx
```

If the ArcGIS Server site was running a Java Web server, then the URLs would be:

Web service catalog:

```
http://www.myserver.com/MyCatalog/Default.jsp
```

Portland map server:

```
http://www.myserver.com/MyCatalog/PortlandMap.jsp
```

Portland geocode server:

```
http://www.myserver.com/MyCatalog/PortlandGeocode.jsp
```

ArcGIS Web services can be consumed from .NET or Java. As a consumer of an ArcGIS Web service, you can use the methods exposed by the Web service by including a reference to the Web service in your .NET or Java project. Web services implemented on a Java Web server can be consumed from a .NET client and vice versa.

Assume you are a .NET developer who wishes to consume the above Web services published by ACME. You can add the following references to your Visual Studio .NET project:

ServiceCatalog:

```
http://www.myserver.com/MyCatalog/Default.aspx?wsdl
```

Portland MapServer Web service:

```
http://www.myserver.com/MyCatalog/PortlandMap.aspx?wsdl
```

Portland GeocodeServer Web service:

```
http://www.myserver.com/MyCatalog/PortlandGeocode.aspx?wsdl
```

The following C# code is an example of using an ArcGIS Server MapServer Web service to draw a map. In this example the Portland Web service has been referenced as *Port*:

```
Port.Portland map = new Port.Portland ();
Port.MapServerInfo mapi = map.GetServerInfo(map.GetDefaultMapName());
Port.MapDescription pMapDescription = mapi.DefaultMapDescription;

Port.ImageType it = new Port.ImageType();
it.ImageFormat  = Port.esriImageFormat.esriImageBMP;
it.ImageReturnType = Port.esriImageReturnType.esriImageReturnMimeData;

Port.ImageDisplay idisp = new Port.ImageDisplay();
idisp.ImageHeight = 400;
idisp.ImageWidth  = 500;
idisp.ImageDPI  = 150;

Port.ImageDescription pID = new Port.ImageDescription();
pID.ImageDisplay = idisp;
pID.ImageType = it;

Port.MapImage pMI = map.ExportMapImage(pMapDescription, pID);
System.IO.Stream pStream = new
System.IO.MemoryStream((byte[])pMI.ImageData);
System.Drawing.Image pImage = Image.FromStream(pStream);
```

The following C# code is an example of using an ArcGIS Server GeocodeServer Web service to locate an address and get the x and y location of the resulting point. In this example the PortlandGeocode Web services has been referenced as *PortGC*:

```
PortGC.PortlandGC gc = new PortGC.PortlandGC();
PortGC.PropertySet ps = new PortGC.PropertySet();
PortGC.PropertySetProperty[] pa = new PortGC.PropertySetProperty[2];
PortGC.PropertySetProperty pr = new PortGC.PropertySetProperty();
pr.Key = "street";
pr.Value = "2111 division st";
pa[0] = pr;

PortGC.PropertySetProperty pr2 = new PortGC.PropertySetProperty();
pr2.Key = "zone";
pr2.Value = "97202";
pa[1] = pr2;
ps.PropertyArray = pa;

PortGC.PropertySet resps = gc.GeocodeAddress(ps,null);
PortGC.PropertySetProperty[] respa = resps.PropertyArray;
PortGC.Point pnt = null;
for (int i = 0;i < respa.Length;i++)
{
  if (respa[i].Key == "Shape")
    pnt = respa[i].Value as PortGC.Point;
  break;
}
double X = pnt.X;
double Y = pnt.Y;
```

The following C# code is an example of using an ArcGIS Server Service Catalog Web service to query the URLs of the Web services in the Web service catalog. In this example the Web service catalog has been referenced as *WSCat*:

```
WSCat.Default sc = new WSCat.Default();
WSCat.ServiceDescription[] wsdesc = sc.GetServiceDescriptions();
WSCat.ServiceDescription sd = null;
string wsURL = null;

for (int i = 1;i < wsdesc.Length;i++)
{
  sd = wsdesc[i];
  wsURL = sd.Url;
  // do something with the URL
}
```

As discussed in the previous sections, developing ArcGIS Server applications is all about remotely programming ArcObjects. As such, the server API is wide and includes a large number of ArcObjects components organized into a series of object libraries. Developers can build any kind of application using the server API, so it is possible to build high-performance, scalable applications, and it is also possible to build extremely slow applications that do not scale.

This section will discuss strategies for avoiding the latter, but it will ultimately be in the hands of the developer and the GIS server administrator to work together to design applications that can perform and scale given available hardware resources.

ArcGIS SERVER AND FINE-GRAINED ArcOBJECTS

ArcGIS Server uses the same ArcObjects that ArcGIS Engine and ArcGIS Desktop use, so if your server application includes GIS functionality that performs poorly in an ArcGIS Engine or ArcGIS Desktop deployment, that same functionality will likely perform poorly in your server deployments. Conversely, if that GIS functionality performs well in an ArcGIS Engine or Desktop deployment, it will also perform well in server deployments if the application is properly tuned with respect to how it makes use of ArcObjects in the server.

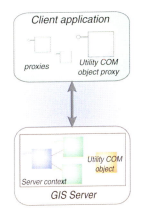

The GIS server can easily be extended to use application-specific COM objects that a developer can write in VB, C++, or .NET. If these COM components are installed on the server object container machines on which your server objects are hosted, then they can be used to do work for your application.

Both coarse-grained calls to remote ArcObjects, such as the methods on the *MapServer* and *GeocodeServer*, and fine-grained calls to remote ArcObjects, such as looping through all the vertices of a polygon, are exposed through the ArcGIS Server API and can be used in your application. However, it's important to note that when making a call against an object running in the server from your Web application, you are making that call across processes. The Web server is running in one process, while the object is running in another process.

Calls to objects across processes are significantly slower than calls to objects in the same process. It is also likely that your Web application is running on a Web server that is actually a different machine from which the object is running on, so the calls are not only cross process, but also cross machine. This performance difference on the scale of a single call or a few 10s of calls is not significant in terms of the overall performance of your application. However, if your application is making 1,000s of fine-grained ArcObjects calls across process or machine to the server, there can be a significant performance impact on the application.

You should minimize the number of round-trips to the GIS server from your application by minimizing the number of fine-grained calls to remote objects. If the nature of your application demands a lot of fine-grained ArcObjects work, one strategy for supporting such functionality, but keeping remote calls to a minimum, is to extend the GIS server with application-specific utility COM objects that you can develop in VB, C++, or .NET.

EXTENDING THE GIS SERVER

The GIS server can easily be extended to use application-specific utility COM objects that a developer can write in VB, C++, or .NET. If these COM components are installed on the server object container machines on which your server objects are hosted, they can be used to do work for your application. For example, you may have a custom network tracing function that you want to use in your server application. You can write the tracing code as a COM object and install it on the server:

```
mylib.TraceUtilities tracer =
pServerContext.CreateObject("mylib.TraceUtilities") as
mylib.TraceUtilities;
result = tracer.DoIsolationTrace(...)
```

In this example, the tracing function may involve thousands of ArcObjects calls, but all of those calls happen in the server where the *TraceUtilities* COM object is running. The coarse-grained *DoIsolationTrace* method is the only method that the Web application needs to call, which means there is only a single remote object call to the server.

To illustrate this in more detail, here is a simple ASP.NET example where a Web application includes a button (*btnTotalAreas*) that reports the total area of polygon features in one of the map's layers that intersect the map extent, and displays the result on a label on the Web form (*lblResult*). In this example, the Web application includes a Map control (*Map1*), from which it gets the *MapServer* and its server context, and includes a TOC control (*Toc1*) that the user of the Web application uses to identify which layer to query.

To do the above, a spatial query is made on the selected layer's feature class using the map's current extent to return a geodatabase cursor. The application then loops through the cursor, gets each feature's geometry, and adds its area to the total. The following is the C# code that would execute when the button is clicked:

```
private void btnTotalAreas_Click(object sender, System.EventArgs e)
{
  using (WebMap webMap = Map1.CreateWebMap() )
  {
    IMapServer map = webMap.MapServer;
    IServerContext ctx = webMap.ServerContext;

    IMapServerObjects mapobj = map as IMapServerObjects;
    IMap fgmap = mapobj.get_Map(map.DefaultMapName);

    // find the selected layer
    IEnumLayer maplayers = fgmap.get_Layers(null,true);
    ILayer lyr;
    string sLayername =
Toc1.GetNodeFromIndex(Toc1.SelectedNodeIndex).Text;

    while ((lyr = maplayers.Next()) != null)
    {
      if (lyr.Name == sLayername)
        break;
    }

    if (lyr == null)
    {
      lblResult.Text = "Layer not found";
      return;
```

```
      }

      // get the feature class and make sure its geometry type is polygon
      IFeatureLayer flyr = lyr as IFeatureLayer;
      IFeatureClass fc = flyr.FeatureClass;

      if (fc.ShapeType != esriGeometryType.esriGeometryPolygon)
       {
        lblResult.Text = "Select a polygon layer";
        return;
       }

      // create the query using the current extent of the map
      ISpatialFilter sf = ctx.CreateObject("esriGeoDatabase.SpatialFilter")
  as ISpatialFilter;
      IMapArea ma = webMap.MapDescription.MapArea;

      sf.SpatialRel = esriSpatialRelEnum.esriSpatialRelIntersects;
      sf.GeometryField = fc.ShapeFieldName;
      sf.Geometry = ma.Extent as IGeometry;

      // execute the query and loop through the results
      IFeature f;
      IArea area;
      double dTotalArea = 0.0;
      IFeatureCursor fcursor = fc.Search(sf,true);
      while ((f = fcursor.NextFeature()) != null)
       {
        area = f.Shape as IArea;
        dTotalArea += area.Area;
       }

      lblResult.Text = dTotalArea.ToString();
    }
  }
```

This code makes approximately eight remote calls on ArcObjects running on the server to get the map from the *MapServer*, get the layers from the map, find the selected layer, check to verify that it is a polygon layer, then create the *QueryFilter* and set its properties. These fine-grained calls into the server do not add up to any significant amount of time.

However, once the query has been executed, three calls are made to the server per feature: one to get the feature, one to get its geometry, one to get the geometry's area. If you consider that (based on the way this application is written) the number of features that may result from the query is indeterminate, the application could potentially loop through thousands of features. If there are 1,000 features intersecting the map extent, then this translates into 3,000 fine-grained calls into the server.

The cost of this number of fine-grained calls does add up and can cause the performance of your application to suffer. To minimize or eliminate this large number of remote calls, you can create a simple COM object that has a method that does the work of looping through all the features and totaling their areas. This aspect of the Web application can then be rewritten to call this single method, reducing this large number of remote calls to a single call to the method on the COM object.

The following VB6 code shows how you could write this COM object. In this example, the method of totaling the areas takes as arguments the feature class and the query filter:

```
Public Function TotalAreas(pFeatureClass As IFeatureClass, pQueryFilter As
IQueryFilter) As Double

  Dim dResult As Double
  dResult = 0#

  ' check to make sure these are polygon features
  If pFeatureClass.ShapeType <> esriGeometryPolygon Then TotalAreas =
dResult

  ' loop through feature and total their areas
  Dim pFeature As IFeature
  Dim pFeatureCursor As IFeatureCursor
  Dim pArea As IArea

  Set pFeatureCursor = pFeatureClass.Search(pQueryFilter, True)
  Set pFeature = pFeatureCursor.NextFeature
  Do Until pFeature Is Nothing
    Set pArea = pFeature.Shape
    dResult = dResult + pArea.Area
    Set pFeature = pFeatureCursor.NextFeature
  Loop

  TotalAreas = dResult

End Function
```

The method returns the total area of all the features that satisfy the query. The return value is tailored to what the calling application (the Web application in this case) wants as an answer (the total area). Once this COM object is installed on the server object container and on the Web server, the code behind the click event of the button in the ASP.NET application can be rewritten as:

```
private void btnTotalAreas_Click(object sender, System.EventArgs e)
{
  using (WebMap webMap = Map1.CreateWebMap() )
  {
    IMapServer map = webMap.MapServer;
    IServerContext ctx = webMap.ServerContext;

    IMapServerObjects mapobj = map as IMapServerObjects;
    IMap fgmap = mapobj.get_Map(map.DefaultMapName);
```

```
    // find the selected layer
    IEnumLayer maplayers = fgmap.get_Layers(null,true);
    ILayer lyr;
    string sLayername =
Toc1.GetNodeFromIndex(Toc1.SelectedNodeIndex).Text;

    while ((lyr = maplayers.Next()) != null)
     {
      if (lyr.Name == sLayername)
        break;
     }

    if (lyr == null)
     {
      lblResult.Text = "Layer not found";
      return;
     }

    // get the feature class and make sure its geometry type is polygon
    IFeatureLayer flyr = lyr as IFeatureLayer;
    IFeatureClass fc = flyr.FeatureClass;

    if (fc.ShapeType != esriGeometryType.esriGeometryPolygon)
     {
      lblResult.Text = "Select a polygon layer";
      return;
     }

    // create the query using the current extent of the map
    ISpatialFilter sf = ctx.CreateObject("esriGeoDatabase.SpatialFilter")
as ISpatialFilter;
    IMapArea ma = webMap.MapDescription.MapArea;

    sf.SpatialRel = esriSpatialRelEnum.esriSpatialRelIntersects;
    sf.GeometryField = fc.ShapeFieldName;
    sf.Geometry = ma.Extent as IGeometry;

    // create out utility COM object on the server
    ServerUtil.clsTotalAreasClass totarea =
ctx.CreateObject("ServerUtil.clsTotalAreas") as
ServerUtil.clsTotalAreasClass;

    double dTotalArea = totarea.TotalAreas(ref fc, ref sf);

    lblResult.Text = dTotalArea.ToString();
   }
 }
```

Using this version of the code, the remote calls when looping through the features, which could total in the 1,000s, have been reduced to a single call. Using this method of pushing fine-grained ArcObjects calls into the server can result in a significant increase in the performance of an operation that requires a large number of fine-grained ArcObjects calls.

The client machine (that is, your Web server) must also install either the COM object or a proxy to the COM object so the application will have access to its interfaces and methods when it's created on the server.

You can create these utility COM objects using Visual Basic, C++, or .NET. When using .NET to create a COM object for use in the GIS server, there are some specific guidelines you need to follow to ensure that you can use your object in a server context and that it will perform well in that environment. For an overview of how to create a COM object using .NET, refer to Appendix D. The guidelines below apply specifically to COM objects you create to run within the server.

- You must explicitly create an interface that your COM class implements. Unlike Visual Basic 6, .NET will not create an implicit interface for your COM class that you can use when creating the object in a server context.

- Your COM class should be marshalled using the Automation marshaller. You specify this by adding the *AutomationProxyAttribute* attribute to your class with a value of true.

- Your COM class should generate a dual class interface. You specify this by adding the *ClassInterfaceAttribute* attribute to your class with a value of *ClassInterfaceType.AutoDual*.

- To ensure that your COM object performs well in the server, it must inherit from *ServicedComponent*, which is in the *System.EnterpriseServices* assembly. This is necessary due to the current COM interop implementation of the .NET framework.

The following C# code shows how you could write the COM object in .NET with the same functionality as previously demonstrated in VB6. Notice the interface definition, the use of class attributes, and the fact that the COM class inherits from *ServicedComponent*:

```csharp
public interface IAreaSum
{
  double sumArea(ref IFeatureClass pFClass, ref IQueryFilter pQFilter);
}

namespace ServerUtilCS
{
  [AutomationProxy(true), ClassInterface(ClassInterfaceType.AutoDual)]
  public class ServerUtil:ServicedComponent, IAreaSum
  {
    public ServerUtil()
    {

    }

    public double sumArea(ref IFeatureClass pFClass, ref IQueryFilter
pQFilter)
    {
```

```
        double dArea = 0;

        // check to make sure these are polygon features
        if (pFClass.ShapeType != esriGeometryType.esriGeometryPolygon)
          return dArea;

        // loop through features and total their areas
        IFeature pFeature = null;
        IArea pArea = null;
        IFeatureCursor pFeatureCursor = pFClass.Search(pQFilter,true);
        while ((pFeature = pFeatureCursor.NextFeature()) != null)
          {
          pArea = pFeature.Shape as IArea;
          dArea += pArea.Area;
          }
        return dArea;
      }
    }
  }
```

The code behind the click event of the button in the ASP.NET application that creates and uses this version of the COM object would look like the following:

```
...
// create out utility COM object on the server
IAreaSum totarea = ctx.CreateObject("ServerUtilCS.ServerUtil") as
IAreaSum;

IQueryFilter qf = sf as IQueryFilter;
double dTotalArea = totarea.sumArea(ref fc, ref qf);
...
```

The following code shows how you could write the COM object in .NET using VB.NET:

```
Public Interface IAreaSum
  Function sumArea(ByRef pFClass As IFeatureClass, ByRef pQFilter As
IQueryFilter) As Double
End Interface

<AutomationProxy(True), ClassInterface(ClassInterfaceType.AutoDual)> _
Public Class ServerUtil
  Inherits ServicedComponent
  Implements IAreaSum

  Public Sub New()
    MyBase.New()
  End Sub
  Public Function sumArea(ByRef pFClass As IFeatureClass, ByRef pQFilter
As IQueryFilter) As Double Implements ServerUtilVBNET.IAreaSum.sumArea
    Dim dResult As Double = 0.0#

    ' check to make sure these are polygon features
    If pFClass.ShapeType <> esriGeometryType.esriGeometryPolygon Then
sumArea = dResult
```

```
' loop through feature and total their areas
Dim pFeature As IFeature = Nothing
Dim pArea As IArea = Nothing
Dim pFeatureCursor As IFeatureCursor = pFClass.Search(pQFilter, True)

pFeature = pFeatureCursor.NextFeature
Do Until pFeature Is Nothing
  pArea = pFeature.Shape
  dResult = dResult + pArea.Area
  pFeature = pFeatureCursor.NextFeature
Loop

sumArea = dResult

  End Function
End Class
```

The code behind the click event of the button in the ASP.NET application that creates and uses this version of the COM object would look like the following:

```
...
// create out utility COM object on the server
IAreaSum totarea = ctx.CreateObject("ServerUtilVBNET.ServerUtil") as
IAreaSum;

IQueryFilter qf = sf as IQueryFilter;
double dTotalArea = totarea.sumArea(ref fc, ref qf);
...
```

If your application makes fine-grained use of ArcObjects, it's not necessary to always extend the server in this way. As discussed above, it comes down to the volume of those calls that your application will make. If your application is written in such a way that it always makes thousands of fine-grained ArcObjects calls, or the number of fined-grained calls is indeterminate based on user interaction with the application, you should consider moving some of the code into the server.

If you design your application such that large volumes of fine-grained ArcObjects calls are not necessary, and your user interface is designed such that your users cannot make requests that result in a large volume of fine-grained ArcObjects calls, then extending the server in this manner is not necessary.

USING POOLED SERVER OBJECTS TO POOL OTHER OBJECTS

If your Web application or Web service does not make use of a *MapServer* or a *GeocodeServer* object for its functionality, you can use empty server contexts for the application or Web service to do its GIS work. In that case, your application will need to create any necessary objects within the empty server context using the *CreateObject* method on the server context.

In some cases, the creation of an object may itself be expensive. A good example of this is a connection to a geodatabase workspace. Making a geodatabase workspace connection involves connecting to a database management system (DBMS) and querying various tables in the database. If each invocation of a Web service started with connecting to a geodatabase workspace, that introduces the fixed cost of making the connection. Additionally, this puts a load on, and can impact the scalability of, the database server.

Ideally your Web service should make use of a pooled workspace connection, such that a small number of connections are made to the database once and shared across invocations of the Web service. One method of doing this is to create a map document that contains a single layer whose source data is a feature class from the workspace that your Web service needs to connect to. You can create a pooled *MapServer* object using this map document as its initialization data. When the instances of the *MapServer* are created to populate the object pool, each will make and hold on to a connection to the workspace. Your Web service can get a reference to one of the pooled instances of the *MapServer*, get the layer from the map, get the feature class from the layer, and ask it for a reference to its workspace. Once the Web service finishes executing and releases the *MapServer*, it and its database connection return to the pool for use by another invocation of the Web service.

Below is an ASP.NET example of how you can use a pooled *MapServer* to get a workspace connection from the first layer:

```
[WebMethod]
public string PooledWSExample()
{
  // connect to the GIS server
  string m_host = "padisha";

  ESRI.ArcGIS.Server.WebControls.ServerConnection connection = new
ESRI.ArcGIS.Server.WebControls.ServerConnection();
  connection.Host = m_host;
  connection.Connect();

  // get a server context
  IServerObjectManager som = connection.ServerObjectManager;
  IServerContext sc = som.CreateServerContext("MyMapServer","MapServer");

  // get the first layer in the map
  IMapServer map = sc.ServerObject as IMapServer;
  IMapServerObjects mapobj = map as IMapServerObjects;

  // get the workspace from the layer's feature class
  IMap fgmap = mapobj.get_Map(map.DefaultMapName);
  IFeatureLayer fl = fgmap.get_Layer(0) as IFeatureLayer;

  IDataset ds = fl.FeatureClass as IDataset;
  IWorkspace ws = ds.Workspace;

  // do something with the workspace
```

```
  sc.ReleaseContext();

  return "Operation complete";
}
```

If you are developing Web services that involve connecting to geodatabase workspaces, you should consider this method of pooling those workspace connections.

LIMITING THE SIZE OF QUERY RESULTS AND OUTPUT

As noted earlier, the server API is wide and includes a large number of ArcObjects components that developers can use to include functionality in their application. This does put responsibility on the developer to build applications that do not allow the user of that application to put either the Web server (in the case of a Web application or Web service) or the database server in a state that results in a denial of service to other users.

Two key areas that will be discussed in detail are writing output and executing and evaluating the results of database queries.

Limiting the size of query results

The GeoDatabase library includes objects that allow you to query data in a database using spatial filters, attribute filters, or a combination of both. At this level of the ArcObjects API, no limits or constraints are put on the nature of the query or the number of records that may be returned by the cursor resulting from executing the query. If, for example, a Web application executes a query that returns 1,000,000 rows and iterates through each row, this can (depending on the nature of the query) tie up the database server while the query is evaluated, then tie up the Web server as the application iterates through the 1,000,000 rows.

These types of queries should be avoided by not allowing users to perform ad hoc queries against the database. Design your application to tightly control the types of queries that users can execute (or that are executed as a result of their interaction with the application). Also, set limits on the number of query results that the application will process for those cases where a large number of query results are returned from the database. You can do this by evaluating a fixed number of maximum rows from result cursors.

The following is an example of executing a query that may return a cursor with a large number of records. In this example, the application will stop evaluating the cursor after the first 100 rows have been retrieved. Assume that *pFeatureClass* is a feature class in the server context:

```
Dim pQueryFilter As IQueryFilter
Set pQueryFilter =
pServerContext.CreateObject("esriGeodatabase.QueryFilter")
pQueryFilter.WhereClause = "ObjectID > 100"

Dim pCursor As ICursor
Set pCursor = pFeatureClass.Search(pQueryFilter, True)

Dim i As Long
i = 1
```

```
Dim pFeature As IFeature
Set pFeature = pCursor.NextRow
While Not pFeature Is Nothing And i < 100
  ' do something with the feature
  i = i + 1
  Set pFeature = pCursor.NextRow
Wend
```

There are some cases when using the coarse-grained methods on the *MapServer* where queries can be executed, but the execution and evaluation of that query takes place in the MapServer. For example, the *QueryFeatureData* method on *IMapServer* returns a fully populated record set containing the results of the query. To ensure that these record sets do not contain a number of rows too large for the system to handle, the MapServer itself has built-in limits to evaluate the results of a query to a maximum record count (logically the same as demonstrated above).

This maximum is set as a property of the MapServer object itself called *MaxRecordCount*. By default, the *MaxRecordCount* is 500, but this property can be modified by the administrator of the GIS server by modifying the value of the *MaxRecordCount* XML tag in the MapServer's configuration file. For more information about the GIS server's configuration files and how to modify them, see Appendix B.

This maximum record count will be applied to the results of the following methods on *IMapServer*:

- *QueryFeatureData*

- *Find*

- *Identify*

The MapServer also allows you to dynamically draw buffers around features by specifying a *SelectionBufferDistanceProperty* on *ILayerDescription* that is greater than 0. If the selection is large, this can greatly increase the resources required to draw a map. The MapServer also has built-in limits to limit the number of features that can be buffered per layer.

This maximum is set as a property of the MapServer object itself called *MaxBufferCount*. By default, the *MaxBufferCount* is 100, but this property can be modified by the administrator of the GIS server by modifying the value of the *MaxBufferCount* XML tag in the MapServer's configuration file. For more information about the GIS server's configuration files and how to modify them, see Appendix B.

The *GeocodeServer* also has built-in limits that prevent requests returning results that are too large. Specifically, the number of records returned by the *FindAddressCandidates* method is limited by the GeocodeServer's *MaxResultSize* property. The default for this is 500, but this property can be modified by the administrator of the GIS server by modifying the value of the *MaxResultSize* XML tag in the GeocodeServer's configuration file. For more information about the GIS server's configuration files and how to modify them, see Appendix B.

The GeocodeServer also has a built-in limit for the number of input records that can be passed into the *GeocodeAddresses* method. This maximum is set as a property of the GeocodeServer itself called *MaxBatchSize* and can be modified by the administrator of the GIS server by modifying the value of the *MaxBatchSize* XML tag in the GeocodeServer's configuration file.

Limiting the size of output

Another area where application developers need to be careful is when their applications write output to a server directory. Files that are large may take a large amount of disk space and a large amount of resources to produce. Developers will have to limit the size of files they write with these considerations in mind.

The MapServer's *ExportMapImage* method takes an *ImageDescription* object that includes the size of image requested. To limit the size of images produced by *ExportMapImage*, the MapServer has built-in limits for the size of images that *ExportMapImage* will produce.

This maximum is set as two properties of the MapServer object itself called *MaxImageWidth* and *MaxImageHeight*, specified in pixels. By default, these are set to 2048, but these properties can be modified by the administrator of the GIS server by modifying the values of the *MaxImageWidth* and *MaxImageHeight* XML tags in the MapServer's configuration file. For more information about the GIS server's configuration files and how to modify them, see Appendix B.

This chapter has covered many details on programming with ArcGIS Server. At the beginning of the chapter, it was stated that programming with ArcGIS Server was about programming ArcObjects remotely and that any ArcObjects developer could become an ArcGIS Server developer given some patterns and rules for programming remote ArcObjects and some knowledge of Internet programming. So far these aspects of programming the server have been covered in great detail. The following is a summary of the programming rules and best practices for developing applications with ArcGIS Server.

CONNECTING TO THE SERVER

- Use the native connection object of the environment in which you are developing your application (.NET , Java, or COM).

- If you are writing Web applications or Web services, you need to use impersonation to fix the identity of your application to a user in the GIS servers users group (*agsusers*).

WORKING WITH SERVER OBJECTS

- You get a server object's context from the server, and the server object from its context.

- A server object exposes a number of coarse-grained methods. You can also access the fine-grained ArcObjects associated with a server object.

- When your request on a server object is finished, you must release the server object back to the server by releasing its context.

MANAGING APPLICATION STATE

- Pooled server objects are intended for stateless use.

- Non-pooled server objects support stateful use.

- There are aspects of your application state you can maintain in .NET or Java session state without making stateful use of the GIS server.

- The keys to scalability are: make stateless use of the GIS server, use pooled server objects, minimize the time your application holds on to a server object, and explicitly release it.

SERVER CONTEXTS

- All ArcObjects in a server application run in server contexts.

- You are responsible for releasing a server context when you are finished working with its objects.

- Always create ArcObjects in a server context using its *CreateObject* method (the exception being the server connection object).

- Objects work best together if they are in the same context.

- Do not directly use objects in a server context with local objects and vice versa.

WEB SERVICES

- ArcGIS Server supports development with both application Web services and ArcGIS Server Web services.

- Application Web services should never return ArcObjects types but should return native .NET or Java types.

WEB CONTROLS

- Web controls handle managing the *MapDescription* in your application's session state.

- The *WebMap* object (.NET) and *AGSWebContext* object (Java) create and release server contexts for you.

- For pooled objects, the Web controls will create and release the server context on each request. For non-pooled objects, the Web controls will hold on to the same context for the duration of the session.

PERFORMANCE TUNING

- Minimize the number of fine-grained calls to remote ArcObjects.

- If large numbers of fine-grained ArcObjects calls are necessary, think about extending the server by creating COM objects that move the fine-grained ArcObjects usage into the server.

- Do not allow application users to perform queries that result in large numbers of rows, and limit the number of results from queries that you process.

- Work with your GIS server administrator to ensure that the built-in limits for queries and output for a particular MapServer or GeocodeServer are configured appropriately for your application.

THE REMAINDER OF THIS BOOK

Now that you have a grasp of the core programming model of ArcGIS Server, the remainder of this book covers the details of the ADFs, several developer scenarios, and detailed object model overviews. Since many of the concepts covered in this chapter are central to developing ArcGIS Server applications, it will be useful to use this chapter as a reference as you read the remainder of this book.

5

Developing Web applications with .NET

ArcGIS Server includes an Application Developer Framework (ADF)—built on top of Microsoft's .NET Framework—that allows you to integrate GIS functionality into your Web applications. The ArcGIS Server ADF for .NET includes a set of custom Web controls and templates—incorporated into Microsoft Visual Studio .NET—that you'll use to build your Web applications. You can start building your Web application with one of several predefined templates, including the Map Viewer template that offers basic map navigation and display, the Search template that finds features by attributes, and the Geocoding template that locates places by address. Alternatively, use the Web controls directly to create your own specialized application in a style that conforms to your existing Web site.

This chapter describes how to create Web applications using the ArcGIS Server ADF for .NET. Many of the topics discussed in this chapter assume you have already read the previous chapters of this book. At a minimum, you should read Chapter 4, 'Developing ArcGIS Server applications'. While it isn't a requirement that you use Microsoft Visual Studio .NET to do so, the examples in this chapter assume you are using this integrated development environment.

Topics covered in this chapter include:

• an overview of the ADF • creating Web applications from templates and Web controls • programming guidelines • Web control reference.

In the previous chapters of this book, you've learned about the ArcGIS Server architecture, administration, and programming practices. You've probably already added a map to your GIS server and previewed it in ArcCatalog. This chapter will show you how to integrate that map—or other server object—into a Web application. You'll find that whatever type of Web application you want to build—from basic map display and query to sophisticated GIS editing and analysis—the ArcGIS Server ADF for .NET allows you to utilize all of ArcObjects in a Web environment.

Creating client applications that access your GIS server ultimately involves programming ArcObjects. The .NET ADF is built on top of Microsoft's .NET Framework, extending the .NET Framework class library with new classes that support a set of custom Web controls and provide remote access to ArcGIS Server and, subsequently, ArcObjects. The diagram to the left shows how the .NET ADF fits into the overall development environment.

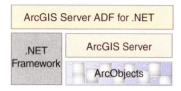

The ArcGIS Server ADF for .NET sits at the top of the various programming components.

You can think of the .NET ADF as a set of:

- Visual Studio .NET templates

- Custom ASP.NET Web controls

- Convenience classes for building client applications that access a GIS server

In the most basic sense, the .NET ADF simply provides you with an additional set of objects to program with. You can use these objects to build desktop and Web applications that access a GIS server; this chapter focuses on creating Web applications. The .NET ADF delivers these objects to you as custom ASP.NET Web controls (also known as server controls) and convenience classes.

The Web controls and convenience classes expose a set of properties, methods, and events that allow you to interact with the GIS server objects, for example, helping you manage connections to the server, access the SOM, and retrieve server objects. Actually, you don't have to use the Web controls at all to create your Web applications. You can directly access ArcGIS Server objects—and thus, ArcObjects—through ASP.NET. However, you'll find that the .NET ADF encapsulates many of the details of programming directly with ArcGIS Server objects and exposes a rich, mapping-centric user interface—through the custom Web controls—that you can place directly on your Web forms.

The Visual Studio .NET templates help you start building your Web applications. Each template incorporates the Web controls into its user interfaces and addresses a particular GIS task—for example, map display and query. While you can use the templates out of the box, they are primarily intended as a starting point for building your own Web application. All the code for the templates is provided to you, so you can easily customize a template to suit your needs or cut and paste code fragments into your own application. The templates also serve as a great learning tool for building your own applications because all the code is there to guide you.

Create your Web application from a predefined mapping template.

VISUAL STUDIO .NET TEMPLATES

When you create a new project in Visual Studio .NET, you're presented with a set of templates that you can use as a starting point to creating an application. In the New Project dialog box, you'll notice that there's a folder that contains ArcGIS Server Projects as well. Within this folder you'll find the set of templates distributed with the .NET ADF. (Note: If you don't see the templates in the New Project dialog box, they probably aren't installed on your system.)

Each template utilizes the set of Web controls that are part of the .NET ADF. For example, those templates that display a map utilize the map control. The primary advantage of building your Web application with one of the templates is that much of the commonly used functionality is already programmed into them, so you don't have to program it yourself. For example, the Map Viewer template displays a toolbar that contains the common map navigation tools for panning and zooming around the map.

As with any template, the look of it simply serves as a starting point for your Web application. You can easily customize the layout of the controls on the template and change elements, such as fonts and colors, to suit your needs. If you plan on integrating the application you create into an existing Web site, you might add other components, such as company logos and site navigation tools, so that it looks like your existing Web pages and integrates seamlessly into your Web site.

Most likely, you'll want to extend the functionality provided in a particular template and incorporate your own custom operations. That's when you'll start programming with the server API and ArcObjects. As mentioned above, each control provides methods that function as entry points into ArcObjects.

Each template included with the .NET ADF is described below. Later in this chapter, you'll see how to use one to build your own Web application in the section called 'Building your first Web applications'.

Map Viewer template

The Map Viewer template provides basic map display capabilities. It consists of a main map, an overview map, a table of contents (legend), a North arrow, and a scale bar. The template also contains a toolbar with built-in tools for panning and zooming. For any map-centric application, the Map Viewer template offers a good starting point.

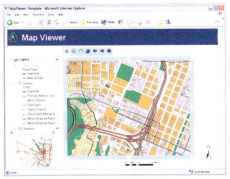

Map Viewer template

Search template

The Search template provides a search-centric interface for finding features on a map. The look of the template is similar to what you might see on the Web for a search engine. Enter a search string and click Go to yield a list of features that match the search string. Click the result you're interested in to get more details about it or to reveal a map that highlights the particular feature.

The Search template searches for matching values in the attribute tables of the layers on the map you

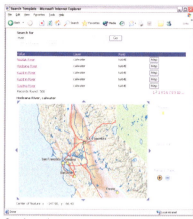

Search template

incorporate into your application. Thus, the list of results returned is restricted to the features on your map. When creating your application from this template, you may want to clearly indicate what type of values can be searched for.

Page Layout template

The Page Layout template displays the entire page layout of a map. It shows all the data frames on the map as well as any map surrounds on the layout, such as the map title, legend, North arrow, and scale bar. This template provides the same view of a map as you'd see in layout view in ArcMap.

The toolbar included in the template allows you to pan and zoom each data frame on the map and also pan and zoom around the page layout itself.

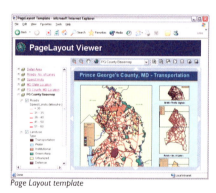

Page Layout template

Geocoding template

The Geocoding template provides an interface for finding map locations by address. Enter the address that you want to find and click Go. You'll be presented with a list of candidates that match the address. Click the result you're interested in to reveal a map that shows the address location. The interface displayed in the template assumes the address style you are using is US Streets or US Streets with Zone.

Geocoding template

You can easily change the interface to conform to other styles of addresses.

Thematic template

The Thematic template adds thematic mapping symbolization capabilities on top of the Map Viewer template. Outwardly, the map display in this template looks the same as that of the Map Viewer template. This template, however, allows the end user to dynamically change how individual layers are drawn by classifying the data in the layer. This template actually modifies the underlying map server object and, thus, requires that you configure it with a non-pooled object.

The Thematic template provides the following classification schemes:

- Natural Breaks

- Equal Interval

- Quantile

Thematic template

Buffer Selection template

The Buffer Selection template allows you to find features in one layer based on their location relative to features in another layer. For instance, suppose you want to determine how many homes a recent flood affected, given that the rivers in the area overflowed their banks by 1,000 meters. An application created from this template can perform the spatial query and identify which residences were affected by the flooding.

Performing this sort of spatial query involves creating a buffer at a specified distance around a set of features—for example, rivers—and finding other features—for example, houses—based on their spatial relationship to the buffered area. The Buffer Selection template provides options for finding features that are completely within or intersect the buffered area.

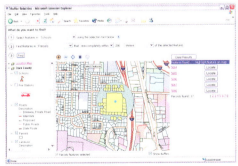

Buffer Selection template

Web Service Catalog template

The Web Service Catalog template provides a way to organize related server objects into groups and make them accessible over the Internet via HTTP as Web services. Use Web service catalogs with ArcGIS Desktop to give people access to the specific server objects they need. For example, you might choose to organize a series of maps used by a particular group of people in a Web service catalog. Alternatively, as a Web developer, you can use the services provided by the Web service catalog in Web applications.

ArcGIS Desktop users can connect to a Web service catalog via the GIS Servers entry in ArcCatalog. When connecting, just provide the URL address of the Web service catalog created through this template. For example, http://www.esri.com/myWebCat/default.aspx. Web developers can reference the Web service catalog with the following: http://www.esri.com/myWebCat/default.aspx?wsdl.

This chapter describes how to create a Web service catalog using the template provided with the .NET ADF. For general information about Web service catalogs, see the section titled 'Programming Web services' in Chapter 4.

The Web Service Catalog template is unlike the other templates above in that it presents no user interface to an end user. Instead, running the template creates a Web service that can be consumed by client applications such as ArcGIS Desktop. When you open the template, you can choose the particular server objects you want to incorporate into your Web service catalog. Once you select the particular server objects, simply build the project to make the Web service catalog available for client access.

WEB CONTROLS AND CONVENIENCE CLASSES

As mentioned earlier in this chapter, the .NET ADF comes with a set of Web controls and convenience classes that you'll use while building your Web applications. The Web controls are analogous to the kinds of controls—such as buttons, labels, and text boxes—you see in any IDE, except in this case, they represent components commonly found on a map, such as the map itself and a legend (also referred to as a table of contents). You can think of the Web control as the user interface component and its associated convenience class the part that does most of the GIS work.

As you might expect, the Web controls contain properties such as height, width, visibility, and border style—everything that relates to arranging the control on a Web form and controlling its appearance. In addition, Web controls generate events based on client interaction, such as when a control is clicked, that the server can respond to. For instance, in the case of a map control, the client action might be the dragging of a box, and the associated server action might be to zoom in to the extent specified by the box or to select all the map features that are contained by the box.

The action executed on the server is typically handled by the control's associated convenience class. Why doesn't the Web control just do all the work? Because there are times—for example, when creating Web services—when no user interface is necessary. By separating the user interface component from the GIS functionality, you can easily program applications with or without a user interface. Both the Web controls and convenience classes expose methods to access their associated object.

The Web controls and convenience classes provide some methods for common mapping operations, such as panning and zooming a map, but they don't attempt to reproduce all of the functionality of ArcObjects. What they do provide, however, is entry points into the ArcGIS Server API and the ArcObjects API. For example, in the code you write, you'll be able to programmatically access a specific map server object from the map control. From there, you'll start programming ArcObjects to implement the specific functionality your application requires—for example, you may want to add new layers to the map or change how the layers are symbolized.

The following text provides a brief description of each Web control and its associated convenience class. You'll find coding examples, along with more detailed information about the interaction of a particular Web control and convenience class, later in this chapter. For a complete description of the properties, methods, and events, see the ArcGIS Developer Help. The .NET ADF Object Model Diagram can be accessed from ArcGIS Developer Help as well.

Map control

The map control displays one particular data frame of the map. As with the ArcMap data view, you can choose which data frame to display in the map control at a given time. To change the data frame, you simply change the *DataFrame* property of the map control. The map control's convenience class is *WebMap*. This class provides methods for panning and zooming the map display. You can also use methods that identify features—returning a list of attributes—and find features by their attributes.

Table of contents control

The table of contents control is equivalent to the table of contents you see in ArcMap. The table of contents lists the layers on the map and shows what the features represent. Checking a layer in the table of contents will draw it on the map or page layout. You can choose to show the table of contents for all data frames in the map server object or just the data frame being displayed in the associated map control. A table of contents is linked to a particular map or page layout control. The convenience class for the table of contents control is *WebToc*.

Overview map control

The overview map control is similar to a map control in that it displays a particular data frame of a map server object. However, the purpose of the overview map is to provide a point of reference for the area displayed on its associated map control. A small box on the overview map represents the currently displayed area on its associated map control. You can interactively move this box around to pan the area displayed in the map control. The overview map control has the same convenience class, *WebMap*, as the map control.

Page layout control

The page layout control displays the layout of a map and is analogous to layout view in ArcMap. The page layout control displays all of the map elements, including data frames and any map surrounds, such as map titles, North arrows, and scale bars. The page layout control's convenience class is *WebPageLayout*. The methods available in this class are similar to those in the *WebMap* class. There are methods for panning and zooming the page and an individual data frame on the layout.

Toolbar control

The toolbar control displays a toolbar on the Web form. Toolbars are associated with one or more map controls or page layout controls. Toolbars contain functions for working with a map or page layout control. For example, you might put common map navigation tools, such as pan and zoom, on a toolbar. The *Toolbar* class implements all the methods it needs; thus, there is no associated convenience class.

North arrow control

The North arrow control displays a North arrow on the Web form. When you add a North arrow to your Web form, you associate it with a particular map control. The North arrow will point in the north direction for the data frame in the map control. You can set the North arrow symbol and size in design time or programmatically at runtime. The *NorthArrow* class implements all the methods it needs; thus, there is no associated convenience class.

Scale bar control

The scale bar control displays the scale bar of a map control. You can change the appearance of the scale bar control, such as the font, color, or size. The *ScaleBar* class implements all the methods it needs; thus, there is no associated convenience class.

impersonation control

The impersonation control allows you to incorporate the appropriate security credentials for accessing a GIS server into your Web applications. The impersonation control allows the Web application to impersonate the identity of a particular user and, thus, has all the privileges that identity is entitled to. As long as the identity being impersonated is part of the ArcGIS Server users group (agsusers), the Web application will be granted access to the GIS server.

GeocodeConnection component

The *GeocodeConnection* component simplifies access to geocode server objects. The *GeocodeConnection* component doesn't have a visual display on a Web form. The primary purpose of this component is to allow you to specify what GIS server you want to connect to and what geocode server object you want to access. The *GeocodeConnection* component's convenience class is called *WebGeocode*. Use this class to locate addresses.

If you've ever worked with any of Microsoft's integrated development environments before, you'll find adding GIS functionality to Web applications in Microsoft Visual Studio .NET to be much the same—you drag and drop controls from a toolbox onto a form (in this case a Web form), set some control properties, and programmatically define how the control works by writing code that responds to events such as mouse clicks. Now, in addition to adding text boxes and buttons, you'll also be able to add things, such as a map and a table of contents, directly to your Web form. If any part of this process seems unfamiliar to you, you should probably take a break from learning about ArcGIS Server and spend some time learning a little more about the .NET Framework, ASP.NET, and Microsoft Visual Studio .NET. This chapter assumes that you are already familiar with this development environment and understand Web forms, Web controls, assemblies, namespaces, and so on.

In this section, you'll start learning about how to build Web applications with the .NET ADF. The following four examples explain in step-by-step fashion how to create the sample applications.

- Creating your first Web application with a template

- Creating a Web service catalog

- Creating a Web application with the Web controls

- Accessing ArcObjects from a Web application

As you build these applications, you'll be exposed to the various classes implemented in the .NET ADF and also see some of the methods, properties, and events exposed on these classes. In addition, Chapter 7, 'Developer scenarios', presents more sophisticated examples of Web applications that focus on a particular GIS activity. For a complete reference of all classes in the .NET ADF, see the online Help incorporated into Visual Studio .NET's help system.

To code and run the examples presented below, you need to have a working GIS server with at least one map server object running on it. See Chapter 3, 'Administering an ArcGIS Server', for information on starting a map server object. In addition, the .NET ADF must be installed on the computer from which you will run Visual Studio .NET. All examples assume you are using this IDE. Although the sample code provided in this chapter is written with C#, you'll find that no matter what language you choose to program in, the process for working with the underlying classes is the same.

When creating Web applications without using a template, you will need to set the BuddyControl *property for the overview map and table of contents controls. This property associates the control with a given map control.*

CREATING YOUR FIRST WEB APPLICATION WITH A TEMPLATE

One of the easiest ways to create a Web application is to start from a template. This example shows you how to create an application from the Map Viewer template provided with the .NET ADF.

1. Start Visual Studio .NET and create a new project. Click the File menu, click New, and click Project.

2. In the New Project dialog box, double-click ArcGIS Server Projects and click Visual C#.

3. On the right side of the dialog box, click Map Viewer Web Application.

4. Type in a location for the Web application and click OK.

   Visual Studio .NET should now show you default.aspx, which contains the following: a map control, an overview map control, a table of contents control, a toolbar control, a scale bar control, and a North arrow control.

5. Set the *Host* property of the map control. Type the name of your GIS server.

6. Set the *ServerObject* property of the map control. Click the ServerObject property dropdown arrow and click a map server object in the list. You can alternatively type in the name of the server object, but be aware that the name is case sensitive.

7. Set the *DataFrame* property of the map control. Click the DataFrame property dropdown arrow and click a data frame on the map.

8. Set the *Host*, *ServerObject*, and *DataFrame* properties of the overview map control to the same values as you set for the map control.

   In general, you can use any map server object for the overview map as long as its geographic extent includes the geographic extent of the server object shown in the map control. If not, you will not see the area-of-interest box on the overview map control.

9. Set the *Identity* property for the impersonation control. Click the ellipsis button to reveal a dialog box and enter a valid username, password, and domain or machine name. The account you specify should be a member of the ArcGIS Server users account. Impersonation is discussed in greater detail later in this chapter in the section titled 'Accessing a GIS server from a Web application'.

10. Click the Start button to run the application.

11. Explore the interface of your first Web application. For example, click the Zoom In tool on the toolbar and drag a rectangle over the map. The extent box on the overview map updates to reflect the current extent.

12. Click the Stop button to stop the application or close the browser window.

As you can see, you can quickly build a Web application from a template. Try experimenting with some of the other templates to see how they work and what functionality they provide. In general, you'll need to set the same properties for the controls in the other templates as you did in this example. Each project you create from a template contains a readme file that describes what you need to do to make an application run from the template. Read this file for more detailed information about a template.

CREATING A WEB SERVICE CATALOG

The Web Service Catalog template provides a way to organize related server objects into groups and make them accessible over the Internet via HTTP as Web services. One of the primary advantages of creating a Web service catalog is that ArcGIS Desktop users can directly connect to a Web service catalog and utilize the server objects exposed through it and, for example, add a map server object to ArcMap. You can include whichever server objects you want to in a Web service catalog. Thus, you might create a Web service catalog for each department in your organization with only the specific server objects each department needs access to.

The .NET ADF Web Service Catalog template allows you to quickly create a Web service catalog. Follow the steps below to create one.

1. Start Visual Studio .NET and create a new project. Click the File menu, click New, and click Project.

2. In the New Project dialog box, double-click ArcGIS Server Projects and click either Visual Basic or Visual C#.

3. On the right side of the dialog box, click Web Service Catalog Application.

4. Type in a location for the Web service catalog and click OK. This is the location that you'll use to connect to the Web service catalog. Avoid using "localhost" as the server name, and instead specify the specific machine name.

5. In the dialog box that appears, type the name of the server you want to connect to and click Connect.

6. Check the server objects you want to include in the Web service catalog and click OK.

7. Select the impersonation control on the Web form and set the *Identity* property. Click the ellipsis button to reveal a dialog box and enter a valid username, password, and domain or machine name. The account you specify should be a member of the ArcGIS Server users account. Impersonation is discussed in greater detail later in this chapter in the section titled 'Accessing a GIS server from a Web application'.

8. Click the Build menu and click Build Solution.

You now have a Web service catalog that you can connect to. In ArcCatalog, specify the URL address for the Web service catalog as shown below:

```
http://myserver/mycatalog1/default.aspx
```

See Chapter 3, 'Administering an ArcGIS Server', for more information on connecting to Web service catalogs from ArcCatalog.

You can also access Web service catalogs programmatically using standard Web protocols. As a developer, you can consume the Web service catalog with the following reference:

```
http://myserver/mycatalog1/default.aspx?wsdl
```

For more information on consuming Web service catalogs from a Web application, see the section titled 'Programming Web services' in Chapter 4.

CREATING A WEB APPLICATION WITH THE WEB CONTROLS

Whether you choose to start building your Web application from a template or from scratch, utilizing the Web controls directly, you'll soon have to write your own custom code to implement the specific functionality you want to incorporate in your application. As you read at the beginning of this chapter, writing custom code means programming with the .NET ADF, ArcGIS Server, and ArcObjects. Through the .NET ADF, you'll be able to access the server objects and, in turn, ArcObjects.

The example presented in this section is divided into two parts. In the first part, you'll see how you can use the map control to access all the map server objects on a specific GIS server. Then, you'll see how you can add tools to interact with the map control.

Connecting to a server and getting a list of server objects

In this section, you'll add a list box and map control to a Web form. The list box will display the list of map server objects running on a particular GIS server machine. Selecting a map from the list will display it on the map control.

1. In Visual Studio .NET, create a new Visual C# project using the ASP.NET Web Application template and provide an appropriate name for the application.

2. Add a list box to the Web form.

3. Add a map control to the Web form. In the Visual Studio .NET Toolbox, you'll notice a category for ESRI Web controls. This category contains the map control delivered with the .NET ADF.

4. Set the *Host*, *ServerObject*, and *DataFrame* properties of the map control to a map server object running on your GIS server.

5. Add an impersonation control to the Web form.

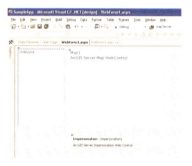

The controls in Visual Studio .NET

6. Set the *Identity* property of the impersonation control. Click the Browse button to reveal a dialog box and enter a valid username, password, and domain or machine name. The account you specify should be a member of the ArcGIS Server users account. Impersonation is discussed in greater detail later in this chapter in the section titled 'Accessing a GIS server from a Web application'.

7. Open the code behind page, WebForm1.aspx.cs, and add the following statement at the top:

```
using ESRI.ArcGIS.Server.WebControls;
```

8. Code the Page_Load event as shown below.

```
private void Page_Load(object sender, System.EventArgs e)
{
  if (!Page.IsPostBack)
  {
    if (Session.IsNewSession)
    {
      ListBox1.DataSource =
        Map1.ServerConnection.GetServerObjectNames("MapServer");
      ListBox1.DataBind();
      ListItem item = ListBox1.Items.FindByText(Map1.ServerObject);
      ListBox1.SelectedIndex = ListBox1.Items.IndexOf(item);
    }
  }
}
```

9. Code the ListBox1_SelectedIndexChanged event as shown below. (Double-click the list box control on the Web form to automatically create an empty function that you can write the code for.)

```
private void ListBox1_SelectedIndexChanged(object sender, System.EventArgs e)
{
  Map1.ServerObject = ListBox1.SelectedValue;
  Map1.DataFrame = null;
  Map1.Draw();
}
```

10. Set the AutoPostBack property of the list box to True.

11. Add the following code to the Session_End event in Global.asax.cs.

```
protected void Session_End(Object sender, EventArgs e)
{
  ESRI.ArcGIS.Server.IServerContext context;
  for (int i = 0; i < Session.Count; i++)
  {
    context = Session[i] as ESRI.ArcGIS.Server.IServerContext;
    if (context != null)
      context.ReleaseContext();
  }
  Session.RemoveAll();
}
```

Upon running the application, you'll see the list of map server objects and the map you selected.

12. Run the application. The list box should show all of the map server objects running on the server. When you click a map server object in the list, the map control should update to show the selected map.

Take a look at how this example works.

When you set the *Host* and *ServerObject* property at design time, you established the GIS server you want to connect to and the map server object to display by default. The list box is populated from this line of code in the Page_Load event:

```
ListBox1.DataSource =
  Map1.ServerConnection.GetServerObjectNames("MapServer");
```

From the map control, Map1, you get the *ServerConnection* object, and from this you can get a list of server objects, in this case, of type *MapServer*. This list of map server objects is then bound to the list box control.

Clicking on an item in the list box executes this code:

```
Map1.ServerObject = ListBox1.SelectedValue;
Map1.DataFrame = null;
Map1.Draw();
```

This code sets the *ServerObject* property of the map control to the item selected in the list box. The *DataFrame* property is set to null, which means the default data frame will be displayed. Last, the map control is drawn.

When you display a non-pooled object from the list of maps, the map control saves the server context for the non-pooled object in session state. Adding code to the Session_End event cleans up the session variable when the application ends and ensures that the server context gets released. The "for" loop cycles through all session variables that were set during the application's lifetime.

```
for (int i = 0; i < Session.Count; i++)
```

If the session variable references a server context, the value of the variable, context, will not be null, and the server context is subsequently released.

```
context = Session[i] as ESRI.ArcGIS.Server.IServerContext;
if (context != null)
    context.ReleaseContext();
```

When working with non-pooled objects, it's good practice to code the Session_End event to release any server contexts held in session state. The code shown in this example is generic. You can use it in any application to release any server context that may be held in session state.

The Session_End event executes when the session times out. You can set the time-out period for your application in the Web.Config file.

Adding a toolbar to zoom in

In this section, you'll add a toolbar control that allows you to zoom in on the map and also return to the full map extent.

1. From the ESRI tab in the toolbox, drag and drop a toolbar control onto your Web form.

2. Set the *BuddyControls* property to specify which map controls the toolbar will work with. This property is actually a collection, so if you have more than one map control on the page, the toolbar could work with all of them. In the BuddyControl Collection Editor, click Add and set the *Name* property to the ID of the map control, Map1.

3. Set the *ToolbarStyle* property to TextOnly.

4. Set the *ToolBarItems* property. This property is a collection of toolbar items. Toolbars can contain four types of items, tools, commands, separators, and spaces. Tools interact with the map control—for example, dragging a box to zoom in—before executing the code behind, whereas commands simply execute the code behind. You can use separators and spaces to group tools and commands on the toolbar. You'll use a tool for the Zoom In operation and a command for the Full Extent operation.

Add a toolbar above the map control.

4a. In the ToolbarItem Collection Editor, click the Add button's dropdown arrow and add a Tool. Set the following properties as shown below:
Text = Zoom In
Name = ZoomInTool
ClientToolAction = DragRectangle
ServerToolActionAssembly = ESRI.ArcGIS.Server.WebControls
ServerToolActionClass = ESRI.ArcGIS.Server.WebControls.Tools.MapZoomIn

4b. In the ToolbarItem Collection Editor, click the Add button's dropdown arrow and add a Command. Set the following properties as shown below:
Text = Full Extent
Name = FullExtent

5. Click OK to dismiss the ToolbarItem Collection Editor.

6. Code the Toolbar1_CommandClick event as shown below. (Double-click the toolbar control on the Web form to automatically create an empty function that you can write the code for.)

```
private void Toolbar1_CommandClick(object sender,
ESRI.ArcGIS.Server.WebControls.ToolbarCommandClickEventArgs e)
{
    using (WebMap webmap = Map1.CreateWebMap())
    {
      webmap.DrawFullExtent();
    }
}
```

7. Run the application. Click the Zoom In tool and drag a box over the map to zoom in to. Click the Full Extent button to return to the full extent of the map server object.

Use the toolbar to navigate the map.

Take a look at how this example works.

As mentioned above, a toolbar can contain two types of items that perform an action—tools and commands. Tools have a client-side action as well as a server-side action. The .NET ADF includes some built-in client- and server-side actions that you can utilize with no extra coding effort on your part.

In the example above, you specified the DragRectangle client-side action and MapZoomIn as the server-side action for the Zoom In tool. The client-side action is implemented as a JavaScript that enables the drawing of a rectangle over the map control. When you set the ClientToolAction property, you may have noticed that there are also other options available to you such as drawing a circle, line, or polygon.

Once the client-side action completes, the server-side action executes. In this case, dragging a box over the map control specifies the area on the map to zoom in to. The coordinates of the box are then passed to the server-side action, which changes the current extent displayed and redraws the map. This server-side action, MapZoomIn, is already implemented for you. However, you can also write your own code that executes on the server.

Toolbar commands only have a server-side action that is implemented through the click event on the toolbar. In this example, there is only one command item and it returns the map display to the full extent. Thus, the code for the toolbar click event is straightforward. The following line of code:

```
using (WebMap webmap = Map1.CreateWebMap())
```

creates a *WebMap* object. *WebMap* is the convenience class associated with the map control and is responsible for getting a server context from the GIS server—essentially giving you a proxy to the map server object you're working with.

When you instantiate a *WebMap* object in your application, your application utilizes a map server object on the server. Until your application releases the object, it is unavailable to other clients who may request it. Thus, it's very important to dispose of the object once you're through using it. For pooled objects, you can ensure the object is disposed of by creating the *WebMap* object within a using statement, as shown in this example, or by calling the *Dispose* method directly. For non-pooled objects, you would additionally call *ReleaseServerContext*, then *Dispose*. Releasing objects is discussed in more detail later in this chapter in the section titled 'Getting and releasing server contexts'.

Only one line of code is needed to return the map to its full extent. The *DrawFullExtent* method resets the displayed extent of the map server object and causes the map control to redraw.

```
webmap.DrawFullExtent();
```

WebMap works differently with respect to pooled and non-pooled objects. With a pooled object, *WebMap* automatically releases the server context at the end of the using block. With a non-pooled object, *WebMap* doesn't release the server context and assumes that your application will use the server context exclusively during the lifetime of the application and explicitly release it, typically at the end of the session. Thus, it's the code in the Session_End event that releases the server context for non-pooled objects.

When creating your own toolbars, you will likely place more than one command on them. For toolbars with multiple command items, you can use a switch statement to conditionally execute code based on which command is clicked. For example, if you want to add another command item to the toolbar in the example, you might write code such as this:

```
private void Toolbar1_CommandClick(object sender,
ESRI.ArcGIS.Server.WebControls.ToolbarCommandClickEventArgs e)
{
    switch (e.CommandName)
    {
        case "FullExtent":
            using (WebMap webmap = Map1.CreateWebMap())
            {
                webmap.DrawFullExtent();
            }
            break;
        case "AnotherCommand":
            //Do something here
            break;
    }
}
```

In the code above, you obtain the name of the command item clicked with the *CommandName* method and use that value as the switch statement.

In general, when working with pooled objects, if you instantiate any WebMap object within a using statement, the server context will be released at the end of the using block. When working with non-pooled objects, you will need to explicitly release the server context when your application is finished using it. This is typically handled in the Session_End event.

ACCESSING ArcObjects FROM A WEB APPLICATION

The .NET ADF exposes a great deal of GIS functionality through its API. In the previous example, you used the *WebMap* class to draw the map at its full extent (*WebMap::DrawFullExtent*). The *WebMap* class has additional methods that, for example, pan and zoom the map, identify features, and find features by an attribute value. All of this functionality is built into the .NET ADF using ArcObjects. While you can write ArcObjects code yourself to do the same thing, the *WebMap* class makes it a bit easier because it hides some of the details in the methods it exposes to you. In a similar way, the *WebPageLayout* class makes it easier to work with the page layout of a map.

The .NET ADF, however, does not duplicate all of the functionality in ArcObjects. To do more than the basic GIS operations provided in the .NET ADF, you'll need to write code using the ArcObjects API. As you'll see in the example below, you'll use the same convenience class—*WebMap*—to access ArcObjects directly.

The goal of this example is to show you how to work with pooled and non-pooled server objects and also how to access the ArcGIS Server and ArcObjects API from the .NET ADF.

The Web form you'll create will contain four list boxes. The first list box displays a list of map server objects running on the server. The second list box displays a list of data frames in the selected map server object. The third list box displays the list of layers in the selected data frame. The fourth list box displays the attribute fields of the selected layer. As you make selections from the list boxes, the other list boxes update accordingly. For example, selecting a new layer in the layers list box updates the fields list box.

1. In Visual Studio .NET, create a new Visual C# project using the ASP.NET Web Application template and provide an appropriate name for the application.

2. Add four list boxes to the Web form. Name the list boxes as shown below:

   lstServerObjs—contains a list of map server objects
   lstDataFrames—contains a list of data frames
   lstLayers—contains the list of layers
   lstFields—contains the list of fields

3. Set the AutoPostBack property to true for lstServerObjs, lstDataFrames, and lstLayers.

4. Add an impersonation control to the Web form.

5. Set the *Identity* property of the impersonation control. Click the Browse button to reveal a dialog box and enter a valid username, password, and domain or machine name. The account you specify should be a member of the ArcGIS Server users group (agsusers). Impersonation is discussed in greater detail later in this chapter in the section titled 'Accessing a GIS server from a Web application'.

Add four list boxes and an impersonation control to the Web form.

6. Open the code behind page, WebForm1.aspx.cs, and add the following statements at the top:

```
using ESRI.ArcGIS.Server.WebControls;
using ESRI.ArcGIS.Carto;
using ESRI.ArcGIS.Geodatabase;
```

7. Code the Page_Load event as shown below, substituting the name of your GIS server for "your_server" below.

```
private void Page_Load(object sender, System.EventArgs e)
{
  if (!Page.IsPostBack)
  {
    if (Session.IsNewSession)
    {
      // Set host machine and save in session state.
      string host = "your_server";
      Session.Add("host", host);
      // Create a server connection.
      ServerConnection svrConnection = new ServerConnection(host, true);
      // Populate server objects and data frames list boxes.
      Get_ServerObjects(svrConnection);
      Get_DataFrames(svrConnection);
      string mapServerObj = lstServerObjs.SelectedValue;
      string dataFrame = lstDataFrames.SelectedValue;
      // Create WebMap and populate layers and fields list boxes.
      WebMap webmap = Get_WebMap(svrConnection,
          host, mapServerObj, dataFrame);
      Get_Layers(webmap);
      Get_Fields(webmap);
      // Keep server context for non-pooled objects.
      Keep_ServerContext(webmap);
    }
  }
}
```

Make sure you substitute the name of your host machine, otherwise the application won't work properly.

8. Code the Get_ServerObjects function as shown:

```
private void Get_ServerObjects(ServerConnection svrConnection)
{
  // Populate list of map server objects on server.
  lstServerObjs.DataSource =
    svrConnection.GetServerObjectNames("MapServer");
  lstServerObjs.DataBind();
  lstServerObjs.SelectedIndex = 0;
}
```

9. Code the Get_DataFrames function as shown:

```
private void Get_DataFrames(ServerConnection svrConnection)
{
  // Populate list of data frames of selected map server object
  string mapServerObj = lstServerObjs.SelectedValue;
  lstDataFrames.DataSource =
    svrConnection.GetDataFramesFromMapServerObject(mapServerObj);
  lstDataFrames.DataBind();
  lstDataFrames.SelectedIndex = 0;
}
```

10. Code the Get_Layers function as shown:

```
private void Get_Layers(WebMap webmap)
{
  // Populate list of layers in selected data frame
  ArrayList arrLayers = new ArrayList();
  IMapServer map = webmap.MapServer;
  IMapServerInfo msi = map.GetServerInfo(webmap.DataFrame);
  IMapLayerInfos layers = msi.MapLayerInfos;
  if (layers.Count != 0)
  {
    for (int i = 0; i < layers.Count; i++)
    {
      IMapLayerInfo oneLayer = layers.get_Element(i);
      arrLayers.Add(oneLayer.Name);
    }
  }
  else
    arrLayers.Add ("No layers in data frame.");
  lstLayers.DataSource = arrLayers;
  lstLayers.DataBind();
  lstLayers.SelectedIndex = 0;
}
```

11. Code the Get_Fields function as shown:

```
private void Get_Fields(WebMap webmap)
{
  // Populate list of fields
  ArrayList arrFields = new ArrayList();
  int layerIndex = lstLayers.SelectedIndex;
  IMapServer map = webmap.MapServer;
  IMapServerInfo msi = map.GetServerInfo(webmap.DataFrame);
  IMapLayerInfos layers = msi.MapLayerInfos;
  if (layers.Count != 0)
  {
    IMapLayerInfo oneLayer = layers.get_Element(layerIndex);
    if (oneLayer.isFeatureLayer)
    {
```

```
              IFields fields = oneLayer.Fields;
              for (int i = 0; i < fields.FieldCount; i++)
                {
                  IField oneField = fields.get_Field(i);
                  arrFields.Add(oneField.Name);
                }
            }
            else
              arrFields.Add ("No fields for this layer.");
        }
        else
          arrFields. Add ("No fields available.");
        lstFields.DataSource = arrFields;
        lstFields.DataBind();
        lstFields.SelectedIndex = 0;
    }
```

12. Code the Get_WebMap function as shown:

```
private WebMap Get_WebMap (ServerConnection svrConnection, string host,
  string mapServerObj, string dataFrame)
{
  if (Session["serverContext"] != null)
  {
    ESRI.ArcGIS.Server.IServerContext svrContext =
      Session["serverContext"] as ESRI.ArcGIS.Server.IServerContext;
    WebMap webmap = new WebMap (svrContext, host, dataFrame);
    return webmap;
  }
  else
    return new WebMap(svrConnection, mapServerObj, dataFrame);
}
```

13. Code the Keep_ServerContext function as shown:

```
private void Keep_ServerContext (WebMap webmap)
{
  // Save ServerContext object in session state for non-pooled object.
  if (!webmap.IsPooled)
    Session.Add("serverContext", webmap.ServerContext);
  webmap.Dispose();
}
```

14. Double-click the lstServerObjects control and code the
    lstServerObjs_SelectedIndexChanged event as shown:

```
private void lstServerObjs_SelectedIndexChanged(object sender,
  System.EventArgs e)
{
  // New Map Server Object selected. Update list boxes.
  string host = Session["host"] as String;
  ServerConnection svrConnection = new ServerConnection(host, true);
  Get_DataFrames(svrConnection);
  string mapServerObj = lstServerObjs.SelectedValue;
  string dataFrame = lstDataFrames.SelectedValue;
```

```
// Remove any ServerContext in session state and release server context.
if (Session["serverContext"] != null)
{
  ESRI.ArcGIS.Server.IServerContext svrContext =
    Session["serverContext"] as ESRI.ArcGIS.Server.IServerContext;
  svrContext.ReleaseContext();
  Session.Remove("serverContext");
}
WebMap webmap = Get_WebMap(svrConnection, host, mapServerObj, dataFrame);
Get_Layers(webmap);
Get_Fields(webmap);
Keep_ServerContext(webmap);
}
```

15. Double-click the lstDataFrames control and code the lstDataFrames_SelectedIndexChanged event as shown:

```
private void lstDataFrames_SelectedIndexChanged(object sender,
  System.EventArgs e)
{
  // New data frame selected. Update list of layers and fields.
  string host = Session["host"] as String;
  string mapServerObj = lstServerObjs.SelectedValue;
  string dataFrame = lstDataFrames.SelectedValue;
  ServerConnection svrConnection = new ServerConnection(host, true);
  WebMap webmap = Get_WebMap(svrConnection, host, mapServerObj, dataFrame);
  Get_Layers(webmap);
  Get_Fields(webmap);
  Keep_ServerContext(webmap);
}
```

As you click on a new server object, data frame, or layer, the appropriate list boxes update.

16. Double-click the lstLayers control and code the lstLayers_SelectedIndexChanged event as shown:

```
private void lstLayers_SelectedIndexChanged(object sender,
  System.EventArgs e)
{
  // New layer selected. Update list of fields.
  string host = Session["host"] as String;
  string mapServerObj = lstServerObjs.SelectedValue;
  string dataFrame = lstDataFrames.SelectedValue;
  ServerConnection svrConnection = new ServerConnection(host, true);
  WebMap webmap = Get_WebMap(svrConnection, host, mapServerObj, dataFrame);
  Get_Fields(webmap);
  Keep_ServerContext(webmap);
}
```

17. Run the application. Each list box should be populated with the appropriate values. For example, as you click a layer name, the list of fields will update.

If you receive an error that the RPC server is unavailable, make sure you substituted the name of your GIS server in the Page_Load event:

```
string host = "your_server";
```

Take a look at how this example works.

Most of the work performed by the application is done in the several functions you coded. The Page_Load function populates each of the four list boxes on the Web form by making a call to these functions. There is a function that populates each list box and additional functions to create a *WebMap* object and save the server context for a non-pooled object. Each function is described in greater detail below.

Get_ServerObjects function

The Get_ServerObjects function calls the *GetServerObjectNames* method on the *ServerConnection* object. This method returns an array that can be bound directly to the list box control.

```
lstServerObjs.DataSource =
    svrConnection.GetServerObjectNames("MapServer");
```

The last thing the function does is select the first element in the list.

```
lstServerObjs.SelectedIndex = 0;
```

This is important because the other functions assume that each list box has a current selected value, and the other functions use this value in code.

Get_DataFrames function

The Get_DataFrames function is almost identical to the Get_ServerObjects function. This function calls the *GetDataFramesFromMapServerObject* method on the *ServerConnection* object to populate the list box control. The method requires the name of a map server object as an argument, which is the selected value in the list of map server objects.

```
string mapServerObj = lstServerObjs.SelectedValue;
lstDataFrames.DataSource =
    svrConnection.GetDataFramesFromMapServerObject(mapServerObj);
```

Get_Layers function

The Get_Layers function requires a little more coding because there is no method in the .NET ADF for obtaining a list of layers in a data frame. Thus, you need to write your own code to populate the list of layers.

The function accepts a *WebMap* object as an argument. The *WebMap* object is used by this function and the Get_Fields function. By passing it in as an argument here, both functions can use the same *WebMap* object and don't have to instantiate their own, which saves getting a second server context from the GIS server.

The *WebMap* class exposes the *MapServer* method, which returns a map server object. This object is the coarse-grained server object that represents the map running on the server. You can use this object as an entry point for accessing other map server objects.

```
ArrayList arrLayers = new ArrayList();
IMapServer map = webmap.MapServer;
```

Through the *IMapServer* interface, the function gets a *MapServerInfo* object. The *IMapServerInfo* interface has properties that describe the map. One property, *MapLayerInfos*, returns a list of layers in the specified data frame.

```
IMapServerInfo msi = map.GetServerInfo(webmap.DataFrame);
IMapLayerInfos layers = msi.MapLayerInfos;
```

If the current data frame has layers, then the function loops through the list of layers, extracting the name of each layer and populating the ArrayList.

```
if (layers.Count != 0)
  {
    for (int i = 0; i < layers.Count; i++)
    {
      IMapLayerInfo oneLayer = layers.get_Element(i);
      arrLayers.Add(oneLayer.Name);
    }
  }
else
  arrLayers.Add ("No layers in data frame.");
```

Last, the function binds the ArrayList to the list box control, which shows the list of layers in the particular data frame, in the particular map server object.

```
LstLayers.DataSource = arrLayers;
LstLayers.DataBind();
LstLayers.SelectedIndex = 0;
```

Get_Fields function

The Get_Fields function populates the last list box control with a list of fields in the selected layer. This function follows the same coding pattern as the Get_Layers function, so you can skip over the beginning section and look specifically at the code that builds the list of fields.

The following code gets the individual layer (*MapLayerInfo* object) using the *get_Element* method and passing in the current selected index number in the list box of layers—recall that the Get_Layers function populated the list box by looping through the list of layers, so the order should match the order in the list box.

```
int layerIndex = lstLayers.SelectedIndex;
...
IMapLayerInfo oneLayer = layers.get_Element(layerIndex);
```

Given the selected layer, the next section of code first checks if the layer is a feature layer—for example, it may be an image layer that doesn't have fields—and if so, builds the list of fields in the layer.

```
if (oneLayer.isFeatureLayer)
{
  IFields fields = oneLayer.Fields;
  for (int i = 0; i < fields.FieldCount; i++)
  {
    IField oneField = fields.get_Field(i);
    arrFields.Add(oneField.Name);
  }
}
```

Last, the array is bound to the list box control.

```
LstFields.DataSource = arrFields;
LstFields.DataBind();
LstFields.SelectedIndex = 0;
```

Get_WebMap function

The Get_WebMap function is a key function in this application. It creates the *WebMap* object and does so differently for pooled and non-pooled server objects. For pooled objects, the Get_WebMap function returns a *WebMap* object based on a server connection. The GIS server hands back a new server context from the pool. For non-pooled objects, the function uses an existing server context— extracted from session state—to create the *WebMap* object. In this case, the server hands back the same server context the application used previously.

The function first checks if there is a session variable that was created by the Keep_ServerContext function.

```
if (Session["serverContext"] != null)
```

If the session variable exists, it means the application is currently accessing a non-pooled object, and that object has been created explicitly for this application and still exists on the server. The function creates a new *WebMap* object by passing in the server context, stored in session state, to the constructor.

```
ESRI.ArcGIS.Server.IServerContext svrContext =
    Session["serverContext"] as ESRI.ArcGIS.Server.IServerContext;
WebMap webmap = new WebMap (svrContext, host, dataFrame);
return webmap;
```

If the session variable is null, it means the application is currently accessing a pooled object or this is the first call to create a non-pooled object from this application. The function then creates a new *WebMap* object from the server connection rather than the server context.

```
return new WebMap(svrConnection, mapServerObj, dataFrame);
```

Keep_ServerContext function

The Keep_ServerContext function saves the server context in session state if the server object accessed by the application is a non-pooled object.

```
if (!webmap.IsPooled)
    Session.Add("serverContext", webmap.ServerContext);
webmap.Dispose();
```

The function always calls the *Dispose* method on the *WebMap* object. For pooled objects, *Dispose* also releases the server context. However, *Dispose* does not release the server context for non-pooled objects.

LstServerObjs_SelectedIndexChanged function

Whenever an end user of this application selects a new map server object, all the list boxes on the Web form must update to reflect the change. To update the list boxes, this function calls the Get_DataFrames, Get_Layers, and Get_Fields functions to populate each list box. The first few lines of the function establish a server connection and retrieve the list of data frames in the newly selected object.

```
// New Map Server Object selected. Update list boxes.
string host = Session["host"] as String;
ServerConnection svrConnection = new ServerConnection(host, true);
Get_DataFrames(svrConnection);
```

The next few lines of code check if a session variable exists. If it does, it means the application was previously working with a non-pooled object and the server context is currently stored in session state.

```
if (Session["serverContext"] != null)
{
  ESRI.ArcGIS.Server.IServerContext svrContext =
    Session["serverContext"] as ESRI.ArcGIS.Server.IServerContext;
  svrContext.ReleaseContext();
  Session.Remove("serverContext");
}
```

The function releases the server context—using the *ReleaseContext* method—since the application is now finished working with the non-pooled object.

Next, the function calls the Get_WebMap function to create a new *WebMap* object and passes that object as an argument to the Get_Layers and Get_Fields functions, which populate their respective list boxes.

```
WebMap webmap = Get_WebMap(svrConnection, host, mapServerObj, dataFrame);
Get_Layers(webmap);
Get_Fields(webmap);
```

The last thing the function does is call Keep_ServerContext.

```
Keep_ServerContext(webmap);
```

If the newly selected server object is a non-pooled object, the Keep_ServerContext function saves the server context in session state. Then, for any subsequent calls to Get_WebMap, such as when a new data frame or layer is selected, a new *WebMap* object is created from the existing server context. However, if the newly selected server object is a pooled object, the function disposes of the *WebMap* object, which also releases the server context.

LstDataFrames_SelectedIndexChanged and LstLayers_SelectedIndexChanged functions

The LstDataFrames_SelectedIndexChanged and LstLayers_SelectedIndexChanged functions are essentially the same. These functions simply make calls to the Get_Layers and Get_Fields functions to repopulate the list boxes when the end user of the application changes the current data frame or layer. The relevant lines of code to do this are:

```
WebMap webmap = Get_WebMap(svrConnection, host, mapServerObj, dataFrame);
Get_Layers(webmap);
Get_Fields(webmap);
Keep_ServerContext(webmap);
```

As you read through this chapter, you're probably already thinking about how to apply what you've learned to your own project. This section describes some of the things you'll need to think about as you begin building your own Web applications. Some of the information presented here is discussed in more detail in other chapters of this book as well; however, it's worth reviewing this information at this point to ensure you have a clear picture of how to build your Web application.

ORGANIZING YOUR DATA AND CREATING SERVER OBJECTS

Presumably, most—if not all—of the GIS functionality you build into your Web applications will be provided through the server objects you host on your GIS server. How many server objects a single Web application utilizes depends on how you organize your server objects and what functionality you want to build into the Web application. Often, a Web application utilizes more than one server object, and these server objects need to have some correlation between them. For example, if you want to build an application that locates places by address, that application will minimally access a geocode server object to find the address location and a map server object to display the location to the user. The geocode and map server objects need to reference the same geographic location, otherwise the geocoded points won't draw on the map. Similarly, the geographic area displayed in an overview map needs to match that of its associated main map.

As map server objects are based on map documents (for example, .mxd files), you have some flexibility as to how you organize the data contained within a given map server object. For instance, suppose you're creating an application that displays three maps. Should you create one map document that results in one map server object with three data frames in it? Or instead, create three map documents and three server objects with one data frame each?

As a general rule, you should organize the data in a map server object according to how your Web applications will use it. For example, you would create one map server object that contains all the data required for a Web application. The server object would have, for example, one data frame for the main map control and one data frame for the overview map. Later, if you build another Web application that can share the overview map with the first application, you may choose to host the data for the overview map in its own server object.

DATA ACCESS CONSIDERATIONS

Ensure that the following holds true for your map server objects.

- The map document and all data must be accessible to all server object container machines.

- The ArcGIS Server Container account you established during the postinstallation should have read access to any shared network directories.

- File-based data (for example, coverages and shapefiles) should be placed on shared networked drives and saved in the map documents using UNC pathnames.

- ArcSDE connections must be saved in the map document before being added to the server.

ACCESSING A GIS SERVER FROM A WEB APPLICATION

You need to consider two levels of security when building a Web application that accesses a GIS server. First, you need to consider who will access your application. Does the application contain sensitive information that is not meant for general consumption? If so, you'll probably want to restrict access to it in some way. Second, you need to consider how the Web application itself accesses and works with the GIS server because the server has its own access control mechanism that your Web application must implement.

To restrict access to a Web application, you need to identify who is attempting to access the Web site. This process, called *authentication*, challenges a user to prove who they are, usually by requiring them to provide some security credentials that only they and the resource they are trying to access know about. For example, to access your bank account through an automated teller machine (ATM), you identify yourself with your ATM card and your personal identification number. Without both, you are not granted access to the account.

For Web applications, security credentials are commonly expressed in the form of a username and password. If the username and password are valid, the user is authenticated and subsequently granted access to the Web site. There are many ways you can authenticate users and grant them access to a Web application. A discussion of these methods is beyond the scope of this book. However, whether you choose to restrict access to your Web application or not, you do need to consider the second level of security, which governs how the Web application accesses the GIS server. This is the main focus of this section.

As you read in the previous chapter, access to a GIS server—and the server objects running on it—is managed by two operating system user groups: ArcGIS Server users (agsusers) and ArcGIS Server administrators (agsadmin). To access the server objects, an application must connect to the server as a user in the ArcGIS Server users group. To additionally administer the GIS server—for example, add and remove server objects—an application must connect as a user in the ArcGIS Server administrators group.

First, consider how you access a GIS server through ArcCatalog. When you connect to your GIS server in ArcCatalog, if the account you're currently logged in as is a member of the ArcGIS Server administrators group, you will see additional menu choices that let you manage the server. For instance, you can add, remove, start, and stop server objects. However, if your account is only a member of the ArcGIS Server users group, then you will not see any menu choices for managing server objects and can instead only view the server objects running on the server. Further, if your account is not a member of either of these user groups, you won't have any access to the server.

Access to a GIS server is managed in a similar way with Web applications. A Web application can connect to the GIS server as a particular user through a process called *impersonation*. With impersonation, the Web application assumes the identity of a particular user and, thus, has all the privileges that user is entitled to. As long as this user is part of the appropriate group—for example, ArcGIS Server users, agsusers—the Web application will have access to the server objects running on the GIS server. So what user should your Web application connect to the server as? The answer to this question depends on how you choose to set up access to the GIS server itself.

If you want to learn more about authentication and impersonation, you can consult almost any book that discusses ASP.NET or visit Microsoft's Developer Network at www.msdn.microsoft.com. Microsoft Visual Studio .NET also contains a thorough discussion of impersonation in the online help.

When incorporating impersonation into your Web application, there are two approaches you can take. You can use the impersonation control or modify the configuration of your application in the Web.config file. No matter how your Web application implements impersonation, most likely the application will impersonate a user who is a member of the ArcGIS Server users group rather than the ArcGIS Server administrators group. As a member of the ArcGIS Server users group, the application will have complete access to the server objects. Thus, you can write code that works with the object in a shallowly stateful or deeply stateful manner. You can do anything except manage the server objects themselves. If you want to write an application that manages the server, the application will have to impersonate a user in the ArcGIS Server administrators group.

Configuring impersonation with the impersonation control

By adding the impersonation control to your Web form at design time, you can set the necessary identity to impersonate. The *Identity* property allows you to specify a username, password, and optional domain or machine name. The identity is encrypted into the application; thus, it can't be read. Once you add an impersonation control to the Web form, all .NET ADF Web controls on the page will utilize the identity established by the impersonation control.

For more information on how the impersonation control works, see the section titled 'Impersonation control' later in this chapter.

Each page of your Web application needs its own impersonation control on it. You can also access the impersonation control in your code through the *Impersonation* class. Thus, you can dynamically change the identity if necessary. Code examples for doing this are presented later in this chapter in the section 'Impersonation control'. Setting impersonation through the control overrides any impersonation set in Web.config, described below.

Configuring impersonation through the Web.config file

Alternatively, you can enable impersonation for your Web application by including the appropriate setting in a configuration file, Web.config, in the application root directory. (When you create a Web application in Visual Studio .NET, the Web.config file is automatically created for you.)

To enable impersonation and to impersonate the authenticated user, include the following line in the application's Web.config file:

```
<identity impersonate="true" />
```

With this setting, the Web application is now configured to impersonate the authenticated user who is accessing the Web application. (Note: You may also need to disable anonymous access to the Web application through the Web server. This can be done through the Microsoft Management Console for IIS.) In the case where the client is accessing the Web application through Internet Explorer, the user identity (Windows account) is automatically passed to the Web application, and thus, the Web application can seamlessly impersonate the authenticated user. In the case where the client is accessing the Web application through Netscape®, the user identity is not automatically passed on to the Web application. Thus, the user will be prompted by a dialog box to enter a username and password for authentication.

To enable impersonation and to impersonate a specific, hard coded user, include the following line in the application's Web.config file:

```
<identity impersonate="true" userName="domain\user" password="mypass" />
```

This allows the application to run with the specified identity as long as the identity information is correct. As you can see, the username and password are stored in clear text. Although this information is not transmitted via IIS, a user on the domain with the proper credentials can see the contents of this file. For increased security, you can encrypt the username and password in the system registry, replacing the clear text versions with the following:

```
userName="registry:HKLM\Software\AspNetIdentity,Name"
password="registry:HKLM\Software\AspNetIdentity,Password"
```

This method of encrypting the username and password in the registry is beyond the scope of this book. For further information on this encryption method, consult Microsoft Knowledge Base Article #329290, entitled HOW TO: Use the ASP.NET Utility to Encrypt Credentials and Session State Connection Strings.

GETTING AND RELEASING SERVER CONTEXTS

As you've already read in this book, the server context provides the means to access a server object and to instantiate new objects that must exist within the same context as the server object. When you add a server object to your GIS server with ArcCatalog, one property you specify is how many instances are available for clients that need to use the object. The number of instances you allow impacts how many client applications can simultaneously access the server object. Of course, the number of instances you can reasonably create depends on, among other things, how many machines comprise your GIS server and how many other server objects your server is hosting. In any event, there is a finite number of instances you can create on your system. You must make sure that your client applications use these instances judiciously so that the server can handle the load you expect.

When creating a server object in ArcCatalog, you can set the maximum usage time, which is the maximum amount of time a client application can hold on to a server context.

Judicious use of server objects means that client applications get and release a server context appropriately. The fundamental difference between an application that utilizes a pooled object versus a non-pooled object is the length of time that the application holds on to the server context. Pooled objects are meant for short-term use. An application that utilizes a pooled object will obtain a new server context for every request it sends to the server, then immediately release it after using it. Non-pooled objects are meant for long term, deeply stateful use of a server object. An application that utilizes a non-pooled object will get and hold on to the server context for the duration of the application. However, it's just as important that the application releases the server context when the application ends because there are a finite number of instances available.

Consider a simple Web application with a map control and a button, where clicking the button zooms in and pans the map. Here's what happens with a pooled object. Clicking the button sends a request to the GIS server, and the server responds by giving the client application a server context. Within that server context, the code to zoom in and pan executes, which generates a new map image that is displayed in the map control. When execution completes, the application releases the server context back to the pool, and it is made available to the next client that requests it. The next time you click the button, the application is again given a server context, but it may not be the same server context used during the last request.

The same Web application, when run with a non-pooled object, functions differently. In this case, the Web application holds on to the server context. Thus, every time you click the button, the application uses the same server context. The GIS server explicitly creates the server context for the application to use during its lifetime. The Web application holds on to the server context by storing the object in session or application state.

Of course, you probably wouldn't use a non-pooled object for a simple Web application such as the one above. However, you can immediately see how you might utilize non-pooled objects. Anything you do to the server object within the server context—for instance, adding a layer to the map, changing a layer's symbolization, or editing a layer's features—you can expect it to be maintained on the server the next time your application sends a request to the server (for example, the new layer will still be there). When the application ends, the server context is released and the server object is deleted.

In the .NET ADF, when you work with the *WebMap*, *WebPageLayout*, or *WebGeocode* class, you are indirectly working with a server context because these classes instantiate a server context. In addition, because these classes are the convenience classes associated with the map control, overview map control, page layout control, and geocode connection component, working with these items also utilizes a server context. By following some simple coding practices, you can ensure that your application properly gets and releases a server context and keeps your GIS server running at optimum performance.

There are a few different coding techniques that you can use in your Web applications to ensure that the application properly releases the server context. The one you implement will depend on whether you are working with a pooled object or a non-pooled object. In general, follow this guideline:

When creating an application that uses a pooled object, always dispose of the WebMap, WebPageLayout, *or* WebGeocode *object. When programming with a non-pooled object, you must also release the server context, then dispose of the object.*

This guideline is illustrated with code samples below. The different coding techniques apply to pooled and non-pooled objects. Although the description and code samples below refer to the *WebMap* class, the concepts apply to the *WebPageLayout* class as well.

Releasing server contexts of pooled objects

When creating an application that uses a pooled object, you need to dispose of any *WebMap* objects you instantiate. In C#, you can create the *WebMap* object within a using statement, which implicitly disposes of the *WebMap* object, or without a using statement and explicitly calling the *Dispose* method on the *WebMap* class in your code. For example:

```
using (WebMap webmap = Map1.CreateWebMap())
{
  // use the WebMap object
}
```

At the end of the using block, the *WebMap* object is implicitly disposed of by the .NET Framework—the framework actually calls the *Dispose* method on the *WebMap* object for you. Alternatively, you can code the same thing explicitly with the following:

The concepts and coding practices described here with WebMap *apply to* WebPageLayout *and* WebGeocode *as well.*

```
WebMap webmap = Map1.CreateWebMap();
try
{
    // use the WebMap object
}
finally
    webmap.Dispose();
```

The recommended method is to use the using statement because you might forget to dispose of the object in your code. Through the using statement, the object will always be disposed of. The *Dispose* method automatically releases the server context if the *WebMap* object references a pooled object. This is done as a convenience to you and also to ensure that pooled server objects are released and made available to other client applications.

Visual Basic does not have the equivalent of a using statement, thus you always need to dispose of the *WebMap* object explicitly.

```
Dim webmap As WebMap = Map1.CreateWebMap()
Try
    ' use the WebMap object
Finally
    webmap.Dispose()
End Try
```

Releasing server contexts of non-pooled objects

When you create an application that uses a non-pooled object, the application should hold on to the server context, typically in session state. When the application ends, all server contexts must be released.

The first time your application creates a *WebMap*, you can save the server context in session state:

```
using (WebMap webmap = Map1.CreateWebMap())
{
    // do something with the WebMap object, then
    // save the server context in session state.
    Session.Add ("serverContext", webmap.ServerContext);
}
```

The next time your Web application needs to use the *WebMap* object, you can re-create it from the server context saved in session state:

IServerContext is part of the ESRI.ArcGIS.Server namespace.

```
string host = Map1.Host;
string dataFrame = Map1.DataFrame;
string serverObject = Map1.ServerObject;
IServerContext context = Session["serverContext"] as IServerContext;
using (WebMap webmap = new WebMap(context, host, serverObject, dataFrame))
{
    // do something with the WebMap object
}
```

When your application ends, you need to release the server context. If you have a *WebMap* object, you can use the *ReleaseServerContext* and *Dispose* methods. For example, you might code an Exit button on your application that includes the following code:

```
webmap.ReleaseServerContext();
webmap.Dispose();
```

Alternatively, if you don't have a *WebMap* object available, you can release the server context directly from the *IServerContext* object stored in session state:

```
IServerContext context = Session["serverContext"] as IServerContext;
context.ReleaseContext();
Session["serverContext"].Remove();
```

Releasing the server context through an Exit button assumes that the user of your Web application actually clicks the button. Unfortunately, this isn't always the case, and you should write code to handle this situation as well. In addition, Web servers clean up session variables after a certain period of inactivity. Thus, if the user of your application steps away from the desk, the session can time-out before the Exit button is ever clicked. In both cases, the GIS server is still allocating a server context for the application, even though the application can no longer access it. This ties up the resource, making it unavailable to any other client application.

In reality, the GIS server does some things to try to free up resources that are no longer being used (see the maximum usage time property on the Server Object Properties dialog box in ArcCatalog). However, there is still some amount of time before this happens during which the resource is unavailable. To ensure that the resource is freed up, add code to the Session_End event of your Web application to release the server context. You can find the Session_End event in the Global.asax.cs file in your project. For example, you can add code to this event to search for your session variable and release the server context.

```
protected void Session_End(Object sender, EventArgs e)
{
  IServerContext context = Session["serverContext"] as IServerContext;
  context.ReleaseContext();
}
```

A more generic version that cleans up all session variables that reference server contexts is shown below.

Releasing server context in the Session_End event

The map control, page layout control, and overview map control internally utilize session state to maintain the server context for non-pooled objects. Thus, when you use these controls in your Web application, you must code the Session_End event to release the server context.

The following code sample is a generic version that will release all server contexts associated with session variables. This will remove those created by the controls and any that you've created in your application.

C# version:

```
protected void Session_End(Object sender, EventArgs e)
{
  IServerContext context;
  for (int i = 0; i < Session.Count; i++)
  {
    context = Session[i] as IServerContext;
    if (context != null)
      context.ReleaseContext();
  }
  Session.RemoveAll();
}
```

Visual Basic version:

```
Sub Session_End(ByVal sender As Object, ByVal e As EventArgs)
  Dim context As IServerContext
  Dim i As Integer
  For i = 0 To Session.Count - 1
    context = Session(i)
    If Not (context Is Nothing) Then
      context.ReleaseContext()
    End If
  Next i
  Session.RemoveAll()
End Sub
```

Always include this Session_End code in any Web application you create that uses non-pooled objects.

WORKING WITH FINE-GRAINED ArcObjects

The .NET ADF provides a set of classes that allow you to create Web applications that incorporate GIS functionality. While the .NET ADF exposes many of the common GIS operations, it doesn't attempt to duplicate all of the functionality of ArcObjects. What the .NET ADF does provide, however, is a mechanism for accessing ArcObjects directly. This is referred to in this book as working with the fine-grained ArcObjects that comprise a server object. This section describes some of the things you need to be aware of when working with fine-grained ArcObjects.

The *WebMap* and *WebPageLayout* classes expose a *MapServer* method that provides you access to the *IMapServer* interface of the *MapServer* coclass. Through the *IMapServerObjects* interface of the *MapServer* coclass, you can access the fine-grained ArcObjects that make up the map document being served up as a map server object on your GIS server.

Additional information about the MapDescription and PageDescription can be found in the 'Map control' and 'Page layout control' sections, respectively, later in this chapter.

When you call any of the methods on the *MapServer* coclass's interfaces, the state of the fine-grained ArcObjects in the server object instance is not permanently changed. Rather, the *MapDescription* or the *PageDescription* is applied for the duration of the method, and the state of the fine-grained ArcObjects returns to the previous state at the end of the call. However, when you work with fine-grained ArcObjects, the state of the server object instance may be permanently changed, depending on what methods you call. For this reason, you should work with non-pooled server objects if you want to make calls on fine-grained ArcObjects that change the state of the object.

You may have noticed that when you work with the ADF .NET Web controls, the Web controls appear to maintain things such as layer visibility and the current map extent. This information is stored in session state as part of the Web application itself. The actual server object instance running on the GIS server is unaltered. Whenever a request gets sent to the server, this state information is reapplied to the server object instance for the duration of the request.

The map and overview map controls store their state in the *MapDescription* object. The *MapDescription* gets applied to the map server object on every draw and query request. The page layout control stores its state in the *PageDescription* object, which also gets applied to the server object on every draw request.

Because the map, overview map, and page layout controls store some properties that define their state, it means that these properties are applied to the map server object and will override any changes you may make directly to the fine-grained ArcObjects in the server object instance. For example, suppose you access the *IActiveView* interface of the data frame you're working with and change *IActiveView::Extent*. If you subsequently call *WebMap::Refresh*, you might expect that the data frame will be drawn with the extent you set on *IActiveView*. This will not happen because the *MapDescription* stored by the map control also has an extent property that you can access via *IMapDescription::MapArea*, and this property overrides the underlying changes to the extent made through *IActiveView*. This behavior applies to every property of the *MapDescription* coclass for the map and overview map controls and every property of the *PageDescription* coclass for the page layout control.

The following steps outline how you can make changes to the fine-grained ArcObjects and also see them reflected in your Web application.

1. Call *WebMap::ApplyMapDescriptionToServer* or *WebPageLayout::ApplyPageDescriptionToServer* to update the server object instance with the properties stored by the Web controls.

2. Make any changes to the server object instance that you need to, for example, adding or removing a layer or changing layer rendering.

3. Call *WebMap::RefreshServerObjects* or *WebPageLayout::RefreshServerObjects* to make the *MapServer* refresh its properties with the current state held by the fine-grained ArcObjects in the server instance. When you call this method with no arguments or with "True" as the argument, the *MapDescription* or *PageDescription* held by the convenience class is updated with the default map or page description from the server.

4. You are now ready to use any of the *WebMap* or *WebPageLayout* methods to draw or query the server object instance.

The following code illustrates this approach:

```
using (WebMap webmap = Map1.CreateWebMap())
{
  webmap.ApplyMapDescriptionToServer();  //apply control state to server
  // work with fine-grained ArcObjects
  IMapServerObjects mapServerObjects = webmap.MapServer as IMapServerObjects;
  IActiveView map = mapServerObjects.get_Map("Layers") as IActiveView;
  IEnvelope extent =
   webmap.ServerContext.CreateObject("esriGeometry.Envelope") as IEnvelope;
  extent.PutCoords(-30,-30,30,30);
  map.Extent = extent;
  // make map server and webmap take note of these changes
  webmap.RefreshServerObjects();
  webmap.Refresh();  // refreshes map control
}
```

CHOOSING CAPABILITIES FOR YOUR WEB SERVICE CATALOG

Web service catalogs may contain both MapServer and GeocodeServer Web services. Each Web service includes functionality that your users can exercise using ArcMap and you can exercise in your applications. The functionality provided by a particular Web service can be limited to, for example, allow consumers of the Web service to draw the map but not to query the data sources of the layers in the map.

The set of functionality that a Web service supports is called its capabilities. The capabilities for MapServer and GeocodeServer Web services are organized into capability groups. The *IRequestHandler::HandleStringRequest* method supported by MapServer and GeocodeServer takes a string parameter named capabilities, whose value is a comma-delimited list of capabilities, such as *HandleStringRequest("map,query,data", <inputSOAP>, <outputSOAP>).*

The capability groups and the methods included in each group for MapServer Web services are:

Map:

- get_DocumentInfo
- get_MapCount
- get_MapName
- get_DefaultMapName
- GetServerInfo
- ExportMapImage
- GetSupportedImageReturnTypes
- GetLegendInfo
- ToMapPoints
- FromMapPoints

Query:

- Identify
- QueryFeatureCount
- QueryFeatureIDs
- QueryHyperlinks
- ComputeScale
- ComputeDistance
- GetSQLSyntaxInfo

Data

- Find
- QueryFeatureData

By default, the Web service catalog template assigns "map" as the capabilities for all MapServer Web services. Therefore, by default, consumers of the MapServer Web service, including ArcMap, will not be able to identify or query layers in the map. You can add additional capability groups (query and data) by editing the Utility.cs class file and changing the value for MAP_CAPABILITIES. Edit the following line of code to include query and data:

```
public const string MAP_CAPABILITIES = "map,query,data";
```

The above line will set the capabilities for all MapServer Web services included in the Web service catalog. If you want to set different capabilities for individual MapServer Web services, edit the aspx.cs file named after the map server object. For example, if you included a map server object named Counties in your Web service catalog, you would edit the Counties.aspx.cs file and search for the following line of code:

```
string responseMessage = svc.QueryService(Utility.MAP_CAPABILITIES,
    request);
```

You can substitute the following line of code, enabling the appropriate capabilities.

```
string responseMessage = svc.QueryService("map,query", request);
```

The capability groups and the methods included in each group for GeocodeServer Web services are:

Geocode:

- GeocodeAddress
- GeocodeAddresses
- StandardizeAddress
- FindAddressCandidates
- GetAddressFields
- GetCandidateFields

- GetIntersectionCandidateFields

- GetStandardizedFields

- GetStandardizedIntersectionFields

- GetResultFields

- GetDefaultInputFieldMapping

- GetLocatorProperties

Reverse geocode:

- ReverseGeocode

By default, the Web service catalog template assigns "geocode,reversegeocode" as the capabilities for all GeocodeServer Web services. You can remove a capability group (geocode or reversegeocode) by editing the Utility.cs class file and changing the value for GC_CAPABILITIES. Edit the following line of code to change the capabilities.

```
public const string GC_CAPABILITIES = "geocode,reversegeocode";
```

The above line of code sets the capabilities for all GeocodeServer Web services. To set the capabilities for an individual GeocodeServer Web service, change the following line of code in the aspx.cs file named after your geocode server object.

```
string responseMessage = svc.QueryService(Utility.GC_CAPABILITIES,
    request);
```

Throughout this chapter, you've seen how the various Web controls that are part of the .NET ADF are used in Web applications. The remaining sections of this chapter describe each control in more detail, examining the classes that each Web control is built with and highlighting those methods that you'll find particularly useful. This chapter, however, does not serve as a complete reference for all the objects in the .NET ADF. You can find a complete description of the properties, methods, and events for all classes in the ArcGIS Developer Help. In addition, you can view the .NET ADF Object Model Diagram through the ArcGIS Developer Help.

EXPLORING THE WEB CONTROL OBJECT MODEL

The .NET ADF custom Web controls are built by extending existing classes in the .NET Framework. The .NET ADF also includes additional convenience classes that help to support the Web controls. All of the .NET ADF Web controls and convenience classes are incorporated into one namespace, ESRI.ArcGIS.Server.WebControls.

The diagram below presents a conceptual object model diagram for the .NET ADF Web controls.

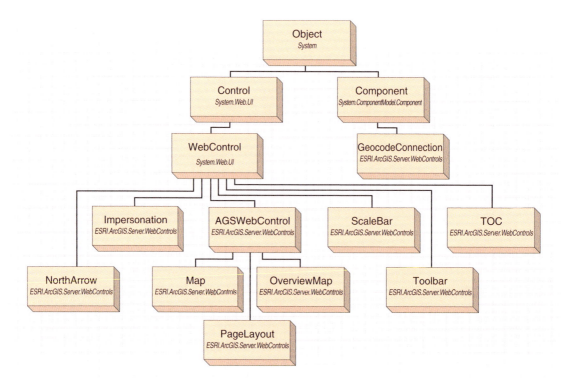

.NET ADF NAMESPACES

Namespaces group classes with similar functionality. Accordingly, the classes distributed with the .NET ADF are grouped into namespaces as well. As a Web application developer, you'll become most familiar with the ESRI.ArcGIS.Server.WebControls namespace, as this namespace contains the custom Web controls that you program with. The following summarizes the three namespaces distributed with the .NET ADF:

• ESRI.ArcGIS.Server.WebControls—contains the set of custom Web controls and classes from which you create your Web applications that access the GIS server.

• ESRI.ArcGIS.Server.WebControls.Design—contains the set of classes that you can use to extend the .NET ADF custom Web controls or create new custom controls similar to those in the .NET ADF.

• ESRI.ArcGIS.Server.WebControls.Tools—contains the set of classes that provide server-side actions for tools on a toolbar control or for *ToolItems* on a map or page layout control. For instance, these classes provide the code that executes on the server for the zoom in, zoom out, and pan operations on a map or page layout control.

The map control provides the display functionality of the ArcMap data view for a map document served using ArcGIS Server. As with data view, a map control displays a single data frame in the map document. Using the map control, you can navigate the data frame, for example, pan and zoom, and access the fine-grained ArcObjects that comprise it.

Map Control Objects

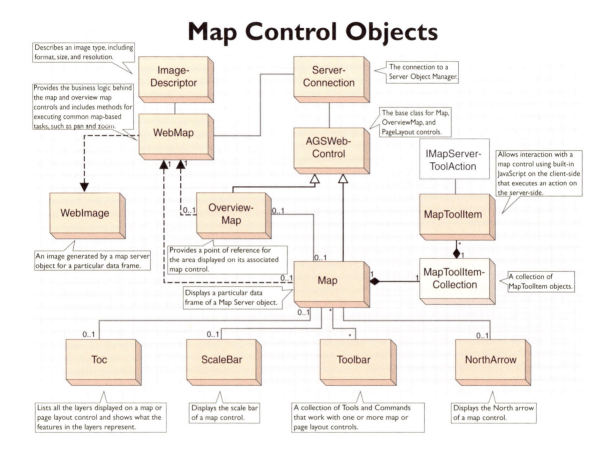

The *Map* class provides access to the properties and methods that govern how the control works. The map control's convenience class is *WebMap*. This class provides the functionality for navigating the data frames and also provides the entry point for accessing fine-grained ArcObjects. You interact with a map control either through a tool on a *Toolbar* or a *MapToolItem*. The toolbar control presents a well-defined user interface comprised of tools and commands that interact with the map control. Alternatively, you can use other user interface elements, such as an HTML button, that activate particular *MapToolItem* objects.

The *Map* class provides the user interface for displaying a data frame from a map server object on a Web form.

Map : AGSWebControl, System.Web.UI.IPostBackEventHandler, System.Web.UI.IPostBackDataHandler	Displays a particular data frame or map server object.
■— CopyrightText: CopyrightText	The copyright text to display on the map.
■■ CurrentToolItem: MapToolItem	The current toolitem that can be used on the control.
■■ DataFrame: System.String	The data frame to display from the current map server object.
■■ Extent: Extent	The current extent of the data frame displayed in the map control.
■■ ImageUrl: System.String	The URL location of the image to display in the map control.
■■ InitialExtent: InitialExtent	Draws the map at either the last saved extent or full extent.
■■ MapDescription: System.String	The string representation of the MapDescription for a particular map server object.
■— ToolItems: MapToolItemCollection	Collection of ToolItems that can be used on the control.
◄— CreateWebMap: WebMap	Creates the WebMap object associated with this control.
◄— Draw: System.Void	Draws the map image using existing settings.
◄— LoadPostData (System.String postDataKey, System.Collections.Specialized.NameVal ueCollection values): System.Boolean	Indicates whether RaisePostDataChangedEvent should be called.
◄— RaisePostBackEvent (System.String eventArgument): System.Void	Handles postbacks. Raises events based on the event argument passed in. Calls the ServerToolAction component of the CurrentToolItem.
◄— RaisePostDataChangedEvent: System.Void	Enables the map control to process an event raised when a form is posted to the server.
◄— Reset: System.Void	Resets the map control to an unbound state.
◁— Circle: MapCircleEventHandler	Occurs when an end user drags a circle on the map control using the Circle client tool action.
◁— DataFrameChanged: DataFrameChangeEventHandler	Occurs when the DataFrame property of the map control has changed.
◁— DragImage: MapToolEventHandler	Occurs when an end user drags the image in the map control using the DragImage client tool action.
◁— DragRectangle: MapToolEventHandler	Occurs when an end user drags a rectangle on the map control using the DragRectangle client tool action.
◁— Line: MapLineEventHandler	Occurs when an end user drags a line on the map control using the Line client tool action.
◁— MapClick: MapPointEventHandler	Occurs when an end user clicks the map.
◁— Oval: MapOvalEventHandler	Occurs when an end user drags an oval on the Map control using the Oval client tool action.
◁— Point: MapPointEventHandler	Occurs when an end user clicks the Map control using the Point client tool action.
◁— Polygon: MapPolygonEventHandler	Occurs when an end user drags a polygon on the Map control using the Polygon client tool action.
◁— Polyline: MapPolylineEventHandler	Occurs when an end user drags a polyline on the Map control using the Polyline client tool action.

The *Map* class contains properties, inherited from the *WebControl* class (System.Web.UI), that specify how the control looks on the Web page, such as the height, width, and border style. Typically, you'll define these properties at design time.

The *CreateWebMap* method is an important method as it gives you access to the map control's corresponding convenience class, *WebMap*. It's the *WebMap* class that provides much of the functionality associated with navigating the data frame. For example, you can enlarge the map extent displayed in a data frame using the *Zoom* method.

The *ToolItems* property gives access to a set of *MapToolItem* objects that can act on a map control. A *MapToolItem* provides a means for interacting with the map control on the client-side using built-in JavaScript, which in turn executes an action on the server-side. The *CurrentToolItem* is the *MapToolItem* that is currently active. See the next section, titled 'Interacting with the map control' for more information.

The *AutoFirstDraw* property, inherited from *AGSWebControl*, allows you to control whether the map will draw automatically when the page is first rendered. When set to true, the control will draw in the *OnInit* method the first time the control is drawn in a session. Setting this property to false allows you to handle the drawing yourself in the Page_Load event, for example:

```
if (!Page.IsPostBack)
{
    using (WebMap webMap = Map1.CreateWebMap())
    {
        webMap.DrawDefaultExtent();
    }
}
```

The *InitialExtent* property, also inherited from *AGSWebControl*, specifies whether the map should draw at the last extent saved in the ArcMap document or at the full extent.

The *MapDescription* property provides access to the string representation of the *MapDescription* object (in the ESRI.ArcGIS.Carto namespace) used to draw the data frame. This string representation of the *MapDescription* is stored in session state and is passed to the MapServer whenever a *WebMap* object is created using *CreateWebMap* and one of its draw or query methods is called. Only changes to the *MapDescription* member are saved by the map control. If any other ArcObjects customizations need to be made, you should be working with a non-pooled server object. See the section titled 'Working with fine-grained ArcObjects' earlier in this chapter for more information.

While you can programmatically get the *MapDescription* object from the string by using *IServerContext::LoadObject*, it's easier to access the object directly through *WebMap::MapDescription*. Conversely, you can go from the *MapDescription* object to the string representation by using *IServerContext::SaveObject*.

With the *UseMIMEData* property, you specify whether the data frame images will be accessed from disk or whether MIME data will be used to get the images as bytes from the map server object and use the images as bytes within the Web tier. You can always access data frame images as MIME data. However, depending on how the server object is configured, you may not always be able to access data frame images from disk.

The *CopyrightText* property allows you to display text over the map control so that the image has the appropriate copyright information on it.

INTERACTING WITH THE MAP CONTROL

Just as you can click a button on a Web form, the map control also allows interaction. When you interact with a map control, some action happens on the client and some action happens on the server. The client-side action is what allows you to, for example, draw a box over the control. The server-side action is what allows you to use the coordinates of the box and, for example, zoom in. During every interaction with a map control, two things happen:

- An end user interacts with the map control on the client.

  The map control allows the user to, for example, draw a box over the map control with the mouse. This client-side action is controlled by a JavaScript function that executes on the client, without any requests being sent to the server. The .NET ADF includes prewritten JavaScript code for common interactions such as drawing a box.

• The server executes an action based on the client interaction.

Once the client-side action is completed, the client sends a request to the server to execute the server-side action. For example, if server-side action is to zoom in, the server calculates the new extent for the data frame—based on the user-defined box—and instructs the map server object to draw the new extent. All of this happens on the server because it's the server that has the ability to access the map control and to draw the new extent. The .NET ADF provides several classes that implement *IMapServerToolAction*. These classes function as server-side actions, for example, *MapZoomIn* and *MapPan*.

As a Web application developer, you build functionality into your Web application by defining the set of client-side actions and associated server-side actions you want to include in your application. You'll use the prewritten JavaScript code and server-side actions included with the .NET ADF or write your own custom server-side actions. Once defined, you need to expose this functionality in the user interface. This is described below.

Using a toolbar to interact with the map control

The toolbar control provides the easiest way of setting up various interactions with the map control. The toolbar control, like any toolbar you'd see in a desktop application, allows you to group a set of related actions together in one place. For example, you might group all the zooming and panning actions on one toolbar. When you add a toolbar to your Web form, you associate it with a map control through the toolbar's *BuddyControls* property. This identifies which controls the toolbar should interact with.

After you associate a toolbar with a map control, you must define what each item on the toolbar will do. Through the *ToolbarItems* property of the toolbar control, you can add new items and edit existing items on the toolbar. Then, for each item on the toolbar, you'll identify what happens when an end user uses it. Some items will execute both a client-side and server-side action; others will only execute a server-side action. For more information on setting up toolbars, see the 'Toolbar control' section in this chapter.

Using the ToolItems property of the map control

While a toolbar provides the easiest way to associate actions with a map control, its user interface may not provide you with the particular look and feel you want in your Web application. The *ToolItems* property of the map control provides an alternative method for associating a set of actions with a map control.

This approach gives you greater control over the user interface design, as it allows you to use other controls—for example, an HTML button—to interact with the map control. However, this approach also requires a bit more coding effort on your part. You will need to more closely manage, through code you write, the actions performed on the map control. You will also need to manage how the user interface exposes these actions to an end user and indicates what action will be performed at any given time. For instance, a toolbar highlights the active tool and indicates to the end user what will happen when the user interacts with the map control. You would need to emulate this behavior in some way.

The *ToolItems* property of the map control exposes a collection of *MapToolItem* objects. As with the toolbar control, you identify a client-side and server-side action for each *MapToolItem*. You can use the same JavaScript functions and .NET ADF classes for the client-side and server-side actions, respectively, that you use with a toolbar control. The *CurrentToolItem* property of the map control allows you to specify which *MapToolItem* is currently active. When the map control draws, it will activate the JavaScript function associated with the *CurrentToolItem*.

You can define a *MapToolItem* through the *ToolItems* property on the map control or programmatically in your code. Accessing the *ToolItems* property of the map control in Visual Studio .NET displays the MapToolItem Collection Editor. Just click Add to add a *MapToolItem*. The figure below shows one entry already added that implements a zoom in functionality. The client-side action is set to *DragRectangle* and the server-side action is set to *MapZoomIn*. The *Name* property is the name you'll use to reference this *MapToolItem*; here, the name is zoomin.

Once you've defined the client-side and server-side actions for the particular *MapToolItem*, you need to make it the map control's *CurrentToolItem*. The following line of code sets the *CurrentToolItem* of a map control.

```
Map1.CurrentToolItem = Map1.ToolItems.Find("zoomin");
```

Typically, you'll have more than one *MapToolItem* associated with the map control. When you create your Web application, you'll probably want to include the above line of code in the Page_Load event to establish a *CurrentToolItem* the first time a page loads. Then, you can expose other controls in your user interface that change the *CurrentToolItem*. For example, you might add a button server control and include the following line of code in the button's click event to change the *CurrentToolItem* to a pan tool.

```
Map1.CurrentToolItem = Map1.ToolItems.Find("pan");
```

Server controls run on the server, and thus, the above line of code actually executes on the server.

Another approach you can take to expose the *MapToolItem* functionality in the user interface is to use HTML controls. An HTML control doesn't require a round-trip to the server to execute. The HTML control you add to your Web form would define the client-side action by calling the JavaScript function directly and use the server-side action defined on the *MapToolItem*. In this case, the JavaScript function you call should correspond to the client-side action you specify on the *MapToolItem*. For example, you can add the following HTML button to your Web form.

```
<INPUT style="Z-INDEX: 102; LEFT: 238px; POSITION: absolute; TOP: 80px"
type="button" value="Zoom In" onclick="MapDragRectangle('Map1', 'zoomin',
true)">
```

In the tag above, the onclick attribute establishes the client-side action the HTML button performs. In this case, the button calls the *MapDragRectangle* JavaScript function directly and passes the name of the map control, the name you specified in the dialog above, and a Boolean value to display a loading image while the page refreshes. The *MapDragRectangle* JavaScript function corresponds to the *DragRectangle* client tool action defined in the MapToolItem Collection Editor above. The name is what links the HTML button to the server-side action. When the JavaScript posts back to the server—after end user interaction with the map control—the *Map* class calls the *ServerAction* method on the *ServerToolAction* class.

It's important to note that if the map control has a toolbar associated with it (through the toolbar's *BuddyControls* property), the settings on the toolbar will override the settings of the map control's *ToolItems* property. For instance, if you define a *MapToolItem* called "MyTool" and you have a similarly named tool on a toolbar associated with the map control, the settings on the toolbar tool take precedence over the settings on the map control's *MapToolItem*.

JavaScripts for client-side interaction

There is a one-to-one relationship between JavaScript function names and *MapClientToolAction* enumerations. Thus, the enumeration name you specify when defining the *MapToolItem* properties should match the JavaScript function name you specify in the HTML button. The list of *MapClientToolAction* enumerations and their corresponding JavaScript functions is presented below.

MapClientToolAction	JavaScript Function
Circle	MapCircle
DragImage	MapDragImage
DragRectangle	MapDragRectangle
Line	MapLine
Oval	MapOval
Polygon	MapPolygon
Polyline	MapPolyline
Point	MapPoint

When invoking the JavaScript functions directly, supply the following three arguments:

- The ID of the map control, for example, Map1.

- The *Name* property of the *MapToolItem* that will handle the server-side action, for example, zoomin. If you plan to write code to handle a tool event, this is the value passed in as the *ToolName* property of the *ToolEventArgs* object.

- A Boolean value indicating whether to show the loading image between postback and page refresh.

MAP CONTROL EVENTS

Events provide another way of executing server-side actions. Each *MapClientToolAction* enumeration maps to a map control event. When the event is triggered, your event code executes.

You can write event handlers that operate in conjunction with any server-side actions you establish through a toolbar or through a *MapToolItem*. Your event handler code will execute *after* the server-side action defined on the toolbar or *MapToolItem*. Alternatively, by omitting a server-side action for a particular item on a toolbar or *MapToolItem*, you can completely control how your application responds to a particular client-side action using an event handler. For example, you might define a client-side action through the *DragRectangle* enumeration (or *MapDragRectangle* JavaScript function) and handle the server-side action through the following event code:

```
private void Map1_DragRectangle(object sender,
  ESRI.ArcGIS.Server.WebControls.ToolEventArgs args)
{
  switch (args.ToolName)
  {
    case "zoomin":
      // Extracts the X,Y screen coordinates passed in as arguments
      RectangleEventArgs rectArgs = args as RectangleEventArgs;
      System.Drawing.Rectangle rectangle = rectArgs.ScreenExtent;
      int xcoord = rectangle.X;
      int ycoord = rectangle.Y;
      //Do something further with the screen coordinates.
       break;
  }
}
```

The above code uses the *ToolEventArgs* object passed to the function and converts it to a *RectangleEventArgs*, which is the specific type of object the *MapDragRectangle* JavaScript function provides. In this example, the code only extracts the X and Y values of the upper left corner and stores them in variables. Your code could use these values to do something further.

The table below shows the relationship between *MapClientToolActions* and *ToolEventArgs*.

Map Control Events	ToolEventArgs Type
Circle	CircleEventArgs
DragImage	PointEventArgs
DragRectangle	RectangleEventArgs
Line	LineEventArgs
Oval	OvalEventArgs
Polygon	PolygonEventArgs
Polyline	PolylineEventArgs
Point	PointEventArgs

WHAT'S STORED IN SESSION STATE

The map control utilizes session state to maintain information about the server object referenced by the control. The names of the session variables can be found in the technical article titled 'Session variables for the .NET Web controls' in ArcGIS Developer Online (*http://arcgisdeveloperonline.esri.com*).

The following objects are stored in session state.

ServerContext (for non-pooled server objects only)

For non-pooled objects, the *ServerContext* is stored in session state so that the same instance of the server object can be used for the entire session.

The *ServerContext* is an object that needs to be released explicitly in order to release server resources. You should release this by calling *IServerContext.ReleaseContext*.

In a deeply stateful Web application that uses a server object for a long period of time, you should release the *ServerContext* at the end of the session. The code in the Session_End method in the Global.asax.cs file gets executed at the end of a session. In this method, you should loop over all the objects stored in Session and check if they are of type *IServerContext*. If they are, you should release the context.

```
IServerContext context;
for (int i = 0; i < Session.Count; i++)
{
   context = Session[i] as IServerContext;
   if (context != null)
      context.ReleaseContext();
}
```

MapDescription

MapDescription is stored in the session because it can be a large string.

ImageURL

ImageURL is stored in the session because it can be a large string.

ToolItems

ToolItems is stored in session because it is a complex object.

CopyrightText

CopyrightText is stored in session because it is a complex object.

Extent

Extent is stored in session because it is a complex object.

CONTROL LIFECYCLE

In addition to what the base class *AGSWebControl* does, here's what the map control does at each point in its lifecycle.

- Instantiation—Initializes member variables with default values.

- OnInit—The first time the control is drawn in a session and if *AutoFirstDraw* is set to true, the map is drawn at the extent specified by the *InitialExtent* property. A copy of the *MapDescription* property is stored to check if the content changes between initialization and just before rendering. Copies of the *Host*, *ServerObject*, and *DataFrame* properties are stored to check if the connection changes between initialization and just before rendering.

- TrackViewState—Begins tracking view-state changes for complex properties, such as the *ToolItems* property.

- LoadViewState—Restores view-state information of complex properties, such as the *ToolItems* property, to what they were at the end of the previous page request. Postback only.

- LoadPostData—Indicates whether RaisePostDataChangedEvent should be called. Always returns false. Postback only.

- OnLoad—Uses the base class implementation.

- RaisePostDataChangedEvent—Does nothing. Postback only.

- RaisePostBackEvent—Fires server-side events based on the postback event argument. Calls the *CurrentToolItem*'s *ServerToolAction* class's *ServerAction* method. Postback only.

- OnPreRender—Checks if *MapDescription*, *Host*, *ServerObject*, or *DataFrame* properties have changed. If the *MapDescription* has changed, it fires the *ContentsChanged* event. If the *Host* or *ServerObject* has changed, it fires the *ConnectionChanged* event. If the *DataFrame* has changed, it fires the *DataFrameChanged* event. Registers JavaScript required for client-side action.

- SaveViewState—Saves view-state information of complex properties such as the *ToolItems* property.

- Render—Renders the HTML required to draw the control. Registers additional JavaScript used to set control-specific properties.

EVENTS

Events fired on RaisePostBackEvent

The following events are fired on RaisePostBackEvent. They are only fired for postbacks.

Event fired on all postbacks to map control:

- MapClick

Events based on client interaction with the map control:

- DragImage—Occurs when the image in a map control has been dragged using the *DragImage* client tool action.

- DragRectangle—Occurs when a rectangle has been dragged on the map control using the *DragRectangle* client tool action.

- Circle—Occurs when a circle has been drawn on the map control using the *Circle* client tool action.

- Line—Occurs when a line has been drawn on the map control using the *Line* client tool action.

- Point—Occurs when the map control has been clicked using the *Point* client tool action.

- Polygon—Occurs when a polygon has been drawn on a map control using the *Polygon* client tool action.

- Polyline—Occurs when a polyline has been drawn on the map control using the *Polyline* client tool action.

- Oval—Occurs when an oval has been drawn on the map control using the *Oval* client tool action.

Events fired on PreRender

The following events are fired on PreRender:

- ConnectionChanged—if the *Host* and *ServerObject* have changed.

- ContentsChanged—if the *MapDescription* has changed.

- DataFrameChanged—if the *DataFrame* has changed.

Generic control events

The following events are generic to all controls. All of these events are fired before the control does its work at the respective cycle. For example, the Init event is fired before the map control does its custom OnInit actions. Here's what you can expect to happen at each of these events:

- Init—Occurs when the server control is initialized, which is the first step in its lifecycle. This is too early in the lifecycle to access any of the map control's properties.

- Load—Occurs when the server control is loaded into the *Page* object. At this stage, all of the map control's initializations are complete and changes saved into view state on the previous request have been applied. This is the stage to access control properties.

- PreRender (inherited from Control)—Occurs when the server control is about to render to its containing *Page* object and before the *ContentsChanged*, *ConnectionChanged*, and *DataFrameChanged* events are fired.

- Unload (inherited from Control)—Occurs when the server control is unloaded from memory. This is where you can check the state after Render.

The page layout control provides the display functionality of the ArcMap layout view for a map document served using ArcGIS Server. Using the page layout control, you can view a map server object's layout, navigate it, navigate the maps within the layout, and access the fine-grained ArcObjects that comprise it.

Page Layout Control Objects

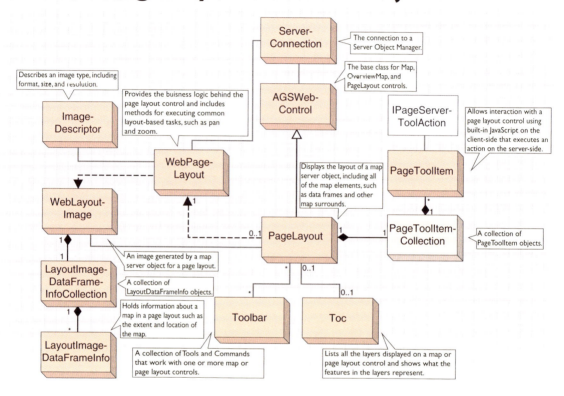

The *PageLayout* class provides access to the properties and methods that govern how the control works. The page layout control's convenience class is *WebPageLayout*. This class provides the functionality for navigating the layout and its data frames and also provides the entry point for accessing fine-grained ArcObjects. You interact with a page layout control either through a tool on a *Toolbar* or a *PageToolItem*. The toolbar control presents a well-defined user interface composed of tools and commands that interact with the page layout control. Alternatively, you can use other user interface elements, such as an HTML button, that activate particular *PageToolItem* objects.

The *PageLayout* class provides the user interface for displaying a map server object's layout on a Web form.

PageLayout : AGSWebControl, System.Web.UI.IPostBackEventHandler, System.Web.UI.IPostBackDataHandler	Displays the layout of a map server object, including all of the map elements, such as data frames and other map surrounds.
■— CopyrightText: CopyrightText	The copyright text to display on the page layout.
■—■ CurrentToolItem: PageToolItem	The current ToolItem that can be used on the control.
■—■ PageDescription: System.String	The string representation of the PageDescription for a particular page layout.
■— ToolItems: PageToolItemCollection	Collection of ToolItems that can be used on the control.
■—■ WebLayoutImage: WebLayoutImage	The image displayed by the page layout control and its properties.
◄— CreateWebPageLayout: WebPageLayout	Creates the WebPageLayout object associated with this control.
◄— Draw: System.Void	Draws the page layout image using existing settings.
◄— LoadPostData (System.String postDataKey, System.Collections.Specialized.NameValueCollection values): System.Boolean	Indicates whether RaisePostDataChangedEvent should be called.
◄— RaisePostBackEvent (System.String eventArgument): System.Void	Handles postbacks. Raises events based on the event argument passed in. Calls the ServerToolAction component of the CurrentToolItem.
◄— RaisePostDataChangedEvent: System.Void	Enables the page layout control to process an event raised when a form is posted to the server.
◄— Reset: System.Void	Resets the page layout to an unbound state.
◄⊢ DragImage: PageToolEventHandler	Occurs when an end user drags the image in the page layout control using the DragImage client tool action.
◄⊢ DragRectangle: PageToolEventHandler	Occurs when an end user drags a rectangle on the page layout control using the DragRectangle client tool action.
◄⊢ MapDragImage: PageToolEventHandler	Occurs when an end user drags a data frame in the page layout control using the MapDragImage client tool action.
◄⊢ MapDragRectangle: PageToolEventHandler	Occurs when an end user drags a rectangle on a data frame in the page layout control using the MapDragRectangle client tool action.
◄⊢ MapPoint: PagePointEventHandler	Occurs when an end user clicks on a data frame in the page layout control using the MapPoint client tool action.
◄⊢ PageClick: PagePointEventHandler	Occurs when an end user clicks the page layout control.
◄⊢ Point: PagePointEventHandler	Occurs when an end user clicks the page layout control using the Point client tool action.

The *PageLayout* class contains properties, inherited from the *WebControl* class (System.Web.UI), that specify how the control looks on the Web page, such as the height, width, and border style. Typically, you'll define these properties at design time.

The *CreateWebPageLayout* method is an important method, as it gives you access to the page layout control's corresponding convenience class, *WebPageLayout*. It's the *WebPageLayout* class that provides much of the functionality associated with navigating the layout. For example, you can enlarge your view of the layout using the *PageZoom* method.

The *ToolItems* property gives access to a set of *PageToolItem* objects that can act on a page layout control. A *PageToolItem* provides a means for interacting with the page layout control on the client-side using built-in JavaScript, which in turn executes an action on the server-side. The *CurrentToolItem* is the *PageToolItem* that is currently active. See the next section, titled 'Interacting with the page layout control' for more information.

The *AutoFirstDraw* property, inherited from *AGSWebControl*, allows you to control whether the page layout will draw automatically when the page is first rendered. When set to true, the control will draw in the OnInit method the first time the control is drawn in a session. Setting this property to false allows you to handle the drawing yourself in the Page_Load event, for example:

```
if ( !Page.IsPostBack )
{
  if ( Session.IsNewSession )
  {
    using (WebPageLayout layout = PageLayout1.CreateWebPageLayout())
    {
      ESRI.ArcGIS.Geometry.IEnvelope env =
        layout.ServerContext.CreateObject("esriGeometry.Envelope") as
        ESRI.ArcGIS.Geometry.IEnvelope;
      env.PutCoords(0,0,2,2);
      layout.PageDrawExtent(env);
    }
  }
}
```

The *InitialExtent* property, also inherited from *AGSWebControl*, specifies whether the page layout should draw at the last extent saved in the ArcMap document or at the full extent.

The *PageDescription* property provides access to the string representation of the *PageDescription* object (in the ESRI.ArcGIS.Carto namespace) used to draw the layout. This string representation of the *PageDescription* is stored in session state and is reapplied to the *MapServer* on every draw. Only changes to the *PageDescription* member are saved by the page layout control. If any other ArcObjects customizations need to be made, you should be working with a non-pooled server object. See the section titled 'Working with fine-grained ArcObjects' earlier in this chapter for more information.

While you can programmatically get the *PageDescription* object from the string by using *IServerContext::LoadObject,* it's easier to access the object directly through *WebPageLayout::PageDescription.* Conversely, you can go from the *PageDescription* object to the string representation by using *IServerContext::SaveObject.*

The *WebLayoutImage* property is the result of a page layout draw operation. It provides access to the image drawn. It also provides information about the image and the data frames in the image. Of special interest is the *Maps* property of the *WebLayoutImage* class. This property provides access to the *LayoutImageDataFrameInfo* objects corresponding to each data frame in the image. The *LayoutImageDataFrameInfo* for a data frame is a required parameter for those *WebPageLayout* methods that draw data frames, for example, *WebPageLayout::MapZoom.*

With the *UseMIMEData* property, you specify whether the page layout images will be written to disk or whether MIME data will be used to get the images as bytes from the map server object and use the images as bytes within the Web tier. You can always access page layout images as MIME data. However, depending on how the server object is configured, you may not always be able to access page layout images from disk.

The *CopyrightText* property allows you to display text over the page layout control, so that the image has the appropriate copyright information on it.

INTERACTING WITH THE PAGE LAYOUT CONTROL

Just as you can click a button on a Web form, the page layout control also allows interaction. When you interact with a page layout control, some action happens on the client and some action happens on the server. The client-side action is what allows you to, for example, draw a box over the control. The server-side action is what allows you to use the coordinates of the box and, for example, zoom in. During every interaction with a page layout control, two things happen:

- An end user interacts with the page layout control on the client.

  The page layout control allows the user to, for example, draw a box over the page layout control with the mouse. This client-side action is controlled by a JavaScript function that executes on the client, without any requests being sent to the server. The .NET ADF includes prewritten JavaScript code for common interactions such as drawing a box.

- The server executes an action based on the client interaction.

  Once the client-side action is completed, the client sends a request to the server to execute the server-side action. For example, if server-side action is to zoom in, the server calculates the new extent for the page layout—based on the user-defined box—and instructs the page layout to draw the new extent. All of this happens on the server because it's the server that has the ability to access the page layout control and to draw the new extent. The .NET ADF provides several classes that implement *IPageServerToolAction*. These classes function as server-side actions, for example, *PageZoomIn* and *PagePan*.

As a Web application developer, you build functionality into your Web application by defining the set of client-side actions and associated server-side actions you want to include in your application. You'll use the prewritten JavaScript code and server-side actions included with the .NET ADF or write your own custom server-side actions. Once defined, you need to expose this functionality in the user interface. This is described below.

Using a toolbar to interact with the page layout control

The toolbar control provides the easiest way of setting up various interactions with the page layout control. The toolbar control, like any toolbar you'd see in a desktop application, allows you to group a set of related actions together in one place. For example, you might group all the zooming and panning actions on one toolbar. When you add a toolbar to your Web form, you associate it with a page layout through the toolbar's *BuddyControls* property. This identifies which controls the toolbar should interact with.

After you associate a toolbar with a page layout control, you must define what each item on the toolbar will do. Through the *ToolbarItems* property of the toolbar control, you can add new items and edit existing items on the toolbar. Then, for each item on the toolbar, you'll identify what happens when an end user uses it. Some items will execute both a client-side and server-side action; others will only execute a server-side action. For more information on setting up toolbars, see the 'Toolbar control' section in this chapter.

Using the ToolItems property of the page layout control

While a toolbar provides the easiest way to associate actions with a page layout control, its user interface may not provide you with the particular look and feel you want in your Web application. The *ToolItems* property of the page layout control provides an alternative method for associating a set of actions with a page layout control.

This approach gives you greater control over the user interface design, as it allows you to use other controls—for example, an HTML button—to interact with the page layout. However, this approach also requires a bit more coding effort on your part. You will need to more closely manage, through code you write, the actions performed on the page layout. You will also need to manage how the user interface exposes these actions to an end user and indicates what action will be performed at any given time. For instance, a toolbar highlights the active tool and indicates to the end user what will happen when he/she interacts with the page layout. You would need to emulate this behavior in some way.

The *ToolItems* property of the page layout control exposes a collection of *PageToolItem* objects. As with the toolbar control, you identify a client-side and server-side action for each *PageToolItem*. You can use the same JavaScript functions and .NET ADF classes for the client-side and server-side actions, respectively, that you use with a toolbar control. The *CurrentToolItem* property of the page layout control allows you to specify which *PageToolItem* is currently active. When the page layout draws, it will activate the JavaScript function associated with the *CurrentToolItem*.

You can define a *PageToolItem* through the *ToolItems* property on the page layout control or programmatically in your code. Accessing the *ToolItems* property of the page layout control in Visual Studio .NET displays the PageToolItem Collection Editor. Just click Add to add a *PageToolItem*. The figure below shows one entry already added that implements a zoom in functionality. The client-side action is set to *DragRectangle* and the server-side action is set to *PageZoomIn*. The *Name* property is the name you'll use to reference this *PageToolItem*; here, the name is zoomin.

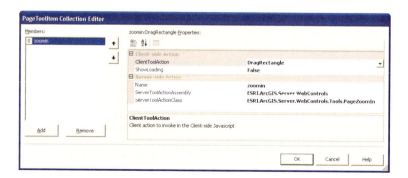

Once you've defined the client-side and server-side actions for the particular *PageToolItem*, you need to make it the page layout's *CurrentToolItem*. The following line of code sets the *CurrentToolItem* of a page layout control.

```
PageLayout1.CurrentToolItem = PageLayout1.ToolItems.Find("zoomin");
```

Typically, you'll have more than one *PageToolItem* associated with the page layout control. When you create your Web application, you'll probably want to include the above line of code in the Page_Load event to establish a *CurrentToolItem* the first time a page loads. Then, you can expose other controls in your user interface that change the *CurrentToolItem*. For example, you might add a button server control and include the following line of code in the button's click event to change the *CurrentToolItem* to a pan tool.

```
PageLayout1.CurrentToolItem = PageLayout1.ToolItems.Find("pan");
```

Server controls run on the server, and thus, the above line of code actually executes on the server.

Another approach you can take to expose the *PageToolItem* functionality in the user interface is to use HTML controls. An HTML control doesn't require a round-trip to the server to execute. The HTML control you add to your Web form would define the client-side action, by calling the JavaScript function directly, and use the server-side action defined on the *PageToolItem*. In this case, the JavaScript function you call should correspond to the client-side action you specify on the *PageToolItem*. For example, you can add the following HTML button to your Web form.

```
<INPUT style="Z-INDEX: 102; LEFT: 238px; POSITION: absolute; TOP: 80px"
type="button" value="Zoom In" onclick="PageDragRectangle('PageLayout1',
'zoomin', true)">
```

In the tag above, the onclick attribute establishes the client-side action the HTML button performs. In this case, the button calls the *PageDragRectangle* JavaScript function directly and passes the name of the page layout control, the name you specified in the dialog box above, and a Boolean value to display a loading image while the page refreshes. The *PageDragRectangle* JavaScript function corresponds to the *DragRectangle* client tool action defined in the PageToolItem Collection Editor above. The name is what links the HTML button to the server-side action. When the JavaScript posts back to the server—after end user interaction with the page layout control—the *PageLayout* class calls the *ServerAction* method on the *ServerToolAction* class.

It's important to note that if the page layout control has a toolbar associated with it (through the toolbar's *BuddyControls* property), the settings on the toolbar will override the settings of the page layout control's *ToolItems* property. For instance, if you define a *PageToolItem* called "MyTool" and you have a similarly named tool on a toolbar associated with the page layout control, the settings on the toolbar tool take precedence over the settings on the page layout control's *PageToolItem*.

JavaScripts for client-side interaction

There is a one-to-one relationship between JavaScript function names and *PageClientToolAction* enumerations. Thus, the enumeration name you specify when defining the *PageToolItem* properties should match the JavaScript function name you specify in the HTML button. The list of *PageClientToolAction* enumerations and their corresponding JavaScript functions is presented below.

PageClientToolAction	JavaScript Function
DragImage	PageDragImage
DragRectangle	PageDragRectangle
MapDragImage	PageMapDragImage
MapDragRectangle	PageMapDragRectangle
MapPoint	PageMapPoint
Point	PagePoint

When invoking the JavaScript functions directly, supply the following three arguments.

- The ID of the page layout control, for example, PageLayout1.

- The *Name* property of the *PageToolItem* that will handle the server-side action, for example, zoomin. If you plan to write code to handle a tool event, this is the value passed in as the *ToolName* property of the *ToolEventArgs* object.

- A Boolean value indicating whether to show the loading image between postback and page refresh.

PAGE LAYOUT EVENTS

Events provide another way of executing server-side actions. Each *PageClientToolAction* enumeration maps to a page layout control event. When the event is triggered, your event code executes.

You can write event handlers that operate in conjunction with any server-side actions you establish through a toolbar or through a *PageToolItem*. Your event handler code will execute *after* the server-side action defined on the toolbar or *PageToolItem*. Alternatively, by omitting a server-side action for a particular item on a toolbar or *PageToolItem*, you can completely control how your application responds to a particular client-side action using an event handler. For example, you might define a client-side action through the *DragRectangle* enumeration (or *PageDragRectangle* JavaScript function) and handle the server-side action through the following event code:

```
private void PageLayout1_DragRectangle(object sender,
  ESRI.ArcGIS.Server.WebControls.ToolEventArgs args,
  ESRI.ArcGIS.Server.WebControls.LayoutImageDataFrameInfo info)
{
  switch (args.ToolName)
  {
    case "zoomin":
    // Extracts the X,Y screen coordinates passed in as arguments
    RectangleEventArgs rectArgs = args as RectangleEventArgs;
    System.Drawing.Rectangle rectangle = rectArgs.ScreenExtent;
    int xcoord = rectangle.X;
    int ycoord = rectangle.Y;
    // Do something further with the screen coordinates.
    break;
  }
}
```

The above code uses the *ToolEventArgs* object passed to the function and converts it to a *RectangleEventArgs*—which is the specific type of object the *PageDragRectangle* JavaScript function provides. In this example, the code only extracts the X and Y values of the upper left corner and stores them in variables. Your code could use these values to do something further.

The table below shows the relationship between *PageClientToolActions* and *ToolEventArgs*.

Page Layout Control Events	ToolEventArgs Type
DragImage	PointEventArgs
DragRectangle	RectangleEventArgs
MapDragImage	PointEventArgs
MapDragRectangle	RectangleEventArgs
MapPoint	PointEventArgs
Point	PointEventArgs

WHAT'S STORED IN SESSION STATE

The page layout control utilizes session state to maintain information about the server object referenced by the control. The names of the session variables can be found in the technical article titled 'Session variables for the .NET Web controls' in ArcGIS Developer Online (*http://arcgisdeveloperonline.esri.com*).

The following objects are stored in session state.

ServerContext

For non-pooled server objects, the *ServerContext* is held in session state so that the same instance of the server object can be used for the entire session.

The *ServerContext* is an object that must be released explicitly in order to release server resources. You should release this by calling *IServerContext.ReleaseContext*.

In a deeply stateful Web application that uses a server object for a long time, you should release the *ServerContext* at the end of the session. The code in the Session_End method in the Global.asax.cs file executes at the end of a session. In this method, you should loop over all the objects stored in session state and check if they are of type *IServerContext*. If they are, you should release the context.

```
IServerContext context;
for ( int i = 0; i < Session.Count; i++)
{
  context = Session[i] as IServerContext;
  if ( context != null )
    context.ReleaseContext();
}
```

PageDescription

PageDescription is stored in the session state because it can be a large string. It is impractical to store the *PageDescription* in view state and have it go back and forth on each request.

WebLayoutImage

WebLayoutImage is also stored in session state for the same reason as *PageDescription*. It can be large depending on the number of data frames in a page layout.

ToolItems

ToolItems is stored in session because it is a complex object.

CopyrightText

CopyrightText is stored in session because it is a complex object.

CONTROL LIFECYCLE

In addition to what the base class *AGSWebControl* does, here's what the page layout control does at each point in its lifecycle.

- Instantiation—Initializes member variables with default values.

- OnInit—The first time the control is drawn in a session and if *AutoFirstDraw* is set to true, the page layout is drawn at its full extent. A copy of the *PageDescription* property is stored to check if the content changes between initialization and just before rendering. Copies of the *Host* and *ServerObject* properties are stored to check if the connection changes between initialization and just before rendering.

- TrackViewState—Begins tracking view-state changes for complex properties such as the *ToolItems* property.

- LoadViewState—Restores view-state information of complex properties, such as the *ToolItems* property, to what they were at the end of the previous page request. Postback only.

- LoadPostData—Indicates whether RaisePostDataChangedEvent should be called. Always returns false. Postback only.

- OnLoad—Uses the base class implementation.

- RaisePostDataChangedEvent—Does nothing. Postback only.

- RaisePostBackEvent—Fires server-side events based on the postback event argument. Calls the *CurrentToolItem*'s *ServerToolAction* class's *ServerAction* method. Postback only.

- OnPreRender—Fires the ContentsChanged event if the *PageDescription* property has changed after LoadViewState for postbacks or after OnInit for nonpostbacks. Fires the ConnectionChanged event if the *Host* or *ServerObject* properties have changed after LoadViewState for postbacks or after OnInit for nonpostbacks. Registers JavaScript required for client-side action.

- SaveViewState—Saves view-state information of complex properties such as the ToolItems property.

- Render—Renders the HTML required to draw the control. Registers additional JavaScript used to set control specific properties.

EVENTS

Events fired on RaisePostBackEvent

The following events are fired on RaisePostBackEvent. They are only fired for postbacks.

Event fired on all postbacks to page layout control:

- PageClick

Events based on client interaction with the whole page displayed in the page layout control:

- DragImage—Occurs when the image in the page layout control has been dragged using the *DragImage* client tool action.

- DragRectangle—Occurs when a rectangle has been dragged on the page layout control using the *DragRectangle* client tool action.

- Point—Occurs when the page layout control has been clicked using the *Point* client tool action.

Events based on client interaction with the maps in the page displayed in the page layout control:

- MapDragImage—Occurs when a data frame in the page layout control has been dragged using the *MapDragImage* client tool action.

- MapDragRectangle—Occurs when a rectangle has been dragged on a data frame in the page layout control using the *MapDragRectangle* client tool action.

- MapPoint—Occurs when a data frame in the page layout control has been clicked upon using the *MapPoint* client tool action.

Events fired on PreRender

The following events are fired on PreRender:

- ConnectionChanged—if the *Host* and *ServerObject* have changed.

- ContentsChanged—if the *PageDescription* has changed.

Generic control events

The following events are generic to all controls. All of these events are fired before the control does its work at the respective cycle. For example, the Init event is fired before the page layout control does its custom OnInit actions. Here's what you can expect to happen at each of these events:

- Init—Occurs when the server control is initialized, which is the first step in its lifecycle. This is too early in the lifecycle to access any of the page layout control's properties.

- Load—Occurs when the server control is loaded into the *Page* object. At this stage, all of the page layout control's initializations are complete and changes saved into ViewState on the previous request have been applied. This is the stage to access control properties other than WebLayoutImage.

- PreRender (inherited from Control)—Occurs when the server control is about to render to its containing *Page* object and before the ContentsChanged and ConnectionChanged events are fired. Between the Load and PreRender phases, the RaisePostBackEvent method could have resulted in refreshing the page layout display. At the PreRender phase, you can assume that the WebLayoutImage property is set to its final value for the page cycle.

- Unload (inherited from Control)—Occurs when the server control is unloaded from memory. This is where you can check the state after Render.

The overview map control is similar to a map control in that it displays a particular data frame of a map server object. However, the purpose of the overview map is to provide a point of reference for the area displayed on its associated map control. A small box on the overview map represents the currently displayed area on its associated map control.

Overview Map Control Objects

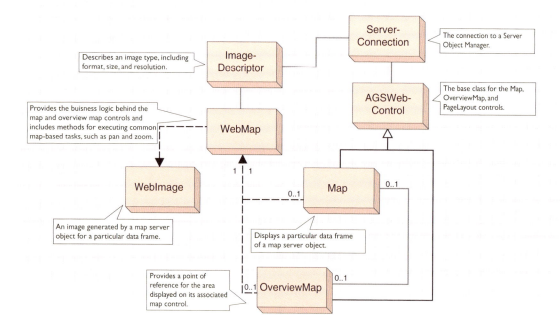

Describes an image type, including format, size, and resolution.

The connection to a Server Object Manager.

The base class for the Map, OverviewMap, and PageLayout controls.

Provides the buisness logic behind the map and overview map controls and includes methods for executing common map-based tasks, such as pan and zoom.

An image generated by a map server object for a particular data frame.

Displays a particular data frame of a map server object.

Provides a point of reference for the area displayed on its associated map control.

The *OverviewMap* class provides access to the properties and methods that govern how the control works. Like the map control, the overview map control's convenience class is *WebMap*. However, unlike the map control, the overview map control doesn't provide the same interactive capabilities as the map control. For example, you can't add a toolbar to your Web form and add tools that interact with the overview map control.

The primary purpose of the overview map control is to provide a visual reference to the area displayed on its associated map control. Thus, the geographic area displayed on an overview map control should match the geographic area displayed on its associated map control.

The *OverviewMap* class provides the user interface for displaying a data frame from a map server object on a Web form.

OverviewMap : AGSWebControl, System.Web.UI.IPostBackEventHandler, System.Web.UI.IPostBackDataHandler	Provides a point of reference for the area displayed in its associated map control.
■–■ AOIExtent: Extent	The extent of the area of interest displayed on the associated map control.
■–■ BuddyControl: System.String	The map control the overview map is linked to.
■–■ DataFrame: System.String	The data frame to display from the map server object.
■–■ Extent: Extent	The current extent of the data frame in the overview map control.
■–■ ImageUrl: System.String	The URL location of the image to display in the overview map control.
■–■ MapDescription: System.String	The string representation of the MapDescription for a particular map server object.
◄– CenterAt (System.Int32 x, System.Int32 y): System.Void	Center the area of interest to the specified screen coordinates.
◄– CreateWebMap: WebMap	Creates the WebMap object associated with this control.
◄– Draw: System.Void	Draws the map image using existing settings.
◄– LoadPostData (System.String postDataKey, System.Collections.Specialized.NameValueCollection values): System.Boolean	Indicates whether RaisePostDataChangedEvent should be called.
◄– RaisePostBackEvent (System.String eventArgument): System.Void	Handles postbacks. Raises events based on the event argument passed in. Calls the ServerToolAction component of the CurrentToolItem.
◄– RaisePostDataChangedEvent: System.Void	Enables the map control to process an event raised when a form is posted to the server.
◄– Reset: System.Void	Resets the map control to an unbound state.
◁– OverviewMapPoint: MapPointEventHandler	Occurs when an end user clicks the control using the Point client tool action.

The *OverviewMap* class contains properties, inherited from the *WebControl* class (System.Web.UI), that specify how the control looks on the Web page, such as the height, width, and border style. Typically, you'll define these properties at design time. The *CreateWebMap* method is an important method as it gives you access to the overview map control's corresponding convenience class, *WebMap*. The *BuddyControl* property links the overview map to a map control.

SETTING THE EXTENT OF THE OVERVIEW MAP CONTROL

When you add an overview map control to a Web form, by default the overview map will display the full extent of the specified data frame. Alternatively, you can set the *InitialExtent* property, inherited from *AGSWebControl*, at design time through the Visual Studio .NET interface. This property allows you to specify whether you want to display the data frame at its full extent or the last extent saved in the ArcMap document that contains it. You can also set the extent programmatically. For instance, you might add the following code to the Page_Load event to explicitly set the extent you want to display in the overview map control.

```
private void Page_Load(object sender, System.EventArgs e)
{
  if (!Page.IsPostBack)
  {
    if (Session.IsNewSession)
    {
      Extent extent = new Extent(2301791, 731414, 2310360, 740087);
      OverviewMap1.Extent = extent;
      // No need to draw control; it will automatically when page draws
    }
  }
}
```

SETTING THE AREA OF INTEREST

The small box that displays on the overview map control represents the area of interest of the associated map control. The overview map automatically updates the area of interest box whenever the extent changes in the associated map control. Thus, if an end user zooms in to an area with the map control, the new extent is automatically reflected by the area of interest box on the overview map control.

If you want, you can programmatically control the area of interest box through the *AOIExtent* property on the overview map control. For example, the following code, attached to the Click event of a server button, sets the area of interest and updates the map control to reflect the new area of interest.

```
private void Button1_Click(object sender, System.EventArgs e)
{
  Extent extent = new Extent(2301791, 731414, 2310360, 740087);
  // Change the area of interest box
  OverviewMap1.AOIExtent = extent;
  OverviewMap1.Draw();
  // Change the extent of the map control to match
  Map1.Extent = extent;
  Map1.Draw();
}
```

WHAT'S STORED IN SESSION STATE

The overview map control utilizes session state to main information about the server object referenced by the control. The names of the session variables can be found in the technical article titled 'Session variables for the .NET Web controls' in ArcGIS Developer Online (*http://arcgisdeveloperonline.esri.com*).

The following objects are stored in session state.

ServerContext (for non-pooled server objects only)

For non-pooled objects, the *ServerContext* is stored in session state so that the same instance of the server object can be used for the entire session.

The *ServerContext* is an object that needs to be released explicitly in order to release server resources. You should release this by calling *IServerContext.ReleaseContext*.

In a deeply stateful Web application that uses a server object for a long period of time, you should release the *ServerContext* at the end of the session. The code in the Session_End method in the Global.asax.cs file gets executed at the end of a session. In this method, you should loop over all the objects stored in Session and check if they are of type *IServerContext*. If they are, you should release the context.

```
IServerContext context;
for (int i = 0; i < Session.Count; i++)
{
  context = Session[i] as IServerContext;
  if (context != null)
    context.ReleaseContext();
}
```

MapDescription

MapDescription is stored in the session because it can be a large string.

ImageURL

ImageURL is stored in the session because it can be a large string.

CONTROL LIFECYCLE

In addition to what the base class *AGSWebControl* does, here's what the map control does at each point in its lifecycle.

- Instantiation—Initializes member variables with default values.

- OnInit—A copy of the *MapDescription* property is stored to check if the content changes between initialization and just before rendering. A copy of the *Host* and *ServerObject* properties is stored to check if the connection changes between initialization and just before rendering.

- LoadViewState—Uses base class implementation. Postback only.

- LoadPostData—Indicates whether RaisePostDataChangedEvent should be called. Always returns false. Postback only.

- OnLoad—Uses the base class implementation.

- RaisePostDataChangedEvent—Does nothing. Postback only.

- RaisePostBackEvent—Fires server-side events based on the postback event argument. Centers the area of interest at the clicked-on location.

- OnPreRender—The first time the control is drawn in a session, if *AutoFirstDraw* is set to true, the overview map is drawn at the extent specified by its *InitialExtent* property. Fires the ContentsChanged event if the *MapDescription* property has changed after LoadViewState for postbacks or after OnInit for nonpostbacks. Fires the ConnectionChanged event if the *Host* or *ServerObject* properties have changed after LoadViewState for postbacks or after OnInit for nonpostbacks. Registers JavaScript required for client-side action.

- Render—Renders the HTML required to draw the control. Registers additional JavaScript used to set control-specific properties.

EVENTS

Events fired on RaisePostBackEvent

The following event is fired on RaisePostBackEvent. It is only fired for postbacks.

• OverviewMapPoint

Events fired on PreRender

The following events are fired on PreRender:

• ConnectionChanged—if the *Host* and *ServerObject* have changed.

• ContentsChanged—if the *MapDescription* has changed.

Generic control events

The following events are generic to all controls. All of these events are fired before the control does its work at the respective cycle. For example, the Init event is fired before the map control does its custom OnInit actions. Here's what you can expect to happen at each of these events:

• Init—Occurs when the server control is initialized, which is the first step in its lifecycle. This is too early in the lifecycle to access any of the overview map control's properties.

• Load—Occurs when the server control is loaded into the *Page* object. At this stage, all of the overview map control's initializations are complete and changes saved into view state on the previous request have been applied. This is the stage to access control properties.

• PreRender (inherited from Control)—Occurs when the server control is about to render to its containing *Page* object and before the *ContentsChanged* and *ConnectionChanged* events are fired.

• Unload (inherited from Control)—Occurs when the server control is unloaded from memory. This is where you can check the state after Render.

The Toc control—or table of contents control—displays the contents of a map and allows an end user to turn layers on and off. The Toc control functions in the same manner as the table of contents in ArcMap. The Toc control presents a hierarchical list of data frames, layers, and symbols displayed on the map.

Toc Control Objects

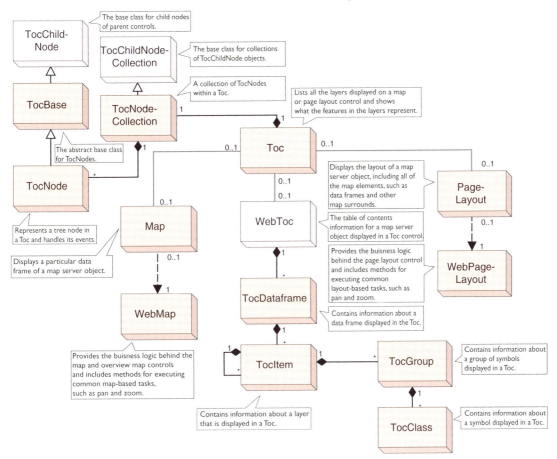

In looking at the above object model diagram, you can conceptually split the diagram in two. There is a set of classes associated with the Toc control—for example, *TocNodeCollection*—that comprises the visual component of the control on a Web form. A Toc control functions as a tree control that contains a set of nodes that you can expand or contract. These nodes represent the individual data frames and layers in a map server object. In addition, there is a set of classes associated with the Toc control's convenience class, *WebToc*. *WebToc* interacts with a map server object to obtain the table of contents information from the map server object that ultimately gets displayed in the Toc control. It uses its associated classes to store the retrieved information.

You can get a *WebToc* object from a *WebMap* or a *WebPageLayout* object. A *WebToc* is comprised of one or more *TocDataFrame* objects that correspond to data frames in a map server object. A *TocDataFrame* has one *TocItem* object for each layer in the data frame. For group layers, a *TocItem* object itself may contain one or more *TocItem* objects, one for each layer in the group. The *TocItem* object provides properties that, for example, allow you to access the layer name and layer visibility.

When you classify a layer in ArcMap, you end up with a group of symbols that might, for instance, represent population values. A *TocGroup* represents this group of symbols used for a particular layer. Each individual symbol in the group is represented with a *TocClass* object.

Toc : PostBackControl	Lists all the layers displayed on a map or page layout control and shows what the features in the layers represent.
■–■ AutoLayerVisibility: System.Boolean	When true, automatically draws the layer when the check box is checked. When false, you can process the event yourself.
■–■ AutoPostBack: System.Boolean	Whether or not to post back to the server on each interaction.
■–■ AutoSelect: System.Boolean	When true, keyboard hovering will automatically select the node.
■–■ BuddyControl: System.String	The map or page layout control the TOC is linked to.
■–■ DefaultStyle: Design.CssCollection	The CSS attributes to apply to the TOC.
■–■ Enabled: System.Boolean	Whether or not the control is enabled.
■–■ EnableViewState: System.Boolean	Whether the control automatically saves its state between round-trips.
■–■ ExpandedImageUrl: System.String	URL of the image to display next to an expanded node.
■–■ ExpandLevel: System.Int32	The minimum number of levels to expand the tree view by default.
■–■ HoverStyle: Design.CssCollection	The CSS attributes to apply when the a node is hovered over.
■–■ ImageFormat: WebImageFormat	Image format to use to generate Web images for TOC symbol swatches.
■–■ ImageUrl: System.String	URL of the image to display next to a node.
■–■ Indent: System.Int32	The number of pixels to indent each level of tree if ShowLines is false.
■– Nodes: TocNodeCollection	Gets the collection of nodes in the control.
■–■ PatchFormat: System.String	Defines how the patches displaying symbology look in the TOC.
■–■ SelectedImageUrl: System.String	URL of the image to display next to a selected node.
■–■ SelectedNodeIndex: System.String	Index of the selected node.
■–■ SelectedStyle: Design.CssCollection	The CSS attributes to apply to a selected node in the TOC.
■–■ SelectExpands: System.Boolean	Expands or collapses a node by clicking on it.
■–■ ShowAllDataFrames: System.Boolean	Indicates whether to display all data frames in a map control.
■–■ ShowLines: System.Boolean	Shows dotted lines connecting the tree hierarchy.
■–■ ShowPlus: System.Boolean	Shows +/- symbols on expandable nodes.
■–■ Target: System.String	The ID of the frame to target upon selecting a node.
■–■ UseMIMEData: System.Boolean	Streams images instead of getting files from the server.
◄– Bind: System.Void	Forces the node to bind immediately, even if it isn't expanded.
◄– Draw: System.Void	Draws the TOC.
◄– GetNodeFromIndex (System.String strIndex): TocNode	Returns the TocNode at the given index location.
◄– SetLayerVisibility (System.String map, System.Int32 layerID, System.Boolean visible): System.Void	Sets the visibility of a layer.
◁– Check: TocClickEventHandler	Occurs when a node's check box is clicked.
◁– Collapse: TocClickEventHandler	Occurs when a node is collapsed.
◁– ContentsChanged: ContentsChangeEventHandler	Occurs when the contents of the Toc control have changed.
◁– Expand: TocClickEventHandler	Occurs when a node is expanded.
◁– SelectedIndexChange: TocSelectEventHandler	Occurs when the selected node changes.

The easiest way to work with a Toc control is to associate it with a map or page layout control, through the *BuddyControl* property. Doing so binds the Toc control to the particular map or page layout control. This allows the Toc control to respond to the *ContentsChanged* event of the map or page layout control and update itself accordingly. However, you can also use a Toc control independently, without any association, by leaving the *BuddyControl* property empty. The Toc control can depict any data that has a tree structure. When using a Toc in this manner, you will need to populate the nodes in it, through the *Nodes* property. The Identify dialog box in the Map Viewer template has an example of a Toc control that has no association with a map or page layout control.

When the Toc control is associated with a map control, you can show all the data frames in the map server object or just a particular data frame. When associated with a page layout control, all data frames are shown in the Toc control.

IMMEDIATE POSTBACK UPON CHECKING A BOX

In ArcMap, when you check a layer in the table of contents, that layer draws immediately. By default, when you associate a Toc control with a map or page layout control on your Web form, you'll see the same behavior in your Web application when it's displayed in a browser. However, with Web applications, you may not want to send a request back to the server every time the end user of your Web application interacts with the Toc control, because round-trips to the server consume server resources and take time to process.

Instead, you can disable this automatic postback behavior by setting the *AutoPostBack* property to false. With this setting, when an end user checks layers on or off, the map display won't update until a postback occurs. The Toc control keeps track of any user interaction, and when a postback does occur, any changes to the layer visibility will be reflected on the map control. For instance, if you disable this property, you might add a server button to refresh the map where the click event contains the following:

```
Map1.Draw();
```

Note, however, when the Toc control is associated with a map control, you can't set the *AutoPostBack* property to false when you show all data frames in the Toc control—when *ShowAllDataFrames* is true. *AutoPostBack* must be set to true when *ShowAllDataFrames* is true.

CONTROLLING LAYER DRAWING

When you check the box next to a layer name in the Toc control, you typically expect the layer to draw. This is, in fact, the default behavior of the Toc control. However, if you need to, you can change this behavior. For example, you may not want an end user to be able to toggle the visibility of a layer.

You control the automatic drawing and clearing of layers through the *AutoLayerVisibility* property of the Toc control. When this property is true, checking a layer on or off will draw or clear the layer, respectively, from the associated map or page layout control. When you set this property to false, nothing will automatically happen when you check or uncheck a layer. In this case, you may also want to code the Toc control's *Check* event. This event is triggered whenever a layer is checked or unchecked. Additionally setting the *AutoPostBack* property to true will execute the event immediately.

The following code emulates the *AutoLayerVisibility* property and draws layers in a map control based on whether or not they are checked in the Toc control. You can easily code the event to do other things.

```
private void Toc1_Check(object sender,
ESRI.ArcGIS.Server.WebControls.TocClickEventArgs args)
{
   TocNode node = Toc1.GetNodeFromIndex(args.NodeIndex);
   using (WebMap webmap = Map1.CreateWebMap())
   {
     IMapDescription mapdesc = webmap.MapDescription;
     ILayerDescriptions layerdesc = mapdesc.LayerDescriptions;
     ILayerDescription onelayerdesc = layerdesc.get_Element(node.LayerID);
     onelayerdesc.Visible = node.Checked;
     webmap.Refresh();
   }
}
```

From the *TocClickEventArgs*, you can get the index of the node that was checked or unchecked and use this index to get a *TocNode* object. With the *TocNode* object, you can obtain the *LayerID* of the particular layer the node represents. With the *LayerID*, you can find the *LayerDescription* object and change its visibility. Notice that the *MapDescription* is obtained from the *WebMap* object. This is the *MapDescription* that is associated with the map control. Thus, the changes you make to this *MapDescription* are reflected on the map control.

If you have both *AutoLayerVisibility* set to true and have coded the *Check* event, the code in the *Check* event executes first. Thus, you can write custom code, yet still allow the control to manage drawing the layers.

WORKING WITH THE TOC TREE

The following code samples illustrate different ways to work with the components of the Toc control. The samples show you some of the things you may want to do when building your own Web applications.

Navigating through the Toc control to find a data frame

The following code gets the name of the data frame the checked layer is contained in. The code is placed in the *Check* event on the Toc control.

```
private void Toc1_Check(object sender,
ESRI.ArcGIS.Server.WebControls.TocClickEventArgs args)
{
   TocNode node = Toc1.GetNodeFromIndex(args.NodeIndex);
   string layername = node.Text;
   // Get the parent of the current node.
   node = node.Parent as TocNode;
   // Use while loop to handle grouped layers
   while (node.Parent is TocNode)
   {
     node = node.Parent as TocNode;
   }
   string dataframename = node.Text;
}
```

Removing check boxes from the Toc control

In some Web applications, you may not want your end users to toggle the visibility of the layers on the map. The following function hides the check boxes on the Toc control. The function calls itself recursively because the Toc control can have many tree levels.

```
private void RemoveCheckbox(TocNodeCollection nodes)
{
   foreach (TocNode node in nodes)
   {
      node.CheckBox = false;
      if (node.Nodes != null) RemoveCheckbox(node.Nodes);
   }
}
```

The function should be called from the *ContentsChanged* event on the Toc control, as shown below.

```
private void Toc1_ContentsChanged(object sender, System.EventArgs e)
{
   RemoveCheckbox(Toc1.Nodes);
}
```

The reason you need to hide the check boxes on every postback is because the Toc control doesn't save this information between postbacks. Thus, by adding code to the *ContentsChanged* event, you can ensure the check boxes remain hidden.

Manually populating a Toc control

The *BuddyControl* property of the Toc allows you to link the control to a map or page layout control. In doing so, you get a lot of built-in functionality, such as the automatic drawing of layers. However, there may be times when you want to manage the Toc control yourself.

The sample code below populates a Toc control with the layer names in the current data frame. When implementing this code, you should not set the *BuddyControl* property of the Toc control.

With a Toc control and a map control on a Web form, add the following code to the Page_Load event:

```
private void Page_Load(object sender, System.EventArgs e)
{
   if ( !Page.IsPostBack )
   {
      using (WebMap webmap = Map1.CreateWebMap())
      {
         WebToc webtoc = webmap.WebToc(WebImageFormat.PNG,false,true, null);
         TocDataFrame dataframe = webtoc.Find(Map1.DataFrame);
         TocNode parent = new TocNode();
         parent.Expanded = true;
         parent.Text = dataframe.Name;
         TocNode node = null;
```

```
// Create a node in the Toc for each layer
foreach (TocItem item in dataframe)
  {
    node = new TocNode();
    node.Text = item.Name;
    parent.Nodes.Add(node);
    // If the layer is a group layer, expand it
    if (item.TocItemCount > 0)
      {
        node.Expandable = ExpandableValue.Always;
         node.Expanded = true;
        GetGroupLayerItems (item, node);
      }
  }
Toc1.Nodes.Add(parent);
  }
 }
}
```

Code the following function. This function gets called from the above Page_Load event and populates the Toc control with group layers. It calls itself recursively because a group layer can contain other group layers.

```
private void GetGroupLayerItems(TocItem item, TocNode parent)
{
  // This function adds layers in a group layer to the Toc
  TocNode node = null;
  int i;
  for (i = 0; i < item.TocItemCount; i++)
  {
    node = new TocNode();
    node.Text = item.GetTocItem(i).Name;
    parent.Nodes.Add(node);
    if (item.GetTocItem(i).TocItemCount > 0)
      {
        node.Expandable = ExpandableValue.Always;
         node.Expanded = true;
        GetGroupLayerItems (item.GetTocItem(i), node);
      }
  }
}
```

WHAT'S STORED IN SESSION

The Toc control utilizes session state to maintain information about the control. The following are stored in session state.

- A HashTable with the list of expanded nodes.

- URL properties: *ImageUrl*, *SelectedImageUrl*, *ExpandedImageUrl*, and *SystemImagesPath* are stored in session because they can potentially be long strings.

- *Nodes* and the style properties, *DefaultStyle*, *HoverStyle*, and *SelectedStyle*, are stored in session state, as they are not simple data types and can be large.

CONTROL LIFECYCLE

Here's what Toc control does at each point in its lifecycle.

- Instantiation—Initializes member variables with default values.

- OnInit—If the *BuddyControl* property is set, the Toc control starts listening to the buddy control's *ContentsChanged* event.

- TrackViewState—Begins tracking view-state changes for complex properties such as *Nodes*, *TocNodeTypes*, *DefaultStyle*, *HoverStyle*, and *SelectedStyle*.

- LoadViewState—Restores view-state information of complex properties such as *Nodes*, *TocNodeTypes*, *DefaultStyle*, *HoverStyle*, and *SelectedStyle*, to what they were at the end of the previous page request. Postback only.

- LoadPostData—At this phase, the control looks at the posted data to determine which node was selected, expanded, collapsed, checked, or unchecked and changes the node accordingly. Postback only.

- OnLoad—*Toc.Draw* is called if it is not a postback.

- RaisePostDataChangedEvent—This IPostBackDataHandler method is implemented to raise events signaling that something about the control has changed. The Toc control raises the *SelectedIndexChanged, Expand, Collapse,* and *Check* events at this phase. If a check box has been checked or unchecked, the visibility of a layer is turned on or off if *AutoLayerVisibility* is set to True. Postback only.

- RaisePostBackEvent—As all the Toc control events are change events, nothing happens in this phase. Postback only.

- OnPreRender—Checks if *HoverNodeIndex* and *SelectedNodeIndex* are valid.

- SaveViewState—Saves view-state information of complex properties such as *Nodes*, *TocNodeTypes*, *DefaultStyle*, *HoverStyle*, and *SelectedStyle*.

- Render—Renders the HTML required to draw the control. RichControl.Render calls RenderUpLevelPath if the browser is at Internet Explorer 5.5 or higher. It calls RenderDownLevelPath if browser is different or the Internet Explorer version is lower than 5.5.

EVENTS

Events fired on RaisePostDataChangedEvent

The following events are fired on RaisePostDataChangedEvent.

- Check—This event is fired when a node in the Toc control is checked or unchecked. In a bound Toc control, only nodes associated with layers have a check box next to them.

- Collapse—This event is fired when a node is collapsed to hide the nodes contained in it.

- Expand—This event is fired when a node is expanded to expose the nodes contained in it.

- SelectedIndexChange—This event is fired when node is selected and the node is different from the previously selected node.

Event fired on PreRender

The following event is fired on PreRender:

• ContentsChanged—The Toc control fires its ContentsChanged event in the OnPreRender method when the user has refreshed the Toc control. When the Toc control is refreshed in response to its buddy control's ContentsChanged event, it fires its ContentsChanged event after refreshing itself. In both of these scenarios, the ContentsChanged event will be fired in the PreRender phase of a page cycle.

The toolbar control displays a toolbar on a Web form. A toolbar allows you to associate a set of actions with one or more map controls or one or more page layout controls. You can place as many toolbars as you like on your Web form to organize the tools in your Web application.

Toolbar Control Objects

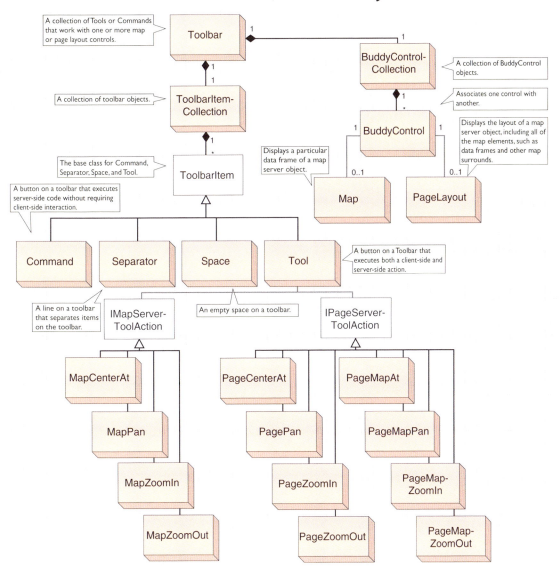

A collection of Tools or Commands that work with one or more map or page layout controls.

Toolbar

A collection of BuddyControl objects.

BuddyControl-Collection

A collection of toolbar objects.

ToolbarItem-Collection

Associates one control with another.

BuddyControl

Displays the layout of a map server object, including all of the map elements, such as data frames and other map surrounds.

The base class for Command, Separator, Space, and Tool.

ToolbarItem

Displays a particular data frame of a map server object.

A button on a toolbar that executes server-side code without requiring client-side interaction.

Map

PageLayout

Command **Separator** **Space** **Tool**

A button on a Toolbar that executes both a client-side and server-side action.

A line on a toolbar that separates items on the toolbar.

IMapServer-ToolAction

An empty space on a toolbar.

IPageServer-ToolAction

MapCenterAt **PageCenterAt** **PageMapAt**

MapPan **PagePan** **PageMapPan**

MapZoomIn **PageZoomIn** **PageMap-ZoomIn**

MapZoomOut **PageZoomOut** **PageMap-ZoomOut**

Unlike some of the other controls in the .NET ADF, a toolbar doesn't have any associated convenience class. This is because the primary purpose of a toolbar is to provide a user interface for a set of actions that can be applied to a map or page layout control. All of the properties and methods that make the toolbar work are built into the *Toolbar* class itself.

Toolbar: System.Web.UI.WebControls.WebControl, System.Web.UI.IPostBackDataHandler, System.Web.UI.IPostBackEventHandler	*A collection of Tools and Commands that work with one or more map or page layout controls.*
Alignment: Alignment	*Specifies how text and images should be aligned.*
BackColor: System.Drawing.Color	*The background color of the control. This property is not used.*
BuddyControls: BuddyControlCollection	*The set of page layout or map controls that this toolbar acts on.*
BuddyControlType: BuddyControlType	*The type of the buddy controls associated with this control, either map or page layout.*
CurrentTool: System.String	*The selected tool on this toolbar.*
Enabled: System.Boolean	*Indicates whether the Web control is enabled. This property is not used.*
EnableViewState: System.Boolean	*Indicates whether the server control persists its view state, and the view state of any child controls it contains, to the requesting client.*
Font: System.Web.UI.WebControls.FontInfo	*The font used in the control. This property is not used.*
ForeColor: System.Drawing.Color	*The foreground color of the control. This property is not used.*
Group: System.String	*The group to which this toolbar belongs.*
Orientation: Orientation	*The orientation of the toolbar, either horizontal or vertical.*
TextPosition: TextPosition	*The text position relative to the image on a button.*
ToolbarItemDefaultStyle: System.Web.UI.WebControls.Style	*The style applied to make Commands and Tools when they are not selected, disabled, or hovered upon.*
ToolbarItemDisabledStyle: System.Web.UI.WebControls.Style	*The style applied to Commands and Tools when they are disabled.*
ToolbarItemHoverStyle: System.Web.UI.WebControls.Style	*The style applied to Commands and Tools when they are hovered upon.*
ToolbarItems: ToolbarItemCollection	*The collection of ToolbarItem objects on the Toolbar.*
ToolbarItemSelectedStyle: System.Web.UI.WebControls.Style	*The style applied to Commands and Tools when they are selected.*
ToolbarStyle: ToolbarStyle	*The style of the toolbar; whether it displays both text and image for a command or tool or only one of the two.*
ToolTip: System.String	*The ToolTip used in the control. This property is not used.*
LoadPostData (System.String postDataKey, System.Collections.Specialized.NameValueCollection values): System.Boolean	*Gets the current tool from the page for postbacks.*
RaisePostBackEvent (System.String eventArgument): System.Void	*Handles postbacks and raises events based on the event argument.*
RaisePostDataChangedEvent: System.Void	*This method is not used.*
CommandClick: ToolbarCommandClickEventHandler	*Occurs when a Command on the Toolbar has been clicked.*

ASSOCIATING A TOOLBAR WITH A CONTROL

You specify which map or page layout controls a toolbar will work with through the toolbar control's *BuddyControls* property. This property allows you to specify one or more map or page layout controls the toolbar will operate on. However, a toolbar can only have buddy controls of one type, either map controls or page layout controls, but not both. Use the *BuddyControlType* property to specify which type the toolbar will work with.

ITEMS ON A TOOLBAR

A toolbar control is composed of many *ToolbarItem* objects. A *ToolbarItem* can be one of several types:

* Tool—A *Tool* executes a client-side action to set the stage for interaction with a map or page layout control. When the interaction completes, an action

executes on the server side. For example, the end user of a Zoom In tool would drag a box over the map to identify an area to zoom in to. Subsequently, the server-side action executes to zoom in to the specified area.

- Command —A *Command* simply executes server-side code and requires no end user interaction. For example, the end user would simply click a full extent command on a toolbar and would not need to interact with the map.

- Space—A *Space* object allows you to make room on your Toolbar for other HTML controls or Web controls. Using the *Size* property of a *Space* object, you can specify how big a space you want on your toolbar. A size of 1 translates to a space of 10 pixels. However, the size of a space can vary with the size of a toolbar control. If you shrink a toolbar control, the size of a space will decrease before other items get truncated. A *Space* does not execute any action.

- Separator—A *Separator* object is a line between items on the toolbar control that allows you to form groups of related items on the toolbar. A *Separator* does not execute any action.

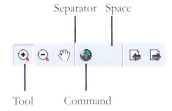

Separator Space

Tool Command

The *ToolBarItems* property on the toolbar control provides access to the *ToolBarItemCollection* class, which is a collection of *ToolBarItem* objects. These are the *Tool*, *Command*, *Separator*, and *Space* objects on the toolbar. You can programmatically add items to the toolbar with the *ToolbarItemCollection::Add* method. In Visual Studio .NET, clicking on the *ToolbarItems* property displays the ToolbarItem Collection Editor.

The *CurrentTool* property of the toolbar identifies the active tool. This is the tool that will be active when the page is first drawn. At runtime, clicking a tool will make it the active tool. There is only one current tool on a toolbar or a set of toolbars that have been grouped together.

GROUPING TOOLBARS

There may be times when you want to have more than one toolbar on your Web form that works with a map or page layout control. For example, you may want to use the toolbar itself as a means for organizing related functions, such as having all the map navigation tools on one toolbar. You might also want to place toolbars in different locations on your Web form, for example, along the top and side of a particular map or page layout control. In both cases, you want the set of toolbars you define to work with the same control. The *Group* property allows one or more toolbars to work together and essentially function as one. Thus, at any given time, there is only one currently active tool on the set of grouped toolbars.

You organize toolbars into groups by using the same value for the *Group* property on each toolbar you want to participate in the group. Setting the *CurrentTool* property on one toolbar will clear the *CurrentTool* property of the other toolbars in the group. When creating a group, you must ensure that you provide a unique name for each *ToolBarItem* across all the grouped toolbars. You are also responsible for ensuring that the *BuddyControls* property is the same for of all the toolbars in the group.

Configuring tools on a toolbar

As described above, a *Tool* performs two actions: a client-side action that enables an end user to interact with the map or page layout control and a server-side action that executes after the end user has finished interacting with the map or page layout control. The ToolbarItems Collection Editor allows you to specify both the client-side and server-side actions thorough the *ClientToolAction*, *ServerToolActionAssembly*, and *ServerToolActionClass* properties.

Client-side actions are controlled by JavaScript functions that execute on the client, without any requests being sent to the server. The .NET ADF includes prewritten JavaScript code for common interactions such as drawing a box. The map and page layout controls each have their own JavaScript functions. For a listing of these JavaScript functions available for each control, see the 'Map control' and 'Page Layout control' sections earlier in this chapter.

You can define a *Tool* (also *Command*, *Space*, and *Separator*) through the ToolbarItem Collection Editor.

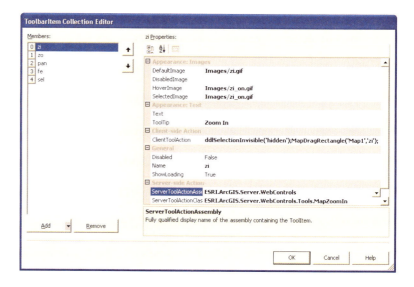

The editor allows you to specify the client-side and server side actions. Use the *ClientToolAction* property to specify the client-side action. With this property, you'll see the list of prewritten JavaScript functions and also a custom option that allows you to write your own function.

Choosing the custom JavaScript option displays a dialog box into which you can enter your JavaScript code. You might simply call your own JavaScript function before calling one of the prewritten JavaScript functions. For example, the following is from the Zoom In tool on the Buffer Selection template.

In a more complex scenario, you might handle all of the client interaction with your custom JavaScript. In this case, your custom *ClientToolAction* JavaScript should set up a function that will be called at the onmousedown event of an HTML element on the form. You would embed the JavaScripts required for your custom *ClientToolAction* into the page itself.

When the interaction finishes, another JavaScript function should post back either to the buddy control or to the toolbar control by calling the __doPostBack() function. One of the arguments in the __doPostBack() functions is the name of the control to post back to. The other is the event argument. If the postback goes to the toolbar control, it will call the *ServerAction* method of the *ServerToolActionClass* with no arguments. If the postback goes to the buddy control, the buddy control will call the *ServerAction* method of the *ServerToolActionClass*. This will happen only if the event argument is the string equivalent of a *PageClientToolAction* enumeration value for page layout controls and if the event argument is the string equivalent of a *MapClientToolAction* enumeration value for map controls.

See Chapter 7, 'Developer scenarios' for examples of tool actions.

Use the prewritten JavaScript functions as an example when writing your own custom *ClientToolActions*. You can find these in the ArcGIS Server installation location under: <ArcGIS Install Location>\DotNet\VirtualRootDir\ aspnet_client\esri_arcgis_server_webcontrols\9_0\JavaScript.

After configuring the client-side action, you need to specify the server-side action. The server-side action executes after the JavaScript has posted back to the buddy control or toolbar control. Two properties control the server-side action: *ServerToolActionAssembly* and *ServerToolActionClass*. The *ServerToolActionAssembly* is the fully qualified display name of the assembly containing the *ToolItem*. The *ServerToolActionClass* property is the fully qualified name of a class. This class is expected to implement *IPageServerToolAction* for toolbars with *BuddyControlType* set to PageLayout; it is expected to implement *IMapServerToolAction* for toolbars with *BuddyControlType* set to Map.

The .NET ADF provides common server-side actions for tools, such as Zoom In, Zoom Out, and Pan. You can also write your own class to handle server-side actions. You would do this by implementing either *IPageServerToolAction* or *IMapServerToolAction* depending on the *BuddyControlType* of the toolbar. If you add a class to your current project you'll have to type in the assembly name for the *ServerToolActionAssembly* property in the ToolbarItem Collection Editor the first time you use it. For the class to be available in the editor, you'll have to build the project after adding the class.

Configuring commands on a toolbar

Commands on a toolbar control execute a server-side action only. You write the code that executes on the server in the *CommandClick* event for a toolbar control. The argument passed into the event handler is of type *ToolbarCommandClickEventArgs*. This argument has a property called *CommandName* through which you can find out which *Command* an end user clicked. The following code shows how you can use a switch statement to execute the appropriate server-side action based on the command clicked.

```
private void Toolbar1_CommandClick(object sender,
  ESRI.ArcGIS.Server.WebControls.ToolbarCommandClickEventArgs e)
{
  switch(e.CommandName)
  {
    case "FullExtent":
      FullExtent();
      break;
    case "ZoomBack":
      ZoomBack();
      break;
    case "ZoomNext":
      ZoomForward();
      break;
    }
}
```

CONTROLLING THE LOOK OF THE TOOLBAR

The *Orientation* property allows you to specify whether the toolbar control is displayed horizontally or vertically. The *ToolbarStyle* allows you to specify whether or not the toolbar will have text and images. You can choose from the following three styles: ImageAndText, TextOnly, and ImageOnly. Use the *TextPosition* and *Alignment* properties to position the text relative to the image.

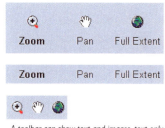

A toolbar can show text and images, text only, or images only.

You control the look of the toolbar with the following four properties. Each style property allows you to set up the background color, font, and so on.

- *ToolbarItemDefaultStyle*—the style applied to *Commands* and *Tools* when they are not disabled; hovered upon; or, in the case of *Tools*, selected.

- *ToolbarItemHoverStyle*—the style applied to *Commands* and *Tools* when the mouse is held over them.

- *ToolbarItemDisabledStyle*—the style applied to *Commands* and *Tools* when their *Disabled* property is set to True.

- *ToolbarItemSelectedStyle*—the style applied to the *CurrentTool* of the toolbar. It's also applied to a *Command* from the time it gets clicked until the page refreshes.

WHAT'S STORED IN SESSION

The following complex properties: *ToolbarItems*, *BuddyControls*, *ToolbarItemDefaultStyle*, *ToolbarItemHoverStyle*, *ToolbarItemDisabledStyle*, and *ToolbarItemSelectedStyle* are stored in session state if *EnableSessionState* is set to True on the Page. Otherwise, they are stored in view state.

CONTROL LIFECYCLE

Here's what the toolbar control does at each point in its lifecycle.

- Instantiation—Initializes member variables with default values.

- OnInit—Toolbox adds any of its *ToolbarItems* that are *Tool* objects to the *ToolItems* collection of each of the buddy controls.

- TrackViewState—Begins tracking view-state changes for complex properties, such as *ToolbarItems*, *BuddyControls*, *ToolbarItemDefaultStyle*, *ToolbarItemHoverStyle*, *ToolbarItemDisabledStyle*, and *ToolbarItemSelectedStyle*.

- LoadViewState—Restores view-state information of complex properties such as *ToolbarItems*, *BuddyControls*, *ToolbarItemDefaultStyle*, *ToolbarItemHoverStyle*, *ToolbarItemDisabledStyle*, and *ToolbarItemSelectedStyle*, to what they were at the end of the previous page request. Postback only.

- LoadPostData—Indicates whether RaisePostDataChangedEvent should be called. Always returns false. Retrieves the selected tool from the hidden field on the Web form used to store it and sets it up as the *CurrentTool* property. Postback only.

- RaisePostDataChangedEvent—Does nothing. Postback only.

- RaisePostBackEvent—Fires server-side events based on the postback event argument. If the event argument is the name of a *Command*, the toolbar control fires the *CommandClick* event passing the name of the *Command* as part of the argument. If the event argument is the name of a *Tool*, the toolbar control calls the Tool's *ServerToolActionClass ServerAction* method with a null argument. Postback only.

- OnPreRender—Sets up the *CurrentTool* of the toolbar as the *CurrentToolItem* for all of the buddy controls. Registers JavaScript required for client-side behavior.

- SaveViewState—Saves view-state information of complex properties such as *ToolbarItems*, *BuddyControls*, *ToolbarItemDefaultStyle*, *ToolbarItemHoverStyle*, *ToolbarItemDisabledStyle*, and *ToolbarItemSelectedStyle*.

- Render—Renders the HTML required to draw the toolbar control. Registers additional JavaScript used to set control specific properties. Registers a hidden field that will keep track of the current *Tool*.

EVENTS

Event fired on RaisePostBackEvent

If the event argument on a postback is the name of a *Command*, the *CommandClick* event is fired.

Generic control events

The following events are generic to all controls. All of these events are fired before the control does its work at the respective cycle. For example, the Init event is fired before the toolbar control does its custom OnInit actions. Here's what you can expect to happen at each of these events:

- Init—Occurs when the server control is initialized, which is the first step in its lifecycle. This is too early in the lifecycle to access any of the toolbar control's properties.

- Load—Occurs when the server control is loaded into the Page object. At this stage, all of the toolbar control's initializations are complete, and changes saved into view state on the previous request have been applied. This is the stage to access control properties.

- PreRender—Occurs when the server control is about to render to its containing Page object and before the *CurrentToolItem* properties of the buddy controls are set.

- Unload—Occurs when the server control is unloaded from memory. This is where you can check the state after Render.

In order to access a GIS server and the server objects running on it, a Web application must connect to the server using an operating system account that has access to the GIS server. The impersonation control helps you set the particular identity, or user account, your Web application will impersonate. The impersonation control accepts a username, password, and a domain or machine name and encrypts this information into the Web application. The control dynamically changes the identity of the Web application from the logged on user account to the impersonated user account.

Impersonation Control Object

Impersonation — Incorporates the appropriate security credentials for accessing a GIS server onto a Web application.

The impersonation control provides a simple mechanism for allowing your Web application to access a GIS server. ASP.NET provides other ways to impersonate a user identity and to manage security in general. A discussion of those methods is beyond the scope of this book. If you have difficulty implementing impersonation in your Web application, you might find the following URL useful:

http://support.microsoft.com/?id=306158

USING THE IMPERSONATION CONTROL

The impersonation control utilizes the Win32 API LogonUser to dynamically perform the impersonation. The LogonUser function verifies that the specific user account is a valid account on the system. If the account is valid, the Web application will successfully impersonate the specified user account. However, that user account must also have access to the GIS server for the Web application to run properly. You need to ensure that you've added this account to the ArcGIS Server users group, called agsusers. For more information on adding a user account to the agsusers group, see the section titled 'Setting up and connecting to a GIS server' in Chapter 3, 'Administering an ArcGIS Server'.

Impersonation : System.Web.UI.WebControls.WebControls	Incorporates the appropriate security credentials for accessing a GIS server into a Web application.
■— CurrentUser: System.String	The current Windows user as returned by the System.Security.Principle.WindowsIdentity object.
■—■ Identity: System.String	The user to impersonate who has access to the GIS server.
■— IsImpersonating: System.Boolean	Returns true if impersonating, false otherwise.
◄— IdentityObject (Identity id): Identity	Sets the identity of a user through the Identity class.
◄— Impersonate: System.Boolean	Dynamically modifies the current windows user with the value in the Identity property.
◄— UndoImpersonation: System.Void	Return to the original Windows identity.

The impersonation control encrypts the identity and stores it in session state on the server for the duration of the session. When creating a Web application, each page of the application should contain an impersonation control. However, only the main page requires that you set a valid identity. This identity is shared by other pages and the controls on those pages.

You can alternatively implement impersonation programmatically in your Web application. For example, suppose you want your Web application to prompt the end user for a username and password. Then, you can use the information he/she enters to set up impersonation. For example, you can add the following code to the Global.asax.cs file of your Web application to enable impersonation on every request.

At the top of the Global.asax.cs file, define the following variable:

```
ESRI.ArcGIS.Server.WebControls.Impersonation impersonate = null;
```

Then code the Application_BeginRequest and Application_EndRequest events, substituting the appropriate values for the username, domain name, and password.

```
protected void Application_BeginRequest(Object sender, EventArgs e)
{
   // Enable impersonation at the beginning of a request.
   impersonate = new ESRI.ArcGIS.Server.WebControls.Impersonation();
   ESRI.ArcGIS.Server.WebControls.Identity id =
      new ESRI.ArcGIS.Server.WebControls.Identity();
   id.UserName = "username";
   id.Domain = "domainname";
   id.Password = "password";
   impersonate.IdentityObject(id);
   impersonate.Impersonate();
}

protected void Application_EndRequest(Object sender, EventArgs e)
{
   // Undo impersonation at the end of a request.
   impersonate.UndoImpersonation();
}
```

In your code, you probably won't hard code the username and password, as shown above, but instead you'll obtain them from an end user.

WHAT'S STORED IN SESSION STATE

The impersonation control stores the *Identity* in application state so that it can be shared between other Web controls in the .NET ADF. Nothing is stored in view state.

CONTROL LIFECYCLE

Here's what the control does at each point in its lifecycle.

- Instantiation—Initializes member variables with default values.

- OnInit—Attempts to log on as the specified identity.

- TrackViewState—Uses the base class implementation.

- LoadViewState—Uses the base class implementation. Postback only.

- LoadPostData—Does nothing.

- OnLoad—Uses the base class implementation.

- RaisePostDataChangedEvent—Does nothing.

- RaisePostBackEvent—Does nothing.

- OnPreRender—Does nothing.

- SaveViewState—Uses the base class implementation.

- Render—Does nothing.

- Unload—If impersonating a validated identity, removes the impersonation and reestablishes the default user.

The North arrow control displays a North arrow on the Web form. A North arrow indicates the orientation of the data frame. The North arrow will point in the north direction for the data frame displayed in the map control.

North Arrow Control Objects

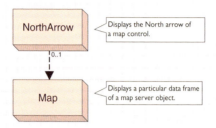

When you add a North arrow to your Web form, you associate it with a particular map control with the *BuddyControl* property. The North arrow control implements all the methods it needs; thus, there is no associated convenience class.

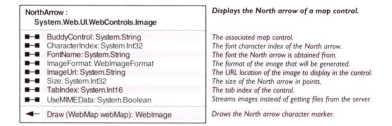

The default *FontName* used by Visual Studio .NET is ESRI North. This font contains the North arrows used in ArcGIS Desktop. Within this font, you specify the particular *CharacterIndex* of the North arrow you want to use. The default *CharacterIndex* is 177. In the ESRI North font, character indexes range from 33 to 125 and 161 to 218.

You can use the Microsoft Windows Character Map to view the fonts available to you.

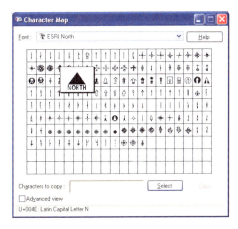

To obtain an appropriate value for *CharacterIndex*, convert the hexadecimal number—displayed in the lower left corner of the dialog box—to decimal and use the decimal value as the *CharacterIndex*. In the figure above, the hexadecimal value of 4E represents a decimal value of 78. You would use 78 as the *CharacterIndex*.

Alternatively, you can use ArcMap to view character indexes directly. Just add a North arrow in ArcMap, then display the properties of the North arrow.

Click the Character button on the dialog box to display the symbols. In the lower left corner, you'll see character index. In the figure below, the *CharacterIndex* is 78.

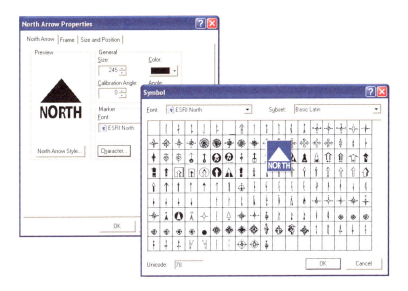

WHAT'S STORED IN SESSION STATE

The North arrow control does not utilize session state unless *UseMIMEData* is set to true. In this case, the *MIMEData* object is stored in session state.

CONTROL LIFECYCLE

Here's what the control does at each point in its lifecycle.

- Instantiation—Initializes member variables with default values.

- OnInit—If the North arrow control is associated with a map control, it listens for the DataFrameChanged event.

- TrackViewState—Uses the base class implementation.

- LoadViewState—Uses the base class implementation. Postback only.

- LoadPostData—Does nothing.

- OnLoad—Uses the base class implementation.

- RaisePostDataChangedEvent—Does nothing.

- RaisePostBackEvent—Does nothing.

- OnPreRender—Draws the North arrow if necessary.

- SaveViewState—Uses the base class implementation.

- Render—Writes the <img> HTML tag.

- Unload—Does nothing.

The scale bar control displays the scale bar of a particular data frame of a map server object. The scale bar provides a visual indication of the sizes of features and the distances between features shown on a data frame. A scale bar is divided into parts and labeled with its ground length, usually in multiples of map units such as tens of kilometers or hundreds of miles. If the scale of the data frame changes, for example, when an end user zooms in, the scale bar control updates accordingly.

Scale Bar Control Objects

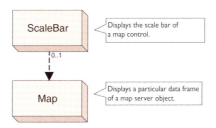

The scale bar control has no convenience class, as all the necessary functionality is encapsulated within the control itself. When you add a scale bar to your Web application, you must set the *BuddyControl* property to associate the scale bar with a particular map control.

ScaleBar : System.Web.UI.WebControls.Image	**Displays the scale bar of a map control.**
■─■ BarColor: System.Drawing.Color	*The bar color of the scale bar image.*
■─■ BarFont: System.Drawing.Font	*The font used for the display of text labels.*
■─■ BarHeight: System.Double	*The scale bar height in points.*
■─■ BarStyle: ScaleBarStyle	*The style of the scale bar.*
■─■ BarUnits: Units	*The measurement units (e.g., miles) of the scale bar.*
■─■ BuddyControl: System.String	*The associated map control.*
■─■ Divisions: System.Int32	*The total number of divisions.*
■─■ DivisionsBeforeZero: System.Int32	*The number of divisions to the left of the zero.*
■─■ ImageFormat: WebImageFormat	*The format of the image that will be generated.*
■─■ ImageUrl: System.String	*The URL location of the image to display in the control.*
■─■ Subdivisions: System.Int16	*The number of subdivisions per major division.*
■─■ TabIndex: System.Int16	*The tab index of the control.*
■─■ UseMIMEData: System.Boolean	*Streams images instead of getting files from the server.*
◀─ Draw (WebMap webMap): WebImage	*Draws the scale bar.*

The properties of the scale bar control allow you to control the visual aspects of the display. You can set the color with *BarColor* property and the font with *BarFont* property. Use *BarStyle* to choose the particular style of scale bar you want to display and *BarUnits* to set the map units.

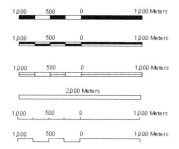

The BarStyle property sets the style of the scale bar. From top to bottom: Alternating, DoubleAlternating, Hollow, SingleDivision, ScaleLine, and SteppedScaleLine.

The scale bar is divided into segments that measure distances. You can control how many segments will appear with the *Divisions* property. For example, the scale bars in the figure to the left have two major divisions. The first division is divided into four subdivisions. The number of subdivisions for the first division is controlled with the *Subdivisions* property.

WHAT'S STORED IN SESSION STATE

The scale bar control does not utilize session state unless *UseMimeData* is set to true. In this case, the *MIMEData* object is stored in session state.

CONTROL LIFECYCLE

Here's what the control does at each point in its lifecycle.

- Instantiation—Initializes member variables with default values.
- OnInit—Listens for the DataFrameChanged or ContentsChanged events on the associated map control.
- TrackViewState—Uses the base class implementation.
- LoadViewState—Uses the base class implementation. Postback only.
- LoadPostData—Does nothing.
- OnLoad—Uses the base class implementation.
- RaisePostDataChangedEvent—Does nothing.
- RaisePostBackEvent—Does nothing.
- OnPreRender—Draws the scale bar if necessary.
- SaveViewState—Uses the base class implementation.
- Render—Writes the <img> HTML tag.
- Unload—Does nothing.

The *GeocodeConnection* component does not present a visual interface to work with in the Visual Studio .NET designer; rather, it provides a server component similar to that of a SQL data provider.

Unlike the other .NET ADF controls, this component derives from System.ComponentModel.Component, not System.Web.UI.WebControls.WebControl.

GeocodeConnection Component Objects

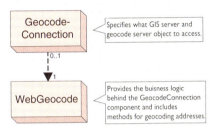

The *GeocodeConnection* component has properties to define the *Host* and *ServerObject* and also a few properties that define how addresses will be matched—*MininumCandidateScore*, *MinimumMatchScore*, and *ShowAllCandidates*. The *GeocodeConnection* component's convenience class is *WebGeocode*. You can create a *WebGeocode* object through the *GeocodeConnection::CreateWebGeocode* method or directly through the *WebGeocode* class itself. Through *WebGeocode*, you can find addresses.

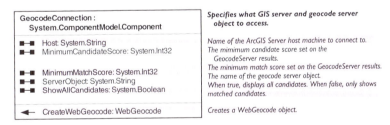

The *GeocodeConnection* object always releases the server context when you dispose of the object (that is, at the end of a using block). This is true for both pooled and non-pooled objects.

FINDING AN ADDRESS

The primary purpose of the *GeocodeConnection* component is to allow you to match addresses using a geocode server object. The example below shows you how to use the *FindAddressCandidates* and *GetCandidateFields* methods on a *WebGeocode* object. This example utilizes the following controls: a *GeocodeConnection* component, an impersonation control, two data grids, and a button.

You will need the following using statements in your code to run the code below.

```
using ESRI.ArcGIS.Server;
using ESRI.ArcGIS.Server.WebControls;
using ESRI.ArcGIS.esriSystem;
```

The following code is attached to the button control's Click event. Although the address information is hard coded into this example, you can easily add user interface components, such as a text input control, to the Web form to prompt the end user for these values.

```
private void Button1_Click(object sender, System.EventArgs e)
{
  geocodeConnection1.Host = "gisserver";
  geocodeConnection1.ServerObject = "RedlandsGeo";
  geocodeConnection1.MinimumCandidateScore = 60;
  geocodeConnection1.MinimumMatchScore = 10;
  geocodeConnection1.ShowAllCandidates = true;
  using (WebGeocode geocodeconn = geocodeConnection1.CreateWebGeocode())
  {
    IServerContext context = geocodeconn.ServerContext;
    IPropertySet pset = context.CreateObject("esriSystem.PropertySet")
      as IPropertySet;
    pset.SetProperty("Street", "380 New York Street");
    pset.SetProperty("Zone", "92373");
    System.Data.DataSet dataset1 =
      geocodeconn.FindAddressCandidates(pset, null, false, false);
    if (dataset1 != null)
     {
       DataGrid1.DataSource = dataset1;
       DataGrid1.DataBind();
     }
    System.Data.DataSet dataset2 = geocodeconn.GetCandidateFields();
    if (dataset2 != null)
     {
       DataGrid2.DataSource = dataset2;
       DataGrid2.DataBind();
     }
  }
}
```

MAPPING AN ADDRESS

Suppose you want to draw an address location on a map. In your Web application, you'll utilize both a geocode server object and a map server object. The geocode server object finds the location of the address you want to draw on the map server object. However, because a geocode server object and a map server object run in separate server contexts, you need to pass the coordinates of the geocoded location to the map server object.

The sample code below geocodes an address and draws a marker symbol on the map at the address location. To create this sample, add the following controls to a Web form: a map control, a *GeocodeConnection* component, an impersonation control, and a button. Set a map server object on the map control and a geocode server object on the *GeocodeConnection* component. The map and geocode server objects should share the same geographic location.

Code the Click event of the button as follows:

```
private void Button1_Click(object sender, System.EventArgs e)
{
    geocodeConnection1.Host = "gisserver";
    geocodeConnection1.ServerObject = "RedlandsGeo";
    using (WebGeocode wgc = geocodeConnection1.CreateWebGeocode())
    {
        IPropertySet pset =
          wgc.ServerContext.CreateObject("esriSystem.PropertySet") as
            IPropertySet;
        pset.SetProperty("Street", "380 New York St");
        pset.SetProperty("Zone", "92373");
        IPoint pt = wgc.GeocodeAddress(pset, null);
        using (WebMap webmap = Map1.CreateWebMap())
        {
            DrawAddress(webmap, pt);
            webmap.Zoom(0.5, pt);
            webmap.Refresh();
        }
    }
}
```

The code above calls the following function. This function draws the address location with a marker symbol that is added to the map's graphic layer.

```
public void DrawAddress (WebMap webmap, IPoint pt)
{
    IServerContext serverContext = webmap.ServerContext;
    IMapDescription mapDescription = webmap.MapDescription;
    // Copy the geocoded address point to the map context
    IPoint point = serverContext.CreateObject("esriGeometry.Point") as
        IPoint;
    IGeometry geometry = point;
    geometry.SpatialReference = webmap.SpatialReference;
    point.PutCoords(pt.X, pt.Y);
    point.Project(mapDescription.SpatialReference);
    if (point == null)
        return;
    // Instantiate a graphics elements collection
    IGraphicElements graphicElements =
        serverContext.CreateObject("esriCarto.GraphicElements") as
        IGraphicElements;
    // Specify symbol RGB color
    IRgbColor rgbColor = Converter.ToRGBColor(serverContext,
        System.Drawing.Color.Red);
    ISimpleMarkerSymbol simpleMarkerSymbol1 =
        serverContext.CreateObject("esriDisplay.SimpleMarkerSymbol") as
        ISimpleMarkerSymbol;
    simpleMarkerSymbol1.Color = rgbColor;
```

```
simpleMarkerSymbol1.Style = esriSimpleMarkerStyle.esriSMSCircle;
simpleMarkerSymbol1.Size = 9;
//Create MarkerElement
IMarkerElement markerElement1 =
  serverContext.CreateObject("esriCarto.MarkerElement") as
  IMarkerElement;
markerElement1.Symbol = simpleMarkerSymbol1;
IElement pointElement1 = markerElement1 as IElement;
//Set the geometry of the point element
pointElement1.Geometry = point;
IGraphicElement markerGraphicElement1 = pointElement1 as
  IGraphicElement;
graphicElements.Add(markerGraphicElement1);
// Pass array of elements to Graphic Container of Map Description
mapDescription.CustomGraphics = graphicElements;
}
```

WHAT'S STORED IN SESSION

The *GeocodeConnection* component doesn't utilize session state.

CONTROL LIFECYCLE

Unlike other ADF Web controls, a server component does not have a control lifecycle; instead, the class is instantiated during Page load and construction in the Page Lifecycle.

Developing Web applications with Java

ArcGIS Server includes an Application Developer Framework (ADF)—built with the

Java 2 Platform Enterprise Edition (J2EE) standard JavaServer Faces (JSF)—that

allows you to integrate GIS functionality into your Web applications. The ArcGIS

Server Java ADF includes a set of custom Web controls and templates that you'll

use to build your Web applications. You can start building your Web application with

one of several predefined templates, including the Map Viewer template that

offers basic map navigation and display, the Search template that finds features

by attributes, and the Geocoding template that locates places by address.

Alternatively, use the Web controls directly to create your own specialized

application in a style that conforms to your existing Web site.

This chapter describes how to create Web applications using the ArcGIS Server

ADF for Java. Many of the topics discussed in this chapter assume you have

already read the previous chapters of this book. At a minimum, you should read

Chapter 4, 'Developing ArcGIS Server applications'.

Topics covered in this chapter include:

• an overview of the ADF • creating Web applications from templates and Web

controls • programming guidelines • Web control reference

The Java ADF, as described in this chapter, was developed with JavaServer Faces (JSF) version 1.0. For the latest information on supported servlet engines and application servers as well as software and documentation updates, visit the ESRI Support Web site at http://support.esri.com. For more information about JSF, visit http://java.sun.com/j2ee/javaserverfaces.

The ArcGIS Server Application Developer Framework for Java sits at the top of the various programming components.

In the previous chapters of this book, you learned about the ArcGIS Server architecture, administration, and programming practices. You've probably already added a map to your GIS server and previewed it in ArcCatalog. This chapter will show you how to integrate that map or a locator into a Web application. You'll find that whatever type of Web application you want to build—from basic map display and query to sophisticated GIS editing and analysis—the ArcGIS Server ADF for Java allows you to utilize all of ArcObjects in a Web environment.

Creating Web applications that access your GIS server ultimately involves programming ArcObjects. The Java ADF is built on top of JSF, using this standard to create new classes that support a set of custom Web controls and provide access to ArcGIS Server and, subsequently, ArcObjects. The diagram to the left shows how the Java ADF fits into the overall development environment.

You can think of the Java ADF as:

- A set of custom Web controls exposed as JavaServer Pages (JSP) tags.

- A set of templates to be used as starting points for your Web application.

- An API for building client applications that access a GIS server.

- A J2EE Connector Architecture (JCA) compliant resource adapter that allows Enterprise JavaBeans (EJB) to call and work with ArcGIS Server objects.

This chapter describes the Web controls, Web templates, and programming practices for building applications that access a GIS server. Information and discussion regarding the JCA resource adapter and EJBs can be found in Appendix C.

In the most basic sense, the Java ADF provides you with an additional set of objects to program with. This chapter focuses on how to create Web applications that access a GIS server with the Java ADF objects. The Java ADF delivers these objects to you as Web controls.

The Web controls expose a set of properties and methods that allow you to interact with the GIS server objects, for example, helping you manage connections to the server, access the SOM, and retrieve server objects. Actually, you don't have to use the Java ADF at all to create your Web applications. You can directly access ArcGIS Server objects—and thus, ArcObjects—through the Java API. However, you'll find that the Java ADF encapsulates many of the details of programming directly with the ArcGIS Server objects and exposes a rich, mapping-centric user interface through the Web controls.

The templates help you start building your Web applications. Each template incorporates the Web controls into its user interface and addresses a particular GIS task—for example, map display and query. While you can use the templates out of the box and simply connect them to your GIS server and server objects, they are primarily intended as a starting point for building your own Web application. All the code for the templates is provided to you, so you can easily customize a template to suit your needs or cut and paste JSP code fragments into your own application. The templates also serve as a great learning tool for building your own applications because all the code is there to guide you.

WEB TEMPLATES

The Java ADF includes a set of templates you can use as a starting point for creating an application.

Each template utilizes a set of Web controls that is part of the Java ADF. For example, those templates that display a map utilize the map control. The primary advantage of building your Web application with one of the templates is that much of the commonly used functionality is already programmed into them so you don't have to program it yourself. For example, the Map Viewer template displays a toolbar that contains the common map navigation tools for panning and zooming around the map.

As with any template, the look of it serves as a starting point for your Web application. You can customize the layout of the controls on the template and change things such as fonts and colors to suit your needs. If you plan to integrate the application you create into an existing Web site, you also might add components, such as company logos and site navigation tools, so that it looks similar to your existing Web pages and integrates seamlessly into your Web site.

Map Viewer template

Most likely, you'll want to extend the functionality provided in a particular template and incorporate your own custom operations. That's when you'll start programming with the server API and ArcObjects.

Each template included with the Java ADF is described below. Later in this chapter, you'll see how to use one to build your own Web application.

Map Viewer template

The Map Viewer template provides basic map display capabilities. It consists of a main map, an overview map, a table of contents (legend), a North arrow, and a scale bar. The template also contains a toolbar with built-in tools for panning and zooming. For any map-centric application, the Map Viewer template offers a good starting point.

Search template

The Search template provides a search-centric interface for finding features on a map. The look of the template is similar to what you might see on the Web for a search engine. Enter a search string and click GO to yield a list of features that match the search string. Click the result you're interested in to get more details about it or to reveal a map that highlights the particular feature.

Search template

The Search template searches for matching values in the attribute tables of the layers on the map you incorporate into your application. Thus, the list of results returned is restricted to the features on your map. When creating your application, you may want to clearly indicate the types of values that can be searched.

Page Layout template

The Page Layout template displays the entire page layout of a map. It shows all the data frames on the map as well as any map surrounds on the layout, such as the map title, legend, North arrow, and scale bar. This template provides the same view of a map as you'd see in layout view in ArcMap. The toolbar included in the template allows you to pan and zoom each data frame on the map and also pan and zoom around the page layout itself.

Page Layout template

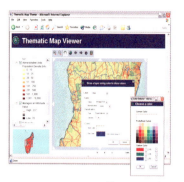

Thematic template

Geocoding template

Buffer Selection template

This chapter describes how to create a Web service catalog using the template provided with the Java ADF. For general information about Web service catalogs, see the section titled 'Programming Web services' in Chapter 4.

Thematic template

The Thematic template adds thematic mapping symbolization capabilities on top of the Map Viewer template. Outwardly, the map display in this template looks the same as that of the Map Viewer template. This template, however, allows the end user to dynamically change how individual layers are drawn by classifying the data in the layer. This template actually modifies the underlying map server object and, thus, requires that you configure it with a non-pooled object.

The Thematic template provides the following classification schemes:

- Natural Breaks
- Equal Interval
- Quantile

Geocoding template

The Geocoding template provides an interface for finding map locations by address. Enter the address that you want to find and click Locate. You'll be presented with a list of candidates that match the address. Click the result you're interested in to reveal a map that shows the address location. The interface displayed in the template changes depending on the address style you specify.

Buffer Selection template

The Buffer Selection template allows you to find features in one layer based on their location relative to features in another layer. For instance, suppose you want to determine how many homes a recent flood affected, given that the rivers in the area overflowed their banks by 1,000 meters. An application created from this template can perform the spatial query and identify which residences were affected by the flooding.

Performing this sort of spatial query involves creating a buffer at a specified distance around a set of features—for example, rivers—and finding other features based on their spatial relationship to the buffered area. The Buffer Selection template provides options for finding features that are completely within or intersect the buffered area.

Web Service Catalog template

The Web Service Catalog template provides a way to organize related server objects into groups and make them accessible over the Internet via HTTP as Web services. Use Web service catalogs with ArcGIS Desktop to give people access to the specific server objects they need. For example, you might choose to organize a series of maps used by a particular group of people in a Web service catalog. Alternatively, as a Web developer, you can use the services provided by the Web service catalog in Web applications.

ArcGIS Desktop users can connect to a Web service catalog via the GIS Servers entry in ArcCatalog. When connecting, provide the URL address of the Web service catalog created through this template, for example, *http://www.esri.com/webcatalog/default.jsp*. Web developers can reference the Web service catalog with the following: *http://www.esri.com/webcatalog/default.jsp?wsdl*.

The Web Service Catalog template is unlike the other templates in that it presents no user interface to an end user. Instead, running the template creates a Web service that can be consumed by client applications such as ArcGIS Desktop. When you build the template, you can choose the particular server objects you want to incorporate into your Web service catalog. Once you select them, simply deploy the template to make the Web service catalog available for client access.

WEB CONTROLS

Data objects are discussed in more detail later in this chapter.

As mentioned earlier, the Java ADF comes with a set of Web controls that you'll use while building your Web applications. The Web controls are analogous to the kinds of controls—such as buttons, labels, and text boxes—you see in any Web application, except in this case, they represent components commonly found on a map, such as the map itself and a legend (also referred to as a table of contents). You can think of the set of Web controls as the user interface component and its associated data object the part that does most of the GIS work.

As you might expect, the Web controls have attributes such as height, width, visibility, border style, and so on. In addition, the Web controls can respond to client actions, such as a control being clicked, and perform an appropriate action. For instance, in the case of a map control, the client action might be the dragging of a box and the associated server-side event might be to zoom in to the extent specified by the box or to select all the map features that are contained by the box. The Java ADF includes tools, command buttons, and listeners. Tools interact with a control, for example, clicking a feature on the map control. A command button directly executes a server-side action. A listener responds to an event passed from a command button. You'll find descriptions of tools, command buttons, and listeners later in this chapter. You can also learn more about the Java ADF listeners in the ArcGIS Server Java ADF Listeners and Actions documentation in the ArcGIS Developer Help.

The Web controls provide listeners and actions for common mapping operations, such as panning and zooming a map or buffering features, but they don't attempt to reproduce all of the functionality of ArcObjects. What they do provide, however, is entry points into the ArcGIS Server and ArcObjects APIs. For example, in the code you write, you'll be able to programmatically access a specific map server object from the map control. From there, you'll start programming ArcObjects to implement the specific functionality your application requires—for example, you may want to add new layers to the map or change how the layers are symbolized.

The following sections provide a brief description of each Web control and a look at how each tag is used in the templates. For a complete description of the tags and attributes, see the Java ADF tag library in the ArcGIS Developer Help. You'll find coding examples along with more detailed information about the interaction of a particular Web control later in this chapter. In addition, view the Java ADF object model diagram accessed from the ArcGIS Developer Help.

Context control

The context control establishes and maintains a connection with the GIS server. In the *context* tag, the *resource* attribute specifies the GIS server machine and the server object. All controls must be nested within a context tag. Nesting other tags within the *context* tag links them together and allows all the controls to share the same context. For example, you put a *map* tag and a *toc* tag in the same context to indicate that they work with each other. The example below is based on the tags in the Map Viewer template's *mapviewer.jsp* file. As you read through the remaining sections that describe the other controls, refer to this code to understand how the tags are structured.

To learn more about the attributes of these tags, refer to the Java ADF tag library documentation in the ArcGIS Developer Help.

```
<ags:context id="mapContext" resource="world@localhost">
  <ags:map ... />
  <ags:overview ... />
  <ags:toc ... />
  <ags:north_arrow ... />
  <ags:scale_bar ... />
</ags:context>
```

Map control

Because the map control displays the map, it will often be the main visual component of your Web application. The map control displays one particular data frame of a map document. Like the ArcMap data view, you can choose which data frame to display in the map control at a given time. You can change the data frame by setting the *dataFrame* attribute of the *map* tag.

The Java ADF provides listeners and actions for panning and zooming the map. Listeners and actions are discussed later in this chapter and described in the ArcGIS Developer Help.

Page layout control

The page layout control displays the layout of a map and is analogous to layout view in ArcMap. The page layout control, represented by the *pageLayout* tag, displays all the map elements, including data frames and any map surrounds. This example is based on the tags in the Page Layout template's *pageLayout.jsp*.

```
<ags:context id="mapContext" resource="world@localhost">
  <ags:pageLayout id="PageLayout0" left="233" top="115" width="535"
    height="408" activeTool="PageMapZoomIn" .../>
  <ags:toc ... />
</ags:context>
```

The Java ADF provides listeners and actions for panning and zooming the page layout or individual data frames on the layout.

Overview control

The overview control is similar to the map control in that it displays a particular data frame of a map server object. However, the purpose of an overview map is to provide a point of reference for the area displayed on its associated map control. The overview control, represented by the *overview* tag, always shows its data frame at full extent. A small area of interest box on the overview map represents the currently displayed area on its associated map control. You can interactively move this box around to pan the area displayed in the map control.

Like the other controls, the *overview* tag should be nested in a *context* tag to link it to a particular map control. But the *overview* tag is different in that it has a *resource* attribute that allows you to specify a different server object (from the map) for the overview map.

Table of contents control

The table of contents control is equivalent to the table of contents you see in ArcMap. The table of contents control, represented by the *toc* tag, lists the layers on the map and shows what the features represent. Checking a layer in the table of contents will draw it on the map or page layout. You can choose to show the table of contents for all data frames in the map server object or just the data frame being displayed in the associated map control.

North arrow control

The North arrow control displays the North arrow of a map control. This control uses a default symbol from the ESRI North TrueType font. You can specify a character index, using the *north_arrow* tag's *charIndex* attribute, if you want to change the North arrow symbol.

Scale bar control

The scale bar control displays the scale bar of a map control. You can change the appearance of the scale bar control, such as the font, color, or size, and set other properties, such as the units and number of divisions, by editing the attributes of the *scale_bar* tag.

Geocode control

The geocode control renders the input fields that allow you to enter address information. This control works with the address style from the locator to determine the input fields to display. For example, if the address style is US Streets with Zone, the control will render an address and ZIP Code text box. The results of a geocode operation are put into a data table.

The example below, based on the Geocoding template's *geocode.jsp* file, uses two *context* tags. The first *context* tag references the geocode server object and the *resourceType* attribute is set to geocode. The second *context* tag includes the *map* and *dataTable* tags. The *dataTable* tag should be nested in the same *context* tag as the *map* tag to display the geocoded address on the map.

```
<ags:context id="geocodeContext" resource="locator@localhost"
  resourceType="geocode">
  <ags:geocode id="geocode1" left="45" top="15" width="688">
    <jsfc:attribute name="fieldAlias:Street" value="Address" />
    <jsfc:attribute name="fieldAlias:Zone" value="ZIP Code" />
  </ags:geocode>
</ags:context>
<ags:context id="mapContext" resource="map@localhost">
  <ags:map .../>
  <!- Geocode result Data Table ->
  <jsfh:dataTable id="geocodeTable"
    binding="#{sessionScope['geocodeContext'].webGeocode.results.dataComponent}"
    rendered=
      "#{sessionScope['geocodeContext'].webGeocode.results.count > 0}"
    rows="5" … >
  </jsfh:dataTable>
  <!- Geocode result Scroller Component ->
  <ags:scroller id="idScroller" dataComponent="geocodeTable" left="300"
    top="250" height="20" width="400" ... />
</ags:context>
```

If you don't want to display the geocoded address on a map, you would put the *geocode* and *dataTable* tags in the same *context* tag. In this example, which is not taken from the Geocoding template, you would only get a list of geocode results.

```
<ags:context id="geocodeContext" resource="locator@localhost"
  resourceType="geocode">
  <ags:geocode id="geocode1" left="45" top="15" width="688">
    <jsfc:attribute name="fieldAlias:Street" value="Address" />
    <jsfc:attribute name="fieldAlias:Zone" value="ZIP Code" />
  </ags:geocode>
  <jsfh:dataTable id="geocodeTable"
    binding="#{sessionScope['geocodeContext'].webGeocode.results.dataComponent}"
    rendered=
      "#{sessionScope['geocodeContext'].webGeocode.results.count > 0}"
    rows=:"5" … >
  </jsfh:dataTable>
  <ags:scroller id="idScroller" dataComponent="geocodeTable" left="300"
    top="250" ... />
</ags:context>
```

Identify results control

The identify results control renders the results of an identify, the action of click-ing a map feature (or features) to return its attributes. You would use an identify tool to perform the action of clicking a map feature.

This sample shows a map, an identify tool, and the *identifyResults* tag.

```
<ags:context id="mapContext" resource="world@localhost">
  <ags:map ... />
  <ags:identifyResults cssClass="identifyClass" />
  <IMG id="imgIdentify" name="imgIdentify" src="images/identify.gif"
    MapPoint('Map0', 'Identify')>
</ags:context>
```

Scroller control

The scroller control implements paging for a data table of results, such as the results from a search or buffer operation. A listener displays search results in a data table. The data table that displays those results has an *id* attribute. You can associate a scroller control with a data table by passing the value of the *id* at-tribute to the *scroller* tag's *dataComponent* attribute.

This example is based on the Search template; it shows the data table for a set of find results and the scroller control attached to this data table.

```
<!- Find results Data table ->
<jsfh:dataTable id="findTable"
  binding="#{sessionScope['searchContext'].
    attributes['esriAGSFindResults'].dataComponent}"
  var="result"
  rows="5"
  rendered="#{sessionScope['searchContext'].
    attributes['esriAGSFindResults'].count > 0}" ...>
  <jsfh:column id="colHeader">
    <jsfc:facet name="header"><jsfh:outputText value="Field" />
        </jsfc:facet>
    <jsfh:outputText value="#{result.field}" />
  </jsfh:column>
  ...
</jsfh:dataTable>
<!- Find result Scroller Component ->
<ags:scroller id="idScroller" dataComponent="findTable" left="300"
  top="240" ... />
```

JSF technology offers the ability to build a Java Web application by dragging and dropping controls to design an interface. Using an IDE that integrates with JSF, you could drag the Java ADF map or page layout control onto a form, the way you would a text box or button. When this book was written, no IDE with these capabilities was available, and thus, there is no drag and drop support with the Java ADF. However, you can use an IDE, such as JBuilder, to write the JSP files for your application.

In this section, you'll start learning about how to build Web applications with the Java ADF. The following four examples explain step-by-step how to create the sample applications.

* Creating your first Web application with a template

* Creating a Web service catalog

* Creating a Web application with the Web controls

* Accessing ArcObjects from a Web application

Each set of steps is followed by a detailed explanation of the code.

When you start using the Java ADF, located at <install location>\DeveloperKit\Templates\Java, you will find a series of folders named after the templates described earlier. The build and deployment tool used by the Java ADF is Another Neat Tool (ANT), a commonly used build and deployment tool for Java applications. ANT is part of the Java ADF installation and is invoked with the arcgisant commands. Do not edit the files associated with the templates found in the <install location>\DeveloperKit\Templates\Java folder. After running the *arcgisant build* command, a Web application with the name you specified is created for you at <install location>\DeveloperKit\Templates\Java\build. It is the JSP file found in this Web application's folder that you will edit.

As you build these sample applications, you'll be exposed to the various controls implemented in the Java ADF and see some of the attributes, actions, events, and listeners that are used. Chapter 7, 'Developer scenarios', presents more sophisticated examples of Web applications that focus on a particular GIS activity. For a complete reference of all the controls in the Java ADF, refer to the ArcGIS Developer Help.

In order to code and deploy the examples yourself, you need to have access to a working GIS server with at least one map server object running on it. See Chapter 3, 'Administering an ArcGIS Server', for information on starting a map server object. In addition, you will need to understand how your application will access the GIS server and server objects before building your first application with the Java ADF. The ArcGIS Server administrator should add your account into the ArcGIS Server users group, agsusers. You will enter that account along with its domain and password when you are creating your Web application. Accessing a GIS server is described in more detail later in this chapter.

Also, verify that you followed the instructions in the ArcGIS Server Java ADF installation guide for configuring the Java ADF. You will need to set some environment variables and edit a common.properties file.

Before deploying Java ADF applications, verify that you edited the common.properties file for your servlet engine or application server. The common.properties file is found in the <installlocation>\DeveloperKit \Templates\Java folder. Information on how to install and configure the Java ADF is found in the ArcGIS Server Java ADF installation guide.

CREATING YOUR FIRST WEB APPLICATION WITH A TEMPLATE

One of the easiest ways to create a Web application is to start from a template. This example shows you how to create an application from the Map Viewer template provided with the Java ADF.

1. Open a command window and navigate to <install_location>\DeveloperKit\Templates\Java.

2. In the command window, type:

   `arcgisant build`

3. In the dialog, type the name of your application. In this example, you can call it "map".

4. Type the name of the GIS server you want to use.

5. Type the domain, username, and password of an account that has access to that GIS server.

6. Click Connect.

The username and password must have previously been granted access to the GIS server. If your machine is not running on a domain, enter the machine name as the domain.

7. Choose the Map Viewer template from the list.

8. Choose a map server object from the list for the Map.

9. Choose a map server object from the list for the Overview Map.

   In general, you can use any map server object for the overview map as long as its geographic extent includes the geographic extent of the server object shown in the map control. If not, you will not see the area of interest box on the overview map.

10. Click OK.

    You can navigate to <install location>\DeveloperKit\Templates\Java\build\map to see your application.

    Once created, you need to deploy your application. The deployment command will depend on the servlet engine you are using.

    In this example, Tomcat is the servlet engine and the Web server's port number is 8080.

Type "arcgisant -projecthelp" to see the help for arcgisant. Unlike the arcgisant build command, the arcgisant deploy command doesn't open a dialog; you'll enter the required information for the deploy into the command window.

11. Type the following command to deploy the application:

    `arcgisant tomcat-deploy`

12. In the command window, type the name of your application. In this example, it is named 'map'.

13. Stop and restart Tomcat.

14. View your application in a Web browser.

This URL assumes that your Web server is running on port 8080.

    `http://myserver:8080/map/index.html`

As you can see, you can quickly build a Web application from a template. Try experimenting with some of the other templates to see how they work and what functionality they provide. In general, you'll follow the same steps to build and deploy the other templates. The next section takes a look at how you would design an application's look and feel.

CUSTOMIZING THE PRESENTATION

When building an application with the arcgisant command, you only set a few properties: a connection to a host and a server object for the map, page, overview map, or geocode controls. Setting these properties is enough to get started with a Java ADF template, but as you proceed with your application development, you will want to change the presentation of the application. For example, you may want to choose a particular data frame to display in the map, set a tooltip for the map, or turn off the overview map. The Java ADF provides three ways to customize the presentation:

- Edit the application's JSP to set new values for tag attributes.

- Edit the variables in the application's EXtensible Stylesheet Language (XSL) file to set style attributes. If a conflicting value is set in the JSP and the XSL, the JSP tag attributes take precedence over the XSL variables.

- Edit the application's Cascading Style Sheet (CSS) file to set the style attributes. If a conflicting value is set in the JSP, XSL, and CSS, the JSP has highest precedence, followed by the XSL, then the CSS.

Although there is some overlap, the JSP, XSL, and CSS do not offer the same set of customization options. Refer to the ArcGIS Server Java ADF Attribute Comparison documentation in ArcGIS Developer Help for a chart showing which aspects of the application's presentation can be changed in the JSP, XSL, or CSS. Editing JSP tags, XSL variables, or CSS attributes is a common practice when customizing Web applications. You may want to reference other materials in addition to this section if these technologies are not familiar to you.

What you get with a Java ADF template

Take a look at what you get when you create an application from a Java ADF template. The files of interest to you are:

- The JSP file, in the main application folder, is the starting point of the template. This JSP includes the tags for each Web control and the HTML tools and command buttons that make up the template's toolbar.

- The two CSS files, in the css folder, define styles used in the template. The CSS named in accordance with the type of template—for example, mapviewer.css—defines the style attributes for the template. The other CSS, webcontrols.css, defines the styles for the controls.

- The XSL files, in the xsl folder, contain variables for setting the display of the template. There is an XSL for each Web control as well as a common.xsl. The common.xsl specifies elements shared by all of the controls.

- The default.xml file in the tools folder defines the default set of tools used by the template. These tools correspond to functions in the JavaScript library provided with the Java ADF.

The other file that you may need to edit is the arcgis_webapps.properties file. This file stores the username, password, and domain of the account for connecting to the GIS server as well as other properties of the controls.

You can navigate to the map application built in the previous example to see the other files that make up a Java ADF template.

Customizing an application's JSP

The JSP file created for applications built from Java ADF templates can be considered the main file of the application. You will find the *context*, *map*, *overview*, *toc*, *pageLayout*, and other Java ADF tags toward the top of the JSP file. Some attributes are already specified for each tag.

Refer to the JSP tag library documentation in the ArcGIS Developer Help for more information on the tags and their attributes.

Take a look at an example of how you would edit the JSP tags. Most of the templates use the map control. The *map* tag, as specified in the Map Viewer template's JSP, is:

```
<ags:map id="Map0" left="233" top="115" ... />
```

If you wanted to add some copyright text to the map, you could add the *copyrightText* attribute and set it to display your text.

```
<ags:map id="Map0" left="233" top="115" ...
    copyrightText="This is the copyright text for my map." />
```

The following sections show you how to use the XSL variables and CSS styles to customize the presentation of the Java ADF templates. Unlike the XSL or CSS which simply change the look and feel of the application, you can edit the JSP to add functionality to your application, such as changing an existing tool or adding a new tool.

A description of how to add a measure tool to your application is in the section titled 'Map control' later in this chapter.

Customizing an application's XSL

The Java ADF uses XSL to convert XML, returned from a processed JSP tag, into HTML for final display in a Web browser. An XSL file can also be regarded as a transformation file where the input XML is transformed, using Extensible Stylesheet Language Transformation (XSLT), into something else, such as another XML or HTML file. Each Web control in the Java ADF has a corresponding XSL file. The goal of this section is to explain how to modify the XSL variables to customize the final display for each Web control.

If you are not familiar with XSL, there are many Web sites where you can learn about it, such as http://www.w3schools.com/xsl/xsl_intro.asp.

Here's how you can edit the XSL of the map control to change the border color and border style of the map and add a tooltip to it. To edit the XSL for the map control, navigate to <install location>\DeveloperKit\Templates\build\<application name>\WEB-INF\classes\xsl and open map.xsl in a text editor. You will see the list of variables at the top of the file. If you want to add a tooltip to the map, which instructs the users of your Web site on how to use the Zoom In tool, you would change the *customToolTip* variable. Suppose you also want to change the color of the map border to blue and its style to dotted. Here are the variables with the suggested changes.

```
<xsl:variable name="customCssClass"></xsl:variable>
<xsl:variable name="customBorderColor">Blue</xsl:variable>
<xsl:variable name="customBorderStyle">Dotted</xsl:variable>
<xsl:variable name="customDragBoxColor"></xsl:variable>
<xsl:variable name="customDragLineWidth"></xsl:variable>
<xsl:variable name="customToolTip">Drag a box on the map</xsl:variable>
```

You would follow a similar practice to change the other variables.

If you want to see the XML created by the controls, you can change the *log_level*

If you make a mistake while editing the XSL, the errors are sent to your servlet engine or application server log.

property in the application's arcgis_webapps.properties file. By setting this property to "FINE", the XML generated by the controls is output to your servlet engine or application server log. Here is an example of the map control's XML:

```
<?xml version="1.0" encoding="UTF-8"?>
<map>
<client>ie</client>
<common-resources-registered>false</common-resources-registered>
<form-id>frmMap</form-id>
<resource-bundle-name>D:\WEB-INF\classes\res\Res.xml</resource-bundle-name>
<first-time>true</first-time>
<id>Map0</id>
<css-class>mapClass</css-class>
<image-url>mimedata?id=map22472173</image-url>
<width>503</width>
<height>376</height>
<left>249</left>
<top>131</top>
<border-width>16</border-width>
<tool>
<tool-key>ZoomIn</tool-key>
<client-action>MapDragRectangle</client-action>
</tool>
</map>
```

In addition to setting variables in the XSL to control the presentation of the Web controls, use the XSL to output to a markup language other than HTML. XML can be transformed into any markup language supported by XSL, such as HTML, XHTML, or PDF.

Refer to the JSP tag library documentation in the ArcGIS Developer Help for more details on the xslFile attribute.

If you want your map in XHTML format, you could write a generateXHTML.xsl that renders the map control into XHTML. When using your own XSL, you would set the *xslFile* attribute of the *map* tag to, for example, generateXHTML.xsl. The ArcGIS Developer Help includes a sample on how to render the contents of a page in XHTML format.

Customizing an application's CSS

CSS is a language for adding display styles to HTML. You can edit the CSS included with a Java ADF application to change the display of the Web controls.

To edit the CSS for the map control, navigate to <install location>\DeveloperKit\Templates\build\<application name>\css and open webcontrols.css in a text editor. The map control is listed at the top of the file:

```
div.mapClass { border-color: #B0C4DE; border-style: Solid; }
```

The color values are in hexadecimal. The value for the map's border color is the hexadecimal value for LightSteelBlue. You can change the map border to a dark red color with the following code:

A Web site for finding hexadecimal color codes is http://www.december.com/html/spec/color.html.

```
div.mapClass { border-color: #B22222; border-style: Solid; }
```

The CSS files specific to each template are also found in the css folder. For a Map Viewer template, open the mapviewer.css, to edit styles for items such as the title and logo.

CREATING A WEB SERVICE CATALOG

The Web Service Catalog template provides a way to organize related server objects into groups and make them accessible over the Internet via HTTP as Web services. One of the primary advantages of creating a Web service catalog is that ArcGIS Desktop users can directly connect to a Web service catalog and utilize the server objects exposed through it and, for example, add a map server object to ArcMap. You can include whichever server objects you want to in a Web service catalog. Thus, you might create a Web service catalog for each department in your organization with only the specific server objects to which they need access.

The Java ADF Web Service Catalog template allows you to quickly create a Web service catalog. Follow the steps below to create one.

1. Open a command window and navigate to <install location>\DeveloperKit\Templates\Java.

2. In the command window, type:

   `arcgisant build`

3. Type the name of your application; in this example, you can call it "webservicecatalog".

4. Type the name of the GIS server you want to use.

5. Type the domain, username, and password of an account that has access to that GIS server.

6. Click Connect.

7. Choose the Web Service Catalog template from the list.

8. Check the server objects you want to include in the Web service catalog.

You can navigate to <install location>\DeveloperKit\Templates\Java\build \webservicecatalog to see your application.

9. Click OK.

10. In the command window, type:

    `arcgisant tomcat-deploy`

11. In the command window, type the name of your application. This is the name you entered in the first input box, Application Name, on the ANT dialog during the build process. For this example, type "webservicecatalog".

12. Stop and restart Tomcat.

You now have a Web service catalog to which you can connect. In ArcCatalog, specify the URL for the Web service catalog as shown below:

    `http://myserver:8080/webservicecatalog/default.jsp`

See Chapter 3, 'Administering an ArcGIS Server', for more information on connecting to Web service catalogs from ArcCatalog.

For more information on adding capabilities to your Web service catalog, see the section titled 'Guidelines for creating your own Web applications' later in this chapter.

You can also access Web service catalogs programmatically using standard Web protocols. As a developer, you can consume the Web service catalog with the following reference:

    `http://myserver:8080/webservicecatalog/default.jsp?wsdl`

For more information on consuming Web service catalogs from a Web application, see the section titled 'Programming Web services' in Chapter 4.

CREATING A WEB APPLICATION WITH THE WEB CONTROLS

Whether you choose to start building your Web application from a template or from scratch utilizing the Web controls directly, you'll soon have to write your own custom code to implement the specific functionality you want to incorporate in your application. As you read at the beginning of this chapter, writing custom code means programming with the Java ADF, ArcGIS Server, and ArcObjects. Through the Java ADF, you'll be able to access the server objects and, in turn, ArcObjects.

In this example, you'll see how you can use the map control to access all the map server objects on a specific GIS server. Then, you'll see how you can add tools to interact with the map control.

In addition to the files included with an application built with the Java ADF Web controls, you will create a JSP file called serverlist.jsp and a Java file called ContextUtil.java. The serverlist.jsp is the page displayed to the Web application user. The page shows an input box and button for connecting to a host, a list box of server objects, a map, and map navigation tools. The Java code in ContextUtil.java is in a bean. The serverlist.jsp file will call the Java code with the *useBean* tag.

Building the Web controls application

You will use the arcgisant command to create an application called *serverlist*.

1. Open a command window and navigate to <install location>\Java\webcontrols.

2. In the command window, type:

   arcgisant build

3. Type the name of the GIS server you want to use.

4. Type the domain, username, and password of an account that has access to that GIS server.

5. Click Check Connection.

6. For the name of the application, type "serverlist".

If you are not using Tomcat, you'll typically use your servlet engine or application server's working directory.

7. In this example, Tomcat is used, so the directory is Apache Group\Tomcat 4.1\webapps. Click OK.

Browse to the Apache Group\Tomcat 4.1\webapps\serverlist directory to review the files that have been created as part of this Java ADF Web controls application. The structure of the Web application is the same as one created from a Java ADF template, but the contents of the Web application are different. You'll notice that you do not have a JSP file in the main folder. You will create serverlist.jsp file in this location. In addition, the css folder only contains a CSS for the Web controls, whereas an application built with a Java ADF template includes a CSS file for the template.

The domain, username, and password entered into the arcgisant build dialog is stored in the arcgis_webapps.properties file. The application has been created for that account. The serverlist.jsp includes an input box and button for connecting to a host. The host can be any GIS server that recognizes this account.

Getting the list of server objects

1. Navigate to your application's WEB-INF directory and create this folder structure: classes\com\esri\arcgis\webcontrols\samples. Create ContextUtil.java in the samples directory.

2. Add the following import statements to the top of the file.

```java
package com.esri.arcgis.webcontrols.samples;
import java.util.*;
import com.esri.arcgis.webcontrols.util.WebUtil;
import com.esri.arcgis.webcontrols.ags.data.AGSWebContext;
import com.esri.arcgis.webcontrols.ags.data.AGSWebMap;
import com.esri.arcgis.webcontrols.ags.data.AGSResource;
import com.esri.arcgis.server.IEnumServerObjectConfigurationInfo;
import com.esri.arcgis.server.IServerObjectConfigurationInfo;
import com.esri.arcgis.carto.IMapLayerInfos;
import com.esri.arcgis.carto.IMapLayerInfo;
import com.esri.arcgis.geodatabase.IFields;
import javax.faces.context.FacesContext;
import java.util.logging.*;
```

3. Declare the class as ContextUtil.

```java
public class ContextUtil {
   ...
}
```

4. Add the following variables:

```java
private static Logger logger =
   Logger.getLogger(ContextUtil.class.getName());
private String host;
private List serverObjects = new ArrayList();
private String serverObject;
```

5. Add the *getChangeOperation* and *changeOperation* methods.

```java
public String getChangeOperation() {
   changeOperation();
   return "";
}
public void changeOperation() {
   Map rMap = FacesContext.getCurrentInstance().
      getExternalContext().getRequestParameterMap();
   String operation = (String)rMap.get("operation");
   if(operation == null || operation.equals(""))
     return;
   if(operation.equals("userHost")) {
     setHost((String)rMap.get("txtHost"));
     return;
    }
   if(operation.equals("serverObject")) {
    setServerObject((String)rMap.get("selServerObject"));
     return;
    }
  }
```

6. Add the methods to set, then return the name of the GIS server.

```java
public void setHost(String value) {
    host = value;
    serverObjects.clear(); getServerObjects();
    setServerObject(serverObjects.size() > 0 ?
      (String)serverObjects.get(0) : "");
}
public String getHost() { return host; }
```

7. Add the methods to set, then return the selected server object.

```java
public void setServerObject(String value) {
    this.serverObject = value;
}
public String getServerObject() { return serverObject; }
```

8. Add the method to return the list of server objects on the GIS server.

```java
public List getServerObjects() {
    if(serverObjects.size() > 0)
      return serverObjects;
    AGSResource resource = null;
    try {
      resource = new AGSResource(host, "");
      IEnumServerObjectConfigurationInfo configs =
        resource.getServerObjectManager().getConfigurationInfos();
      configs.reset();
      IServerObjectConfigurationInfo config;
      while ((config = configs.next()) != null) {
        if (!config.getTypeName().equalsIgnoreCase("MapServer"))
            continue;
        serverObjects.add(config.getName());
      }
      return serverObjects;
    }
    catch (Exception e) {
      logger.log(Level.WARNING, "Unable to get server objects.", e);
      return serverObjects;
    }
    finally {
      if (resource != null)
        resource.destroy();
    }
}
```

9. Save ContextUtil.java, compile it, and verify that the class file is in the WEB-INF\classes\com\esri\arcgis\webcontrols\samples directory.

Take a look at ContextUtil.java.

You have added methods to the bean to get the name of the GIS server and the list of server objects running on that host.

getChangeOperation and changeOperation methods

The *getChangeOperation* method is called when the page is loaded. The *getChangeOperation* method calls the *changeOperation* method to determine if the host, server object, data frame, or layer has been set. The first time this application is run, the *changeOperation* method does nothing. However, if the application user enters a new host name or selects a new server object, the *changeOperation* method calls the *setHost* and *setServerObject* methods. This case of finding a previously entered host is only valid for the same browser session. Once the browser is closed, the connection to the host is released.

setHost and getHost methods

The *setHost* method sets the name of the GIS server. The *getHost* method returns the name of the GIS server.

setServerObject, getServerObject, and getServerObjects methods

The *setServerObject* method sets the selected server object. The *getServerObject* method returns the name of the currently selected server object. The *getServerObjects* method returns the list of all the map server objects on the GIS server.

Displaying the list of server objects

In this section, you'll add an input box, a button, a list box, a context control, and a map control to the application. The input box is for entering the name of a GIS server. The list box will display the map server objects running on the GIS server. Selecting a server object from the list will display it on the map control.

1. Browse to the Apache Group\Tomcat 4.1\webapps\serverlist folder and create a file named serverlist.jsp.

2. Add this tag to set the content type and character set for the page.

   ```
   <%@ page contentType="text/html; charset=ISO-8859-1" %>
   ```

3. Add references to the following tag libraries. These tag libraries are required for any Java ADF application.

   ```
   <%@ taglib uri="http://www.esri.com/arcgis/webcontrols" prefix="ags" %>
   <%@ taglib uri="http://java.sun.com/jsf/core" prefix="jsfc" %>
   <%@ taglib uri="http://java.sun.com/jsf/html" prefix="jsfh" %>
   ```

4. Add a reference to the JavaServer Pages Standard Tag Library (JSTL). JSTL is a commonly used tag library in JSP applications.

   ```
   <%@ taglib uri="http://java.sun.com/jstl/core" prefix="c" %>
   ```

5. Add the *useBean* and *setProperty* tags to reference the bean in the ContextUtil.java file. If localhost is not your GIS server, replace localhost with the name of your GIS server.

   ```
   <jsp:useBean id="contextUtil"
     class="com.esri.arcgis.webcontrols.samples.ContextUtil"
     scope="session">
     <jsp:setProperty name=
       "contextUtil" property="host" value="localhost" />
   </jsp:useBean>
   ```

6. Add opening and closing tags for the *view* tag and add the *getProperty* tag.

```
<jsfc:view>
   <jsp:getProperty name="contextUtil" property="changeOperation" />
   ...
</jsfc:view>
```

7. Add opening and closing tags for the *html*, *head*, *title*, *htmlBase*, and *noCache* tags to the *view* tag.

```
<html>
<head>
<title>Getting Started with the Java ADF</title>
<ags:htmlBase />
</head>
<ags:noCache />
...
</html>
```

8. Add the JavaScript *changeOperation* function inside the *noCache* tag. This function is called when a server object is selected from the list.

```
<script language="Javascript">
  function changeOperation(operation) {
        var theForm = document.forms[0];
        theForm.operation.value = operation;
        theForm.submit();
  }
</script>
```

9. Add opening and closing tags for the *body* and *form* tags. The *form* tag encloses all controls that display or collect data.

```
<body>
<jsfh:form id="frmServerList">
...
</jsfh:form>
</body>
```

10. Add the input box and button for connecting to a host and a list box for the server objects. When a server object is selected from the list, the *changeOperation* JavaScript function is called.

```
<div id="host" style="position:absolute;left:50px;top:50px;">
   <b>Host </b>
   <input name="txtHost" value="<c:out
     value='${sessionScope.contextUtil.host}'/>" />
   <input type="BUTTON" value="Connect"
     onclick="javascript:changeOperation('userHost');"/>
</div>

<div id="serverobject" style="position:absolute;left:50px;top:100px;">
   <input type="HIDDEN" name="operation"><b>Server Objects </b>
   <select name="selServerObject" onchange=
     "javascript:changeOperation('serverObject');">
     <c:forEach var="item" items=
       "${sessionScope.contextUtil.serverObjects}">
```

```
        <option value='<c:out value="${item}"/>'
        <c:if test='${sessionScope.contextUtil.serverObject ==
          item}'>selected="true"</c:if>><c:out value='${item}'/></option>
      </c:forEach>
    </select>
  </div>
```

11. Add the *context* tag. ContextUtil.java has two methods, *getServerObject* and *getHost*, which are used to set the *resource* attribute of the *context* tag.

```
<ags:context id="sampleContext"
resource='<%=(contextUtil.getServerObject() + "@" +
  contextUtil.getHost())%>' >

...

</ags:context>
```

12. Add the *map* tag to the *context* tag. The *map* tag will display a 400 x 400 map with a solid border.

```
<ags:context id="sampleContext"
resource='<%=(contextUtil.getServerObject() + "@" +
  contextUtil.getHost())%>' >

  <ags:map id="Map0" left="450" top="100" width="400" height="400" />

...

</ags:context>
```

13. Save serverlist.jsp.

Take a look at serverlist.jsp.

As mentioned above, the first three tag libraries referenced in this application are required for any Java ADF application.

```
<%@ taglib uri="http://www.esri.com/arcgis/webcontrols" prefix="ags" %>
<%@ taglib uri="http://java.sun.com/jsf/core" prefix="jsfc" %>
<%@ taglib uri="http://java.sun.com/jsf/html" prefix="jsfh" %>
```

The first tag library, *http://www.esri.com/arcgis/webcontrols*, is the core ArcGIS Web controls tag library. It includes such controls as the context, map, and TOC. The next two tag libraries reference JSF. The core JSF tag library is http://java.sun.com/jsf/core. It contains the essential tags for building any JSF Web application. The standard JSF components tag library is http://java.sun.com/jsf/html. These standard components are rendered as HTML widgets, such as forms, text fields, or check boxes. They are associated with their server-side components by the JSF technology.

The fourth tag library is the JavaServer Pages Standard Tag Library (JSTL).

```
<%@ taglib prefix="c" uri="http://java.sun.com/jstl/core" %>
```

JSTL is a commonly used tag library in JSP applications. In this application, JSTL is used to iterate a collection of server objects in order to populate a list.

This may be your first time using the *view* tag from the core JSF tag library.

```
<jsfc:view>

...

</jsfc:view>
```

When you reference JSF tags from within a JSP page, you must enclose them in the *view* tag. You can enclose other content, such as HTML and other JSP tags, within the *view* tag as well.

The next section of code uses the *html/Base* tag.

```
<ags:htmlBase />
```

The *html/Base* tag is from the core ArcGIS Web controls tag library. Before a JSF application can launch a JSP page, the Web application must invoke the *FacesServlet* to begin the JSF application's lifecycle. To do this, "faces" must be included in the URL path. The URL for this serverlist application is http://localhost/serverlist/faces/serverlist.jsp. With this as the URL, the default base of the page is http://localhost/serverlist/faces/. If any image or any other resource on the page is accessed as <img src="images/image.gif">, the browser will resolve this as http://localhost/serverlist/faces/images/image.gif. However, there is no physical resource that corresponds to "faces", so this URL would show a broken link. To prevent this problem, the utility *html/Base* tag has been included with the Java ADF, which will explicitly set the base of the page to the root of the Web application, that is, http://localhost/serverlist.

When a new server object is selected from the list, the *changeOperation* JavaScript function is called. This function calls the bean's *getServerObject* and *getHost* methods to set the name of the GIS server and the selected server object.

The *context* and *map* tags, from the ArcGIS Web controls tag library, expose the context and map controls. When a map server object is chosen from the list, the map control displays the selected map.

Adding a Full Extent button and Zoom In and Pan tools

In this section, you'll add a command button to display the map at its full extent and tools for zooming in and panning the map.

1. Open serverlist.jsp.

2. Add Zoom In and Pan tools within the *context* tag.

```
<div id="button" style="position:absolute;left:550px;top:50px;">
  <input type="button" value="Zoom In" title="Zoom In"
    onclick="MapDragRectangle('Map0', 'ZoomIn');" />
  <input type="button" value="Pan" title="Pan"
    onclick="MapDragImage('Map0','Pan');" />
  ...
</div>
```

3. Add a Full Extent button to the *div* tag added in the previous step. This button is associated with the *ZoomFullExtentListener*.

```
<jsfh:commandButton id="idZoomFullExtent" value="Zoom Full Extent"
  title="Full Extent">
  <jsfc:actionListener type=
  "com.esri.arcgis.webcontrols.ags.faces.event.ZoomFullExtentListener"
  />
</jsfh:commandButton>
```

The *context* tag should look like this:

```
<ags:context id="sampleContext"
resource='<%=(contextUtil.getServerObject() + "@" +
  contextUtil.getHost())%>' >
<div id="button" style="position:absolute;left:550px;top:50px;">
  <input type="button" value="Zoom In" title="Zoom In"
    onclick="MapDragRectangle('Map0', 'ZoomIn');" />
```

```
<input type="button" value="Pan" title="Pan"
  onclick="MapDragImage('Map0','Pan');" />
<jsfh:commandButton id="idZoomFullExtent" value="Zoom Full Extent"
  title="Zoom Full Extent">
  <jsfc:actionListener type=
  "com.esri.arcgis.webcontrols.ags.faces.event.ZoomFullExtentListener"
  />
</jsfh:commandButton>
</div>
  <ags:map id="Map0" left="450" top="100" width="400" height="400" />
</ags:context>
```

4. Save serverlist.jsp.

5. Stop and restart Tomcat.

6. If you do not want to use localhost as your GIS server, verify that you added your GIS server to the *value* attribute of the *setProperty* tag.

```
<jsp:useBean id="contextUtil"
  class="com.esri.arcgis.webcontrols.samples.ContextUtil"
  scope="session">
  <jsp:setProperty name=
    "contextUtil" property="host" value="localhost" />
</jsp:useBean>
```

From the application, you can connect to any GIS server that recognizes the account you used when building the application with the *arcgisant build* command.

In a Web browser, type "http://localhost/serverlist/faces/serverlist.jsp".

The application you created looks like this:

You can click the Zoom In tool and drag a box over the map to zoom in. Click the Pan tool and drag the map image to a new location. Click the Full Extent button to return to the full extent of the map.

Take a look at these additions to serverlist.jsp.

A toolbar can contain two types of items—tools and commands. Tools have a client-side action as well as a server-side action. Tools interact with the map control—for example, dragging a box to zoom in—before executing a server-side event. Commands directly execute a server-side event. Full Extent is implemented as a JSF command button, but Zoom In and Pan are tools.

In the example above, you specified *MapDragRectangle* as the JavaScript function for the Zoom In tool.

```
<input type="button" value="Zoom In" title="Zoom In"
onclick="MapDragRectangle('Map0', 'ZoomIn');" />
```

MapDragRectangle enables the drawing of a rectangle over the map control. When this client-side action completes, a request is sent to the server passing the arguments of the *MapDragRectangle* function: the ID of the control and the tool key. In this example, the ID of the control is "Map0" and the tool key is "ZoomIn". The map control parses these arguments and delegates processing to the server-side action class associated with the tool's key. The Zoom In tool is associated with the *ZoomInToolAction* class.

Commands, on the other hand, directly execute server-side actions. JSF has standard command components, such as buttons and hyperlinks. These command components are associated with listeners. When a command is clicked, a server-side event is triggered, calling the associated listeners to process the event. The Java ADF includes a series of listeners for common mapping operations, such as displaying the map at full extent or panning the map in a particular direction. You can learn more about the Java ADF listeners in the ArcGIS Developer Help.

ACCESSING ArcObjects FROM A WEB APPLICATION

The Java ADF exposes a great deal of GIS functionality through its API. In the previous example, you used the *map* tag with the *ZoomFullExtentListener* to display the map at its full extent and the Zoom In and Pan tools for zooming in to and panning the map. Similar tools are available for changing the extent of the page layout. The Java ADF has additional tools for identifying or selecting features.

The Java ADF does not duplicate all of the functionality in ArcObjects. To do more than the basic GIS operations provided in the Java ADF, you'll need to write code directly against ArcObjects. The goal of this example is to show you how to access the ArcGIS Server and ArcObjects API from the Java ADF.

This example describes how to add code to the ContextUtil.java and serverlist.jsp files created in the previous example in order to display the data frames, layers, and fields of a server object. The first additional list box displays the list of data frames for a selected map server object. The next list box displays the list of layers in the selected data frame. The final list box displays the attribute fields of the selected layer. As you make selections from the list boxes, the other list boxes will update to reflect the current selection. For example, if you select a new data frame, the layers and fields list boxes will update to reflect the change.

Getting the data frames, layers, and fields for a server object

1. Open ContextUtil.java created in the previous example.

2. Add four variables to the list of variables.

```java
private static Logger logger =
Logger.getLogger(ContextUtil.class.getName());
private String host;
private List serverObjects = new ArrayList();
private String serverObject;
private Map layers = new Hashtable();
private String layer;
private List fields = new ArrayList();
private String contextId;
```

3. Add the conditional checks to the *changeOperation* function for the data frames and layers.

```java
public void changeOperation() {
    Map rMap = FacesContext.getCurrentInstance().getExternalContext()
        .getRequestParameterMap();
    String operation = (String)rMap.get("operation");
    if(operation == null || operation.equals(""))
      return;
    if(operation.equals("userHost")) {
     setHost((String)rMap.get("txtHost"));
      return;
     }
    if(operation.equals("serverObject")) {
     setServerObject((String)rMap.get("selServerObject"));
      return;
     }
    if(operation.equals("dataFrame")) {
```

```
          setDataFrame((String)rMap.get("selDataFrame"));
          return;
        }
        if(operation.equals("layer")) {
          setLayer((String)rMap.get("selLayer"));
          return;
        }
      }
    }
```

4. Add code to set the data frame to the *setServerObject* method.

```
public void setServerObject(String value) {
    this.serverObject = value;
    setDataFrame(null);
}
```

Insert this code after the getServerObjects method.

5. Add the methods to set and return the selected data frame.

```
public void setDataFrame(String value) {
    if(value != null) {
      AGSWebContext agsContext = getAGSWebContext();
      ((AGSWebMap)agsContext.getWebMap(true)).setFocusMapName(value);
      agsContext.refresh();
    }
    layers.clear(); setLayer(null);
}
public String getDataFrame() {
    try {
      return
      ((AGSWebMap)getAGSWebContext().getWebMap(true)).getFocusMapName();
    }
    catch(Exception e) {
      logger.log(Level.WARNING, "Unable to find the current data frame.",
      e);
      return null;
    }
}
```

6. Add the method to return the collection of data frames.

```
public Collection getDataFrames() {
    return getAGSWebContext().getDataFrames();
}
```

7. Add the methods to set the selected layer, return the selected layer, and return the list of layers for a data frame.

```
public void setLayer(String value) {
    this.layer = value;
    fields.clear();
}
public String getLayer() { return layer; }
public Map getLayers() {
    if(layers.size() > 0)
      return layers;
    if(getDataFrame() == null) {
      layer = null; layers.clear(); fields.clear();
```

```
      return layers;
    }
    try {
      IMapLayerInfos infos =
        ((AGSWebMap)getAGSWebContext().getWebMap(true)).
        getFocusMapServerInfo().getMapLayerInfos();
      IMapLayerInfo info;
      for (int i = 0; i < infos.getCount(); i++) {
        info = infos.getElement(i);
        layers.put(Integer.toString(info.getID()), info.getName());
      }
      if(layers.size() > 0)
        setLayer((String)layers.keySet().iterator().next());
        return layers;
      }
    catch (Exception e) {
      logger.log(Level.WARNING, "Unable to get layers", e);
      return layers;
    }
  }
```

8. Add the methods to return the fields for the selected layer.

```
public List getFields() {
    if(fields.size() > 0)
      return fields;
    if(layer == null) {
      fields.clear();
      return fields;
    }
    try {
      IFields iFields = ((AGSWebMap)getAGSWebContext().getWebMap(true)).
        getFocusMapServerInfo().getMapLayerInfos().
        getElement(Integer.parseInt(layer)).getFields();
      if (iFields != null) {
        for (int i = 0; i < iFields.getFieldCount(); ++i)
          fields.add(iFields.getField(i).getName());
      }
      return fields;
    }
    catch (Exception e) {
      logger.log(Level.WARNING, "Unable to get fields.", e);
      return fields;
    }
}
```

9. Add the *getContextId*, *setContextId*, and *getAGSWebContext* methods.

```
public String getContextId() { return contextId; }
public void setContextId(String value) { contextId = value; }
private AGSWebContext getAGSWebContext() {
return
(AGSWebContext)WebUtil.getWebContext(FacesContext.getCurrentInstance(),
  contextId);
}
```

10. Save ContextUtil.java and compile it.

Displaying the list of server objects, data frames, layers, and fields

You'll add list boxes for the list of data frames, layers, and fields to serverlist.jsp.

1. Add code to set the *contextId* property to the value used in the *context* tag.

```
<jsp:useBean id="contextUtil" class="com.esri.arcgis.webcontrols
  .samples.ContextUtil" scope="session">
  <jsp:setProperty name=
    "contextUtil" property="host" value="localhost" />
  <jsp:setProperty name="contextUtil" property="contextId"
    value="sampleContext" />
</jsp:useBean>
```

2. Open serverlist.jsp in a text editor. Add this code to the *context* tag.

```
<ags:context id="sampleContext"
resource='<%=(contextUtil.getServerObject() + "@" +
contextUtil.getHost())%>'>
<div id="dataFrame" style="position:absolute;left:50px;top:150px;">
    <b>DataFrames </b>
    <select name="selDataFrame"
      onchange="javascript:changeOperation('dataFrame');">
      <c:forEach var="item"
        items="${sessionScope.contextUtil.dataFrames}">
        <option value='<c:out value='${item}'/>'
          <c:if test="${sessionScope.contextUtil.dataFrame ==
          item}">selected='true'</c:if> > <c:out value="${item}"/>
          </option>
        </c:forEach>
      </select>
    </div>
    <div id="layer" style="position:absolute;left:50px;top:200px;">
      <b>Layers </b>
      <select name="selLayer"
        onchange="javascript:changeOperation('layer');">
        <c:forEach var="item" items="${sessionScope.contextUtil.layers}">
          <option value='<c:out value="${item.key}"/>'
            <c:if test="${sessionScope.contextUtil.layer ==
            item.key}">selected="true"</c:if> >
            <c:out value="${item.value}"/></option>
          </c:forEach>
        </select>
      </div>
      <div id="fields" style="position:absolute;left:50px;top:250px;">
        <b>Fields </b><br>
        <c:forEach var="item" items="${sessionScope.contextUtil.fields}">
          <c:out value="${item}"/><br>
        </c:forEach>
      </div>
      <div id="button" style="position:absolute;left:550px;top:50px;">
        <input type="button" value="Zoom In" title="Zoom In"
          onclick="MapDragRectangle('Map0', 'ZoomIn');" />
        <input type="button" value="Pan" title="Pan"
          onclick="MapDragImage('Map0','Pan');" />
```

```
<jsfh:commandButton id="idZoomFullExtent" value="Zoom Full Extent"
    title="Zoom Full Extent">
  <jsfc:actionListener type=
      "com.esri.arcgis.webcontrols.ags.faces.event.
      ZoomFullExtentListener" />
</jsfh:commandButton>
</div>
<ags:map id="Map0" left="450" top="100" width="400" height="400" />
</ags:context>
```

3. Save serverlist.jsp.

4. Stop and restart your servlet engine. You are ready to view the application. In a Web browser, type "http://localhost/serverlist/faces/serverlist.jsp".

The application you just created looks like this:

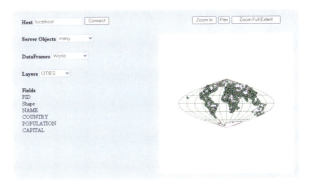

Take a look at how this example works.

setDataFrame, getDataFrame, and getDataFrames methods

The *setDataFrame* method calls the *AGSWebMap's setFocusMapName* method to set the selected data frame. Similarly, the *getDataFrame* method calls the *AGSWebMap's getFocusMapName* method and returns the selected data frame. The *getDataFrames* method returns the collection of data frames.

setLayer, getLayer, and getLayers methods

The *setLayer* and *getLayer* methods respectively set, then return the selected layer. The *getLayers* method calls the *getMapLayerInfos* method on *IMapServerInfo* to return the layers of the focus map.

```
IMapLayerInfos infos =
  ((AGSWebMap)getAGSWebContext().getWebMap(true)).
  getFocusMapServerInfo().getMapLayerInfos();
```

getFields method

The *getFields* method, based on the selected layer, returns the list of fields by calling the *getFields* method on *IMapLayerInfo*.

```
IFields iFields = ((AGSWebMap)getAGSWebContext().getWebMap(true)).
  getFocusMapServerInfo().getMapLayerInfos().
  getElement(Integer.parseInt(layer)).getFields();
if (iFields != null) {
  for (int i = 0; i < iFields.getFieldCount(); ++i)
    fields.add(iFields.getField(i).getName());
  }
return fields;
```

getContextId, setContextId, and getAGSWebContext methods

The *getContextId* and *setContextId* methods respectively set, then return the ID of the context. The *getAGSWebContext* method is a convenience method to get the *AGSWebContext*.

As you read through this chapter, you're probably already thinking about how to apply what you've learned to your own project. This section describes some of the things you'll need to think about as you begin building your own Web applications. Some of the information presented here is discussed in more detail in other chapters of this book; however, it's worth reviewing this information to ensure you have a clear picture of how to build your Web application.

ORGANIZING YOUR DATA AND CREATING SERVER OBJECTS

Presumably, most—if not all—of the GIS functionality you build into your Web applications will be provided through the server objects you host on your GIS server. How many server objects a single Web application utilizes depends on how you organize your server objects and what functionality you want to build into the Web application. Often, a Web application utilizes more than one server object and these server objects need to have some correlation between them. For example, if you want to build an application that locates places by address, that application will at minimum access a geocode server object to find the address location and a map server object to display the location. The map and geocode server object need to reference the same geographic location; otherwise the geocoded points won't draw on the map. Similarly, the geographic area displayed in an overview map needs to match that of its associated main map.

As map server objects are based on map documents (.mxd files), you have some flexibility as to how you organize the data contained within a given map server object. For instance, suppose you're creating an application that displays three maps. Should you create one map document resulting in one map server object with three data frames in it? Or should you create three map documents—and three server objects—with one data frame each?

As a general rule, you should organize the data in a map server object according to how your Web applications will use it. For example, you would create one map server object that contains all the data required for a Web application. The server object would have, for example, one data frame for the main map control and one data frame for the overview map. Later, if you build another Web application that can share the overview map with the first application, you may choose to host the data for the overview map in its own server object.

DATA ACCESS CONSIDERATIONS

Ensure that the following holds true for your map server objects.

- The map document and all data must be accessible to the SOM machine and all container machines.

- The ArcGIS Server Container account you established during the postinstallation should have read access to any shared network directories.

- File-based data (for example, coverages and shapefiles) should be placed on shared networked drives and saved in the map documents using UNC pathnames.

- ArcSDE connections must be saved in the map document before being added to the server.

ACCESSING A GIS SERVER FROM YOUR WEB APPLICATION

You need to consider two levels of security when building a Web application that accesses a GIS server. First, you need to consider who will access your application. Does the application contain sensitive information that is not meant for general consumption? If so, you'll probably want to restrict access to it in some way. Second, you need to consider how the Web application itself accesses and works with the GIS server because the server has its own access control mechanism that your Web application must implement.

In order to restrict access to a Web application, you need to identify who is attempting to access the Web site. This process, called *authentication*, challenges a user to prove who they are, usually by requiring them to provide some security credentials that only they and the resource they are trying to access know about. For example, to access your bank account through an ATM, you identify yourself with your ATM card and your personal identification number. Without both, you are not granted access to the account.

For Web applications, security credentials are commonly expressed in the form of an account and a password. If the account and password are valid, the user is authenticated and subsequently granted access to the Web site. There are many ways you can authenticate users and grant them access to a Web application. A discussion of these methods is beyond the scope of this book. However, whether you choose to restrict access to your Web application or not, you do need to consider the second level of security, which governs how the Web application accesses the GIS server. This is the main focus of this section.

As you read in the previous chapters, access to a GIS server—and the server objects running on it—is managed by two operating system user groups: ArcGIS Server users (agsusers) and ArcGIS Server administrators (agsadmin). To access the server objects, an application must connect to the server as an account in the agsusers group. To additionally administer the GIS server—for example, add and remove server objects—an application must connect as an account in the agsadmin group.

First, take a look at how you access a GIS server through ArcCatalog. When you connect to your GIS server in ArcCatalog, if the account you're currently logged in as is a member of the ArcGIS Server administrators group, you will see additional menu choices that let you manage the server. For instance, you can add, remove, start, and stop server objects. However, if your account is only a member of the ArcGIS Server users group, you will not see any menu choices for managing server objects and can instead only view the server objects running on the server. Further, if your account is not a member of either of these user groups, you won't have any access to the server.

Access to a GIS server is managed in a similar way with Web applications. A Web application can connect to the GIS server as a particular account through a process called *impersonation*. With impersonation, the Web application assumes the identity of a particular account and, thus, has all the privileges the account is entitled to. As long as this account is part of the appropriate group, for example, agsusers, the Web application will have access to the server objects running on the GIS server. So what account should your Web application connect to the server as? The answer to this question depends on how you choose to set up access to the GIS server itself.

Your application will connect to the GIS server as a specific account. You essentially hard code the account into your Web application. You can view the username and encrypted password that your Web application uses in the arcgis_webapps.properties file, located at <install location>\DeveloperKit\Templates\Java\build\<Web application name>\WEB-INF\classes. The ArcGIS Server administrator will either add your account or give you an existing account to use. Either way, the account is a member of the agsusers group.

Most likely the application will impersonate an account that is a member of the agsusers group rather than the agsadmin group. As a member of the agsusers group, the application will have complete access to the server objects. Thus, you can write code that works with the object in a stateless or stateful manner. You can do anything except manage the server objects themselves. If you want to write an application that manages the server, the application will have to impersonate an account in the agsadmin group.

Refer to the 'Context control' section later in this chapter for more guidelines on working with the server context, pooled versus non-pooled objects, and making fine-grained ArcObjects calls in your Java ADF applications.

CHOOSING CAPABILITIES FOR YOUR WEB SERVICE CATALOG

Web service catalogs may contain both MapServer and GeocodeServer Web services. Each Web service includes functionality that your users can exercise using ArcMap and you can exercise in your applications. The functionality provided by a particular Web service can be limited to, for example, allow consumers of the Web service to draw the map but not to query the data sources of the layers in the map.

The set of functionality that a Web service supports is called its capabilities. The capabilities for MapServer and GeocodeServer Web services are organized into capability groups. The *IRequestHandler::HandleStringRequest* method supported by MapServer and GeocodeServer takes a string parameter named capabilities, whose value is a comma-delimited list of capabilities, such as *HandleStringRequest("map,query,data", <inputSOAP>, <outputSOAP>)*.

The capability groups and the methods included in each group for MapServer Web services are:

Map:

- get_DocumentInfo
- get_MapCount
- get_MapName
- get_DefaultMapName
- GetServerInfo
- ExportMapImage
- GetSupportedImageReturnTypes
- GetLegendInfo

- ToMapPoints
- FromMapPoints

Query:

- Identify
- QueryFeatureCount
- QueryFeatureIDs
- QueryHyperlinks
- ComputeScale
- ComputeDistance
- GetSQLSyntaxInfo

Data:

- Find
- QueryFeatureData

By default, the Web service catalog template assigns "map" as the capabilities for all MapServer Web services. Therefore, by default, consumers of the MapServer Web service, including ArcMap, will not be able to identify or query layers in the map. You can add additional capability groups, query, and data by editing the JSP file created for each server object in the Web service catalog. For example, if two MapServer objects named canada and mexico were added to the Web service catalog, and you wanted to add query and data capabilities, you would edit the canada_MapServer.jsp and mexico_MapServer.jsp files. Edit the following line of code to include query and data:

```
String agsCapabilities="map,query,data";
```

The capability groups and the methods included in each group for GeocodeServer Web services are:

Geocode:

- GeocodeAddress
- GeocodeAddresses
- StandardizeAddress
- FindAddressCandidates
- GetAddressFields
- GetCandidateFields
- GetIntersectionCandidateFields
- GetStandardizedFields
- GetStandardizedIntersectionFields
- GetResultFields
- GetDefaultInputFieldMapping

- GetLocatorProperties

Reverse geocode:

- ReverseGeocode

By default, the Web service catalog template assigns "geocode,reversegeocode" as the capabilities for all GeocodeServer Web services. You can remove a capability group, geocode, or reversegeocode by editing the JSP file created for each server object in the Web service catalog. For example, if a GeocodeServer object named locator was added to the Web service catalog, and you wanted to remove the reverse geocoding capability, you would edit the locator_GeocodeServer.jsp file. Edit the following line of code to remove the reversegeocode capability:

```
String agsCapabilities="geocode";
```

Throughout this chapter, you've seen how the various Java ADF Web controls get used in Web applications. The remaining sections of this chapter describe each control in more detail, examining the classes with which each Web control is built and highlighting those methods that you'll find particularly useful. In addition, you will see how the Java ADF works with the JSF standard, how to implement new classes for tool actions and listeners, and the basics of the Java ADF object model. This chapter, however, does not serve as a complete reference for all the objects in the Java ADF. You can find a complete description of the packages, classes, and methods in the ArcGIS Developer Help. In addition, you can view the Java ADF object model diagram in the ArcGIS Developer Help.

EXPLORING THE WEB CONTROL OBJECT MODEL

The Java ADF Web controls extend the JSF Framework. Controls that accept events extend the *UICommand* class and implement *ActionListener*, enabling them to respond to action events and invoke listeners. The other controls extend from *UIComponentBase*. In addition to the JSF classes mentioned, other concepts from JSF will be referred to throughout this section. If you are not familiar with JSF, you may want to visit http://java.sun.com/j2ee/javaserverfaces.

The diagram below presents a conceptual object model diagram of the Java ADF Web controls.

Control Objects Overview

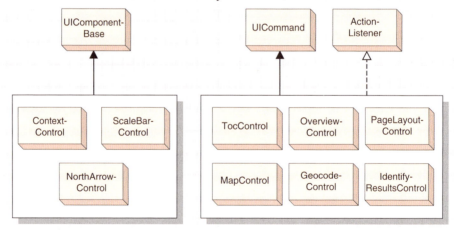

JAVA ADF PACKAGES

Packages group classes with similar functionality. Accordingly, the classes distributed with the Java ADF are organized as Java packages. The base package of the Java ADF is com.esri.arcgis.webcontrols. This section lists the packages that comprise this base package.

faces.component package

All of the Java ADF Web controls are contained in the faces.component package. A Web control is a JSF custom component. If the component responds to client actions, it is created by extending the UICommand class and implementing ActionListener. Otherwise, the component extends the UIComponentBase class. The context, North arrow, and scale bar controls are examples of controls that do not respond to client-side action events.

faces.event package

The faces.event package contains event argument objects for some of the Web controls. It also contains:

• *ClientActionArgs* subclasses representing client actions such as dragging a rectangle or an image.

• *IMapToolAction* and *IPageLayoutToolAction* interfaces that you can use to implement your own map or page layout tools.

• Classes that act as descriptors for the map and page layout tools.

faces.renderkit.xml package

The faces.renderkit.xml package contains classes for the custom renderers associated with each Web control. These renderers output XML. This XML then goes through an XSLT transformation to a markup language, such as another XML or HTML, specified in an XSL file. The Java ADF's processing of XML is described in more detail in the section titled 'Building your first Web applications' earlier in this chapter.

faces.taglib package

All the tag handler classes that wrap the Web controls are stored in the faces.taglib package. For example, the *MapTag* class exposes the functionality of the map control. These tag handler classes extend from the JSF *UIComponentTag* class.

faces.validator package

The faces.validator package contains custom validator classes that validate user inputs. When an input is considered invalid, a validator throws a ValidatorException with a FacesMessage. This message shows the reason the input wasn't validated. When an input is considered invalid, no events in that request are executed.

data package

The data package contains the data classes that work with the Web controls. Some examples of classes in this package include *WebMap*, *WebOverview*, and *WebPageLayout*. These classes are typically abstract and require concrete implementations for use with their respective Web controls. The default implementations of these abstract classes, to be used with a GIS server, can be found in the com.esri.arcgis.webcontrols.ags.data package. Each of these classes stores properties that reflect the type of control they represent. The *WebMap* class, for example, can retrieve information about the map such as width, height, data frame, and image format.

ags.data package

The ags.data package comprises the default implementations of the abstract business classes defined in the *com.esri.arcgis.webcontrols.data* package and contains objects that store data in the ADF. For example, *AGSBuffer* stores buffer data and *AGSMapResource* stores information pertaining to the server connection. Classes that represent a GIS server's resources and other helper classes are also included in this package.

ags.faces.event package

The ags.faces.event package has implementations of the Java ADF server-side tool actions and listeners.

ags.util package

The ags.util package contains the AOUtil class, which provides ArcObjects helper methods.

util package

The util package contains utility helper classes. The *WebUtil* class gives you quick access to the *WebContext* and information about your *WebApplication* and *WebSession*. The *ApplicationProperties* class reads and stores information pertaining to the arcgis_webapps.properties file. The *XMLUtil* class provides access to XML functions for tasks such as creating documents or elements. This package also has classes that implement the MIME data servlet, the servlet context listener, and the session time-out servlet filter.

The context control establishes a working environment with a GIS server. By specifying a value for the *context* tag's *resource* and *resourceType* attributes, a connection to a GIS server is established and maintained.

Context Control Objects

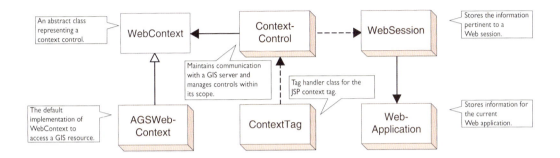

The diagram above illustrates how the *ContextControl* works with an abstract *WebContext* class. By default, the *ContextControl* provides a concrete subclass of *WebContext* called *AGSWebContext* that has been designed to work with a GIS server. The *ContextTag* class wraps the functionality of the *ContextControl* and enables the use of the *context* tag in a JSP page.

The *AGSWebContext* works with the *ContextControl* to take care of various tasks essential to working with a GIS server. *AGSWebContext* does the following:

- Establishes and maintains a connection with a GIS server.

- Detects the pooling method of a server object.

- When working with non-pooled server objects, connects to a GIS server when the session starts and persists the server context for the entire session.

- When working with pooled server objects, establishes a server context with a GIS server when a request is received and releases the server context once the request has been serviced.

- When working with pooled server objects, serializes the changed *MapDescription* and persists this even after the server context has been released. When the server context is released, the client's view of the server context must be persisted. For example, if a map is zoomed, the changed extent needs to be persisted on the client. The server context is re-created when a connection to the GIS server is reestablished on receipt of the next request and the *MapDescription* is reapplied.

- Maintains relationships with all the controls within its scope.

WORKING WITH THE CONTEXT CONTROL

The *ContextControl* acts as a container for all other controls. The first task of the context control is to create a *WebContext* for the current *WebSession* by passing in the resource, the resource type (map or geocode), the default data frame, and a class name for the implementation of the abstract *WebContext*. The default implementation of *WebContext* is *AGSWebContext*. This new *WebContext* object is stored in session scope and is accessible by invoking the following code. In this example, CONTEXTID is the user-assigned identifier for the context control.

```
WebUtil.getWebContext(facesContext, CONTEXTID)
```

During the initialization phase of *AGSWebContext*, new objects are created based on the resource type of the context. If the type is map, a new *AGSMapResource* and *AGSWebMap* are created. These objects work with the map control. If the type is geocode, new *AGSGeocodeResource* and *AGSWebGeocode* objects are created to work with the geocode control. The *WebMap* and *WebGeocode* objects can be retrieved by calling *WebContext.getWebMap* or *WebContext.getWebGeocode*.

You can ask for data objects used by the other controls, for example, *WebPageLayout*, *WebToc*, and *WebOverview*. These objects are automatically created and stored on the *WebContext* when the controls are placed on the page.

All other data objects are stored as request attributes to the *WebContext* and are only accessible by calling the *getAttribute* method. These objects are instantiated on an as-needed basis and are classified as managed context attributes. For example, the identify tool uses a data object called *AGSWebIdentifyResults* to store the results of an identify. To use *AGSWebIdentifyResults* you must first initialize the object so that you can call the *setLocation* method. The *setLocation* method sets the x,y point to use for the identify action. Once the *setLocation* method has been called, call *fetchResults* to return the results of the identify. The code for an identify is:

```
AGSWebIdentifyResults results =
  (AGSWebIdentifyResults)context.getAttribute
  (AGSWebIdentifyResults.WEB_CONTEXT_ATTRIBUTE_NAME);
  results.setLocation(new AGSPoint
   (wMap.toMapPoint(clientPoint.getX(), clientPoint.getY())));
  results.fetchResults();
```

The first line of code in this example asks for the *AGSWebIdentifyResults* object stored under its *WEB_CONTEXT_ATTRIBUTE_NAME*. Because this is the first time you are calling the *getAttribute* method, the attribute does not currently exist. The Java ADF checks to see if this object is a managed context attribute specified in the WEB-INF\classes\managed_context_attributes.xml file. If it is listed in this file, a new *AGSWebIdentifyResults* object is instantiated and stored under the appropriate attribute name.

The data objects of the other controls are described in the following sections.

If you need to add a custom object to the managed_context_attributes.xml file, you would use the following syntax:

```
<managed-context-attribute>
  <name>esriWebIdentifyResults</name>
  <attribute-class>
    com.esri.arcgis.webcontrols.ags.data.AGSWebIdentifyResults
  </attribute-class>
  <description>
    Results as a result of the identify tool operation.
  </description>
</managed-context-attribute>
```

On initialization or change of the WebContext, objects are automatically created or updated. In order to have objects that are managed context attributes updated by the context control, the objects must implement an interface called *WebContextInitialize* and provide an implementation of the *init* method.

Another important interface is *WebContextObserver*. This interface requires an implementation of the *update* method. When the current context is refreshed, the *WebContext* calls the *update* method for all registered observers. To become a registered observer, call the *WebContext.addObserver* method. The *refresh* method can also take an optional *args* argument. You would use the *args* argument if, for example, you wanted the observer to refresh only if the map has changed.

As mentioned earlier, a *WebContext* is part of a *WebSession*. A *WebSession* contains a *WebApplication*. The responsibilities of the *WebApplication* are:

- Read the arcgis_webapps.properties file in order to initialize impersonation.

- Load the collection of tool items.

- Load the collection of managed context attributes.

- Provide access to the resource bundle, enabling the lookup of localized strings.

- Specify a directory for file caching of MIME data.

METHODS

Every control works with an abstract class and a concrete implementation of this abstract class to enable access to the business logic. The *ContextControl* can retrieve the associated class by calling the *getWebContext* method, which then returns a *WebContext* object. The default implementation of *WebContext* is *AGSWebContext*. The *AGSWebContext* contains methods to:

- Create server objects.

- Retrieve the server context, map and layer descriptions, and data frames.

- When working with pooled server objects, load and save objects to a string.

- Retrieve the business objects associated with the *AGSWebContext* by calling methods such as *getWebMap* or *getWebOverview*.

- Refresh the *WebContext* and all controls registered as a *WebContextObserver* to the context.

The diagrams below show the methods available from *WebContext* and *AGSWebContext*.

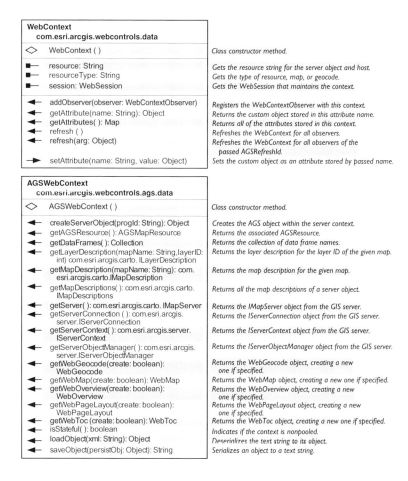

WebContext com.esri.arcgis.webcontrols.data	
◇ WebContext ( )	Class constructor method.
■— resource: String	Gets the resource string for the server object and host.
■— resourceType: String	Gets the type of resource, map, or geocode.
■— session: WebSession	Gets the WebSession that maintains the context.
◄— addObserver(observer: WebContextObserver)	Registers the WebContextObserver with this context.
◄— getAttribute(name: String): Object	Returns the custom object stored in this attribute name.
◄— getAttributes( ): Map	Returns all of the attributes stored in this context.
◄— refresh ( )	Refreshes the WebContext for all observers.
◄— refresh(arg: Object)	Refreshes the WebContext for all observers of the passed AGSRefreshId.
►— setAttribute(name: String, value: Object)	Sets the custom object as an attribute stored by passed name.

AGSWebContext com.esri.arcgis.webcontrols.ags.data	
◇ AGSWebContext ( )	Class constructor method.
◄— createServerObject(progId: String): Object	Creates the AGS object within the server context.
◄— getAGSResource( ): AGSMapResource	Returns the associated AGSResource.
◄— getDataFrames( ): Collection	Returns the collection of data frame names.
◄— getLayerDescription(mapName: String,layerID: int) com.esri.arcgis.carto. ILayerDescription	Returns the layer description for the layer ID of the given map.
◄— getMapDescription(mapName: String): com. esri.arcgis.carto.IMapDescription	Returns the map description for the given map.
◄— getMapDescriptions( ): com.esri.arcgis.carto. IMapDescriptions	Returns all the map descriptions of a server object.
◄— getServer( ): com.esri.arcgis.carto. IMapServer	Returns the IMapServer object from the GIS server.
◄— getServerConnection ( ): com.esri.arcgis. server.IServerConnection	Returns the IServerConnection object from the GIS server.
◄— getServerContext( ): com.esri.arcgis.server. IServerContext	Returns the IServerContext object from the GIS server.
◄— getServerObjectManager( ): com.esri.arcgis. server.IServerObjectManager	Returns the IServerObjectManager object from the GIS server.
◄— getWebGeocode(create: boolean): WebGeocode	Returns the WebGeocode object, creating a new one if specified.
◄— getWebMap(create: boolean): WebMap	Returns the WebMap object, creating a new one if specified.
◄— getWebOverview(create: boolean): WebOverview	Returns the WebOverview object, creating a new one if specified.
◄— getWebPageLayout(create: boolean): WebPageLayout	Returns the WebPageLayout object, creating a new one if specified.
◄— getWebToc (create: boolean): WebToc	Returns the WebToc object, creating a new one if specified.
◄— isStateful( ): boolean	Indicates if the context is nonpooled.
◄— loadObject(xml: String): Object	Deserializes the text string to its object.
◄— saveObject(persistObj: Object): String	Serializes an object to a text string.

Take a look at some examples on how you would use *AGSWebContext*. In these examples, you'll see that each must start with this line of code that retrieves the context from the JSF Faces context. In this example, CONTEXTID should be replaced with the identifier of the context control.

In this code example as well as others in the remainder of this chapter, CONTEXTID is a placeholder for the identifier of the context control. For example, if the context tag specifies the id="mapContext", then CONTEXTID would be replaced with "mapContext".

```
AGSWebContext context = (AGSWebContext)WebUtil.
  getWebContext(FacesContext.getCurrentInstance(), CONTEXTID);
```

In a Java ADF application, if the overview control extent changes, the map control refreshes to that new extent as well because the map is registered as an observer to the context. In this code example, the *WebContext* and all the observer controls are refreshed.

```
AGSWebContext context = (AGSWebContext)WebUtil.
  getWebContext(FacesContext.getCurrentInstance(), CONTEXTID);
context.refresh();
```

This example shows how to retrieve objects for the SOM, server context, and server object's map descriptions and data frames.

```
AGSWebContext context = (AGSWebContext)WebUtil.
  getWebContext(FacesContext.getCurrentInstance(), CONTEXTID);
IServerObjectManager som = context.getServerObjectManager();
IServerContext srvContext = context.getServerContext();
IMapDescriptions descriptions = context.getMapDescriptions();
Collection dataFrames = context.getDataFrames();
```

In this example, a new server object is created within the server context.

```
AGSWebContext context = (AGSWebContext)WebUtil.
  getWebContext(FacesContext.getCurrentInstance(), CONTEXTID);
IPolyline pLine = new
  IPolylineProxy(context.createServerObject(Polyline,getClsid()));
```

In this example, a request attribute is stored in the *WebContext*.

```
AGSWebContext context = (AGSWebContext)WebUtil.
  getWebContext(FacesContext.getCurrentInstance(), CONTEXTID);
context.setAttribute("newAttribute", MyObject)
```

ACCESSING OBJECTS THROUGH REQUEST SCOPE

The *WebContext* is stored in session scope and is available for the duration of the client session. Earlier you saw how to retrieve the *WebContext* in Java code by calling the *getWebContext* method. Now take a look at how to retrieve the *WebContext* from session within a JSP page. The previous example stored a request attribute on the *WebContext* called *newAttribute* and set the value to *MyObject*.

To retrieve the *WebContext* and access a property on it, use the following JSTL syntax:

```
${sessionScope['CONTEXTID'].attributes['newAttributeName'].myProperty}
```

If you want to use this request attribute within a JSF component, you must follow the JSF value binding expression syntax. The syntax of a value binding expression is syntactically the same as JSP 2.0 variable references defined in their expression language. A value binding must be referenced using #{Value}. To change the previous example to use value binding expression syntax, change the "$" to a "#" to reference the attribute stored in the *WebContext*.

```
#{sessionScope['CONTEXTID'].attributes['newAttribute'].myProperty}
```

In this example, variables are set using JSTL to store the resource and resource type used in the *WebContext*.

```
<c:set var="res" value="${sessionScope['CONTEXTID'].resource}"/>
<c:set var="resType" value="${sessionScope['CONTEXTID'].resourceType}"/>
```

In this example, the value of the context's resource string is output using the JSF *outputText* tag.

```
<jsfh:outputText value="#{sessionScope['CONTEXTID'].resource }"
```

The other Java ADF Web controls are discussed in the remainder of this chapter. For each control, look for a section titled 'Accessing objects through WebContext' to see examples of how to access other objects stored in the *WebContext*.

As mentioned, the context control establishes a working environment with a GIS server. Now that you have learned how to use the context control, the following sections describe how the context control takes care of using pooled and non-pooled server objects. Also included in the remainder of this section is a discussion on how the context control provides access to ArcObjects components that allow more advanced customization of your Web applications.

GETTING AND RELEASING SERVER CONTEXTS

In the Java ADF, all Web controls work within the context control. The context control takes care of working with the server context. This means you do not need to make coding considerations for pooled versus non-pooled objects; the release of the server context is taken care of by the context control.

How the context control works with the server context

If you want to write a new class that requires saving state between requests for a pooled object, then you will need to implement the *WebLifecycle* interface and provide *activate* and *passivate* methods. The *WebContext* calls *activate* and *passivate* for any object that is an instance of *WebLifecycle* if it is stored as an attribute to the *WebContext*.

Take a look at the coding considerations you need to make when writing a new class.

When you work with *AGSWebMap*, *AGSWebPageLayout*, or *AGSWebGeocode*, you are indirectly working with a server context because these classes are associated with an *AGSResource* or *AGSGeocodeResource* object. *AGSResource* establishes communication with a MapServer context based on the pooling model. *AGSResource* utilizes the *WebLifecycle* interface and implements two important methods: *activate* and *passivate*. The *activate* method does the following:

- Creates a new *ServerConnection* and connects to the server

- Creates a new server context

- Retrieves the server object from the server context

The *passivate* method releases the server context.

If the server object is pooled, *activate* is called when a request is received and *passivate* is called when the response has been generated. If the server object is non-pooled, *activate* is called once and maintained for the duration of the session or until explicitly released by calling *passivate*.

Maintaining state using WebLifecycle

In addition, *WebLifecycle* can be used to maintain any state needed between requests. The *WebContext* calls the *activate* and *passivate* methods as needed for every object registered. In the case of *AGSWebMap*, the state of the map is maintained between requests by storing the *MapDescription* and *ImageDescription* objects.

This code shows how the *activate* and *passivate* methods save these objects:

```
public void activate() {
  try {
    if (null != xImageDesc)
```

```
      imageDesc = new
      IImageDescriptionProxy(agsContext.loadObject(xImageDesc));
   if (null != xCopyrightSymbol)
      copyrightSymbol = new
      IFormattedTextSymbolProxy(agsContext.loadObject(xCopyrightSymbol));
   }
   catch(Exception e) {
    logger.log(Level.SEVERE, "Cannot activate map.", e);
   }
   finally {
    xImageDesc = null;
    xCopyrightSymbol = null;
   }
  }
 public void passivate() {
   try {
    if (null != imageDesc)
     xImageDesc = agsContext.saveObject(imageDesc);
    if (null != copyrightSymbol)
     xCopyrightSymbol = agsContext.saveObject(copyrightSymbol);
   }
   catch(Exception e) {
    logger.log(Level.SEVERE, "Cannot passivate map.", e);
   }
   finally {
    imageDesc = null;
    copyrightSymbol = null;
   }
  }
}
```

WORKING WITH FINE-GRAINED ArcObjects

The Java ADF provides a set of classes that allow you to create Web applications that incorporate GIS functionality. While the Java ADF exposes many of the common GIS operations, it doesn't attempt to duplicate all of the functionality of ArcObjects. What the Java ADF does provide, however, is a mechanism for accessing ArcObjects directly. This is referred to in this book as working with the fine-grained ArcObjects that comprise a server object. This section describes some of the things you need to be aware of when working with fine-grained ArcObjects.

AGSWebContext and *AGSWebMap* expose a *getServer* method that provides access to the *IMapServer* interface of the *MapServer* coclass. Through *IMapServer*, you can access the fine-grained ArcObjects that make up the map document being served on your GIS server. When you call any of the methods on the *MapServer* coclass's interface, the state of the fine-grained ArcObjects in the server object instance is not permanently changed. Rather, the *MapDescription* is applied for the duration of the method, and the state of the fine-grained ArcObjects returns to the previous state at the end of the request. The same holds true for the *PageDescription* when using *IMapServerLayout*, retrieved using the *getServerLayout* method on *AGSWebPageLayout*. However, when you work with fine-grained ArcObjects, the state of the server object instance may be permanently changed, depending on

AGSWebMap *and* AGSWebPageLayout *are discussed in detail later in this chapter.*

what methods you call. For this reason, you should work with non-pooled server objects if you want to make calls to fine-grained ArcObjects that change the state of the object.

You may have noticed that when you work with the Java ADF Web controls, the Web controls appear to maintain things such as layer visibility and the current map extent. This information is stored in session state as part of the Web application itself. The actual server object instance running on the GIS server is unaltered. Whenever a request gets sent to the server, this state information is reapplied to the server object instance for the duration of the request.

You can learn more about MapDescription in Chapter 4, 'Developing ArcGIS Server applications'.

The map and overview controls store their state in the *MapDescription* object. The *MapDescription* gets applied to the map server object on every draw and query request. The page layout control stores its state in the *PageDescription* object, which also gets applied to the map server object on every draw request.

Because the map, overview, and page layout controls store some properties that define their state, it means that these properties get applied to the map server object and will override any changes you may make directly to the fine-grained ArcObjects in the server object instance.

For example, suppose you access the *IActiveView* interface to the data frame you're working with and change *IActiveView.setExtent*. If you subsequently call *WebContext.refresh*, you might expect that the data frame will be drawn with the extent you set on *IActiveView*. This will not happen because the *MapDescription* stored by the map control also has an *extent* property, which you can access via *MapDescription.setMapArea*, and this property overrides the underlying changes to the extent made through *IActiveView*. This behavior applies to every property of the *MapDescription* coclass for the map and overview controls and every property of the *PageDescription* coclass for the page layout control.

The following steps outline how you can make changes to the fine-grained ArcObjects and also see them reflected in your Web application.

1. Call *AGSWebContext.applyDescriptions* to apply the current state of the map, page, and layer descriptions to the underlying server object

2. Make any changes to the server object instance that you need to, for example, adding or removing a layer, modifying the extent, or changing layer rendering.

3. Call *AGSWebContext.reloadDescriptions* to refresh the map description and reload the Web controls with these changes held by the fine-grained ArcObjects in the server instance.

The following code illustrates this approach:

```
AGSWebContext context =
  (AGSWebContext)WebUtil.getWebContext(event.getComponent());
AGSWebMap webMap = (AGSWebMap)context.getWebMap();
 // Step 1: Apply the descriptions.
context.applyDescriptions();
// Step 2: Modify the extent.
IActiveView map =
  new IActiveViewProxy(mapServerObjects.getMap(webMap.getFocusMapName()));
IEnvelope extent =
```

```
new IEnvelopeProxy(context.createServerObject(Envelope.getClsid()));
extent.putCoords(6837437,1854359,6849844,1842728);
map.setExtent(extent);
// Step 3: Refresh the map description and reload the Web controls.
context.reloadDescriptions();
```

The map control provides the display functionality of the ArcMap data view for a map document served using ArcGIS Server. As with data view, a map control displays a single data frame in the map document. Using the map control, you can navigate the data frame, for example, pan and zoom, and access the fine-grained ArcObjects that comprise it.

Map Control Objects

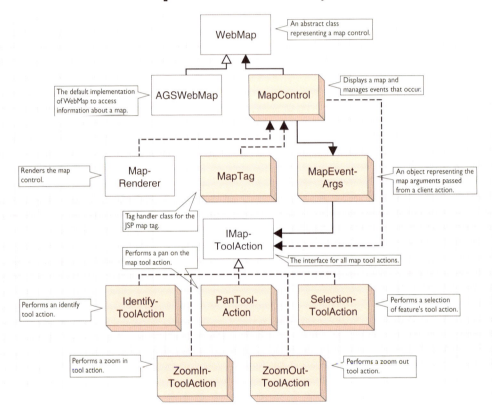

The diagram above illustrates the relationships between the map control and its associated business objects. The *MapControl* works with an abstract *WebMap* object. The *AGSWebContext* creates a new *AGSWebMap* object, which is a sub-type of the *WebMap*. The *MapControl* then associates itself with the *WebMap* object. The *MapRenderer* class renders the control to the appropriate markup, and the *MapTag* class exposes the control as a JSP custom tag named *map*. In addition, events are exposed through *MapEventArgs,* and various map tool actions can be performed on the control.

The majority of the business logic is exposed through *AGSWebMap*, which performs the following tasks:

- Render the map.

- Expose an API that provides access to the details of the map.

- Set up the map for client-side actions. These client-side actions could be dragging a rectangle on the map for zoom operations, clicking a map feature for identify operations, or drawing complex vector features, such as a circle or a polyline.

- Associate the client-side tool actions to server-side action classes.

- Set up a framework to implement new tool actions.

- Execute server-side listeners.

THE MODEL VIEW CONTROLLER PATTERN

Unlike the context control, the map control is a visual JSF component. The map control renders a view, controls events, and implements the business logic. To better understand the architecture of the map control, take a look at a popular design pattern known as the Model View Controller (MVC).

The Java ADF Web controls follow the MVC design pattern. Knowing this pattern will help you understand how the map control responds to tool actions and listeners, accesses the business logic, and renders the map. You can then apply your knowledge to customize the Java ADF and plug in your own tools or listeners.

The MVC pattern distinctly separates the different behaviors of the components: view, controller, and model.

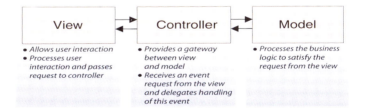

In the MVC pattern:

- The view provides a visual representation of the component. The client interacts with this view and submits events to the controller.

- The controller delegates business processing to the other entities, which make use of the model objects to carry out tasks. When events are queued on a component, the component can delegate the handling of these events to the listeners registered on it. The listeners, in turn, communicate with the business objects to process these events.

- The model objects maintain the data and provide methods that aid in the actual business processing.

- Once the business task is finished, the view once again renders the control to the client.

The diagram below illustrates how the MVC paradigm maps to the Java ADF:

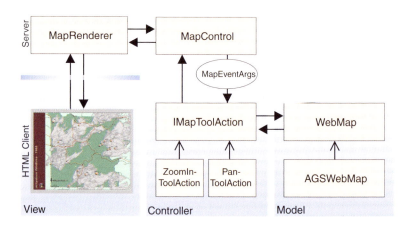

View

In a JSP usage, the map control is placed on the page using the following code:

```
<ags:context id="mapContext" resource="ServerObject@MyHost" >
  <ags:map id="Map0" left="0" top="0" width="400" height="300" />
<ags:context />
```

The *map* tag is invoked using the *MapTag* handler class. The *MapTag* uses the *MapRenderer* to generate the markup—XML translated to HTML—and to display the map. On the client-side, the map is set up for various JavaScript operations, such as zoom, pan, and identify. After a client-side action is performed, the *MapRenderer* decodes the request and passes it on to the *MapControl*.

Controller

The *MapControl* processes the request. First, the *MapControl* creates an event argument object, *MapEventArgs*, to represent the event in an object form. The *MapControl* has tool actions associated with it. Based on the operation that the client performed, the map control next delegates the business processing to the appropriate tool action object. For example, if the client performed a *ZoomIn* action, the map control would forward the *MapEventArgs* to the *ZoomInToolAction* class. The tool action then derives the necessary information from the event arguments—in the case of *ZoomIn*, the x,y coordinates of the zoom rectangle. The mapping business objects, such as *WebMap* and *AGSWebMap*, use this information to perform the task. Finally, on completion of the task, the *MapRenderer* displays the changed map and the whole cycle repeats.

Model

The business objects, *WebMap* and *AGSWebMap*, interface with the GIS server and the necessary ADF objects to perform GIS tasks for the map control. On initialization of *AGSWebContext*, a new *AGSWebMap* is created, which is a sub-type of the *WebMap*. The *MapControl* associates itself with this *WebMap* object.

This diagram shows the methods available from *WebMap* and *AGSWebMap*.

WebMap com.esri.arcgis.webcontrols.data	
◇ WebMap(context: WebContext)	Class constructor method.
■–■ copyrightText: String	Gets/sets the copyright text on the map.
■–■ dpi: int	Gets/sets the dots per inch of the map image.
■–■ height: int	Gets/sets the height of the map image.
■–■ imageFormat: String	Gets/sets the image format of the map.
■–■ width: int	Gets/sets the width of the map image.
◄– getWebContext( ): WebContext	Returns the WebContext associated with this WebMap.
◄– isMimeData( ): boolean	Indicates whether the control returns MIME data.
◄– isVisible( ): boolean	Indicates whether the map is visible.
►– setUseMIMEData(useMIMEData: boolean)	Sets the control to return MIME data.
►– setVisible(visible: boolean)	Sets the visibility of the map.

AGSWebMap com.esri.arcgis.webcontrols.ags.data	*The default implementation of WebMap.*
◇ AGSWebMap(agsContext: AGSWebContext)	Class constructor method
■–■ focusMapExtent: com.esri.arcgis. geometry.IEnvelope	Gets/sets the extent of the focus map.
■–■ focusMapName: String	Gets/sets the name of the focus map.
◄– createServerObject(progId: String): Object	Creates the AGS object represented by the progId.
◄– getFeatureLayers(mapName: String): Hashtable	Returns the feature layers for a given map.
◄– getFocusMapDescription( ):com.esri. arcgis.carto.IMapDescription	Returns the focus map's map description.
◄– getFocusMapFeatureLayers( ): Hashtable	Returns the feature layers of the focus map.
◄– getFocusMapFullExtentScale( ): double	Returns the scale of the focus map when it is at full extent.
◄– getFocusMapScale( ): double	Returns the current scale of the focus map.
◄– getFocusMapServerInfo ( ): com.esri.arcgis.carto.IMapServerInfo	Returns the focus map's IMapServerInfo object.
◄– getFocusMapUnits( ): int	Returns the units of the focus map.
◄– getFocusMapUnitsString( ): String	Returns the units of the focus map as descriptive text.
◄– getImageDescription( ): com.esri.arcgis.carto.IImageDescription	Returns the image description for the map.
◄– getImageUrl ( ): String	Returns the URL of the map to be displayed by the map control.
◄– getMapFullExtentScale(mapName: String): double	Returns the scale of the given map when it is at full extent.
◄– getMapScale(mapName: String): double	Returns the current scale of the given map.
◄– getMapUnitsOptions( ): ArrayList	Calls AOUtil.getUnits() to return the options for map units.
◄– getServer( ): com.esri.arcgis.carto. IMapServer	Returns the IMapServer object from the server.
►– setHeight(height: int)	Sets the height of the map.
►– setImageFormat(imageFormat: String)	Sets the image format of the returned map image.
►– setWidth(width: int)	Sets the width of the map.
◄– fromMapPoint(point: com.esri.arcgis. geometry.IPoint): int[ ]	Converts a map point to screen coordinates.
◄– toMapPoint(x: int, y: int): com.esri.arcgis. geometry.IPoint	Converts the screen coordinates to a map point.

The *AGSWebMap* is the gateway to information about the map displayed in the map control.

Take a look at some code that works with *AGSWebMap*.

The following code shows how to access information about the map.

```
AGSWebMap wMap = (AGSWebMap)context.getWebMap();
Double mapScale = wMap.getMapScale("dataframe1");
System.out.println("The map scale is: " + mapScale);
ArrayList featureLayers = wMap.getFeatureLayers();
for (int i=0; i<featureLayers.size(); i++)
  System.out.println("Layer name: " + featureLayers.get(i));
```

The following code shows how to change the current extent of the map. Use the *toMapPoint* method to convert from screen coordinates to map coordinates.

```
AGSWebMap wMap = (AGSWebMap)context.getWebMap();
IEnvelope extent = wMap.getFocusMapExtent();
extent.centerAt(wMap.toMapPoint(20), wMap.toMapPoint(20));
wMap.setFocusMapExtent(extent);
context.refresh
  (com.esri.arcgis.webcontrols.ags.data.AGSRefreshId.MAP_OPERATION);
```

The following code shows how to change the image format for the map.

```
AGSWebMap wMap = (AGSWebMap)context.getWebMap();
wMap.setImageFormat("JPG");
```

The following code shows how to access objects on the GIS server with *AGSWebMap*.

```
AGSWebMap wMap = (AGSWebMap)context.getWebMap();
IMapServerInfo serverInfo = wMap.getFocusMapServerInfo();
int mapUnits = serverInfo.getMapUnits();
System.out.println("Map units: " + AOUtil.getUnitsString(mapUnits,
  agsContext));
```

INTERACTING WITH THE MAP CONTROL USING TOOLS

When you interact with a map control, an action happens on the client that triggers an action on the server. The client-side action is what allows you to, for example, draw a box over the control. The server-side action is what allows you to use the coordinates of the box and, for example, zoom in.

During every interaction with a map control, two things happen:

• An end user interacts with the map control on the client.

  The client-side action is controlled by a JavaScript function that executes on the client without any requests being sent to the server. The Java ADF includes JavaScript code for common interactions, such as drawing a box.

• The server executes an action based on the client interaction.

  Once the client-side action is completed, the client sends a request to the server to execute the server-side action. The *MapRenderer* decodes the request and passes the *FacesContext* to the *MapControl*. The *FacesContext* contains the request parameters and the tool key, for example, *ZoomIn*. The *MapControl* then creates a *MapEventArgs* object that holds the tool action and the client arguments for that tool action. Last, an *ActionEvent* is fired and the map control's *processAction* method retrieves the appropriate *IMapToolAction*, for example,

ZoomInToolAction, and calls the execute method with the *MapEventArgs*. The *ZoomInToolAction* retrieves the arguments and performs the zoom in.

A *ToolItemCollection* consists of one or more *ToolItem* objects. *ToolItem* stores the tool key, the client-side action, and the server-side action class, *IToolAction*. The diagram below shows the methods on the *ToolItem* object.

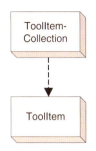

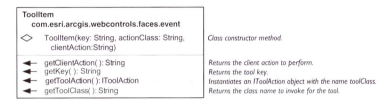

A *ToolItemCollection* consists of one or more *ToolItem* objects.

As a Web application developer, you build functionality into your Web application by defining a set of client-side actions and associated server-side actions.

By default, the Java ADF initializes the *ToolItemCollection* object with the following map tool items:

Key	Server ToolAction	JavaScript method
ZoomIn	ZoomInToolAction	MapDragRectangle
ZoomOut	ZoomOutToolAction	MapDragRectangle
Pan	PanToolAction	MapDragImage
Identify	IdentifyToolAction	MapPoint
Selection	SelectionToolAction	MapDragRectangle

The tool actions are found in the com.esri.arcgis.webcontrols.ags.faces.event package, and the JavaScript is located in the Web application's js directory. You can change an existing tool provided by the Java ADF or add a new tool.

Tools are listed in the default.xml file found in your application's WEB-INF\classes\tools directory. You can edit the existing default.xml file or write your own XML file. If you choose to write your own, you either need to name your file default.xml or set the *map* tag's *toolItemCollection* attribute to the name of your new XML file. The only requirement is that the file be placed in the WEB-INF\classes\tools directory.

Adding a tool

To add a tool, a new tool item must be added to the default.xml file. The code below is a tool item for a measure tool.

```
<tool-item>
  <key>Measure</key>
  <action-class>
  com.esri.arcgis.webcontrols.ags.faces.event.MeasureToolAction</action-class>
  <client-action>MapPolyline</client-action>
</tool-item>
```

This new tool item sets the client-side action and the server-side action class. The Java ADF JavaScript library has a *MapPolyline* client action. The *MapPolyline* client action allows the drawing of multiple line segments on the map and finds the total distance of those segments. For the server-side action, the next section describes how you would write an action class called *MeasureToolAction*.

Creating a server-side action class

A common task when working with the Java ADF will be to create a new server-side tool action class. In the previous section, you added a tool item for a measure tool. Now, implement the code for the *MeasureToolAction* class.

Create MeasureToolAction.java and add the following code.

```
package com.esri.arcgis.webcontrols.ags.faces.event;

import com.esri.arcgis.webcontrols.faces.event.*;
import com.esri.arcgis.webcontrols.ags.data.AGSWebMap;
import com.esri.arcgis.webcontrols.ags.data.AGSWebContext;
import com.esri.arcgis.geometry.*;

public class MeasureToolAction implements IMapToolAction {
  public static final String MEASURE_REQUEST_ATTRIBUTE_NAME =
    "esriMeasureDistance";

  public void execute(MapEventArgs args) throws Exception {
    PolylineArgs polyLineArgs = (PolylineArgs)args.getClientActionArgs();
    ClientPoint[] points = polyLineArgs.getPoints();

    AGSWebContext agsContext = (AGSWebContext)args.getWebContext();
    AGSWebMap wMap = (AGSWebMap)agsContext.getWebMap();

    IPolyline pLine = new
      IPolylineProxy(wMap.createServerObject("Polyline.getClsid()));
    IPointCollection collection = new IPointCollectionProxy(pLine);
    for(int i = 0; i < points.length; ++i) {
      collection.addPoint(wMap.toMapPoint(points[i].getX(),
        points[i].getY()), null, null);
    }

      agsContext.setAttribute(MEASURE_REQUEST_ATTRIBUTE_NAME,
        Double.toString(pLine.getLength()));
  }
}
```

Look at this code in more detail.

All map tool action classes must implement *IMapToolAction* and the *execute* abstract method.

```
public class MeasureToolAction implements IMapToolAction {
        public void execute(MapEventArgs args) throws Exception {
        }
}
```

As mentioned previously, the *MapEventArgs* object was provided to the *MapControl* by the *MapRenderer*; the *MapRenderer* received this information from the client's action request.

To implement the *execute* method, follow these steps:

1. Retrieve the client action arguments from the *MapEventArgs* object. In the case of the measure tool, the client action arguments are the line segments drawn on the map to measure a distance.

   The Java ADF has several classes representing the available *ClientActionArgs*. The table below describes the default action mappings for the map control.

Client Action	ClientActionArgs
MapPoint	PointArgs
MapDragImage	DragImageArgs
MapDragRectangle	DragRectangleArgs
MapPolyline	PolylineArgs
MapLine	LineArgs
MapPolygon	PolygonArgs
MapCircle	CircleArgs
MapOval	OvalArgs

   You can add to the default collection of action argument classes by calling this code:

   ```
   ClientActionArgs.addClientActionArgs("actionName", "argsClassName");
   ```

   In this example, the *PolylineArgs* class, from the default collection, is used because the client-side action is *MapPolyline*.

   ```
   PolylineArgs polyLineArgs = (PolylineArgs)args.getClientActionArgs();
   ClientPoint[] points = polyLineArgs.getPoints();
   ```

   Now, all the points are accessible to write some ArcObjects code to calculate the length of the polyline.

2. Retrieve the *WebMap* object.

   ```
   AGSWebContext agsContext = (AGSWebContext)args.getWebContext();
   AGSWebMap wMap = (AGSWebMap)agsContext.getWebMap();
   ```

3. Use the *createServerObject* method on the *WebContext* to create any new server objects within the server context.

   ```
   IPolyline pLine = new
       IPolylineProxy(agsContext.createServerObject("Polyline.getClsid()));
   ```

4. Convert the points from screen coordinates to map points by using the *toMapPoint()* method on the *WebMap*.

   ```
   IPointCollection collection = new IPointCollectionProxy(pLine);
   for(int i = 0; i < points.length; ++i) {
       collection.addPoint(wMap.toMapPoint(points[i].getX(),
       points[i].getY()), null, null);
   }
   ```

A collection of points is stored in the *pLine* object. The length of the line can be determined by using *pLine.getLength*. The length of the polyline is then stored as an attribute to the *WebContext* so it can be retrieved in the JSP page.

```
agsContext.setAttribute(MEASURE_REQUEST_ATTRIBUTE_NAME,
    Double.toString(pLine.getLength()));
```

Now that you have completed the code for MeasureToolAction.java, navigate to your application's WEB-INF directory and create this structure: classes\com\esri\arcgis\webcontrols\ags\faces\event. Compile the file and place the class file in the event directory.

To add the measure tool to the JSP, edit your JSP file to include this code that will display the measure tool as well as text that reports the measure distance.

```
<ags:context id="mapContext" resource="world@localhost">
  <ags:map ... />
  <input type="button" value="Measure" onclick="MapPolyline('Map0',
    'Measure');" />
  <div style="position:absolute;left:233px;top:600px;">
    Measured Distance: <font color="red"><c:out value="${sessionScope
    ['mapContext'].attributes['esriMeasureDistance']}"/></font>
  </div>
</ags:context>
```

To use the measure tool, click Measure and click the map at the points you want to measure. Double-click to end measuring; the measure distance, reported in map units, will be displayed under the map.

INTERACTING WITH THE MAP CONTROL USING LISTENERS

Not all actions performed on the map require a client-side action. In some cases, you may want to implement custom server-side functionality invoked by a user command. JSF ships with some standard command components, such as buttons and hyperlinks. These command components can be associated with listeners. When the command component is clicked, the JSF framework triggers a server-side event and calls the associated listeners to process the event.

The Java ADF provides these listeners as part of the com.esri.arcgis.webcontrols.ags.faces.event package for the map:

- DirectionalPanListener

- FixedZoomListener

- ZoomRelativeToFullExtentListener

- ZoomFullExtentListener

You can get more details about the listeners provided with the Java ADF in the ArcGIS Developer Help.

To program a listener in JSF, you must implement the *javax.faces.event.ActionListener* interface. This interface requires an implementation of the following abstract method:

```
public void processAction(ActionEvent event)
```

This example shows the JSP code for the *FixedZoomListener*. The purpose of this listener is to zoom a map in or out by a fixed factor. In this example, the zoom factor is 50 percent.

```
<jsfh:commandButton id="zoom" image="images/zoom.gif">
  <jsfc:actionListener type=
  "com.esri.arcgis.webcontrols.ags.faces.event.FixedZoomListener" />
  <jsfc:attribute name="factor" value="0.5" />
</jsfh:commandButton>
```

The JSF *commandButton* component, when clicked, sends the event to the listener associated with the *actionListener* tag. The *factor* attribute is passed to the *FixedZoomListener*, and the server-side event is handled.

This is the implementation of the listener code.

```java
package com.esri.arcgis.webcontrols.ags.faces.event;
import javax.faces.event.*;
import com.esri.arcgis.webcontrols.util.WebUtil;
import java.util.logging.*;
import com.esri.arcgis.webcontrols.data.*;
import com.esri.arcgis.geometry.*;
import com.esri.arcgis.webcontrols.ags.data.*;


public class FixedZoomListener implements ActionListener {
  private static Logger logger =
    Logger.getLogger(FixedZoomListener.class.getName());
  public static final String FACTOR = "factor";
  public void processAction(ActionEvent event) throws
    javax.faces.event.AbortProcessingException {
      try {
        AGSWebMap wMap =
          (AGSWebMap)WebUtil.getWebContext(event.getComponent()).getWebMap();
        double factor =
          Double.parseDouble((String)event.getComponent().getAttribute(FACTOR));
        IEnvelope extent = wMap.getFocusMapExtent();
        extent.expand(factor, factor, true);

        IEnvelope fullExtent = wMap.getFocusMapServerInfo().getFullExtent();
        if(extent.getWidth() > fullExtent.getWidth() || extent.getHeight() >
          fullExtent.getHeight())
        extent = fullExtent;

        wMap.setFocusMapExtent(extent);
        wMap.getWebContext().refresh(AGSRefreshId.MAP_OPERATION);
      }
      catch(Exception e) {
        logger.log(Level.WARNING, "Unable to process action for
          FixedZoomListener.", e);
      }
  }
}
```

The first step in implementing a listener is to implement from the *ActionListener* interface.

```java
public class FixedZoomListener implements ActionListener {
}
```

As mentioned, *ActionListener* requires the implementation of the abstract method, *processAction*. This method gets invoked from the client's command action. Refer to the code comments for more details about this code.

```java
public static final String FACTOR = "factor";
public void processAction(ActionEvent event) throws
javax.faces.event.AbortProcessingException {
  try {
```

```
// The WebMap, retrieved from the WebContext, can return
// the AGSWebMap object using the getWebMap method on WebContext.
AGSWebMap wMap =
  (AGSWebMap)WebUtil.getWebContext(event.getComponent()).getWebMap();
// Get the factor attribute provided to the listener.
double factor =
  Double.parseDouble((String)event.getComponent().getAttribute(FACTOR));
// Get the focus map extent from the WebMap and store this extent.
IEnvelope extent = wMap.getFocusMapExtent();
// Using ArcObjects IEnvelope expand method, stretch the
// current extent by the factor specified.
extent.expand(factor, factor, true);
// If the new extent is greater than the full extent of
// the focus map, set the new extent to the full extent.
IEnvelope fullExtent = wMap.getFocusMapServerInfo().getFullExtent();
if(extent.getWidth() > fullExtent.getWidth() || extent.getHeight() >
  fullExtent.getHeight())
 extent = fullExtent;
// Set the calculated extent to the new extent.
wMap.setFocusMapExtent(extent);
// Update the context and tell all the observers of the context
// that the map has changed and they may need to update.
wMap.getWebContext().refresh(AGSRefreshId.MAP_OPERATION);
}
 catch(Exception e ) {
   logger.log(Level.WARNING, "Unable to process action for
     FixedZoomListener.", e);
 }
 }
```

The JSF listeners (working within the JSF framework) eliminate the need for you, as the Web developer, to take care of when and how the server-side event is triggered.

INTERACTING WITH THE MAP CONTROL USING ACTION METHODS

A JSF component can trigger events by associating itself to a JavaBeans method. The Java ADF refers to these as action methods. For a list of the available action methods see the ArcGIS Server Java ADF Listeners and Actions documentation in the ArcGIS Developer Help.

Now take a look at an example of how you would create one of these action methods. In the previous section, a *FixedZoomListener* was created. Now you will implement this functionality using an action method.

1. Implement the bean class *AGSFixedZoom* and the *doFixedZoom* action method.

```
package com.esri.arcgis.webcontrols.ags.data;

import com.esri.arcgis.webcontrols.data.WebContext;
import com.esri.arcgis.webcontrols.data.WebContextInitialize;
import java.util.logging.Logger;
import java.util.logging.Level;
import com.esri.arcgis.geometry.*;
```

```
// Implement WebContextIntialize to update this class with
// the current WebContext.
public class AGSFixedZoom implements WebContextInitialize {
  private static Logger logger =
    Logger.getLogger(AGSFixedZoom.class.getName());
  private AGSWebContext m_agsContext;
  // Set a default zoom factor.
  double factor = .5;

  // Update the value of WebContext on initialization or change.
  public void init(WebContext agsContext){
    if(agsContext == null || !(agsContext instanceof AGSWebContext))
      throw new IllegalArgumentException
        ("WebContext is null or is not an instance of AGSWebContext.");
    this.m_agsContext = (AGSWebContext)agsContext;
  }
  // Sets the zoom factor.
  public void setFactor(double factor){
    this.factor = factor;
  }
  // Returns the zoom factor.
  public double getFactor(){
    return this.factor;
  }
  // Implement the doFixedZoom action method.
  // Retrieve the WebMap and determine the extent of the focus map.
  //  Expand the IEnvelope by the set factor in both directions. Change
  //  the WebMap to use the new map extent and refresh the WebContext.
  public String doFixedZoom(){
    try {
    AGSWebMap wMap = (AGSWebMap)m_agsContext.getWebMap();
    IEnvelope extent = wMap.getFocusMapExtent();
    extent.expand(factor, factor, true);
    IEnvelope fullExtent =
      wMap.getFocusMapServerInfo().getFullExtent();
    if(extent.getWidth() > fullExtent.getWidth() ||
      extent.getHeight() > fullExtent.getHeight())
      extent = fullExtent;
    wMap.setFocusMapExtent(extent);
    wMap.getWebContext().refresh(AGSRefreshId.MAP_OPERATION);
    }
    catch(Exception _) {
    logger.log(Level.WARNING, "Unable to do fixed zoom", _);
    }
    return null;
  }
}
```

Managed context attributes are discussed in more detail in the 'Working with the context control' section earlier in this chapter.

2. Add the *AGSFixedZoom* class as a managed context attribute. Edit the Web application's managed_context_attributes.xml file with the entry below. The new bean is referenced as esriAGSFixedZoom.

```
<managed-context-attribute>
```

```
<name>esriAGSFixedZoom</name>
<attribute-class>com.esri.arcgis.webcontrols.ags.data.AGSFixedZoom
   </attribute-class>
<description>Performs a fixed zoom on the map.</description>
</managed-context-attribute>
```

3. Write JSP code to access the *doFixedZoom* action method and dynamically set the factor using a JSF inputText box.

```
<ags:context id="mapContext" resource="world@localhost">
  <ags:map id="Map0" left="233" top="115" width="535" height="408" ... />
  <table cellpadding="0" cellspacing="1" class="tblToolBar">
  <tr>
    <td>
     <jsfh:inputText id="factor" value="#{sessionScope['mapContext']
        .attributes['esriAGSFixedZoom'].factor}" />
    </td>
    <td>
     <jsfh:commandButton id="cmdFixed" value="Fixed Zoom"
        action="#{sessionScope['mapContext']
        .attributes['esriAGSFixedZoom'].doFixedZoom}" />
    </td>
  </tr>
  </table>
</ags:context>
```

ACCESSING OBJECTS THROUGH WEBCONTEXT

As mentioned in the section titled 'Context control', the *WebMap* object is accessible through the *WebContext* and can be retrieved using the following Java code:

```
AGSWebContext context = (AGSWebContext)
  WebUtil.getWebContext(FacesContext.getCurrentInstance(), CONTEXTID);
AGSWebMap map = (AGSWebMap)context.getWebMap();
// The WebPageMap, retrieved from the WebContext,
// can be used to retrieve the name of the focus map.
String focusMapName = map.getFocusMapName();
```

The name of the focus map can also be retrieved in JSP code:

```
${requestScope['esriWebContext'].webMap.focusMapName}:
<c:set var="focusMapName"
value="${sessionScope['CONTEXTID'].webMap.focusMapName}" />
```

The above code sets a *focusMapName* variable, with the value retrieved from the *getFocusMapName* method on *AGSWebMap*.

Take a look at a few more examples that relate to the buffer operation on a map.

The following code shows how to display a JSF inputText box to set a buffer distance. A buffer distance must be set before a buffer operation can be performed.

```
<jsfh:inputText value="#{sessionScope['CONTEXTID'].attributes
  ['esriAGSBuffer'].bufferDistance}"/>
```

The following code uses the JSF *selectitems* tag to display a dropdown list of choices for setting the map units. The *selectOneMenu* tag automatically calls *AGSBuffer.setBufferUnits* with the value selected.

```
<jsfh:selectOneMenu id="selBufferUnit" value=
  "#{sessionScope['CONTEXTID'].attributes['esriAGSBuffer'].bufferUnits}">
  <jsfc:selectItems id="selBufferUnitOption" value=
    "#{sessionScope['CONTEXTID'].webMap.mapUnitsOptions}"/>
</jsfh:selectOneMenu>
```

The following code shows how to display the details of a buffer result by using the JSF *dataTable*. First, the *ArrayList* of details is associated to the *value* attribute. Then for each result detail, the field and value are accessed via the item variable. The *rendered* attribute of the *dataTable* ensures that there are buffer result details available to print out.

```
<jsfh:dataTable id="detailsTable"
  value="#{sessionScope['CONTEXTID']
    .attributes['esriAGSBufferResults'].detailsResult.details}"
  var="item"
    rendered="#{sessionScope['CONTEXTID']
      .attributes['esriAGSBufferResults'].detailsResult != null}" … >
  <jsfh:column>
      <jsfc:facet name=
        "header"><jsfh:outputText value="Field" /></jsfc:facet>
      <jsfh:outputText value="#{item[0]}" />
  </jsfh:column>
  <jsfh:column>
      <jsfc:facet name=
        "header"><jsfh:outputText value="Value" /></jsfc:facet>
    <jsfh:outputText value="#{item[1]}" />
  </jsfh:column>
</jsfh:dataTable>
```

The page layout control provides the display functionality of the ArcMap layout view for a map document served using ArcGIS Server. Using the page layout control, you can view a map server object's layout, navigate it, navigate the maps within the layout, and access the fine-grained ArcObjects that comprise it.

Page Layout Control Objects

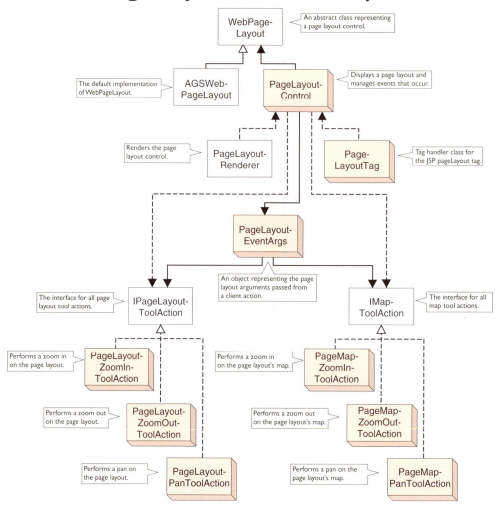

The diagram above illustrates the relationships between the page layout control and its associated business objects. The *PageLayoutControl* works with an abstract *WebPageLayout* object. The actual *WebPageLayout* object is created by the parent context control, and by default, a new *AGSWebPageLayout* is initialized. The *PageLayoutControl* also accepts map tool actions for the maps on the page. In addition, the *PageLayoutControl* responds to events on the page itself by executing actions of type *IPageLayoutToolAction*. The *PageLayoutRenderer* class renders the control to the appropriate markup, and the *PageLayoutTag* class exposes the control as a JSP custom tag named *pageLayout*. *AGSWebPageLayout* performs these important functions for the *PageLayoutControl*:

- Renders the page layout.

- Exposes an API that provides access to the details of the page layout and map.

- Sets up the page layout for client-side actions on the map and the page. These client-side actions could be dragging a rectangle to do a page zoom or dragging an image to pan the page.

- Associates the client-side tool actions to server-side action classes.

- Sets up a framework to implement new tool actions.

- Executes server-side listeners.

The diagram below shows some of the methods available on *WebPageLayout*.

WebPageLayout com.esri.arcgis.webcontrols.data	
◇ WebPageLayout(context: WebContext)	*Class constructor method.*
■–■ dpi: int	*The dots per inch (DPI) of the page layout.*
■–■ height: int	*The height of the page layout.*
■–■ imageFormat: String	*The image format of the page layout.*
■–■ width: int	*The width of the page layout.*
◄— getWebContext( ): WebContext	*Returns the WebContext associated with this WebPageLayout.*
◄— isMimeData( ): boolean	*Indicates whether the control returns MIME data.*
◄— isVisible( ): boolean	*Indicates whether the page layout is visible.*
—► setUseMIMEData(useMIMEData: boolean)	*Sets whether the control returns MIME data.*
—► setVisible(visible: boolean)	*Sets the visibility of the page layout.*

The diagram below shows some of the methods available on *AGSWebPageLayout*.

AGSWebPageLayout com.esri.arcgis.webcontrols.ags.data	The default implementation of WebPageLayout.
◇ AGSWebPageLayout(agsContext: AGSWebContext)	Class constructor method.
■—■ pageExtent: com.esri.arcgis. geometry.IEnvelope	The extent of the page.
◄— createServerObject(progId: String): Object	Creates the AGS object represented by the progId.
◄— getDataFrameName(index: int): String	Returns the data frame name based on an integer index.
◄— getFullExtent( ): AGSExtent	Returns the full extent of the page layout.
◄— getImageDescription( ): com.esri.arcgis. carto.IImageDescription	Returns the image description for the page layout.
◄— getImageRectangle(mapName: String): ImageRectangle	Returns the data frame location in screen coordinates.
◄— getImageRectangles( ): ImageRectangle[ ]	Return all data frame locations in screen coordinates.
◄— getImageUrl( ): String	Returns the URL of the page layout.
◄— getMapDescriptions( ): com.esri.arcgis. carto.IMapDescriptions	Returns the map descriptions for the page layout.
◄— getMapExtent(mapName: String): com. esri.arcgis.geometry.IEnvelope	Returns the extent of the given data frame.

◄—   getMapServerInfo(mapName: String):
   com.esri.arcgis.carto.IMapServerInfo
	Returns the IMapServerInfo object based on data
	   frame name.

◄— getPageDescription( ): com.esri.arcgis.carto.IPageDescription	Returns the page description for the page layout.
◄— getServerLayout( ): com.esri.arcgis.carto.IMapServerLayout	Returns the IMapServerLayout object from the server.
► setHeight(height: int)	Sets the height of the page layout.
► setImageFormat(imageFormat: String)	Sets the image format for the page layout.
► setMapExtent(extent: com.esri.arcgis. geometry.IEnvelope, mapName: String)	Sets the extent for the given data frame name.
► setWidth(width: int)	Sets the width of the page layout.
◄— fromMapPoint(point: com.esri.arcgis. geometry.Ipoint, mapName: String): int[ ]	Converts a map point to screen coordinates.
◄— toMapPoint(x: int, y: int, mapName: String): com.esri.arcgis.geometry.IPoint	Converts a screen coordinate to a map point.
◄— fromPagePoint(point: com.esri.arcgis. geometry.IPoint): int[ ]	Converts a page point to screen coordinates.
◄— toPagePoint(x: int, y: int): com.esri.arcgis.geometry.IPoint	Converts the screen coordinates to a page point.

The *AGSWebPageLayout* provides you with an API to access information about the page layout and control its behavior. The examples below illustrate some common usages.

The following code shows how to change the current page extent of the layout and refresh all observers of the *WebContext*.

```
AGSWebPageLayout wPage =
  (AGSWebPageLayout)context.getWebPageLayout();
IEnvelope extent = wPage.getPageExtent();
extent.centerAt(wPage.toPagePoint(50),wPage.toPagePoint(50));
wPage.setPageExtent(extent);
context.refresh(AGSRefreshId.PAGE_LAYOUT_OPERATION);
```

The following code shows how to access objects on a GIS server with *AGSWebPageLayout*. This code also retrieves the map images contained in the layout and prints out the URL to the first map.

```
AGSWebPageLayout wPage =
  (AGSWebPageLayout)context.getWebPageLayout();
IPageDescription pageDesc = wPage.getPageDescription();
IImageDescription imageDesc = wPage.getImageDescription();
```

```
ILayoutImage srvLayout = wPage.getServerLayout()
  .exportLayout(pageDesc,imageDesc);
IMapImages images = srvLayout.getMapImages();
System.out.println("The first map: " + images.getElement(0).getURL());
```

The following code shows how to perform operations on a map in the page layout. This code sets the map to its full extent and refreshes the display.

```
AGSWebPageLayout wPage =
  (AGSWebPageLayout)context.getWebPageLayout();
wPage.setMapExtent(wPage.getMapServerInfo("dataframe1").
  getFullExtent(), "dataframe1");
context.refresh(AGSRefreshId.MAP_OPERATION);
```

The following code shows how to query the page layout to determine if the image is returned as a URL or as MIME data. If the server object is set up as MIME Only, the *isMIMEData* method will return True even if *PageLayoutControl.setUseMIMEData* set it to False.

```
AGSWebPageLayout wPage =
  (AGSWebPageLayout)context.getWebPageLayout();
System.out.println("MIME is set to: " + wPage.isMIMEData());
```

INTERACTING WITH THE PAGE LAYOUT CONTROL USING TOOLS

The page layout control interacts with tools in the same way that the map control interacts with tools. The only difference between the two is that the page layout control can process both map and page tool actions.

In addition to the tool items for the map control, the *ToolItemCollection* object is populated with tool items for the page layout control. Below is the listing of the tool items specific to the page layout control and their client-side and server-side mappings.

Key	Server ToolAction	JavaScript method
PageLayoutZoomIn	PageLayoutZoomInToolAction	PageDragRectangle
PageLayoutZoomOut	PageLayoutZoomOutToolAction	PageDragRectangle
PageLayoutPan	PageLayoutPanToolAction	PageDragImage
PageMapZoomIn	PageMapZoomInToolAction	PageMapDragRectangle
PageMapZoomOut	PageMapZoomOutToolAction	PageMapDragRectangle
PageMapPan	PageMapPanToolAction	PageMapDragImage

The *PageLayoutRenderer* decodes the request and passes the *FacesContext* to the *PageLayoutControl*. The *FacesContext* contains the request parameters and the tool key for that request. The *PageLayoutControl* then creates an *EventArgs* object that holds the tool action and the tool action's client arguments. If the *ToolAction* is for a map, it creates a *MapEventArgs*. If the *ToolAction* is for a page, it creates a *PageEventArgs*. An *ActionEvent* is fired, and the page layout control's *processAction* method retrieves the appropriate *IToolAction* and calls the *execute* method with the *EventArgs*. In the case of a zoom in on the page, the *PageLayoutZoomInToolAction* parses the arguments and performs the zoom in on the page.

Look at creating a new tool for the page layout. The steps here are similar to the steps taken to create a new tool for the map control. In this example, the new tool's client-side action is *PagePoint*. The server-side action takes the point from the client and zooms in relative to that point by a factor of 50 percent.

1. Change the application's default.xml file to add a new tool, *PageFixedZoomIn*.

    ```
    <tool-item>
      <key>PageFixedZoomIn</key>
      <action-class>
      com.esri.arcgis.webcontrols.ags.faces.event.PageFixedZoomInToolAction
      </action-class>
      <client-action>PagePoint</client-action>
    </tool-item>
    ```

    The *ClientActionArgs* mapping for the page layout control is below. The *PagePoint* action maps to the *PointArgs* object on the server.

Client Action	ClientActionArgs
PagePoint	PointArgs
PageDragRectangle	DragRectangleArgs
PageDragImage	DragImageArgs
PageMapPoint	PointArgs
PageMapDragRectangle	DragRectangleArgs
PageMapDragImage	DragImageArgs

2. Create *PageFixedZoomInToolAction.java*.

3. Add the following code to *PageFixedZoomInToolAction.java*. The code below has comments describing the implementation.

    ```java
    package com.esri.arcgis.webcontrols.ags.faces.event;

    import com.esri.arcgis.webcontrols.faces.event.*;
    import com.esri.arcgis.webcontrols.ags.data.AGSWebPageLayout;
    import com.esri.arcgis.webcontrols.ags.data.AGSWebContext;
    import com.esri.arcgis.webcontrols.ags.data.AGSRefreshId;
    import com.esri.arcgis.geometry.*;

    // Implement from the IPageLayoutToolAction interface.
    public class PageFixedZoomInToolAction implements IPageLayoutToolAction {

    // Zoom in by a fixed factor of 50%.
      protected static final double ZOOM_IN_FACTOR = 0.5;

    // Implement the execute method to perform the task.
      public void execute(PageLayoutEventArgs args) throws Exception {

    // The PagePoint method maps to PointArgs. Extract the client action
    // arguments and cast it to PointArgs.
      PointArgs pointArgs = (PointArgs)args.getClientActionArgs();

    // Retrieve the WebContext and WebPageLayout.
      AGSWebContext agsContext = (AGSWebContext)args.getWebContext();
      AGSWebPageLayout wPage =
        (AGSWebPageLayout)agsContext.getWebPageLayout(true);

    // Get the current page's extent from the WebPageLayout.
      IEnvelope extent = wPage.getPageExtent();
    ```

```
// Extract the x,y coordinate point from PointArgs. Convert both of these
// screen points to page points using the toPagePoint method.
// Center the new extent of the page at the new point and expand it 50%.
  extent.centerAt(wPage.toPagePoint(pointArgs.getPoint().getX(),
    pointArgs.getPoint().getY()));
  extent.expand(ZOOM_IN_FACTOR, ZOOM_IN_FACTOR, true);

// Set the new page extent to the expanded extent.
  wPage.setPageExtent(extent);

// Refresh the context and update those controls that listen for
// changes to the page layout.
    agsContext.refresh(AGSRefreshId.PAGE_LAYOUT_OPERATION);
    }
}
```

4. Now that you have completed the code for PageFixedZoomInToolAction.java, navigate to your application's WEB-INF directory and create this directory structure:

   classes\com\esri\arcgis\webcontrols\faces\event

   Compile the file, and place the class file in the event directory.

5. To add the button for zooming the page layout to the JSP, edit your JSP file to include this code that will display the new button. The button's *onclick* event is set to execute the *PagePoint* JavaScript method and pass the identifier of the page layout control and the tool key.

```
<ags:context id="mapContext" resource="world@localhost">
  <ags:pageLayout ... />
  <input type="button" value="Page Fixed Zoom"
    onclick="PagePoint('PageLayout0', 'PageFixedZoomIn', 'false');" />
</ags:context>
```

The third parameter is an optional Boolean to specify whether you want to see the loading image while the request is processing; the default is true.

INTERACTING WITH THE PAGE LAYOUT CONTROL USING LISTENERS

Like the map control, listeners can be associated with JSF command components. Refer to the section titled 'Interacting with the map control using listeners' for details on how to implement a new listener and attach listeners to commands.

The Java ADF provides these listeners, as part of the com.esri.arcgis.webcontrols.ags.faces.event package, for the page layout:

* PageLayoutDirectionalPanListener

* PageLayoutZoomFullExtentListener

* PageLayoutZoomFullSizeListener

Refer to the ArcGIS Server Java ADF Listeners and Actions documentation in the ArcGIS Developer Help for more details on each listener.

ACCESSING OBJECTS THROUGH WEBCONTEXT

The *WebPageLayout* can be retrieved through the *WebContext* by using the following Java code:

```
AGSWebContext context = (AGSWebContext)
    WebUtil.getWebContext(FacesContext.getCurrentInstance(), CONTEXTID);
AGSWebPageLayout page = (AGSWebPageLayout)context.getWebPageLayout();
// The WebPageLayout, retrieved from the WebContext,
// can be used to retrieve the width and height of the page.
int width = page.getWidth();
int height = page.getHeight();
```

The width and height of the page can also be retrieved using the following JSP code:

```
<c:set var="pageWidth" value=
    "${sessionScope['CONTEXTID'].webPageLayout.width}" />
<c:set var="pageHeight" value=
    "${sessionScope['CONTEXTID'].webPageLayout.height}" />
```

The overview control is similar to a map control in that it displays a particular data frame of a map server object. However, the purpose of the overview map is to provide a point of reference for the area displayed on its associated map control. The overview control can shows its data frame at full extent or the last extent the file was saved at. A small box on the overview map represents the currently displayed area on its associated map control. You can interactively move this box around to pan the area displayed in the map control.

Overview Control Objects

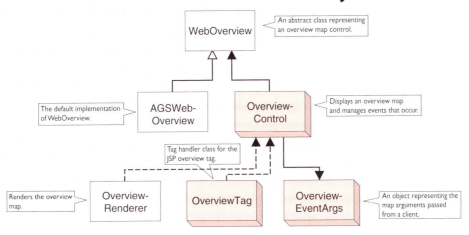

The diagram above illustrates the relationships between the overview control and its associated business objects, *WebOverview* and *AGSWebOverview*. The overview control works with an abstract *WebOverview*. The actual *WebOverview* object is created by the parent context control, and by default, a new *AGSWebOverview* is initialized. The *OverviewEventArgs* object is used by the control to extract the point on the overview map to center at. The *OverviewRenderer* class renders the control to the appropriate markup, and the *OverviewTag* class exposes the control as a JSP custom tag named *overview*. *AGSWebOverview* performs these important functions for the overview control:

- Renders the overview map

- Exposes an API to retrieve information about the overview map

- Sets up the overview map to handle events for panning the area of interest

- Updates the map based on an overview map event

The diagram below shows some of the methods available from *WebOverview* and *AGSWebOverview*.

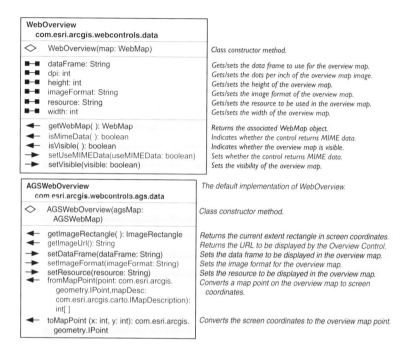

The *AGSWebOverview* provides you with an API to access information about the overview and control its behavior. The examples below illustrate some common usages.

The following code shows how to generate the overview map image and retrieve a reference to it; this reference could be a URL or a MIME data string.

```
AGSWebOverview wOverview = (AGSWebOverview)context.getWebOverview();
String ovMap = wOverview.getImageUrl();
```

The following code shows how to determine the width and height of the area of interest box on the overview map.

```
AGSWebOverview wOverview =(AGSWebOverview)context.getWebOverview();
ImageRectangle rect = wOverview.getImageRectangle();
System.out.println("The width: " + rect.getX1() - rect.getX2());
System.out.println("The height: " + rect.getY1() - rect.getY2());
```

The following code shows how to set a new data frame to display in the overview map.

```
AGSWebOverview wOverview =(AGSWebOverview)context.getWebOverview();
wOverview.setDataFrame("newDataFrame");
context.refresh();
```

INTERACTING WITH THE OVERVIEW CONTROL

The overview control automatically handles two events on the client. These functions can be found in the overview_functions.js file included with Java ADF applications.

- *OVClick*—A single mouse click to center the area of interest.

- *OVDragUp*—A drag rectangle up event to define a new area of interest.

After either of these two events occurs on the client, the center x,y value is posted back to the server and received by the *OverviewRenderer*. The renderer decodes the center coordinate and passes these values to the *OverviewControl* for creation of an *OverviewEventArgs* object. An action event is then fired on the control to update the map control accordingly. Below is the code for the *handleEvent* method on *AGSWebOverview*:

```
public void handleEvent(OverviewEventArgs args) throws Exception {
        IEnvelope extent = agsMap.getFocusMapExtent();
        extent.centerAt(toMapPoint(args.getCenterX(), args.getCenterY()));
        agsMap.setFocusMapExtent(extent);
        agsMap.getWebContext().refresh(AGSRefreshId.MAP_OPERATION);
}
```

The *OverviewEventArgs* object is passed to the *handleEvent* method by the framework, and the x,y coordinate is extracted using *getCenterX* and *getCenterY*. The x,y center is converted to map units, and the new extent is set on the *WebMap*. The *WebContext* is refreshed to update all observers that the map extent has changed.

ACCESSING OBJECTS THROUGH WEBCONTEXT

The *WebOverview* can be retrieved through the *WebContext* using the following Java code:

```
AGSWebContext context = (AGSWebContext)
    WebUtil.getWebContext(FacesContext.getCurrentInstance(), CONTEXTID);
AGSWebOverview overview = (AGSWebOverview)context.getWebOverview();
// The WebOverview, retrieved from the WebContext, can be
// used to retrieve the resource string used with the overview map.
String resource = overview.getResource();
```

The same information can be accessed in JSP code using:

```
<c:set var="resource" value=
  "${sessionScope['CONTEXTID'].webOverview.resource}"/>
```

The above code sets a variable, *resource*, with the value retrieved from the *WebOverview.getResource* method.

The Toc control—or table of contents control—displays the contents of a map and allows an end user to turn layers on and off. The Toc control functions in the same manner as the table of contents in ArcMap: it presents a hierarchical list of data frames, layers, and symbols displayed on the map.

Table of Contents Objects

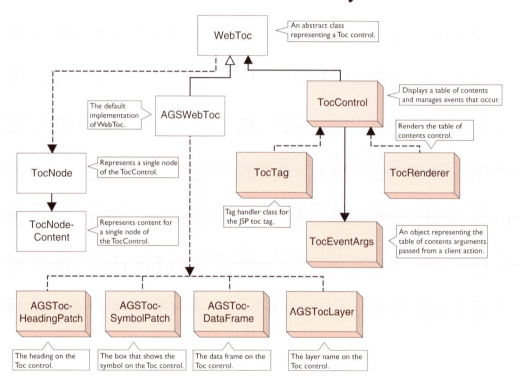

The diagram above shows the relationships between the *TocControl* and its business objects, *WebToc* and *AGSWebToc*. The *TocControl* works with an abstract *WebToc* object. The actual *WebToc* object is created by the parent context control, and by default, a new *AGSWebToc* is initialized. The *TocRenderer* class renders the control to the appropriate markup, and the *TocTag* class exposes the control as a JSP custom tag named *toc*. The *AGSWebToc* performs the following tasks:

- Renders the table of contents

- Exposes an API to retrieve information about the table of contents

- Sets up the table of contents to handle events for toggling layer visibility and changing data frames

- Updates the map or page layout based on a Toc event

WORKING WITH THE TOC TREE

The table of contents is made up of a collection of *TocNode* objects. A *TocNode* stores content, maintains hierarchical relationships, and responds to an event that occurs at that node. To store content, a *TocNode* requires a *TocNodeContent* object to encapsulate the content at that node. A *TocNode* handles its own expand or collapse event and delegates the check box event—for layer visibility—and the node operation—when node content is clicked—to the *TocNodeContent's* event handling methods:

```
public void handleCheckedEvent(TocEventArgs args)
public void handleNodeEvent(TocEventArgs args)
```

The diagram below shows you some of the methods available from *WebToc* and *AGSWebToc*.

| WebToc |
| com.esri.arcgis.webcontrols.data |

◇	WebToc(map: WebMap)	Class constructor method.
■-■	expandLevel: int	The number of levels for the TOC to be expanded.
■-■	imageFormat: String	The format of the TOC images.
■-■	showAllDataFrames: boolean	Indicates whether to show all data frames or only the default data frame.
■-■	useMIMEData: boolean	Indicates whether MIME data is used.
■-■	visible: boolean	Indicates whether the TOC is visible.
◄-	findNode(key: String): TocNode	Returns a TocNode from the TOC tree.
◄-	getWebMap( ): WebMap	Returns the associated WebMap object.
◄-	isMimeData( ): boolean	Indicates whether the control returns MIME data.
◄-	isShowAllDataFrames( ): boolean	Indicates whether to show all data frames or only the default data frame.
◄-	isVisible( ): boolean	Indicates whether the TOC is visible.
-►	setShowAllDataFrames (showAllDataFrames: boolean)	Sets whether to show all data frames or only the default data frame.
-►	setUseMIMEData(useMIMEData: boolean)	Sets whether the control returns MIME data.
-►	setVisible(visible: boolean)	Sets the visibility of the TOC.

The default implementation of WebToc.

| AGSWebToc |
| com.esri.arcgis.webcontrols.ags.data |

◇	AGSWebToc(agsMap: AGSWebMap)	Class constructor method.
◄-	getDataFrames( ): Collection	Gets the TocNodes for the top-level data frames.
◄-	reload( )	Reloads the TOC control.
-►	setImageFormat(imageFormat: String)	Sets the image format for the TOC control.

In general, you will only work with these methods if you want to change the way the existing *TocControl* works in order to write a new tree control. An example of this in the Java ADF is the *IdentifyResultsControl*, which uses the Toc objects to display the results of an identify in tree form. When a result is clicked in the tree, the attributes are shown.

The *AGSWebToc* provides you with an API to access information about the table of contents and control its behavior. The *AGSWebToc* utilizes additional data objects to build the TOC structure. Each of these data objects extends *TocNodeContent*.

- *AGSTocDataFrame*—Represents the content for a data frame

- *AGSTocHeadingPatch*—Represents the content of the symbol patch node

- *AGSTocLayer*—Stores the content for the layer in the TocNode

The examples below illustrate how to use these business objects for the TOC.

The following code shows how to loop through all the data frames in the TOC and collapse all the nodes.

```java
AGSWebToc wToc = (AGSWebToc)context.getWebToc();
Object obj;
for(java.util.Iterator dataFrames = wToc.getDataFrames().iterator();
    dataFrames.hasNext();) {
    obj = dataFrames.next();
    if(!(obj instanceof TocNode))
      continue;
    ((TocNode)obj).setExpanded(false);
}
```

This diagram shows the methods available from *TocNodeContent*.

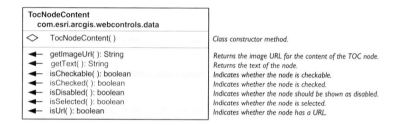

The following code shows how to find a node in the TOC and retrieve information about its content.

```java
AGSWebToc wToc = (AGSWebToc)context.getWebToc();
TocNode firstNode = wToc.findNode("0");
System.out.println("The text on the node: " +
  firstNode.getContent().getText());
System.out.println("Is the node checked? " +
  firstNode.getContent().isChecked());
```

The following code shows how to create a data frame node and add the layers of the data frame as children (*TocNodeContent*) to that node.

```java
TocNode dataFrameNode;
TocNode node;
IMapLayerInfos lInfos;
IMapLayerInfo info;
String mapName = "map1";
IMapServer server = null;
AGSWebToc wToc = (AGSWebToc)context.getWebToc();
AGSWebMap agsMap = (AGSWebMap)context.getWebMap();
server = agsMap.getServer();
lInfos = server.getServerInfo(mapName).getMapLayerInfos();

// Create data frame node.
dataFrameNode = new TocNode(new AGSTocDataFrame(mapName, agsMap), 0);
```

```
// Add layers to the data frame node.
for(int j = 0; j < lInfos.getCount(); ++j) {
    info = lInfos.getElement(j);
    if(info.getParentLayerID() != -1)
      continue;
    node = dataFrameNode.addChild
      (new AGSTocLayer(info, mapName, agsMap));
    addLayerChildren(node, info, mapName, lInfos);
}
```

The following code shows how to reload the TOC.

```
AGSWebToc wToc = (AGSWebToc)context.getWebToc();
wToc.reload();
```

INTERACTING WITH THE TOC CONTROL

The *TocRenderer* controls how the *TocControl* is displayed on the page and manages the events for the table of contents. The *TocRenderer* decodes the following request parameters from the *FacesContext*:

- *nodeKey*—The key of the node responsible for the operation

- *nodeOperation*—The operation to be performed on the node, which is either click or expandCollapse

The *TocRenderer* also retrieves all of the available node key parameter names and builds a *HashMap* of checked and unchecked key values. The *TocRenderer* then calls the *TocControl's setEventArgs* method with the *nodeKey*, *nodeOperation*, and *checkedNodeKeys*. A new *TocEventArgs* is constructed with this information, and an action event is fired on the control to process the event. The *processAction* event for the control does the following using the *TocEventArgs* object:

- Retrieves the node where the event occurred

- Calls *node.handleNodeEvent* with the arguments

- Iterates through all the checked nodes and calls each node's *handleCheckedEvent* method

By default, the *TocNodeContent* can respond to events two ways:

The autoPostBack attribute is discussed in the next section.

- *AGSTocDataFrame*—The *handleNodeEvent* method receives the click event for changing a data frame. A new focus map is set and the TOC is redrawn.

- *AGSTocLayer*—The *handleCheckedEvent* method toggles the layer visibility and the map is redrawn automatically if the *autoPostBack* attribute is set to true.

IMMEDIATE POSTBACK UPON CHECKING A BOX

In ArcMap, when you check on a layer in the table of contents, that layer draws immediately. By default, when you associate a *TocControl* with a map or page layout control, you'll see the same behavior in your Java ADF application. However, with Web applications, you may not want to send a request back to the server every time the end user of your Web application interacts with the *TocControl*. Round-trips to the server consume server resources and take time to process.

Instead, you can disable this automatic postback behavior by setting the *autoPostBack* attribute to false in the *toc* tag. With this setting, when an end user checks layers on or off, the map display won't update until a postback occurs. The TOC keeps track of any user interaction when a postback does occur, and any changes to the checked layers will be reflected on the map control. When the autoPostBack attribute is set to false, you might add a JSF command button and attach a listener to that button to refresh the context for the current view of the TOC. For example, the *processAction* method of the listener would look like this:

```
public void processAction(ActionEvent event) throws
   javax.faces.event.AbortProcessingException {
   try {
    WebUtil.getWebContext(event.getComponent()).getWebMap().refresh();
   }
   catch(Exception e) {
   logger.log(Level.WARNING, "Unable to process action for
      RefreshTocListener.", e);
   }
   }
```

CONTROLLING LAYER DRAWING

A *TocControl* is associated with a map or page layout control by nesting it within the same context control. You can customize the presentation or client-side actions of the *TocControl* by modifying the toc.xsl. For example, if you would like to remove the ability to toggle a layer's visibility, you would remove the presentation logic that allows you to check that box. To accomplish this, there are two places in toc.xsl where code needs to be removed.

First, remove this entire section of code:

```
<!- Add the checkbox based on checked attribute ->
<xsl:call-template name="fcheckbox">
<xsl:with-param name="ischecked" select="content/@checked"/>
<xsl:with-param name="isdisabled" select="content/@disabled"/>
<xsl:with-param name="level" select="level"/>
<xsl:with-param name="id" select="$id"/>
<xsl:with-param name="key" select="$key"/>
<xsl:with-param name="autopostback" select="/toc/auto-post-back"/>
</xsl:call-template>
```

Second, remove the *fcheckbox* function:

```
<xsl:template name="fcheckbox">

      ...

</xsl:template>
```

After this change, your Web application will no longer display checkboxes for controlling the layer visibility.

ACCESSING OBJECTS THROUGH WEBCONTEXT

The *WebToc* can be retrieved through the *WebContext* by using the following Java code:

```
AGSWebContext context = (AGSWebContext)
  WebUtil.getWebContext(FacesContext.getCurrentInstance(), CONTEXTID);
AGSWebToc toc = (AGSWebToc)context.getWebToc();
// The WebToc has been retrieved from the WebContext in order to
// test if the TOC is visible.
boolean isVisible = toc.isVisible();
```

And the *WebToc* can be retrieved through the *WebContext* to test if the TOC is visible by using the following JSP code:

```
<c:set var="initialVisibility"
  value="${sessionScope['CONTEXTID'].webToc.visible}"/>
```

The above code sets a variable, *initialVisibility*, with the value retrieved from the *WebToc.isVisible* method.

The North arrow control displays a North arrow on the page. The North arrow matches the North direction for the data frame in the map control. You can set the North Arrow symbol and size.

North Arrow Control Objects

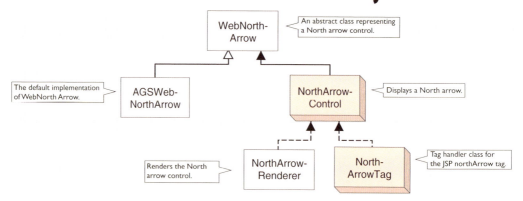

The diagram above shows the relationship between the associated North arrow objects to the *NorthArrowControl*. The *NorthArrowControl* works with an abstract *WebNorthArrow* object. The default implementation of *WebNorthArrow* is *AGSWebNorthArrow*. The *NorthArrowRenderer* class renders the control, and the *NorthArrowTag* class exposes the control as a JSP custom tag named *northArrow*.

The diagram below shows the methods available from *WebNorthArrow*.

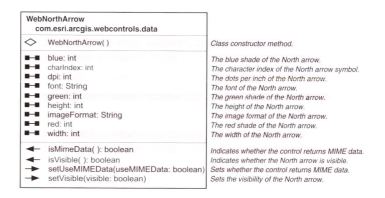

The diagram below shows the methods available from *AGSWebNorthArrow*.

| AGSWebNorthArrow | |
com.esri.arcgis.webcontrols.ags.data	
◇ AGSWebNorthArrow( )	
◄ getImageUrl(): String	
► init (agsContext: WebContext)	
► setCharIndex(charIndex: int)	
► setFont(font: String)	
► setImageFormat(imageFormat: String)	

The default implementation of WebNorthArrow.

Class constructor method.

Returns the URL of the North arrow.
Initializes the object with the associated WebContext.
Sets the character index of the North arrow symbol.
Sets the font of the North arrow.
Sets the image format of the North arrow.

AGSWebNorthArrow performs the following tasks:

- Displays the map's North arrow

- Exposes an API to control the appearance of the North arrow

The *AGSWebNorthArrow* provides you with an API to access information about the North arrow and control its appearance on the page. The diagrams above show some of the methods on *AGSWebNorthArrow* and *WebNorthArrow*.

The *WebNorthArrow* is stored as a request attribute of the *WebContext*. The code below retrieves the *WebNorthArrow* from the *WebContext* by asking for the object stored in *WebNorthArrow.WEB_CONTEXT_ATTRIBUTE*.

```
(AGSWebNorthArrow)context.
    getAttribute(WebNorthArrow.WEB_CONTEXT_ATTRIBUTE_NAME);
```

Take a look at what can be accomplished using the *WebNorthArrow* API.

The following code shows how to retrieve information about the appearance of the North arrow.

```
AGSWebNorthArrow northarrow = (AGSWebNorthArrow)context.
    getAttribute(WebNorthArrow.WEB_CONTEXT_ATTRIBUTE_NAME);
int arrowSize = northarrow.getSize();
int characterIndex = northarrow.getCharIndex();
int size = northarrow.getImageFormat();
```

The following code shows how to change the character index, color, and size of the North arrow.

```
AGSWebNorthArrow northarrow = (AGSWebNorthArrow)context.
    getAttribute(WebNorthArrow.WEB_CONTEXT_ATTRIBUTE_NAME);
northarrow.setSize(60);
northarrow.setCharIndex(177);
northarrow.setFont("ESRI North");
context.refresh();
```

The default font for the North arrow is ESRI North. This font contains the North arrows used in ArcGIS Desktop. Within this font, you specify the particular charIndex of the North arrow you want to use. The default charIndex is 177. In the ESRI North font, character indexes range from 33 to 125 and 161 to 218.

You can use the Microsoft Windows Character Map to view the fonts.

To obtain an appropriate value for charIndex, convert the hexadecimal number—displayed in the lower left corner of the dialog box—to decimal and use the decimal value as the charIndex. In the figure above, the hexadecimal value of 4E represents a decimal value of 78. You would use 78 as the charIndex.

Alternatively, you can use ArcMap to view character indexes directly. Just add a North arrow in ArcMap, then display the properties of the North arrow.

Click the Character button on the dialog box to display the symbols. In the lower left corner, you'll see character index. In the figure below, the charIndex is 78.

ACCESSING OBJECTS THROUGH WEBCONTEXT

The previous code examples showed you how to access *WebNorthArrow* through the *WebContext* using:

```
AGSWebContext context = (AGSWebContext) WebUtil
  .getWebContext(FacesContext.getCurrentInstance(), CONTEXTID);
(AGSWebNorthArrow)
  context.getAttribute(WebNorthArrow.WEB_CONTEXT_ATTRIBUTE_NAME);
```

The example below shows you how to access information about the size, character index, and image format of the North arrow using JSP code:

```
<c:set var="size" value="${sessionScope['CONTEXTID'].
  attributes['esriWebNorthArrow'].size}"/>
<c:set var="charIndex" value="${sessionScope['CONTEXTID'].
  attributes['esriWebNorthArrow'].charIndex}"/>
<c:set var="imageFormat" value="${sessionScope['CONTEXTID'].
  attributes['esriWebNorthArrow'].imageFormat}"/>
```

The scale bar control displays the scale bar of a map control. You can change the appearance of the scale bar control, such as the font, color, or size.

Scale Bar Control Objects

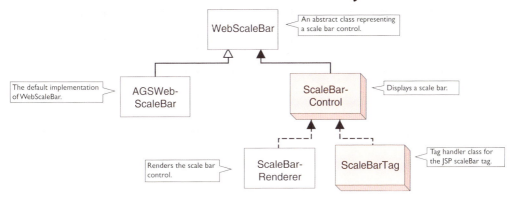

The diagram above shows the relationship between the associated scale bar objects to the *ScaleBarControl*. The scale bar control works with an abstract *WebScalebar* object. The default implementation of *WebScalebar* is *AGSWebScaleBar*. The *ScaleBarRenderer* class renders the control, and the *ScaleBarTag* class exposes the control as a JSP custom tag named *scaleBar*.

The diagram below shows the methods available from *WebScaleBar*.

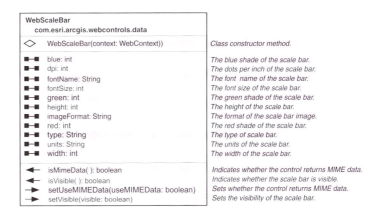

The diagram below shows the methods available from *AGSWebScaleBar*.

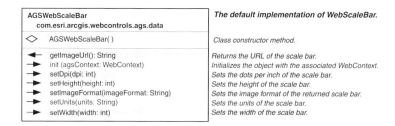

AGSWebScaleBar com.esri.arcgis.webcontrols.ags.data	**The default implementation of WebScaleBar.**
◇ AGSWebScaleBar( )	*Class constructor method.*
◄ getImageUrl(): String	*Returns the URL of the scale bar.*
► init (agsContext: WebContext)	*Initializes the object with the associated WebContext.*
► setDpi(dpi: int)	*Sets the dots per inch of the scale bar.*
► setHeight(height: int)	*Sets the height of the scale bar.*
► setImageFormat(imageFormat: String)	*Sets the image format of the returned scale bar.*
► setUnits(units: String)	*Sets the units of the scale bar.*
► setWidth(width: int)	*Sets the width of the scale bar.*

AGSWebScaleBar performs the following tasks:

- Displays the map's scale bar

- Registers itself as a *WebContextObserver* and updates itself for any map changes

- Exposes an API to control the appearance of the scale bar

The *AGSWebScaleBar* provides you with an API to access information about the scale bar and control its appearance on the page.

The *WebScaleBar* is stored as a request attribute of the *WebContext*. The code below retrieves the *WebScaleBar* from the *WebContext* by asking for the object stored in *WebScaleBar.WEB_CONTEXT_ATTRIBUTE*.

```
(AGSWebScaleBar)context.getAttribute
  (WebScaleBar.WEB_CONTEXT_ATTRIBUTE_NAME);
```

Take a look at what can be accomplished using the *WebScaleBar* API.

The following code shows how to get information about the appearance of the scale bar.

```
AGSWebScaleBar scalebar = (AGSWebScaleBar)context.
  getAttribute(WebScaleBar.WEB_CONTEXT_ATTRIBUTE_NAME);
int fontSize = scalebar.getFontSize();
String fontName = scalebar.getFontName();
int format = scalebar.getImageFormat();
```

The following code shows how to change the units, type, color, and text font (bold) for the scale bar.

```
AGSWebScalebar scalebar = (AGSWebScaleBar)context.
  getAttribute(WebScaleBar.WEB_CONTEXT_ATTRIBUTE_NAME);
scalebar.setUnits("Kilometers");
scalebar.setType("SteppedScaleLine");
scalebar.setFontBold();
scalebar.setRed(255);
scalebar.setGreen(0);
scalebar.setBlue(0);
context.refresh();
```

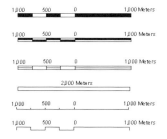

The type attribute sets the style of the scale bar. From top to bottom: Alternating, DoubleAlternating, Hollow, SingleDivision, ScaleLine, and SteppedScaleLine.

ACCESSING OBJECTS THROUGH WEBCONTEXT

The previous examples showed how to access *WebScaleBar* through the *WebContext* using:

```
AGSWebContext context = (AGSWebContext) WebUtil
  .getWebContext(FacesContext.getCurrentInstance(), CONTEXTID);
(AGSWebScaleBar)context.getAttribute(WebScaleBar.WEB_CONTEXT_ATTRIBUTE_NAME);
```

The example below shows you how to access information about the font size, font name, and image format of the scale bar using JSP code:

```
<c:set var="fontSize" value="${sessionScope['CONTEXTID'].
  attributes['esriWebScaleBar'].fontSize}"/>
<c:set var="fontName" value="${sessionScope['CONTEXTID'].
  attributes['esriWebScaleBar'].fontName}"/>
<c:set var="imageFormat" value="${sessionScope['CONTEXTID'].
  attributes['esriWebScaleBar'].imageFormat}"/>
```

The geocode control renders the input fields to be displayed for entering the address information. This control works with the specified address style to determine which input fields to display. For example, if the address style is US Streets with Zone, the control will render an address and ZIP Code text box. If the address style is simply US Streets, only an address input box will be rendered.

As with the other controls, the geocode control must work within a context control. The geocode control is different in that the context control specifies a locator resource instead of a map resource. The geocode control performs the geocode operation and stores the results in the *AGSGeocodeResults* object.

Geocode Control Objects

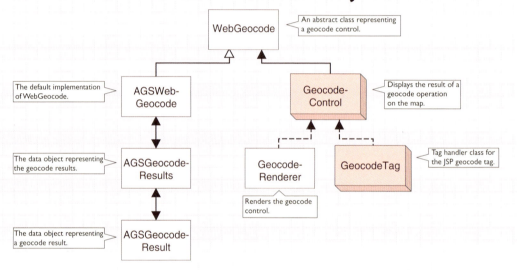

WebGeocode — An abstract class representing a geocode control.

The default implementation of WebGeocode. — AGSWeb-Geocode

GeocodeControl — Displays the result of a geocode operation on the map.

The data object representing the geocode results. — AGSGeocodeResults

GeocodeRenderer — Renders the geocode control.

GeocodeTag — Tag handler class for the JSP geocode tag.

The data object representing a geocode result. — AGSGeocodeResult

The diagram above shows the relationship between the associated geocode objects to the *GeocodeControl*. The *GeocodeControl* works with an abstract *WebGeocode* object. The default implementation of *WebGeocode* is *AGSWebGeocode*. The *GeocodeRenderer* class renders the control, and the *GeocodeTag* class exposes the control as a JSP custom tag named *geocode*.

The diagrams below show the methods available from *WebGeocode* and *AGSWebGeocode*.

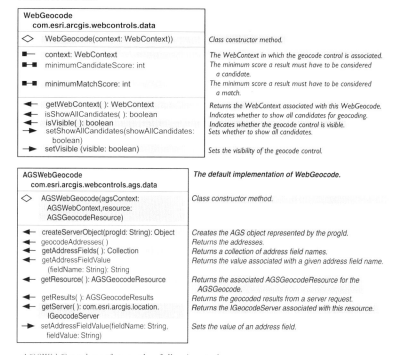

AGSWebGeocode performs the following tasks:

- Establishes a connection with the GeocodeServer

- Retrieves the address fields

- Retrieves the locator properties, such as minimum candidate score and minimum match score

- Performs a geocode operation

- Exposes an API to manage the results of a geocode operation

The *AGSWebGeocode* provides you with an API to geocode a location and access the results. The *WebGeocode* object is accessible through the *WebContext* by calling the *getWebGeocode* method.

Take a look at what can be accomplished using the *WebGeocode* API.

The following code shows how to perform a geocode operation and set properties to control the type of results returned. The code assumes that the address fields and values to geocode have already been set for *AGSWebGeocode* to access.

```
AGSWebGeocode geocode = (AGSWebGeocode)context.getWebGeocode();
geocode.setMinimumMatchScore(75);
geocode.setMinimumCandidateScore(75);
geocode.setShowAllCandidates(true);
geocode.geocodeAddresses();
```

The following code shows how to access the geocode server object through *AGSWebGeocode* and get information about the locator.

```
AGSWebGeocode geocode = (AGSWebGeocode)context.getWebGeocode();
IGeocodeServer server = geocode.getServer();
// Get the minimum candidate score property.
int score =
Integer.parseInt(properties.getProperty("MinimumCandidateScore").toString());
// Fetch all the address fields.
IFields fields = server.getAddressFields();
    int count = fields.getFieldCount();
    String fieldName;
    for(int i = 0; i < count; ++i) {
    fieldName = fields.getField(i).getName();
    System.out.println("Field name: " + fieldname);
    }
```

INTERACTING WITH THE GEOCODE CONTROL

The geocode control renders input fields for geocoding and a button to submit the geocode request. The request is passed to the *GeocodeRenderer*, which retrieves the field values set in the control. A collection of field name and value pairs are set using *WebGeocode.setAddressFieldValue (fieldName, fieldValue)*. An *ActionEvent* is then fired on the geocode control, and *WebGeocode.geocodeAddresses* is invoked. Then the *AGSGeocodeResults* are available for display.

WORKING WITH THE RESULTS OF A GEOCODE

As mentioned before, the geocode control populates the *AGSGeocodeResults* object with the results of a geocode operation. The diagram below shows the methods available on *AGSGeocodeResults* and *AGSGeocodeResult*.

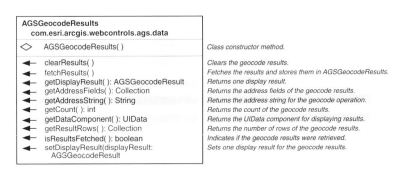

AGSGeocodeResults performs the following tasks:

- Renders a list of results for an associated geocode control

- For each geocode result, stores the x,y location and the field name/value pairs

- Maintains a list of geocode results and information about those results

The *AGSGeocodeResults* provides you with an API to access the results of a geocode operation. The diagram above shows the methods available from *AGSGeocodeResults*. The *AGSGeocodeResults* object is accessible through the *WebContext* by calling *WebContext.getAttribute*.

Look at how to work with the *AGSGeocodeResults* API.

The following code shows how to retrieve the results of a geocode operation and iterate through all the address fields and results.

```
AGSGeocodeResults results =
 (AGSGeocodeResults)context.
 getAttribute(AGSGeocodeResults.WEB_CONTEXT_ATTRIBUTE_NAME);
int totalRows = results.getCount();
Collection rows = results.getResultRows();
// Get all the address fields.
Collection addressFields = results.getAddressFields();
for (Iterator iterator = addressFields.iterator(); iterator.hasNext();)
 System.out.println("Address field is: " + (String) iterator.next());

// Get all the values for each address field.
String value;
AGSGeocodeResult[] result =
 (AGSGeocodeResult[])rows.toArray(new AGSGeocodeResult[0]);
for(int i = 0; i < totalRows; i++) {
 for(Iterator iterator = addressFields.iterator(); iterator.hasNext();) {
 val = (String)iterator.next();
 System.out.println
  ("Value of address field: " + result[i].getValue(value));
 }
}
```

Binding geocode results to a data table

A data table can be used to display AGSFindResults and AGSBufferResults as well. For more information on how to use a data table, refer to the JSF documentation.

You can use a data table to access information about *AGSGeocodeResults* and display the geocode results. The following code shows the *dataTable* tag:

```
<ags:context id="geocodeContext" resource="locator@localhost"
 resourceType="geocode">
 <ags:geocode id="geocode1" left="45" top="15" width="688">
  <jsfc:attribute name="fieldAlias:Street" value="Address" />
  <jsfc:attribute name="fieldAlias:Zone" value="ZIP Code" />
 </ags:geocode>
 <jsfh:dataTable id="geocodeTable" binding=
   "#{sessionScope['geocodeContext'].webGeocode.results.dataComponent}"
   rendered=
     "#{sessionScope['geocodeContext'].webGeocode.results.count > 0}"
   rows="5"...>
 </jsfh:dataTable>
```

```
<ags:scroller id="idScroller" dataComponent="geocodeTable" left="300"
  top="250" height="20" width="400" ... />
</ags:context>
```

The *dataTable* tag allows you to access the properties of *AGSGeocodeResults* directly using the following syntax:

```
#{sessionScope['geocodeContext'].results}
```

In this example, geocodeContext is the ID of the context control which established a connection to the locator.

The binding attribute of the *dataTable* tag must point to a *UIData* component accessible through the *getDataComponent* method on *AGSGeocodeResults*. This *UIData* component is populated by the *fetchResults* method when *AGSWebGeocode* performs the geocode operation. If you want the geocode results to display on the map, you place the *dataTable* tag within a map context.

The following line shows the exact syntax:

```
binding="#{sessionScope['geocodeContext'].webGeocode.results.dataComponent}"
```

The *rendered* attribute of the *dataTable* tag allows you to control when the data table is rendered. In the example below, *AGSGeocodeResults* is checked to determine if it has any entries.

```
rendered="#{sessionScope['geocodeContext'].webGeocode.results.count > 0}"
```

Other attributes of the *dataTable* tag allow you to control the table output. For example, the *rows* attribute can be set to control the number of rows displayed. If the *rows* attribute is used in conjunction with the *scroller* tag, you can implement paging of results. The *scroller* tag takes the ID of the data table in its *dataComponent* attribute and automatically works with data table to page the results. The *dataTable* tag exposes attributes to set the style of the header and rows.

ACCESSING OBJECTS THROUGH WEBCONTEXT

The *WebGeocode* can be retrieved through the *WebContext* using the following Java code:

```
AGSWebContext context = (AGSWebContext)
  WebUtil.getWebContext(FacesContext.getCurrentInstance(), CONTEXTID);
AGSWebGeocode geocode = (AGSWebGeocode)context.getWebGeocode();
// The WebGeocode, retrieved from the WebContext, can be
// used to retrieve the address string.
String addressStr = geocode.getResults().getAddressString();
```

Now, the same information can be accessed in JSP code using:

```
<c:set var="addressStr" value=
  "${sessionScope['CONTEXTID'].webGeocode.results.addressString}" />
```

The above code sets a variable, addressStr, with the value retrieved from *WebGeocode.results.addressString*.

The identify results control renders results of an identify, the action of clicking a map feature (or features) to return the feature's attributes. Initially, the control renders the features found for the topmost layer. A dropdown list at the top of the identify dialog allows you to identify features in the topmost layer, visible layers, all layers, or a specified layer. After clicking a map feature, the identify results control lists the feature and its layer name.

Identify Results Objects

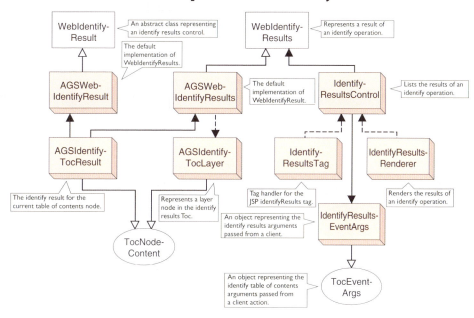

The diagram above shows the relationship between the associated identify results objects to the *IdentifyResultsControl*. The *IdentifyResultsControl* works with a *WebIdentifyResults* object, which is a collection of *WebIdentifyResult* objects. The default implementation of *WebIdentifyResults* is *AGSWebIdentifyResults*. The *IdentifyResultsRenderer* class renders the control, and the *IdentifyResultsTag* class exposes the control as a custom JSP tag named *identifyResults*.

The diagram below shows the methods available from *WebIdentifyResults* and *AGSWebIdentifyResults*.

```
WebIdentifyResults
com.esri.arcgis.webcontrols.data
```
◇   WebIdentifyResults( )   *Class constructor method.*

▬   visible: boolean   *Stores whether the identify results are visible.*
▬   WEB_CONTEXT_ATTRIBUTE_NAME: String   *The name by which this object is referenced in webContext.*
▬▬   identifyOption: int   *The current identify option.*
▬▬   selectedResult: WebIdentifyResult   *The currently selected identify result.*

◄   fetchResults( )   *Populates the nodes of the identify results tree.*
◄   findNode(key: String): TocNode   *Finds the TocNode identified by this key.*
◄   isVisible( ): boolean   *Indicates whether the identify results are visible.*
◄   setVisible(visible: boolean)   *Sets the visibility of the identify results..*

```
AGSWebIdentifyResults
com.esri.arcgis.webcontrols.ags.data
```
The default implementation of WebIdentifyResults.

◇   AGSWebIdentifyResults( )   *Class constructor method.*

▬▬   location: AGSPoint   *The location point used to identify features.*

◄   clearResults( )   *Clears the identify results.*
◄   fetchResults( )   *Fetches the identify results.*
◄   getIdentifyOptions( ): Map   *Returns the available identify options.*
◄   getLayers( ): Collection   *Returns the top-level layer nodes for the identify results.*
◄   getMapName( ): String   *Returns the map name.*
◄   getWebContext( ): AGSWebContext   *Returns the associated WebContext.*
►   init (agsContext: WebContext)   *Initializes the object with the associated WebContext.*

AGSWebIdentifyResults performs the following tasks:

- Renders a list of results from the identify tool action or event on the control

- Maintains a list of results and information about those results

- Exposes an API to manipulate the results

The *AGSWebIdentifyResults* provides you with an API to access the results of an identify. The diagram above shows some of the methods available on *AGSWebIdentifyResults*. The *WebIdentifyResults* object is accessible through the *WebContext* by retrieving the attribute stored in request scope.

Take a look at how to work with the *WebIdentifyResults* API.

The following code shows how to implement the identify tool action's *execute* method and adds a loop through the selected identify result.

```java
public void execute(MapEventArgs args) throws Exception {
    // Retrieve the x,y coordinate of the identify tool action.
    PointArgs pointArgs = (PointArgs)args.getClientActionArgs();
    ClientPoint clientPoint = pointArgs.getPoint();

    // Retrieve the WebContext and WebMap from the map's event arguments.
    AGSWebContext wContext = (AGSWebContext)args.getWebContext();
    AGSWebMap wMap = (AGSWebMap)wContext.getWebMap();

    // Retrieve the identify results from the WebContext
    // stored in request scope.
    AGSWebIdentifyResults results = (AGSWebIdentifyResults)wContext.
      getAttribute(WebIdentifyResults.WEB_CONTEXT_ATTRIBUTE_NAME);
```

```
// Set the location of the client's point for the result set.
results.setLocation(new AGSPoint(wMap.toMapPoint(clientPoint.getX(),
    clientPoint.getY())));
// Change the identify option to use all visible layers
// (default is topmost).
results.setIdentifyOption(new Integer
    (esriIdentifyOption.esriIdentifyVisibleLayers));

// Populate AGSWebIdentifyResults with the matching
// results based on the identify option and client point.
// Fetch results calls results.clearResults() to clear
// any previous results.
results.fetchResults();

// Loop through the selected identify result.
WebIdentifyResult result = results.getSelectedResult();
if(result != null) {
    String[] field;
    // Retrieve the collection of field names and values and iterate.
    for (Iterator iterator = result.getProperties().iterator();
        iterator.hasNext(); ) {
        field = (String[])iterator.next();
        System.out.println("Field name is: " + field[0]);
        System.out.println("Value of field is: " + field[1]);
        }
    }
}
```

INTERACTING WITH THE IDENTIFY RESULTS CONTROL

The identify results control is set up to handle three operations:

- Clicking a layer name in the tree to expand or collapse the display of identified features

- Displaying the selected identify result's field name and value by clicking a feature result in the tree

- Changing the identify option to select features and fetching a list of new results

The identifyResults.xsl utilizes the *identifyNodeFunc* JavaScript method—in the identify_functions.js—to pass information about the node key representing the selected layer and the node operation (click or expandCollapse).

The identifyResults.xsl also defines the dropdown list of identify options and sets the value of the *_identifyOption* parameter. The *_hide* parameter can also be set to control the current visibility of the control.

The *IdentifyResultsControl* creates a new *IdentifyResultsEventArgs* and passes a *WebIdentifyResult* to perform the node operation for the selected layer or the changing of the identify option. An *ActionEvent* is then fired on the control, which does the following:

- Retrieves the TocNode for the event. If there is an event to perform, it calls the *handleNodeEvent* method on *TocNode* with the necessary arguments.

- Retrieves the current identify option and fetches the results based on this option.

- Checks the *_hide* parameter. If true, it sets the visibility of the control to false.

ACCESSING OBJECTS THROUGH WEBCONTEXT

The *WebIdentifyResults* can be retrieved through the *WebContext* with the following Java code:

```
AGSWebContext context = (AGSWebContext)
  WebUtil.getWebContext(FacesContext.getCurrentInstance(), CONTEXTID);
AGSWebIdentifyResults results = (AGSWebIdentifyResults)wContext.
  getAttribute(WebIdentifyResults.WEB_CONTEXT_ATTRIBUTE_NAME);
// The WebIdentifyResults, retrieved from the WebContext, can
// be used to retrieve the map name for the identified results.
results.getMapName();
```

The same information can be accessed in JSP code using:

```
<c:set var="mapName" value="${sessionScope['CONTEXTID']
  .attributes['esriWebIdentifyResults'].mapName}"/>
```

7

Developer scenarios

Previous chapters of this book introduced several programming concepts, patterns, and new APIs. This chapter contains walk-throughs of example developer scenarios of building applications using ArcGIS Server that apply these concepts and use these APIs. Each scenario is available with the ArcGIS Developer samples installed as part of the ArcGIS Server developer kits.

The developer scenarios included are:

• creating an application Web service (.NET and Java) • extending a Web application template (.NET and Java) • developing an ArcGIS Server Web service client (.NET and Java) • extending the GIS server with utility COM objects

Rather than walk through this scenario, you can get the completed Web application from the samples installation location. The sample is installed as part of the ArcGIS developer samples.

This walk-through is for developers who need to build and deploy a .NET Web application that extends one of the application templates installed as part of the .NET Application Developer Framework SDK. The application incorporates focused geodatabase editing capabilities using the ArcGIS Server API, .NET Web controls, and ArcObjects. It describes the process of building, deploying, and consuming the *Extending_a_template* sample, which is part of the ArcGIS Developer samples.

You can find this sample in:
<install_location>\DeveloperKit\Samples\Developer_Guide_Scenarios\
        ArcGIS_Server\Extending_a_web_application_template

PROJECT DESCRIPTION

The MapViewer Web application template is installed as part of the .NET Application Developer Framework SDK.

The purpose of this scenario is to create an ASP.NET Web application using Visual Studio .NET to extend the MapViewer ArcGIS Server Web application project template. The application uses ArcObjects to manage a geodatabase edit session and allows users to start editing, create new features, undo, redo, and save their edits.

The following is how the user will interact with the application:

1. Open a Web browser and specify the URL of the Web application.

2. Click *Start Editing* to start an edit session.

3. Click the *Add Conservation Plan* tool on the toolbar and digitize the new conservation plan polygon on the map.

4. Additional conservation plan features can be created, and users can click the *Undo* and *Redo* buttons to undo and redo their edits.

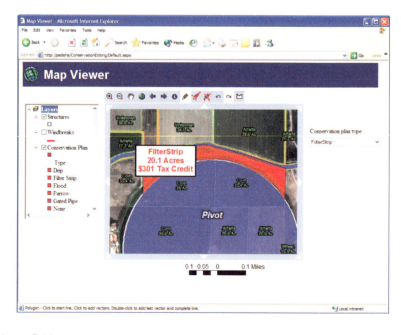

5. Once finished, users can either click *Stop Editing and Save* to save their edits or *Stop Editing and Discard* to discard their edits.

If the users choose to save their edits, the geodatabase edit session is ended and their changes are saved. If the users choose to not save edits, then the geodatabase edit session is ended and their edits are discarded.

CONCEPTS

This application is designed to directly edit the version of the geodatabase that the map server object is connected to. Therefore, all users of the application are editing the same version. This can be augmented with version management capabilities to, for example, create a new version for each session and have all edits go into that version.

While the design of this example allows the editing of personal geodatabases, one could only deploy this application to support multiple editors if the geodatabase was a multiuser geodatabase managed by ArcSDE.

The application templates provided with the ArcGIS Server Application Developer Framework SDK provide a good starting point for developers to create Web applications with advanced GIS functionality. Developers will extend these applications using remote ArcObjects that are exposed through the ArcGIS Server API.

Both coarse-grained calls to remote ArcObjects, such as the methods on the MapServer and GeocodeServer, as well as fine-grained calls to remote ArcObjects, such as creating new geometries, are exposed through the ArcGIS Server API and can be used in your application. With the functionality of ArcObjects available to the Web application developer, the template applications can be extended to include a wide variety of GIS functionality. This functionality includes what is possible using ArcObjects, including analysis; query; display; and, as in this application, data maintenance.

DESIGN

This Web application is an example of a deeply stateful application in that it's designed to make stateful use of the GIS server. Since this Web application supports edit sessions that span multiple requests for operations, such as creating new features and supporting undo, redo, and the ability to stop editing and discard your edits, the application must use the same workspace object and geodatabase edit session throughout a Web application session. To do this, the Web application must make use of the same server context throughout the user session, then release that context only when the session has ended.

It is possible to create an ArcGIS Server application that includes editing functionality and is stateless. In such an application, each request to make an edit would be its own edit session and the application would not support undo, redo, and the option to save or not save edits. This example includes undo and redo functionality and, therefore, must make use of a non-pooled server object.

Because each user session needs its own server context and server object dedicated to it, the server object cannot be shared across multiple user sessions and, therefore, cannot be pooled. When this application is deployed, there will be an instance of a running map server object for each concurrent user of the application.

To provide the actual functionality, the application uses the toolbar Web control and events on the map Web control to expose commands for managing the edit session (Start Editing, Stop Editing, Undo, Redo) and tools to get a polygon from the user and to use that polygon and ArcObjects to update the geodatabase. To support this application, a non-pooled map server object must be added to ArcGIS Server using ArcCatalog.

The Web application will use the Web controls to manage the connection to the GIS server, and the MapViewer Web application template will provide the basic mapping functionality that is required for this application. You will add new tools and commands to the toolbar and map control that allow users to manage edit sessions and click the map as input to creating new features.

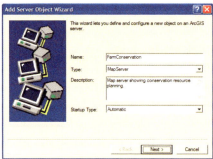

The Add Server Object wizard

REQUIREMENTS

The requirements for working through this scenario are that you have ArcGIS Server and ArcGIS Desktop installed and running. The machine on which you develop this Web application must have the ArcGIS Server .NET Application Developer Framework SDK installed.

You must have a map server object configured and running on your ArcGIS Server that uses the Conservation.mxd map document installed with the ArcGIS Developer Samples. The map document references a personal geodatabase with feature classes of farm data. It also references a QuickBird satellite imagery courtesy of DigitalGlobe.

In ArcCatalog, create a connection to your GIS server, and use the Add Server Object command to create a new server object with the following properties:

Name: FarmConservation

Type: MapServer

Description: Map server showing conservation resource planning.

Map document: <install_location>\DeveloperKit\Samples\

Data\ServerData\Conservation\Conservation.mxd

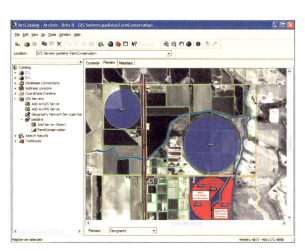

Output directory: Choose from the output directories configured in your server.

Pooling: The Web service makes stateful use of the server object. Click Not pooled for the pooling model and accept the defaults for the max instances (4).

Accept the defaults for the remainder of the configuration properties.

After creating the server object, start it and click the Preview tab to verify that it is correctly configured and that the map draws.

You can refer to Chapter 3 for more information on using ArcCatalog to connect to your server and create a new server object. Once the server object is configured and running, you can begin to code your Web application.

ArcCatalog is used for managing your spatial data holdings, defining your geographic data schemas, and managing your ArcGIS Server. Once you have created your FarmConservation server object, preview it using ArcCatalog to verify it is correctly configured.

The following ArcObjects .NET assemblies will be used in this example:

- ESRI.ArcGIS.Carto
- ESRI.ArcGIS.Geodatabase
- ESRI.ArcGIS.Geometry
- ESRI.ArcGIS.Server
- ESRI.ArcGIS.Server.WebControls
- ESRI.ArcGIS.esriSystem

The development environment does not require any ArcGIS licensing; however, connecting to a server and using a map server object require that the GIS server is licensed to run ArcObjects in the server. None of the assemblies used require an extension license.

The IDE used in this example is Visual Studio .NET 2003. This Web application can be implemented with other .NET IDEs.

IMPLEMENTATION

In this scenario, you will use the MapViewer template project that is installed as part of the ArcGIS Server Application Developer Framework's SDK to provide some basic mapping functionality, and you will extend this template with your own functionality. The code for this scenario will be written in C#; however, you can also write this Web application using VB.NET.

The first step is to create the new project.

Create a new project

1. Start Visual Studio .NET.

2. Click *File*, click *New*, then click *Project*.

The New Project dialog box

3. In the New Project dialog box, under Project Types, click the *ArcGIS Server Projects* category, then the *Visual C#* category. Under *Templates*, click *MapViewer Web Application*.

4. For the Web application name, type "http://localhost/ConservationWebApp".

5. Click *OK*. This will create a new project that contains all of the functionality included in the MapViewer template.

Set the necessary properties on the Web controls

The template includes a *Map* control, an *OverviewMap* control, and an *Impersonation* control. The map and impersonation controls require properties to be set, specifically the GIS server name and the MapServer object that the map control will use and the user account that the Web application will run as for the impersonation controls. This application does not need the overview control, so you will delete it from the application.

If the FarmConservation map server object is not listed, verify that the server object is started.

1. In the Solution Explorer, double-click *Default.aspx*. This will open the Web form in design mode.

2. On the Web form, click the *Map1* map control.

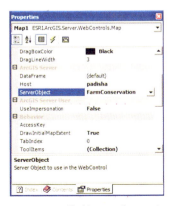

The Map control's properties

3. In the Properties for the map control, type the name of the GIS server for the *host* property, then click the *ServerObject* dropdown and click *FarmConservation* for the server object.

4. On the Web form, click the *Impersonation1* impersonation control.

5. In the Properties for the impersonation control, click the *Identity* property and click the Browse button. This will open the Identity dialog box.

6. Type the username, password, and domain for the account that your Web application will run as, then click OK.

The Impersonation control's Identity dialog box

For your Web application to successfully connect to the GIS server, the account you specify in the impersonation control's properties must be a member of the ArcGIS Server users group on the GIS server. Since the Impersonation control sets impersonation at the Web page level, there is an impersonation control on both the Default.aspx and Identify.aspx pages.

It is possible to build a Web application that uses a non-pooled server object for stateful use without using the Web controls. In such an application, it would be the developer's responsibility to keep a reference to the same server context for the duration of the Web application session.

7. On the Web form, right-click the *OverviewMap1* control, and click Delete.

Get a reference to the workspace from the map

Now that the basic properties of the application template have been set, you need to add code to execute at session startup and session end to get the workspace from the FarmConservation server object that you will be using throughout the application. Since you need to keep the workspace around, you will use the *SetObject* method on the server context to add the workspace to the server context's object dictionary. When you need to use the workspace throughout the application, you can use the *GetObject* method to get the workspace out of the server context.

Since the *FarmConservation* map server object is a non-pooled object, the map control will ask the GIS server for an instance of that map server object and hold on to it for the duration of the session. This means that each time you get the server object and server context from the map control during the application session, you know you are always getting the same one.

In this example, you will get the workspace from the first layer in the map. You do this by getting a reference to the map server object from the WebMap's server context. You can then use the *IMapServerObjects* interface to access the map server's fine-grained ArcObjects to get a reference to the first layer in the map.

1. In the Solution Explorer, double-click *Default.aspx*. This will open the Web form in design mode.

2. Right-click the Web form and click *View Code*. This will open the code behind for *Default.aspx*.

3. Add using statements for the additional assemblies used in this project. At the top of the code window, add the following using statement:

```
using ESRI.ArcGIS.Geodatabase;
```

4. Scroll in the code window until you find the following line:

```
if ( Session.IsNewSession )
```

5. Click the plus sign to expand the *New Session Startup* code region.

6. Add the following lines of code to the *New Session Startup* region:

```
using (WebMap webMap = Map1.CreateWebMap())
{
  // get the workspace from the first layer and set it in the context
  IMapServerObjects mapo =  webMap.MapServer as IMapServerObjects;
  IMap map = mapo.get_Map(webMap.DataFrame);

  IFeatureLayer fl = map.get_Layer(0) as IFeatureLayer;
  IDataset ds =  fl.FeatureClass as IDataset;
  IWorkspace ws = ds.Workspace;
  IServerContext sc = webMap.ServerContext;
  sc.SetObject("theWorkspace",ws);
}
```

Your code for the *New Session Startup* code region should now look like the following:

```
// Is this a PostBack or just started?
if ( !Page.IsPostBack )
{
  // Is this a new session?
  if ( Session.IsNewSession )
  {
    #region New Session Startup - - - TODO:Add new session startup code
here
    // Set default tool to ZoomIn
    Map1.CurrentToolItem = Map1.ToolItems.Find("ZoomIn");
    // Save extent history to Session
    m_extenthistory = new ArrayList();
    m_extenthistory.Add(Map1.Extent);
    Session.Add("extenthistory", m_extenthistory);
    Session.Add("index", -1);

    m_lastextent = Map1.Extent;

    using (WebMap webMap = Map1.CreateWebMap())
    {
      // Get the workspace from the first layer and set it in the context
      IMapServerObjects mapo = webMap.MapServer as IMapServerObjects;
      IMap map = mapo.get_Map(webMap.DataFrame);

      // Get workspace from first layer
      IFeatureLayer fl = map.get_Layer(0) as IFeatureLayer;
      IDataset ds = fl.FeatureClass as IDataset;
      IWorkspace ws = ds.Workspace;
      IServerContext sc = webMap.ServerContext;
      sc.SetObject("theWorkspace",ws);
    }
    #endregion
  }
}
```

Close any open edit sessions on session end

A server context contains an object dictionary that serves as a convenient place for you to store references to commonly used objects. This example uses the SetObject method on ISeverContext to add the workspace to the object dictionary. As you will see later, since you are using the same server context through the application session, you can get to the workspace by using GetObject in various parts of the application.

During the application session, users may start but not stop an edit session before the session times out. This can happen if users start editing and close their browser without stopping editing or if their session times out while their edit session is active.

This will keep the geodatabase edit session open even though the user is no longer using the Web application. You want to safeguard against this by explicitly stopping the edit session if the session times out and it is still active. The template application includes code that executes when the session ends. You will add code to include stopping any active geodatabase edit session.

1. In the Solution Explorer, double-click *Global.asax*.

2. Click the *click here to switch to code view* button.

The session time-out defines how long the session stays active after there has been no user interaction. This includes users closing their Web browsers. The session time-out is a configurable property. By default, the template application's session time-out is set to 20 minutes, but you can change this by changing the time-out property in the session state settings in the application's Web.config file.

3. In the code view, scroll down to the *Session_end* method.

4. Add the following lines of code to the if (context != null) statement:

```
// need to close any open edit session
ESRI.ArcGIS.Geodatabase.IWorkspaceEdit wse =
context.GetObject("theWorkspace") as
ESRI.ArcGIS.Geodatabase.IWorkspaceEdit;

if (wse.IsBeingEdited())
  wse.StopEditing(false);
```

Your *Session_End* method should now look like this:

```
protected void Session_End(Object sender, EventArgs e)
{
  IServerContext context;
  for ( int i = 0; i < Session.Count; i++)
  {
    context = Session[i] as IServerContext;
    if ( context != null )
    {
      // need to close any open edit session
      ESRI.ArcGIS.Geodatabase.IWorkspaceEdit wse =
context.GetObject("theWorkspace") as
ESRI.ArcGIS.Geodatabase.IWorkspaceEdit;

      if (wse.IsBeingEdited())
        wse.StopEditing(false);

      context.ReleaseContext();
    }
  }
  Session.RemoveAll();
}
```

Add editing commands to your toolbar

This application allows the user to start and stop edit sessions, create features, and undo and redo edit operations. To provide the necessary tools and commands for the user to do this, you will add them using the toolbar control.

The template includes a toolbar control that already has a number of tools (*zoom in*, *zoom out*, *pan*, *identify*, and so forth). You will add the following commands to the toolbar:

Start Editing: starts a new edit session

Stop Editing and Save Edits: stops editing and saves the edits

Stop Editing and Discard Edits: stops editing and discards the edits

Undo: undoes an edit operation

Redo: redoes an edit operation

You will add the following tool:

Create Conservation Plan: creates a new conservation plan feature

The MapViewer template's toolbar already contains tools and commands for navigating the map (zoom in, zoom out, pan) and for identifying features in the map.

Before adding the commands, you will need to copy a set of images that you will use for the commands and tools in the toolbar. Copy the following image files from <install_location>\DeveloperKit\Samples\Data\ServerData\Conservation to your application's Images folder (this will be c:\inetpub\wwwroot\VegetationWebApp\Images):

- polygon.gif
- polygonU.gif
- polygonD.gif
- Redo.gif
- RedoD.gif
- RedoU.gif
- StartEditing.gif
- StartEditingD.gif
- StartEditingU.gif
- StopEditingDiscard.gif
- StopEditingDiscardD.gif
- StopEditingDiscardU.gif
- StopEditingSave.gif
- StopEditingSaveD.gif
- StopEditingSaveU.gif
- Undo.gif
- UndoD.gif
- UndoU.gif

Now you will add the commands to the toolbar.

In the Solution Explorer, double-click *Default.aspx* to open the Web form in design mode.

1. Click the toolbar control.

2. In the properties for the toolbar control, click the *ToolbarItemsCollection* property and click the Browse button. This will open the *Toolbar Item Collection Editor*.

3. Click the *Add* dropdown, then click *Command*. This will add a new command to the toolbar collection.

4. Click the *Name* property for the new command and type "tbStartEditing" for the name.

5. Click the *Text* property and type "Start Editing" for the text.

6. Click the *ToolTip* property and type "Starts an edit session" for the tool tip.

7. Click the *DefaultImage* property and type "Images\StartEditing.gif" for the default image.

The Toolbar control's properties

8. Click the *HoverImage* property and type "Images\StartEditingU.gif" for the hover image.

9. Click the *SelectedImage* property and type "Images\StartEditingD.gif" for the click image.

10. Click OK.

The ToolbarItem Collection Editor dialog box

11. Repeat steps 3 to 10 to add a command with the following properties:

   Name: tbStopEditingandSave

   Text: Stop Editing and Save Edits

   ToolTip: Stop editing and save your edits

   DefaultImage: Images\StopEditingSave.gif

   HoverImage: Images\StopEditingSaveU.gif

   SelectedImage: Images\StopEditingSaveD.gif

   Disabled: True

12. Repeat steps 3 to 10 to add a command with the following properties:

   Name: tbStopEditingandDiscard

   Text: Stop Editing and Discard Edits

   ToolTip: Stop editing and discard your edits

   DefaultImage: Images\StopEditingDiscard.gif

   HoverImage: Images\StopEditingDiscardU.gif

   SelectedImage: Images\StopEditingDiscardD.gif

   Disabled: True

13. Repeat steps 3 to 10 to add a command with the following properties:

   Name: tbUndo

   Text: Undo

   ToolTip: Undoes the edit operation

   DefaultImage: Images\Undo.gif

   HoverImage: Images\UndoU.gif

   SelectedImage: Images\UndoD.gif

   Disabled: True

14. Repeat steps 3 to 10 to add a command with the following properties:

   Name: tbRedo

   Text: Redo

   ToolTip: Redoes the edit operation

   DefaultImage: Images\Redo.gif

HoverImage: Images\RedoU.gif

SelectedImage: Images\RedoD.gif

Disabled: True

Now that the commands have been added to the toolbar, you will add the code to execute when they are clicked.

The new commands that you add through the Toolbar Collection Editor will appear on the toolbar.

15. Double-click the toolbar. This will open the code window in the toolbar's *CommandClick* event.

You will see the code that executes when the user clicks one of the toolbar commands that was included as part of the template. You will add additional code to handle the new commands that you created.

16. Add the following case statements:

```
case "tbStartEditing":
  break;
case "tbStopEditingandSave":
  break;
case "tbStopEditingandDiscard":
  break;
case "tbUndo":
  break;
case "tbRedo":
  break;
```

In the first case statement you will add code to handle the user clicking the *Start Editing* button. This code will get the workspace you set in the server context and start an edit session. When the edit session is started, you will also enable the *Stop Editing* commands and disable the *Start Editing* command. You will add more code later to this case block to enable your edit tool, but you have not created the tool yet.

17. In the *tbStartEditing* case block, add the following lines of code:

```
case "tbStartEditing":
  using (WebMap webMap = Map1.CreateWebMap())
  {
    // get the workspace from the server context
    IServerContext ctx = webMap.ServerContext;
    IWorkspaceEdit wse = ctx.GetObject("theWorkspace") as IWorkspaceEdit;

    // check to see if an edit session is already started
    if (!wse.IsBeingEdited())
      wse.StartEditing(true);

    // enable stop editing command and add conservation feature tool
    ToolbarItemCollection tbcol = Toolbar1.ToolbarItems;
    Command cmd = tbcol.Find("tbStopEditingandSave") as Command;
    cmd.Disabled = false;
    cmd = tbcol.Find("tbStopEditingandDiscard") as Command;
    cmd.Disabled = false;
```

The objects and interfaces used for managing geodatabase edit sessions can be found in the GeoDatabase object library. To learn more about geodatabase edit session objects, see the online developer documentation.

```
    // disable start editing
    cmd = tbcol.Find("tbStartEditing")as Command;
    cmd.Disabled = true;

    // TODO enable AddConservationPlan tool
    }
  break;
```

In the second case statement, you will add code to handle the user clicking the *Stop Editing and Save* button. This button will call stop editing on the workspace and saves the edits the user has made. Once you add the tool to create conservation plan features, you will revisit this code to also disable that tool when the *Stop Editing and Save* button is clicked.

18. In the *tbStopEditingandSave* case block, add the following lines of code:

```
case "tbStopEditingandSave":
  using (WebMap webMap = Map1.CreateWebMap())
  {
    // get the workspace from the server context
    IServerContext ctx = webMap.ServerContext;
    IWorkspaceEdit wse = ctx.GetObject("theWorkspace") as IWorkspaceEdit;

    // stop editing and save edits
    wse.StopEditing(true);

    // disable stop editing commands, undo and redo commands,
    // new conservation tool and enable start editing command
    ToolbarItemCollection tbcol = Toolbar1.ToolbarItems;
    Command cmd = tbcol.Find("tbStopEditingandSave") as Command;
    cmd.Disabled = true;
    cmd = tbcol.Find("tbStopEditingandDiscard") as Command;
    cmd.Disabled = true;
    cmd = tbcol.Find("tbUndo") as Command;
    cmd.Disabled = true;
    cmd = tbcol.Find("tbUndo") as Command;
    cmd.Disabled = true;

    // TODO disable AddConservationPlan tool

    cmd = tbcol.Find("tbStartEditing") as Command;
    cmd.Disabled = false;
  }
  break;
```

In the third case statement, you will add code to handle the user clicking the *Stop Editing and Discard* button. This button will call stop editing on the workspace and not save the edits the user has made. Once you add your tool to create conservation plan features, you will revisit this code to also disable that tool when the *Stop Editing and Discard* button is clicked.

19. In the *tbStopEditingandDiscard* case block, add the following lines of code:

```
case "tbStopEditingandDiscard":
 using (WebMap webMap = Map1.CreateWebMap())
  {
   // get the workspace from the server context
   IServerContext ctx = webMap.ServerContext;
   IWorkspaceEdit wse = ctx.GetObject("theWorkspace") as IWorkspaceEdit;

   // stop editing and don't save edits
   wse.StopEditing(false);

   // disable stop editing commands, undo and redo commands,
   // new conservation tool and enable start editing command
   ToolbarItemCollection tbcol = Toolbar1.ToolbarItems;
   Command cmd = tbcol.Find("tbStopEditingandSave") as Command;
   cmd.Disabled = true;
   cmd = tbcol.Find("tbStopEditingandDiscard") as Command;
   cmd.Disabled = true;
   cmd = tbcol.Find("tbUndo") as Command;
   cmd.Disabled = true;
   cmd = tbcol.Find("tbUndo") as Command;
   cmd.Disabled = true;

   // TODO disable AddConservationPlan tool

   cmd = tbcol.Find("tbStartEditing") as Command;
   cmd.Disabled = false;

   webMap.Refresh();
  }
 break;
```

The *tbUndo* case statement will contain the code to execute when the user clicks the *Undo* button. This button will only be enabled if there are edit operations on the edit stack that can be undone. When this button is clicked, you must enable the *Redo* button, and if this button undoes the last edit operation, the *Undo* button must be disabled.

20. In the *tbUndo* case block, add the following lines of code:

```
case "tbUndo":
 using (WebMap webMap = Map1.CreateWebMap())
  {
   // get the workspace from the server context
   IServerContext ctx = webMap.ServerContext;
   IWorkspaceEdit wse = ctx.GetObject("theWorkspace") as IWorkspaceEdit;

   // undo the last edit operation
   wse.UndoEditOperation();

   // enable Redo command
   ToolbarItemCollection tbcol = Toolbar1.ToolbarItems;
   Command cmd = tbcol.Find("tbRedo") as Command;
```

```
    cmd.Disabled = false;

    // check to see if there are still operations that can be undone
    bool bhasUndos = false;
    wse.HasUndos(ref bhasUndos);
    if (!bhasUndos)
     {
      cmd = tbcol.Find("tbUndo") as Command;
      cmd.Disabled = true;
     }
     webMap.Refresh();
   }
  break;
```

The *tbRedo* case statement will contain the code to execute when the user clicks the *Redo* button. This button will only be enabled if there are edit operations on the edit stack that can be redone. When this button is clicked, you must enable the *Undo* button, and if the *Redo* button redoes the last edit operation, the *Redo* button must be disabled.

21. In the *tbRedo* case block, add the following lines of code:

```
  case "tbRedo":
   using (WebMap webMap = Map1.CreateWebMap())
   {
    // get the workspace from the server context
    IServerContext ctx = webMap.ServerContext;
    IWorkspaceEdit wse = ctx.GetObject("theWorkspace") as IWorkspaceEdit;

    // undo the last edit operation
    wse.RedoEditOperation();

    // enable Undo command
    ToolbarItemCollection tbcol = Toolbar1.ToolbarItems;
    Command cmd = tbcol.Find("tbUndo") as Command;
    cmd.Disabled = false;

    // check to see if there are still operations that can be redone
    bool bhasRedos = false;
    wse.HasRedos(ref bhasRedos);
    if (!bhasRedos)
     {
      cmd = tbcol.Find("tbRedo") as Command;
      cmd.Disabled = true;
     }
     webMap.Refresh();
   }
   break;
```

Adding the New Conservation Plan tool

Now that you have added commands to handle managing your edit session, you will add the tool that will actually make edits on the database. This application allows the user to digitize new conservation plan features in the geodatabase by digitizing a polygon on the map control. To allow the user to do this, you will add a new tool to the toolbar's tool collection.

In this example, you will add a single tool for creating new conservation plan polygons of a user-specified type. The types of possible conservation plans will be available from a dropdown list that the user will pick from.

The first step is to add the dropdown list that will contain the list of conservation plan types to your application:

1. In the Solution Explorer, double-click *Default.aspx* to open the Web form in design mode.

2. In the Microsoft Visual Studio .NET toolbox, click the *Web Forms* tab to display the Web Forms tools.

3. In the toolbox, click *Label* and drag a label onto the form next to the map control.

4. In the label's properties, type "Conservation plan type:" for the *Text* property.

5. In the toolbox, click *DropDownList* and drag a dropdown list onto the form under the label you just added.

6. In the dropdown list's properties, type "drpTypes" for the *ID* property.

The VegetationWebApp project in the Visual Studio .NET IDE

Now that you have added the dropdown list control to the Web form, you need to add code to populate it with the different conservation plan types. The feature class in the geodatabase that stores the conservation plan features contains a type field. The type field has a domain associated with it that defines the valid types of conservation plans. You will add code to the session startup to read the values from this domain and add them to the dropdown list.

7. Right-click the Web form and click *ViewCode*. This will open the code behind for *Default.aspx*.

8. Scroll in the code window until you find the following line:

```
if ( Session.IsNewSession )
```

9. Click the plus sign to expand the *New Session Startup* code region.

10. Add the following lines of code to the *New Session Startup* region in the same Using block that you created when getting the workspace and setting it into the server context:

```
// populate the dropdown list with the conservation types
IFeatureWorkspace fws = ws as IFeatureWorkspace;
IFeatureClass fc = fws.OpenFeatureClass("ConservationPlan");
```

```
IField fld = fc.Fields.get_Field(fc.FindField("TYPE"));
ICodedValueDomain cv = fld.Domain as ICodedValueDomain;
for (int i = 0; i < cv.CodeCount; i++)
{
  drpTypes.Items.Add(cv.get_Name(i));
}
```

Your code for the New Session Startup code region should now look like the following:

```
// Is this a PostBack or just started?
if ( !Page.IsPostBack )
{
  // Is this a new session?
  if ( Session.IsNewSession )
  {
    #region New Session Startup  - - TODO:Add new session startup code
here
    // Set default tool to ZoomIn
    Map1.CurrentToolItem = Map1.ToolItems.Find("ZoomIn");
    // Save extent history to Session
    m_extenthistory = new ArrayList();
    m_extenthistory.Add(Map1.Extent);
    Session.Add("extenthistory", m_extenthistory);
    Session.Add("index", -1);

    m_lastextent = Map1.Extent;

    using (WebMap webMap = Map1.CreateWebMap())
     {
      // Get the workspace from the first layer and set it in the context
      IMapServerObjects mapo =  webMap.MapServer as IMapServerObjects;
      IMap map = mapo.get_Map(webMap.DataFrame);

      // Get workspace from first layer
      IFeatureLayer fl = map.get_Layer(0) as IFeatureLayer;
      IDataset ds =  fl.FeatureClass as IDataset;
      IWorkspace ws = ds.Workspace;
      IServerContext sc = webMap.ServerContext;
      sc.SetObject("theWorkspace",ws);

      // Populate the dropdown list with the conservation types
      IFeatureWorkspace fws = ws as IFeatureWorkspace;
      IFeatureClass fc = fws.OpenFeatureClass("ConservationPlan");
      IField fld = fc.Fields.get_Field(fc.FindField("TYPE"));
      ICodedValueDomain cv = fld.Domain as ICodedValueDomain;
      for (int i = 0; i < cv.CodeCount; i++)
      {
        drpTypes.Items.Add(cv.get_Name(i));
      }

     }
```

```
    #endregion
}
```

Now you can add the tool to your toolbar that will actually let users digitize new conservation plan polygons on the map. Before adding the tool to the toolbar, you must create a new class that implements the *IMapServerToolAction* interface and defines the functionality for the tool.

Create the NewConservationPlan class

The first step is to add the new class to the project.

1. In the Solution Explorer, right-click the *ConservationWebApp* project, click *Add*, then click *Add New Item*.

2. In the *Add New Item* dialog box, under *Templates*, click *Class*.

3. For the *Name*, type "NewConservationPlan.cs".

4. Click Open.

This will add a new class (*NewConservationPlan*) to your project and will open the code for the class with some autogenerated code.

5. Add using statements for the assemblies you will use in this class. At the top of the code window, add the following using statements:

```
using ESRI.ArcGIS.Server;
using ESRI.ArcGIS.Server.WebControls;
using ESRI.ArcGIS.Server.WebControls.Tools;
            using ESRI.ArcGIS.Geometry;
            using ESRI.ArcGIS.Geodatabase;
```

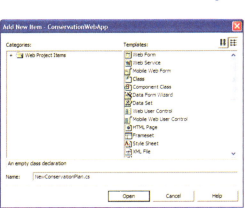

The Add New Item dialog box

Implement the NewConservationPlan class

Now you are ready to implement the *NewConservationPlan* class that contains the code to execute when a new polygon is digitized by the user on the map. Since this class is a map server tool action, it must implement the *IMapServerToolAction* interface.

1. Change the following line:

```
public class NewConservationPlan
```

to:

```
public class NewConservationPlan: IMapServerToolAction
```

2. In the Class View window, expand the class list to the *Bases and Interfaces* of the *NewConservationPlan* class.

3. Right-click the *IMapServerToolAction* interface, click *Add*, then click *Implement Interface*.

Visual Studio stubs out the members of *IMapServerToolAction* in the code window automatically, bracketing the stubs within a region named *IMapServerToolAction Members*.

The *IMapServerToolAction* has a single method to implement called *ServerAction*. This method is where you will put the code to execute when the user clicks the map. The following code will be added to your class:

```
#region IMapServerToolAction Members
```

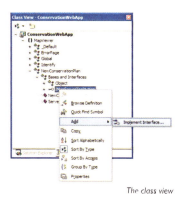

The class view

```
public void ServerAction(ToolEventArgs args)
{
  // TODO: Add NewConservationPlan.ServerAction implementation
}
```

#endregion

The remainder of the code will be added to this method.

In order to create a new feature, you need to use the workspace object you saved in the server context to start and stop an edit operation. You will get the map server's context from the map control's *WebMap* object and create the ArcObjects you will use to perform the edit operation within that context. Since the *FarmConservation* map server object is a non-pooled server object, the WebMap will hand you a reference to the same context in which you saved the workspace object. You will scope the use of the WebMap within a *Using* block.

The *args* object passed into the *ServerAction* method includes a reference to the map control.

4. Add the following code to your ServerAction method:

```
if (args.Control is ESRI.ArcGIS.Server.WebControls.Map)
{
  Map map = args.Control as Map;
  using (WebMap webMap = map.CreateWebMap() )
  {
  }
}
```

You need to get a reference to the WebMap's server context; you can then use the *IServerContext* interface to get the workspace that you set in the context when the session began.

5. Add the following lines of code to your using block:

```
// get the server object and server context from the Web map,
// and get the workspace from the context
IServerObject so = (IServerObject) webMap.MapServer;
IServerContext soc = webMap.ServerContext;
IWorkspaceEdit wse = soc.GetObject("theWorkspace") as IWorkspaceEdit;
```

The *args* object also contains the *VectorEventArgs* collection of points and vectors that describe the polygon the user digitized in screen coordinates. You will use these points with the *Converter* class to convert the set of points to a new polygon geometry in the server context.

6. Add the following lines of code to your using block:

```
// get the polygon from the event
VectorEventArgs vargs = args as VectorEventArgs;
System.Drawing.Point[] pts = vargs.Vectors;
IGeometry geom = Converter.ToPolygon(webMap,pts) as IGeometry;
```

Next, you will use the coded value domain to look up the value associated with the domain description that the user selected in the dropdown list on the Web form. To do this, you will use the workspace object to open the *ConservationPlan* feature class and get the type field and its domain.

7. Add the following lines of code to your using block:

Because the ConservationPlan feature class is a layer in the map server object's map document, it was opened when the map server object was created at the beginning of the session. The call to open the ConservationPlan feature class made against the workspace will actually get the feature class that is already opened in the map.

```
// open the feature class
IFeatureWorkspace fws = wse as IFeatureWorkspace;
IFeatureClass fc = fws.OpenFeatureClass("ConservationPlan");

// get the type from the dropdown list and look up the value from the domain
System.Web.UI.WebControls.DropDownList drpDomain =
map.Page.FindControl("drpTypes") as
System.Web.UI.WebControls.DropDownList;
string sDomainDesc = drpDomain.SelectedItem.Text;
string sDomainValue = "";
int lTypeFld = fc.FindField("TYPE");
IField fld = fc.Fields.get_Field(lTypeFld);
ICodedValueDomain cv = fld.Domain as ICodedValueDomain;
for (int i = 0; i < cv.CodeCount; i++)
{
  if (cv.get_Name(i) == sDomainDesc)
  {
    sDomainValue = cv.get_Value(i) as string;
  }
}
```

Now that you have the geometry for the new conservation plan feature and the type, you can create your new conservation plan feature. To do this, you will use the workspace object to start an edit operation. Within this edit operation, you will create a new feature, set the geometry for the new feature to be your polygon, and set the type to be the domain value associated with the description the user chose.

Once you have set these new properties for the feature, you will use Store to store it in the geodatabase and stop the edit operation.

8. Add the following lines of code to your using block:

An edit operation may be thought of as a short transaction nested within the long transaction corresponding to the edit session. Edit operations also define collections of edits that correspond to a single undo or redo. All related changes to objects in the database within an edit session should be grouped into edit operations. To learn more about geodatabase edit session objects, see the online developer documentation.

```
// start an edit operation and create the new feature
wse.StartEditOperation();
IFeature f = fc.CreateFeature();
IFeatureSimplify fs = (IFeatureSimplify) f;
fs.SimplifyGeometry(geom);
f.Shape = geom;

// initialize default values
IRowSubtypes subt = f as IRowSubtypes;
subt.SubtypeCode = 0;
subt.InitDefaultValues();

// set the type as specified by the user
f.set_Value(lTypeFld, sDomainValue);
f.Store();
wse.StopEditOperation();
```

Now that the new feature is created, you need to update a couple of aspects of the application. Specifically, you will enable the *Undo* command on the toolbar, because there is now an edit operation to undo, and refresh the map to draw the new feature.

9. Add the following lines of code to your Using block:

```
// enable undo
Toolbar tb = map.Page.FindControl("Toolbar1") as Toolbar;
ToolbarItemCollection tcoll = tb.ToolbarItems;
Command cmd = tcoll.Find("tbUndo") as Command;
cmd.Disabled = false;

// refresh the map
webMap.Refresh();
```

Now that the class is defined, you must compile your project to add your new *NewConservationPlan* class to the .NET assembly list.

10. Click *Build*, then click *Build Solution*.

11. Fix any errors.

By compiling the project, this allows you to pick MapView.NewConservationPlan for the ServerToolActionClass in the Toolbar CollectionEditor.

Add the New Conservation Plan tool to the toolbar

Now that you have implemented the class, you will add the tool to your toolbar, which will allow the user to digitize a new polygon and execute the code in the *NewConservationPlan* class.

1. In the Solution Explorer, double-click *Default.aspx* to open the Web form in design mode.

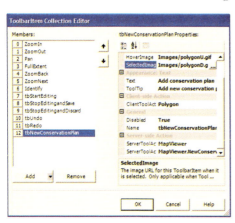

The ToolbarItem Collection Editor dialog box

2. Click the toolbar control.

3. In the properties for the toolbar control, click the *ToolbarItemsCollection* property and click the Browse button. This will open the *Toolbar Item Collection Editor*.

4. Click the *Add* dropdown, then click *Tool*. This will add a new tool to the toolbar collection.

5. Click the *Name* property for the new tool and type "tbNewConservationPlan" for the name.

6. Click the *Text* property and type "Add Conservation Plan" for the text.

7. Click the *ToolTip* property and type "Add new conservation plan" for the tool tip.

8. Click the *ClientToolAction* property dropdown list and click *Polygon*.

9. Click the *ServerToolActionAssembly* property and type "ConservationWebApp" for the assembly name.

10. Click the *ServerToolActionClass* property dropdown list and click *ConservationWebApp.NewConservationPlan* for the class (this is the class you just created).

11. Click the *Disabled* property dropdown list and click true.

The ClientToolAction specifies what code is executed in the client (the Web browser). In this case, JavaScript for drawing a polygon on the map control is the client tool action.

The ServerToolAction is the code in the server that is executed when the client tool action has completed. In this case, the server tool action is defined by the NewConservationPlan class.

The new tool that you add through the Toolbar Collection Editor will appear on the toolbar.

12. Click the *DefaultImage* property and type "Images\polygon.gif" for the default image.

13. Click the *HoverImage* property and type "Images\polygonU.gif" for the hover image.

14. Click the *SelectedImage* property and type "Images\polygonD.gif" for the selected image.

15. Click OK. This will add the new tool to your toolbar.

Finally, you will revisit the code in the toolbar's *CommandClick* event to enable and disable this tool appropriately.

16. Double-click the *Toolbar1* control. This will open the code window in the toolbar's *CommandClick* event.

17. In the *tbStartEditing* case block, change the following line:

```
// TODO enable AddConservationPlan tool
```

to:

```
Tool t = tbcol.Find("tbNewConservationPlan") as Tool;
t.Disabled = false;
```

18. In the *tbStopEditingandSave* and *tbStopEditingandDiscard* case blocks, change the following line:

```
// TODO disable AddConservationPlan tool
```

to:

```
Tool t = tbcol.Find("tbNewConservationPlan") as Tool;
t.Disabled = true;
```

Your Web application is now ready to be tested. Compile the project (*Build/Build Solution*), and fix any errors.

Test the Web application

If you run the Web application from within Visual Studio, it will open a browser and connect to the application's startup page (*Default.aspx*).

1. Click *Debug*, then click *Start*.

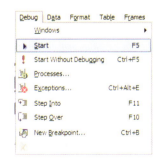

2. Click the Start Editing button on the toolbar.

3. Click the *conservation type* dropdown list and click a conservation type to create.

4. Click the *New Conservation Plan* tool.

5. Click the map to digitize your new conservation plan polygon. Once you have added all the points, double-click.

The new conservation plan polygon will be drawn on the map.

6. Click the *Undo* command to undo the edit.

7. Click the *Redo* command to redo the edit.

8. Click either the *Stop Editing and Save* command or the *Stop Editing and Discard* command to stop editing.

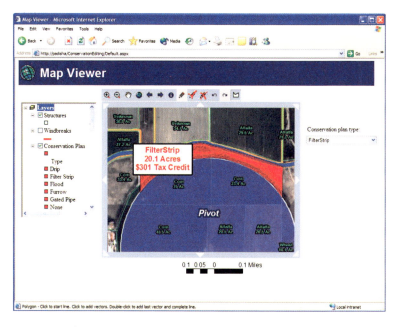

DEPLOYMENT

Presumably you developed this Web application using your development Web server. To deploy this Web application on your production Web server, you can use the built-in Visual Studio .NET tools to copy the project.

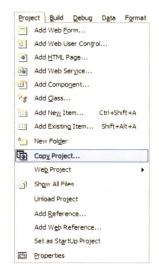

1. In the Solution Explorer, click the *ConservationEditing* project.

2. Click *Project*, then click *Copy Project*.

3. In the *Copy Project* dialog box, specify the location on your Web server to copy the project to.

4. Click OK.

ADDITIONAL RESOURCES

This scenario includes functionality and programming techniques covering a number of different aspects of ArcObjects, the ArcGIS Server API, .NET application templates, and Web controls.

You are encouraged to read Chapter 4 of this book to get a better understanding of core ArcGIS Server programming concepts such as stateful versus stateless server application development. Chapter 4 also covers concepts and programming guidelines for working with server contexts and ArcObjects running within those contexts.

This scenario makes use of a Web application template and the ArcGIS Server .NET ADF Web controls to provide the majority of the user interface for this Web application. To learn more about this Web application template and other template applications that are included with the .NET ADF, see Chapter 5 of this book. Chapter 5 also includes detailed descriptions and examples of using the

.NET Web controls, including the map and toolbar Web controls that you made use of while programming this Web application. If you are unfamiliar with ASP.NET Web development, it's also recommended that you refer to your .NET developer documentation to become more familiar with Web application development.

ArcGIS Server applications exploit the rich GIS functionality of ArcObjects. This application is no exception. It includes the use of ArcObjects to work with the components of a MapServer, manage a geodatabase edit session, create and set properties of features in the geodatabase, and manipulate geometries. To learn more about these aspects of ArcObjects, refer to the online developer documentation on the Carto, GeoDatabase, and Geometry object libraries.

The Java ADF, as described in this scenario, was developed with JavaServer Faces version 1.0. For the latest information on supported servlet engines and application servers as well as software and documentation updates, visit the ESRI Support Web site at http://support.esri.com. For more information about JSF, visit http:// java.sun.com/j2ee/javaserverfaces.

This walk-through is for developers who need to build and deploy a Web application that extends one of the application templates installed as part of the Java Application Developer Framework SDK. The application incorporates focused geodatabase editing capabilities using the ArcGIS Server API, Java Web controls, and ArcObjects. It describes the process of building, deploying, and consuming the *Extending_a_template* Sample, which is part of the ArcGIS developer samples.

You can find this sample in:

<install_location>\DeveloperKit\Samples\Developer_Guide_Scenarios\
        ArcGIS_Server\Extending_a_web_application_ template

PROJECT DESCRIPTION

The purpose of this scenario is to create a JSP Web application by extending the Java ADF MapViewer template. The application uses ArcObjects to manage a geodatabase edit session and allows the user to start editing; create new features; and undo, redo, and save their edits.

Rather than walk through this scenario, you can get the completed Web application from the samples installation location. The sample is installed as part of the ArcGIS Developer samples.

The following is how the user will interact with the application:

1. Open a Web browser and specify the URL of the Web application.

The MapViewer Web application template is installed as part of the Java ADF SDK.

2. Click *Start Editing* to start an edit session.

3. Click the *Add Conservation Plan* tool on the toolbar and digitize the new conservation plan polygon on the map.

4. Additional conservation plan features can be created, and users can click the *Undo* and *Redo* buttons to undo and redo their edits.

5. Once finished, users can either click *Stop Editing and Save* to save their edits or *Stop Editing and Discard* to discard their edits.

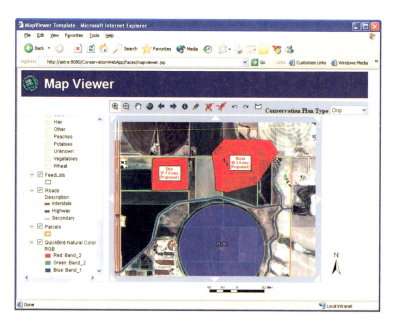

If users choose to save their edits, the geodatabase edit session is ended and their changes are saved. If users choose to not save edits, then the geodatabase edit session is ended and their edits are discarded.

CONCEPTS

The application templates provided with the ArcGIS Server Application Developer Framework SDK provide a good starting point for developers to create Web applications with advanced GIS functionality. Developers will extend these applications using remote ArcObjects that are exposed through the ArcGIS Server API.

Both coarse-grained calls to remote ArcObjects, such as the methods on the MapServer and GeocodeServer, as well as fine grained calls to remote ArcObjects, such as creating new geometries, are exposed through the ArcGIS Server API and can be used in your application. With the functionality of ArcObjects available to the Web application developer, the template applications can be extended to include a wide variety of GIS functionality. This functionality includes what is possible using ArcObjects, including analysis; query; display; and, as in this application, data maintenance.

DESIGN

This Web application is an example of a deeply stateful application in that it's designed to make stateful use of the GIS server. Since this Web application supports edit sessions that span multiple requests for operations, such as creating new features, supporting undo and redo, and the ability to stop editing and discard your edits, the application must use the same workspace object and geodatabase edit session throughout a Web application session. To do this, the Web application must make use of the same server context throughout the user session and release that context only when the session has ended.

Because each user session needs its own server context and server object dedicated to it, the server object cannot be shared across multiple user sessions and, therefore, cannot be pooled. When this application is deployed, there will be an instance of a running map server object for each concurrent user of the application.

To provide the actual functionality, the application uses command buttons to work with the events on the map Web control to expose and manage the edit session (Start Editing, Stop Editing, Undo, Redo) and tools to get a polygon from the user and use that polygon and ArcObjects to update the geodatabase. In order to support this application, a non-pooled map server object must be added to the ArcGIS Server using ArcCatalog.

The Web application will use the Web controls to manage the connection to the GIS server, and the MapViewer Web application template will provide the basic mapping functionality that is required for this application. You will add new tools and commands to the tools collection and map control that allow users to manage their edit session and click the map as input to creating new features.

This application is designed to directly edit the version of the geodatabase that the map server object is connected to. Therefore, all users of the application are editing the same version. This can be augmented with version management capabilities to, for example, create a new version for each session and have all edits go into that version.

While the design of this example allows the editing of personal geodatabases, one could only deploy this application to support multiple editors if the geodatabase was a multiuser geodatabase managed by ArcSDE.

It is possible to create an ArcGIS Server application that includes editing functionality and is stateless. In such an application, each request to make an edit would be its own edit session and the application would not support undo, redo, and the option to save or not save edits. This example includes undo and redo functionality and, therefore, must make use of a non-pooled server object.

The Add Server Object wizard.

REQUIREMENTS

The requirements for working through this scenario are that you have ArcGIS Server and ArcGIS Desktop installed and running. The machine on which you develop this Web application must have the ArcGIS Server Java Application Developer Framework SDK installed.

You must have a map server object configured and running on your ArcGIS Server that uses the *Conservation.mxd* map document installed with the ArcGIS developer samples. The map document references a personal geodatabase with feature classes of farm data. It also references a QuickBird satellite image courtesy of DigitalGlobe.

In ArcCatalog, create a connection to your GIS server and use the Add Server Object command to create a new server object with the following properties:

Name: FarmConservation

Type: MapServer

Description: Map server showing conservation resource planning

Map document: <install_location>\DeveloperKit\Samples\
Data\ServerData\Conservation\Conservation.mxd

Output directory: Choose from the output directories configured in your server.

Pooling: The Web service makes stateful use of the server object. Click Not pooled for the pooling model and accept the defaults for the max instances (4).

Accept the defaults for the remainder of the configuration properties.

After creating the server object, start it and click the Preview tab to verify that it is correctly configured and that the map draws.

You can refer to Chapter 3 for more information on using ArcCatalog to connect to your server and create a new server object. Once the server object is configured and running, you can begin to code your Web application.

ArcCatalog is used for managing your spatial data holdings, defining your geographic data schemas, and managing your ArcGIS Server. Once you have created your FarmConservation server object, preview it using ArcCatalog to verify it is correctly configured.

The following ArcObjects—Java packages will be used in this example:

- com.esri.arcgis.carto

- com.esri.arcgis.geodatabase

- com.esri.arcgis.geometry

- com.esri.arcgis.server

- com.esri.arcgis.webcontrols

- com.esri.arcgis.system

The development environment does not require any ArcGIS licensing; however, connecting to a server and using a map server object require that the GIS server is licensed to run ArcObjects in the server. None of the assemblies used require an extension license.

The IDE used in this example is JBuilder 9 Enterprise Edition, and all IDE specific steps will assume that is the IDE you are using. This Web application can be implemented with other Java IDEs.

IMPLEMENTATION

In this scenario, you will use the MapViewer template project that is installed as part of the ArcGIS Server Application Developer Framework's SDK to provide some basic mapping functionality, and you will extend this template with your own functionality.

The first step is to create an application based on the MapViewer template, which you will extend.

Create an application

1. Open a command window and navigate to the folder *<install_location>\ArcGIS\DeveloperKit\Templates\Java*

2. At the command prompt, type "arcgisant build".

3. In the Input dialog box, type "ConservationWebApp" for the *Application name*.

4. In the Connect to a GIS server section, for *GIS Server*, *Domain*, *Username*, and *Password* enter information pertaining to your GIS server and the account that you want the Web application to connect to the GIS server as.

5. Click *Connect*.

6. In the *Choose template and set properties* section, click the *Name* dropdown and click *Map Viewer*.

7. Click the *Map* dropdown and click *FarmConservation*.

8. Click the Overview map dropdown and click *FarmConservation*.

9. Click OK.

The template Web application deployment Input dialog box

If the FarmConservation map server object is not listed, verify that the server object is started.

This will create the folder *ConservationWebApp* under *<install_location>\ArcGIS\DeveloperKit\Templates\Java\build*. This is a fully functional Web application that you will extend with your editing functionality. First, you will load the application into JBuilder.

Create a working directory

1. Create a new folder, *editing_projects*. This is where your Web application project files will be copied.

2. Copy the folder ConservationWebApp from *<install_location>\ArcGIS\DeveloperKit\Templates\Java\build* to your *editing_projects* folder.

3. Create a folder, *editing*, under the *editing_projects* folder.

4. Create a folder, *src*, under the *editing* folder.

5. Copy the contents of *ConservationWebApp\WEB-INF\classes* into the location *editing_projects\editing\src*.

Your directory structure should look like the following:

```
+---editing_projects
|   +---ConservationWebApp
|   |   +---*.jsp, *.html
|   |   +---css
|   |   +---images
|   |   +---js
|   |   +---WEB-INF
|   |       +---classes
|   |       |   +---arcgis_webapps.properties
|   |       |   +---managed_context_attributes.xml
|   |       |   +---tools
|   |       |   +---xsl
|   |       +---lib
|   +---editing
|       +---src
|           +---res
|           +---tools
|           +---xsl
|           +---arcgis_webapps.properties
|           +---managed_context_attributes.xml
```

Create a new project

1. Start JBuilder.

2. Click *File*, then click *New Project* to open the Project Wizard window.

3. In the Project Wizard dialog box, type "editing" for the *Name*, then type the path to the *editing_projects* folder you created above for the *Directory*.

4. Click *Finish*.

Add references to the Java Application Developer Framework library

The Project wizard

If you have already created the library in one of the other developer scenario examples, you can add the existing library to your project.

In order to use the Web controls, you need to add references to the Java ADF libraries. These libraries are installed when you install the Java ADF. To add these references, you will create a new library and add it to your project.

1. Click *Tools*, then click *Configure Libraries*.

2. In the Configure Libraries dialog box, click *New* to create a new library.

The Configure Libraries dialog box

3. In the New Library wizard, type "webcontrols_lib" for the *Name*.

The New Library wizard

4. Click *Add*, and browse to the location and select *<install_location>/ArcGIS/ java/webcontrols/WEB-INF/lib*, and click OK.

The Directory browser dialog box

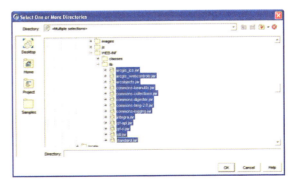

5. Click *OK* to close the Configure Libraries dialog box.

Now you will add the new library to your project.

6. Click *Project*, then click *Project Properties*.

7. In the Project Properties dialog box, click the *Required Libraries* tab.

8. Click *Add*.

9. Under *User Home*, click *webcontrols_lib* and click OK.

The Configure Libraries dialog box

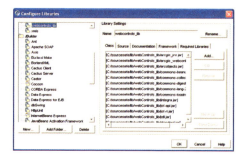

10. Click *OK*.

Add a new Web application project

The Object Gallery dialog box

1. Click *File*, then click *New*.

2. In the Object Gallery dialog box, click the *Web* tab and click *Web Application*.

3. Click *OK*.

4. In the Web Application wizard, type "ConservationWebApp" for *Name*.

5. Click the Browse button and browse to your *editing_projects/ ConservationWebApp* folder for the *Directory*.

6. Click *OK*.

The Web Application wizard

The project properties dialog box

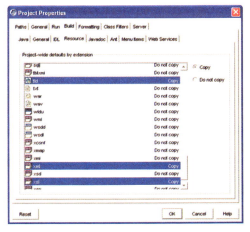

The Project Properties dialog box

It is possible to build a Web application that uses a non-pooled server object for stateful use without using the Web controls. In such an application, it would be the developer's responsibility to keep a reference to the same server context for the duration of the Web application session.

Configure the Web application project

The last step to set up the new project is to configure the deployment descriptors and enable copying of resources to the project.

1. In the Project pane, right-click *ConservationWebApp*, then click *Properties*.

2. Click the *Directories* tab, and check the *Include regular content in WEB-INF and subdirectories* check box.

3. Click *OK*.

4. Click *Project*, then click *Project Properties*.

5. Click the *Build* tab, then click the *Resource* tab.

6. In the file types list, select *xml*, *xsl*, and *properties*.

7. Click the *Copy* option.

8. Click *OK*.

9. Click *Project*, then click *Rebuild project "editing.jpx"* to build the application.

Get a reference to the workspace from the map

Now that the basic properties of the application template have been set, you need to add code to execute at session startup and session end to get the workspace from the *FarmConservation* server object that you will be using throughout the application. Since you need to keep the workspace around, you will use the *SetObject* method on the server context to add the workspace to the server context's object dictionary. When you need to use the workspace throughout the application, you can use the *GetObject* method to get the workspace out of the server context.

Since the *FarmConservation* map server object is a non-pooled object, the context control will ask the GIS server for an instance of that map server object and hold on to it for the duration of the session. This means that each time you get the server object and server context from the context control during the application session, you know you are always getting the same one.

To do this, you will create a JavaBean to store all of the business data.

1. Click *File*, then click *New Class*.

2. In the Class wizard, type "conservationPlan" for the *Package* and type "ConservationPlan" for the *Class Name*.

3. Click *OK*.

This will add a new class to your project and will open the code for the class with some autogenerated code. The autogenerated code will include the name of the package (*conservationPlan*) and a stubbed out class definition for *ConservationPlan*.

```
package conservationPlan;
public class ConservationPlan {
  public ConservationPlan() {
  }
}
```

4. Add import statements for the additional Java packages you will use in this project. At the top of the code window, add the following import statements:

```
import java.io.*;
import java.util.*;
import java.util.logging.*;

import javax.faces.model.*;
import com.esri.arcgis.webcontrols.ags.data.*;
import com.esri.arcgis.webcontrols.data.*;

import com.esri.arcgis.server.*;
import com.esri.arcgis.carto.*;
import com.esri.arcgis.geodatabase.*;
```

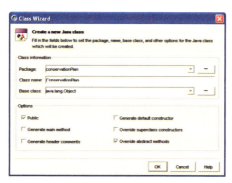

The Class wizard

For the *ConservationPlan* class to be part of the request response life cycle of the Web application it must implement *WebLifeCycle*. By implementing *WebLifeCycle*, you can add actions to *ConservationPlan* that need to be performed when the session is created and destroyed. You also want the context control to manage the instantiation of our *ConservationPlan* class. To achieve this the *ConservationPlan* class should implement *WebContextInitialize*.

5. Change the following line:

```
public class ConservationPlan {
```

to:

```
public class ConservationPlan implements WebLifecycle,
WebContextInitialize {
```

You must add the *WebLifeCycle* and *WebContextInitialize* methods to your class.

Add the following lines of code to your *ConservationPlan* class.

```
public void init(WebContext ctx) {
}

public void destroy() {
}

public void activate() {
}

public void passivate() {
}
```

Your code should now look like the following:

```
package conservationPlan;

import java.io.*;
import java.util.*;
import java.util.logging.*;

import javax.faces.model.*;
import com.esri.arcgis.webcontrols.ags.data.*;
```

```
import com.esri.arcgis.webcontrols.data.*;

import com.esri.arcgis.server.*;
import com.esri.arcgis.carto.*;
import com.esri.arcgis.geodatabase.*;

public class ConservationPlan implements WebLifecycle,
WebContextInitialize {
  public void init(WebContext ctx) {
  }

  public void destroy() {
  }

  public void activate() {
  }

  public void passivate() {
  }
}
```

In this example, you will get the workspace from the first layer in the map. You do this by getting a reference to the map server object from *AGSWebContext*. You can then use the *IMapServerObjects* interface to access the map server's fine-grained ArcObjects to get a reference to the first layer in the map.

6. Add the following attributes to your class.

```
Logger logger = Logger.getLogger(ConservationPlan.class.getName());
  public static final String WEB_CONTEXT_ATTRIBUTE_NAME =
"esriConservationPlan";
  private static final String WS_NAME = "theWorkspace";
  private AGSWebContext agsContext;
```

The code to get the workspace from the map will be added to the init method. The code in the init method will be executed when this class is initialized by the context control.

7. Add the following code to your class.

```
public void init(WebContext ctx) {
  this.agsContext = (AGSWebContext)ctx;
  try {
    AGSWebMap agsMap = (AGSWebMap)agsContext.getWebMap();
    IMapServerObjects mapo = new
IMapServerObjectsProxy(agsMap.getServer());
    IMap map = new IMapProxy(mapo.getMap(agsMap.getFocusMapName()));

    // get workspace from first layer
    IFeatureLayer fl = new IFeatureLayerProxy(map.getLayer(0));
    IDataset ds = new IDatasetProxy(fl.getFeatureClass());
    IWorkspace ws = new IWorkspaceProxy(ds.getWorkspace());
    IServerContext sc = agsContext.getServerContext();
    sc.setObject(WS_NAME, ws);
  }
```

A server context contains an object dictionary that serves as a convenient place for you to store references to commonly used objects. This example uses the SetObject method on IServerContext to add the workspace to the object dictionary. As you will see later, since you are using the same server context through the application session, you can get to the workspace by using GetObject in various parts of the application.

```
  catch (Exception e) {
    throw new IllegalStateException("Cannot create conservation plan.\n\t"
+ e);
  }
}
```

To facilitate getting the workspace from the server context later in the application, you will add a help method to your *ConservationPlan* class to return the workspace.

8. Add the following code to your *ConservationPlan* class.

```
public IWorkspace getWorkspace() {
  try {
    IServerContext sc = agsContext.getServerContext();
    IWorkspace ws = new IWorkspaceProxy(sc.getObject(WS_NAME));
    return ws;
  }
  catch(Exception _) {
    throw new IllegalStateException("Cannot get workspace.\n\t" + _);
  }
}
```

Close any open edit sessions on session end

During the application session, users may start, but not stop, an edit session before the session times out. This can happen if users start editing and close their browser without stopping editing, or if their session times out while their edit session is active.

This will keep the geodatabase edit session open even though the user is no longer using the Web application. You want to safeguard against this by explicitly stopping the edit session if the session times out and it is still active. Since the *ConservationPlan* class implements *WebLifeCycle*, its destroy method will be called when the session times out. You will add code to the destroy method to stop any active edit sessions.

1. Add the following lines of code to your destroy method.

```
// need to close any open edit session
try {
  IWorkspaceEdit wse = new IWorkspaceEditProxy(getWorkspace());
  if (wse.isBeingEdited()) {
    wse.stopEditing(false); // stop editing and don't save edits
  }
}
catch (IOException ex) {
}
```

You don't have to add any code to the activate and passivate methods. The code for the *ConservationPlan* class should look like the following:

```
package conservationPlan;
import java.io.*;
import java.util.*;
import java.util.logging.*;
import javax.faces.model.*;
```

```java
import com.esri.arcgis.webcontrols.ags.data.*;
import com.esri.arcgis.webcontrols.data.*;
import com.esri.arcgis.server.*;
import com.esri.arcgis.carto.*;
import com.esri.arcgis.geodatabase.*;

public class ConservationPlan implements WebLifecycle,
WebContextInitialize {
  Logger logger = Logger.getLogger(ConservationPlan.class.getName());
  public static final String WEB_CONTEXT_ATTRIBUTE_NAME =
"esriConservationPlan";
  private static final String WS_NAME = "theWorkspace";
  private AGSWebContext agsContext;

  public void init(WebContext ctx) {
    this.agsContext = (AGSWebContext)ctx;
    try {
      AGSWebMap agsMap = (AGSWebMap)agsContext.getWebMap();
      IMapServerObjects mapo = new
IMapServerObjectsProxy(agsMap.getServer());
      IMap map = new IMapProxy(mapo.getMap(agsMap.getFocusMapName()));

      // get workspace from first layer
      IFeatureLayer fl = new IFeatureLayerProxy(map.getLayer(0));
      IDataset ds = new IDatasetProxy(fl.getFeatureClass());
      IWorkspace ws = new IWorkspaceProxy(ds.getWorkspace());
      IServerContext sc = agsContext.getServerContext();
      sc.setObject(WS_NAME, ws);
      }

    catch (Exception e) {
      throw new IllegalStateException("Cannot create conservation
plan.\n\t" + e);
      }
    }

  public IWorkspace getWorkspace() {
      try {
      IServerContext sc = agsContext.getServerContext();
      return new IWorkspaceProxy(sc.getObject(WS_NAME));
      }
      catch(Exception e) { throw new IllegalStateException("Cannot get
workspace.\n\t" + e); }
    }


  public void destroy() {
    // need to close any open edit session
    try {
      IWorkspaceEdit wse = new IWorkspaceEditProxy(getWorkspace());
      if (wse.isBeingEdited()) {
```

```
                        wse.stopEditing(false); // stop editing and don't save edits
                      }
                    }
                  catch (IOException ex) {
                    }

                }

            public void activate() {
            }

            public void passivate() {
            }
        }
```

Add editing commands to your toolbar

This application allows the user to start and stop edit sessions, create features, and undo and redo edit operations. To support these functions you will add code to the *ConservationPlan* class.

1. You will add code to get the workspace you saved in the server context. Add the following code to your class:

```
private IWorkspaceEdit getWorkspaceEdit() {
  try { return new IWorkspaceEditProxy(getWorkspace()); }
  catch (IOException e) {
    throw new IllegalStateException("Cannot get workspace edit.\n\t" + e);
  }
}
```

Next, define action methods that will be called when users of the application click buttons to control the edit session, as well as the enabled and disabled state of the buttons, on the JSP page. JSF expects action methods to return a string outcome. This string outcome is used for page navigation. Since our application does not require navigation to another page, the methods return null.

The objects and interfaces used for managing geodatabase edit sessions can be found in the GeoDatabase object library. To learn more about geodatabase edit session objects, see the online developer documentation.

2. Add the following code to your class:

```
public String start() {
  try {
    getWorkspaceEdit().startEditing(true);
  }
  catch (IOException e) {
    logger.log(Level.WARNING, "Unable to start editing.", e);
  }
  return null;
}

public String save() {
  try {
    getWorkspaceEdit().stopEditing(true);
    agsContext.refresh();
  }
```

```
     catch (IOException e) {
       logger.log(Level.WARNING, "Unable to save editing.", e);
     }
     return null;
   }

   public String discard() {
     try {
       getWorkspaceEdit().stopEditing(false);
       agsContext.refresh();
     }
     catch (IOException e) {
       logger.log(Level.WARNING, "Unable to discard editing.", e);
     }
     return null;
   }

   public String undo() {
     try {
       getWorkspaceEdit().undoEditOperation();
       agsContext.refresh();
     }
     catch (IOException e) {
       logger.log(Level.WARNING, "Unable to undo editing.", e);
     }
     return null;
   }

   public String redo() {
     try {
       getWorkspaceEdit().redoEditOperation();
       agsContext.refresh();
     }
     catch (IOException e) {
       logger.log(Level.WARNING, "Unable to redo editing.", e);
     }
     return null;
   }
```

3. To determine whether the Undo and Redo buttons should be enabled, add the
   following methods to your class, which will be called by the JSP page:

```
   public boolean isCanUndo() {
     try {
       boolean[] hasUndos = new boolean[]{ true };
       getWorkspaceEdit().hasUndos(hasUndos);
       return hasUndos[0];
     }
     catch (Exception e) {
       logger.log(Level.WARNING, "Unable to determine if can undo.", e);
     }
```

```
      return true;
    }

    public boolean isCanRedo() {
      try {
        boolean[] hasRedos = new boolean[]{ true };
        getWorkspaceEdit().hasRedos(hasRedos);
        return hasRedos[0];
      }
      catch (Exception e) {
        logger.log(Level.WARNING, "Unable to determine if can redo.", e);
      }
      return true;
    }

    public boolean isEditing() {
      try {
        return getWorkspaceEdit().isBeingEdited();
      }
      catch (Exception e) {
        logger.log(Level.WARNING, "Unable to determine if editable.", e);
      }
      return false;
    }
```

Add the New Conservation Plan tool

Now that you have added commands to handle managing your edit session, you will add the tool that will actually make edits on the database. This application allows the user to digitize new conservation plan features in their geodatabase by digitizing a polygon on the map control. To allow the user to do this, you will add a new tool to the tools collection.

In this example, you will add a single tool for creating new conservation plan polygons of a user-specified type. The types of possible conservation plans will be available from a dropdown list that the user will pick from based on a domain assigned to the Type field in the *ConservationPlan* feature class. The first step is to add code to get the list of valid domain values to the *ConservationPlan* class.

Add the following code to your *ConservationPlan* class.

```
    private ArrayList domainOptions;
    public ArrayList getDomainOptions() {
      if (domainOptions != null) return domainOptions;
      // populate the dropdown list with the conservation types
      try {
        IFeatureWorkspace fws = new IFeatureWorkspaceProxy(getWorkspace());
        IFeatureClass fc = new
IFeatureClassProxy(fws.openFeatureClass("ConservationPlan"));
        IField fld = new
IFieldProxy(fc.getFields().getField(fc.findField("TYPE")));
        ICodedValueDomain cv = new ICodedValueDomainProxy(fld.getDomain());
        domainOptions = new ArrayList(cv.getCodeCount());
```

```
    String cvName;
    for (int i = 0; i < cv.getCodeCount(); i++) {
      cvName = cv.getName(i);
      domainOptions.add(new SelectItem(cvName, cvName, cvName));
     }
     return domainOptions;
   }
  catch (IOException e) { throw new IllegalStateException("Cannot get
domain options.\n\t" + e); }
}

private String domainOption;

public void setDomainOption(String domainOption) {
  this.domainOption = domainOption;
}

public String getDomainOption() {
  return domainOption;
}
```

Add the class to the collection of managed beans of the application. The Context controls will instantiate the classes added to this collection by calling their init method.

1. In the Project pane, expand <Project Source>double-click managed_context_attribute.xml to open it in the code window.

2. Add the following lines to the end of the file before the tag </*managed-context-attributes*>:

```
<managed-context-attribute>
<name>esriConservationPlan</name>
<attribute-class>conservationPlan.ConservationPlan</attribute-class>
<description>Performs Edit Operations.</description>
</managed-context-attribute>
```

The next step is to create a new class that implements the *IMapToolAction* interface and defines the functionality for the tool.

1. Click *File*, then click *New Class*.

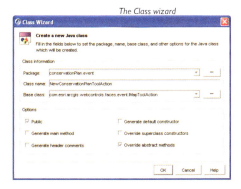
The Class wizard

2. In the Class wizard, type "conservationPlan.event" for the *Package*, "NewConservationPlanToolAction" for the *Class Name*, and "com.esri.arcgis.webcontrols.faces.event.IMapToolAction" for the *Base Class*.

3. Click *OK*.

This will add a new class to your project and will open the code for the class with the following autogenerated code.

```
package conservationPlan.event;

import com.esri.arcgis.webcontrols.faces.event.IMapToolAction;
import com.esri.arcgis.webcontrols.faces.event.MapEventArgs;
```

```
public class NewConservationPlanToolAction implements IMapToolAction {
  public void execute(MapEventArgs parm1) throws java.lang.Exception {
    /**@todo Implement this
com.esri.arcgis.webcontrols.faces.event.IMapToolAction method*/
    throw new java.lang.UnsupportedOperationException("Method execute()
not yet implemented.");
  }
}
```

4. Edit the auto-generated code to change the variable name parm1 to args and remove the UnsupportedOperationException. Your code should look like the following:

```
package conservationPlan.event;

import com.esri.arcgis.webcontrols.faces.event.IMapToolAction;
import com.esri.arcgis.webcontrols.faces.event.MapEventArgs;

public class NewConservationPlanToolAction implements IMapToolAction {
  public NewConservationPlanToolAction() {
  }
  public void execute(MapEventArgs args) throws java.lang.Exception {
  }
}
```

5. Add the following import statements to add the packages you will use in this class.

```
import com.esri.arcgis.geodatabase.*;
import com.esri.arcgis.geometry.*;
import com.esri.arcgis.webcontrols.ags.data.*;
import com.esri.arcgis.webcontrols.faces.event.*;
import conservationPlan.*;
```

Implement the NewConservationPlanToolAction class

Now you are ready to implement the *NewConservationPlanToolAction* class that contains the code to execute when a new polygon is digitized by the user on the map. Since this class is a map tool action, it must implement the *IMapToolAction* interface. The *IMapToolAction* has a single method to implement called *execute*. This method is where you will put the code to execute when the user clicks the map. The remainder of the code will be added to this method.

In order to create a new feature, you need to use the workspace object you saved in the server context to start and stop an edit operation. You will get the map server's context from the *AGSWebContext* object and create the ArcObjects you will use to perform the edit operation within that context. Since the *FarmConservation* map server object is a non-pooled server object, the *AGSWebContext* will hand you a reference to the same context in which you saved the workspace object.

The *args* object passed into the *execute* method includes a reference to the map control.

1. Add the following code to your *execute* method:

```
AGSWebMap wMap = (AGSWebMap) args.getWebContext().getWebMap();
AGSWebContext context = (AGSWebContext) args.getWebContext();
```

```
ConservationPlan plan =
  (ConservationPlan)context.getAttribute(ConservationPlan.WEB_CONTEXT_ATTRIBUTE_NAME);
```

You need to get a reference to the workspace that you set in the context when the session began.

2. Add the following line of code to your *execute* method:

```
IWorkspaceEdit wse = new IWorkspaceEditProxy(plan.getWorkspace());
```

The *args* object also contains the *ClientActionArgs* collection of points and vectors that describe the polygon the user digitized in screen coordinates. You will use these points with the *ToMapPoint* method on the *AGSWebMap* to get the set of points in the spatial reference of the map. You will then use these points to create a new polygon geometry in the server context.

3. Add the following lines of code to your *execute* method:

The objects and interfaces used for creating and working with geometries can be found in the Geometry object library. To learn more about geometry objects, see the online developer documentation.

```
PolygonArgs pargs = (PolygonArgs) args.getClientActionArgs();
ClientPoint[] pts = pargs.getPoints();
IPointCollection pcol = new
IPointCollectionProxy(context.createServerObject(Polygon.getClsid()));
for (int i = 0; i < pts.length; i++) {
  IPoint pt = wMap.toMapPoint(pts[i].getX(), pts[i].getY());
  pcol.addPoint(pt, null, null);
}
IGeometry geom = new IGeometryProxy(pcol);
```

Next, you will use the coded value domain to look up the value associated with the domain description that the user selected in the dropdown list on the Web form. To do this, you will use the workspace object to open the *ConservationPlan* feature class and get the type field and its domain.

4. Add the following lines of code to your *execute* method:

Because the ConservationPlan feature class is a layer in the map server object's map document, it was opened when the map server object was created at the beginning of the session. The call to open the ConservationPlan feature class made against the workspace will actually get the feature class that is already opened in the map.

```
IFeatureWorkspace fws = new IFeatureWorkspaceProxy(wse);
IFeatureClass fc = new
IFeatureClassProxy(fws.openFeatureClass("ConservationPlan"));
String sDomainDesc = plan.getDomainOption();
String sDomainValue = null;
int lTypeFld = fc.findField("TYPE");
if (lTypeFld != -1) {
  IField fld = fc.getFields().getField(lTypeFld);
  ICodedValueDomain cv = new ICodedValueDomainProxy(fld.getDomain());
  for (int i = 0; i < cv.getCodeCount(); i++) {
    if (cv.getName(i).equals(sDomainDesc)) {
      sDomainValue = (String) cv.getValue(i);
      break;
    }
  }
}
```

A coded value domain is a valid value list that can be associated with a field for a class or subtype in a geodatabase. The coded value domain consists of value/description pairs that allow user interfaces to display user understandable text strings to describe valid database values. To learn more about coded value domain objects, see the online developer documentation.

Now that you have the geometry for the new conservation plan feature and the type, you can create the new conservation plan feature. To do this, you will use the workspace object to start an edit operation. Within this edit operation, you will create a new feature, set the geometry for the new feature to be your polygon, and the type to be the domain value associated with the description the user chose.

An edit operation may be thought of as a short transaction nested within the long transaction corresponding to the edit session. Edit operations also defined collections of edits that correspond to a single undo or redo. All related changes to objects in the database within an edit session should be grouped into edit operations. To learn more about geodatabase edit session objects, see the online developer documentation.

Once you have set these new properties for the feature, you will use Store to store it in the geodatabase and stop the edit operation.

5. Add the following lines of code to the *execute* method:

```
wse.startEditOperation();
IFeature f = fc.createFeature();
IFeatureSimplify fs = new IFeatureSimplifyProxy(f);
fs.simplifyGeometry(geom);
f.setShapeByRef(geom);

IRowSubtypes subt = new IRowSubtypesProxy(f);
subt.setSubtypeCode(0);
subt.initDefaultValues();
if (lTypeFld != -1) {
  f.setValue(lTypeFld, sDomainValue);
}
f.store();
wse.stopEditOperation();
context.refresh();
```

Add the New Conservation Plan tool to the tools collection

Now that you have implemented the class, you will add the tool to your tools collection, which will allow the user to digitize a new polygon and execute the code in the *NewConservationPlan* class.

1. In the Project pane, click *tools*, then double-click *default.xml* to open it in the code window.

2. Add the following lines to the end of the file before the tag *</tool-item-collection>*:

```
<tool-item>
<key>NewConservationPlan</key>
<action-class>conservationPlan.event.NewConservationPlanToolAction</action-class>
<client-action>MapPolygon</client-action>
</tool-item>
```

The tag *<key>* will be used in the JSP page to identify the tool inside the application. The tag *<action-class>* is the class you created in the previous step that will be executed on the server in response to this tool. The tag *<client-action>* represents the name of the JavaScript function that will be executed when the user interacts with the tool.

Add functionality to the JSP page

Now that you have created your business object *ConservationPlan* and the map tool *NewConservationPlanToolAction*, you can add commands to the *mapviewer.jsp* page.

The template includes a collection of tools and commands (*zoom in*, *zoom out*, *pan*, *identify*, and so forth). You will add the following commands to this collection:

Start Editing: starts a new edit session

Stop Editing and Save Edits: stops editing and saves the edits

Stop Editing and Discard Edits: stops editing and discards the edits

Undo: undoes an edit operation

Redo: redoes an edit operation

You will add the following tool:

Create Conservation Plan: creates a new conservation plan feature

1. In the Project pane, click *ConservationWebApp*, click *Root directory*, then double-click *mapviewer.jsp*.

The application does not need the overview map control, so you will delete it from the application.

2. Delete the following lines of code from *mapviewer.jsp*. The text "[OvMapServerObject]@[Server]" will be populated with information that you chose when creating your application.

```
<!-- Overview map control -->
<ags:overview resource="[OvMapServerObject]@[Server]" left="16" top="430"
width="204" height="124" borderWidth="2" cssClass="overviewClass" />
```

3. Before adding the commands, you will need to copy a set of images that you will use for the commands and tools in the toolbar. Copy the following image files from *<install_location>\DeveloperKit\Samples\Data\ServerData\Conservation* to your application's Images folder:

- Redo.gif
- RedoD.gif
- RedoU.gif
- StartEditing.gif
- StartEditingD.gif
- StartEditingU.gif
- StopEditingDiscard.gif
- StopEditingDiscardD.gif
- StopEditingDiscardU.gif
- StopEditingSave.gif
- StopEditingSaveD.gif
- StopEditingSaveU.gif
- Undo.gif
- UndoD.gif
- UndoU.gif

4. Add the following lines of code after the identify tool to add the commands and tools to the JSP page.

```
<td>
<!-- Start editing button -->
<jsfh:commandButton id="start" image="images/StartEditing.gif"
title="Start Editing" alt="Start Editing"
```

```
action="#{sessionScope['mapContext'].attributes['esriConservationPlan'].start}"

disabled="#{sessionScope['mapContext'].attributes['esriConservationPlan'].editing}"
       onmousedown="this.src='images/StartEditingD.gif'"
       onmouseover="this.src='images/StartEditingU.gif'"
       onmouseout="this.src='images/StartEditing.gif'"/>
<!-- Stop editing Discard button -->
<jsfh:commandButton id="discard" image="images/StopEditingDiscard.gif"
title="Stop Editing Discard" alt="Stop Editing Discard"

action="#{sessionScope['mapContext'].attributes['esriConservationPlan'].discard}"

disabled="#{!sessionScope['mapContext'].attributes['esriConservationPlan'].editing}"
       onmousedown="this.src='images/StopEditingDiscardD.gif'"
       onmouseover="this.src='images/StopEditingDiscardU.gif'"
       onmouseout="this.src='images/StopEditingDiscard.gif'"/>
<!-- Stop editing Save button -->
<jsfh:commandButton id="save" image="images/StopEditingSave.gif"
title="Stop Editing Save" alt="Stop Editing Save"

action="#{sessionScope['mapContext'].attributes['esriConservationPlan'].save}"

disabled="#{!sessionScope['mapContext'].attributes['esriConservationPlan'].editing}"
       onmousedown="this.src='images/StopEditingSaveD.gif'"
       onmouseover="this.src='images/StopEditingSaveU.gif'"
       onmouseout="this.src='images/StopEditingSave.gif'"/>
<!-- Undo editing button -->
<jsfh:commandButton id="undo" image="images/Undo.gif" title="Undo Editing"
alt="Undo Editing"

action="#{sessionScope['mapContext'].attributes['esriConservationPlan'].undo}"

disabled="#{!sessionScope['mapContext'].attributes['esriConservationPlan'].canUndo}"
       onmousedown="this.src='images/UndoD.gif'"
       onmouseover="this.src='images/UndoU.gif'"
       onmouseout="this.src='images/Undo.gif'"/>
<!-- Redo editing button -->
<jsfh:commandButton id="redo" image="images/Redo.gif" title="Redo Editing"
alt="Redo Editing"

action="#{sessionScope['mapContext'].attributes['esriConservationPlan'].redo}"

disabled="#{!sessionScope['mapContext'].attributes['esriConservationPlan'].canRedo}"
       onmousedown="this.src='images/RedoD.gif'"
       onmouseover="this.src='images/RedoU.gif'"
       onmouseout="this.src='images/Redo.gif'"/>
<!-- Polygon editing button -->
<jsfh:commandButton id="planTool" image="images/polygon.gif" title="New
Conservation Plan" alt="New Conservation Plan"
       onclick="this.src='images/polygonD.gif';MapPolygon('Map0',
'NewConservationPlan');return false;"

disabled="#{!sessionScope['mapContext'].attributes['esriConservationPlan'].editing}"
```

```
        onmouseover="this.src='images/polygonU.gif'"
        onmouseout="ButtonOut('planTool', 'Map0', 'NewConservationPlan',
'images/polygon.gif', 'images/polygonD.gif')" />

<!-- Select Box for Domain types -->
<b>Conservation Plan Type</b>
<jsfh:selectOneMenu
value="#{sessionScope['mapContext'].attributes['esriConservationPlan'].domainOption}">
<jsfc:selectItems
value="#{sessionScope['mapContext'].attributes['esriConservationPlan'].domainOptions}"/
>
</jsfh:selectOneMenu>
</td>
```

Test the Web application

The JavaScript functions will not work correctly inside the JBuilder browser, so you will modify the IDE options to start the application. Once started, you can use another browser to test the application.

The IDE Options dialog box

1. Click *Tools*, then *IDE options*.

2. On the IDE Options dialog box, click the *Web* tab.

3. Click the *Do not use Web View, always launch separate process* option and click OK.

4. In the Project pane, click *ConservationWebApp*, then click *Root directory*.

5. Right-click index.html, and click *Web Run using defaults*.

   This will execute the application. After executing index.html, open an external browser, such as Internet Explorer or Netscape, and type the URL "http://localhost:8080/ConservationWebApp/index.html".

6. Click the Start Editing button on the toolbar.

7. Click the *Conservation Plan Type* dropdown list and choose a conservation type to create.

8. Click the *New Conservation Plan* tool.

9. Click on the map to digitize your new conservation plan polygon. Once you have added all the points, double-click.

10. The new conservation plan polygon will be drawn on the map.

11. Click the *Undo* command to undo the edit.

12. Click the *Redo* command to redo the edit.

13. Click either the *Stop Editing and Save* command or the *Stop Editing and Discard* command to stop editing.

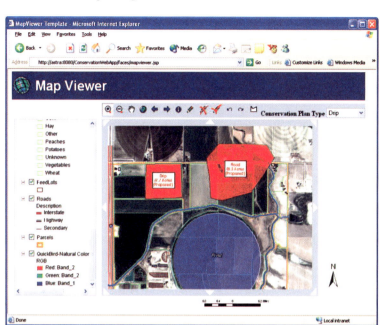

DEPLOYMENT

The Web application was developed with JBuilder's built-in Web server. To deploy this Web application on your production Web server, use the Web archive file ConservationWebApp.war at $/editing_projects and follow the deployment procedure for your Web server.

ADDITIONAL RESOURCES

This scenario includes functionality and programming techniques covering a number of different aspects of ArcObjects, the ArcGIS Server API, Java application templates, and Web controls.

You are encouraged to read Chapter 4 of this book to get a better understanding of core ArcGIS Server programming concepts such as stateful versus stateless server application development. Chapter 4 also covers concepts and programming guidelines for working with server contexts and ArcObjects running within those contexts.

This scenario makes use of a Web application template and the ArcGIS Server's Java ADF Web controls to provide the majority of the user interface for this Web application. To learn more about this Web application template and other template applications that are included with the Java ADF, see Chapter 6 of this book. Chapter 6 also includes detailed descriptions and examples of using the Java Web controls, including the map control that you made use of while programming this Web application. If you are unfamiliar with Java Web development, it's also recommended that you refer to your Java developer documentation to become more familiar with Web application development.

ArcGIS Server applications exploit the rich GIS functionality of ArcObjects. This application is no exception. It includes the use of ArcObjects to work with the components of a MapServer, manage a geodatabase edit session, create and set properties of features in the geodatabase, and manipulate geometries. To learn more about these aspects of ArcObjects, refer to the online developer documentation on the Carto, GeoDatabase, and Geometry object libraries.

Rather than walk through this scenario, you can get the completed Web service from the samples installation location. The sample is installed as part of the ArcGIS developer samples.

This walk-through is for developers who need to build and deploy a .NET application Web service incorporating geocoding and spatial query functionality using the ArcGIS Server API. It describes the process of building, deploying, and consuming the *Application_web_service* sample, which is part of the ArcGIS developer samples.

You can find this sample in: <install_location>\DeveloperKit\Samples\ Developer_Guide_Scenarios\ArcGIS_Server\Application_web_service

PROJECT DESCRIPTION

The purpose of this scenario is to create an ASP.NET Web service using Visual Studio .NET that uses ArcObjects to locate all of the toxic waste sites within a specified distance of a specified address. The Web service returns a .NET array of application-defined toxic waste site objects.

This Web service is intended to be called by other programs, and an example of such a client program is a .NET Windows application. This scenario will also provide an example of how a client application would consume this Web service.

CONCEPTS

A Web service is a set of related application functions that can be programmatically invoked over the Internet. The function can be one that solves a particular application problem, as in this example; a Web service that finds all of the toxic waste sites within a certain distance of an address; or performs some other type of GIS function. Web services can be implemented using the native Web service framework of your Web server such as ASP.NET Web service (WebMethod) or Java Web service (Axis).

When using native frameworks, such as ASP.NET and J2EE, to create and consume your application Web services, you need to use native or application-defined types as both arguments and return values from your Web methods. Clients of the Web service will not be ArcObjects applications, and as such, your Web service should not expect ArcObjects types as arguments and should not directly return ArcObjects types.

To learn more about WSDL, refer to http://www.w3.org.

Any development language that can use standard HTTP to invoke methods can consume this Web service. The Web service consumer can get the methods and types exposed by the Web service through its Web Service Description Language (WSDL). As you walk through this scenario, you will see where special attributes need to be added to your methods and classes such that they can be expressed in WSDL and serialized as XML.

DESIGN

This Web service is designed to make stateless use of the GIS server. It uses ArcObjects on the server to locate an address and query a feature class. In order to support this application, you need to add a pooled geocode server object to your ArcGIS Server using ArcCatalog.

One key aspect of designing your application is whether it is stateful or stateless. You can make either stateful or stateless use of a server object running within the GIS server. A stateless application makes read-only use of a server object, meaning the application does not make changes to the server object or any of its associated objects. A stateful application makes read–write use of a server object where the application does make changes to the server object or its related objects.

Web services are, by definition, stateless applications.

Both coarse-grained calls to remote ArcObjects, such as the methods on the MapServer and GeocodeServer, as well as fine-grained calls to remote

ArcObjects, such as creating new geometries, are exposed through the ArcGIS Server API and can be used in your Web service. The Web service will connect to the GIS server and use an instance of the geocode server object to locate the address supplied to the Web method from the calling application. In order to then buffer the resultant point and use that buffered geometry to query toxic waste sites, you will use the geocode server's server context. Since the geodatabase has been designed such that the address locator is stored in the same geodatabase as the feature class containing the toxic waste sites, you can use the fine-grained objects associated with the geocode server to get a reference to that workspace.

REQUIREMENTS

The requirements for working through this scenario are that you have ArcGIS Server and ArcGIS Desktop installed and running. The machine on which you develop this Web service must have the ArcGIS Server .NET Application Developer Framework installed.

The Add Server Object wizard

You must have a geocode server object configured and running on your ArcGIS Server that uses the Portland.loc locator installed with the ArcGIS Developer Samples. In ArcCatalog, create a connection to your GIS server and use the Add Server Object command to create a new server object with the following properties:

Name: PortlandGeocode

Type: GeocodeServer

Description: Geocode server object for metropolitan Portland

Locator:
<install_location>\DeveloperKit\Samples\Data\ServerData\Toxic\Portland.loc

Batch size: 10 (default)

Pooling: The Web service makes stateless use of the server object. Accept the defaults for the pooling model (pooled server object with minimum instances = 2, max instances = 4).

Accept the defaults for the remainder of the configuration properties.

After creating the server object, start it and right-click to verify that it is correctly configured and that the geocoding properties are displayed.

You can refer to Chapter 3 for more information on using ArcCatalog to connect to your server and to create a new server object. Once the server object is configured and running, you can begin to code your Web service.

ArcCatalog is used for managing your spatial data holdings, defining your geographic data schemas, and managing your ArcGIS Server. Once you have created your PortlandGC server object, open its properties using ArcCatalog to verify it is correctly configured.

The following ArcObjects assemblies will be used in this example:

- ESRI.ArcGIS.Geodatabase
- ESRI.ArcGIS.Geometry
- ESRI.ArcGIS.Location
- ESRI.ArcGIS.Server
- ESRI.ArcGIS.Server.WebControls
- ESRI.ArcGIS.esriSystem

The development environment does not require any ArcGIS licensing; however, connecting to server and using a geocoding server object does require that the GIS server is licensed to run ArcObjects in the server. None of the assemblies used require an extension license.

The IDE used in this example is Visual Studio .NET 2003, and all IDE specific steps will assume that is the IDE you are using. This Web service can be implemented with other .NET IDEs.

IMPLEMENTATION

All code written in this example is in C#; however, you can write this Web service using VB.NET. To begin, you must create a new project in Visual Studio .NET.

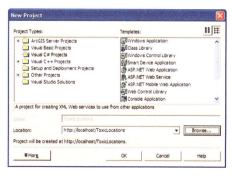

The New Project dialog box

Create a new project

1. Start Visual Studio .NET

2. Click *File*, click *New*, then click *Project*.

3. In the New Project dialog box, click the *Visual C#* category and click *ASP.NET Web Service*.

4. For the Web service name, type "http://localhost/ToxicLocations".

5. Click OK. This will create a blank Web service application.

6. In the Solution Explorer, right-click *Service1.asmx* and click *Rename*. Type "ToxicLocations.asmx" as the new name.

Add references to ESRI assemblies to your project

In order to program using ArcGIS Server, you need to add references to the ESRI assemblies that contain proxies to the ArcObjects components that the Web service will use. These assemblies were installed when you installed the ArcGIS Server .NET Application Developer Framework.

1. In the Solution Explorer, right-click *References* and click *Add Reference*.

2. In the Add Reference dialog box, double-click the following assemblies:

- ESRI.ArcGIS.Geodatabase
- ESRI.ArcGIS.Geometry
- ESRI.ArcGIS.Location
- ESRI.ArcGIS.Server

A proxy object is a local representation of a remote object. The proxy object controls access to the remote object by forcing all interaction with the remote object to be via the proxy object. The supported interfaces and methods on a proxy object are the same as those supported by the remote object. You can make method calls on, and get and set properties of, a proxy object as if you were working directly with the remote object.

The Solution Explorer

The Add Reference dialog box

- ESRI.ArcGIS.Server.WebControls
- ESRI.ArcGIS.System

3. Click OK.

This Web service does not use any of the Web controls, but the *ESRI.ArcGIS.Server.WebControls* assembly contains the .NET *ServerConnection* object required to connect to the GIS server.

Add using statements to add these assemblies' namespaces to your application.

4. In the Solution Explorer, double-click *ToxicLocations.asmx*.

5. In the design window, right-click and click *View Code*. The code for the implementation of this Web service will appear.

6. At the top of the code window, add the following using statements:

```
using ESRI.ArcGIS.Geodatabase;
using ESRI.ArcGIS.Geometry;
using ESRI.ArcGIS.Location;
using ESRI.ArcGIS.Server;
using ESRI.ArcGIS.Server.WebControls;
using ESRI.ArcGIS.esriSystem;
```

Modify the automatically generated code

The code generated by Visual Studio .NET defaults the name of the Web service and contains an example *HelloWorld* Web method. You will rename the Web service and delete the sample Web method.

1. Rename the class and its constructor to "ToxicSiteLocator".

Change the following line:

```
public class Service1 : System.Web.Services.WebService
```

to:

```
public class ToxicSiteLocator: System.Web.Services.WebService
```

Change the following line:

```
public Service1()
```

to:

```
public ToxicSiteLocator()
```

2. Delete the following lines of code:

```
// WEB SERVICE EXAMPLE
// The HelloWorld() example service returns the string Hello World
// To build, uncomment the following lines then save and build the
// project
// To test this Web service, press F5

//    [WebMethod]
//    public string HelloWorld()
//    {
//      return "Hello World";
//    }
```

Your code should look like the following:

```
using System;
using System.Collections;
using System.ComponentModel;
using System.Data;
using System.Diagnostics;
using System.Web;
using System.Web.Services;
using ESRI.ArcGIS.Geodatabase;
using ESRI.ArcGIS.Geometry;
using ESRI.ArcGIS.Location;
using ESRI.ArcGIS.Server;
using ESRI.ArcGIS.esriSystem;

namespace ToxicLocations
{
  /// <summary>
  /// Summary description for Service1.
  /// </summary>
  public class ToxicSiteLocator: System.Web.Services.WebService
  {
    public ToxicSiteLocator()
      {
      // CODEGEN: This call is required by the ASP.NET Web Services
      // Designer
      InitializeComponent();
      }

    Component Designer generated code here
  }
}
```

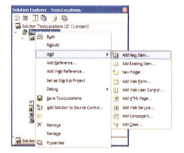

The Solution Explorer

The point of this Web service is to expose a method that is accessible via HTTP-based SOAP requests and that returns all of the toxic site locations within a specified distance of a specified address. These types of methods must be declared as public and support the *[WebMethod]* attribute.

Create the toxic site class

Before you create your new method, first define your toxic waste class that will be returned as a result of the method.

1. In the Solution Explorer, right-click the *ToxicLocations* project, click *Add*, then click *Add New Item*.

2. In the *Add New Item* dialog box, under *Templates*, click *Class*.

3. For the *Name*, type "ToxicSite.cs".

4. Click Open.

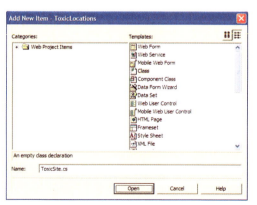

The Add New Item dialog box

This will add a new class (*ToxicSite*) to your project and will open the code for the class with some autogenerated code.

The code view for the *ToxicSite* class should look like the following:

```
using System;

namespace ToxicLocations
{
  /// <summary>
  /// My custom toxic site class
  /// </summary>
  public class ToxicSite
  {
    public ToxicSite()
    {
      //
      // TODO: Add constructor logic here
      //
    }
  }
}
```

Classes that have the XmlInclude attribute can be serialized into XML. Any custom type that is returned by a Web service must have the XmlInclude attribute.

Results returned from Web services must be serializable in XML. The *ToxicSite* type must be marked as to be serialized as XML by including the *XmlInclude* attribute. Because the *XmlInclude* attribute is defined in the *System.Xml.Serialization* namespace, you need to add a using statement for that assembly.

5. Add the following using statement to your code:

```
using System.Xml.Serialization;
```

6. Add the following attribute to your class declaration:

```
[XmlInclude(typeof(ToxicSite))]
```

Your code should now look like this:

```
using System;
using System.Xml.Serialization;

namespace ToxicLocations
{
  /// <summary>
  /// My custom toxic site class
  /// </summary>
  [XmlInclude(typeof(ToxicSite))]
  public class ToxicSite
  {
    public ToxicSite()
    {
      //
      // TODO: Add constructor logic here
      //
    }
  }
}
```

You are interested in returning the type of toxic site, the name of the organization associated with the toxic site, and the x, y coordinates of the toxic site feature. So, you will add four public fields to your *ToxicSite* class and an overloaded constructor to set the data.

7. Add the following lines of code to your *ToxicSite* class:

```
public string Name;
public string Type;
public double X;
public double Y;
```

8. To add the overloaded constructor, type the following in your class definition:

```
public ToxicSite(string sName, string sType, double dX, double dY)
{
  Name = sName;
  Type = sType;
  X = dX;
  Y = dY;
}
```

The code for your *ToxicSite* class should now look like the following:

```
using System;
using System.Xml.Serialization;

namespace ToxicLocations
{
  /// <summary>
  /// My custom toxic site class
  /// </summary>
  [XmlInclude(typeof(ToxicSite))]
  public class ToxicSite
  {
    public ToxicSite()
    {
      //
      // TODO: Add constructor logic here
      //
    }
    public ToxicSite(string sName, string sType, double dX, double dY)
    {
      Name = sName;
      Type = sType;
      X = dX;
      Y = dY;
    }

    public string Name;
    public string Type;
    public double X;
    public double Y;
  }
}
```

Create the Web service method

Now that you have defined your *ToxicSite* class, you will implement your Web service method. As described above, the point of this Web service is to expose a method that is accessible via HTTP-based SOAP requests and that returns all of the toxic site locations within a specified distance of a specified address. These types of methods must both be declared as public and support the *[WebMethod]* attribute.

You will create a method called *FindToxicLocations* that takes as arguments an address, ZIP Code, and a search distance and returns an array of *ToxicSite* objects.

This method will ultimately open a cursor on a feature class in a geodatabase. In order to ensure that your reference to the cursor is released at the end of each request, you will use the *WebObject* object to explicitly release the cursor. You will scope the use of the *WebObject* in a using block.

1. In the Solution Explorer, double-click *ToxicLocations.asmx*.

2. In the design window, right-click and click *View Code*. The code for the implementation of this Web service will appear.

3. Add the following lines of code:

```
[WebMethod]
public ToxicSite[] FindToxicLocations(string Address, string ZipCode,
double Distance)
{
  using (WebObject webObj = new WebObject())
  {
    return null;
  }
}
```

Your code should now look like the following:

```
using System;
using System.Collections;
using System.ComponentModel;
using System.Data;
using System.Diagnostics;
using System.Web;
using System.Web.Services;
using ESRI.ArcGIS.Geodatabase;
using ESRI.ArcGIS.Geometry;
using ESRI.ArcGIS.Location;
using ESRI.ArcGIS.Server;
using ESRI.ArcGIS.esriSystem;

namespace ToxicLocations
{
  /// <summary>
  /// Summary description for Service1.
  /// </summary>
  public class ToxicSiteLocator: System.Web.Services.WebService
  {
```

.NET methods within a class that have the [WebMethod] attribute set are called XML Web service methods and are callable from remote Web clients.

The FindToxicLocations Web method returns an array of ToxicSite objects, rather than a generic collection, such as an ArrayList. If the method did not have a return type of ToxicSite, then the custom class wouldn't be expressed in the Web service's WSDL.

References to COM objects in .NET are not released until garbage collection kicks in. Some objects, such as geodatabase cursors, lock or use resources that the object frees only in its destructor. You can use the WebObject object to explicitly release references to COM objects when it is disposed. For more information on releasing COM objects and the WebObject object, refer to Chapter 4.

```
public ToxicSiteLocator()
{
    // CODEGEN: This call is required by the ASP.NET Web Services
Designer
    InitializeComponent();
}

    Component Designer generated code here
    [WebMethod]
    public ToxicSite[] FindToxicLocations(string Address, string ZipCode,
double Distance)
    using (WebObject webObj = new WebObject())
    {
      return null;
    }
  }
}
```

Validate input parameters

The first thing the Web method will do is verify the provided parameters are valid and, if not, return null.

Type the following in your using block method (note all code additions from this step on are made to this using block):

```
if (Address == null || ZipCode == null || Distance == 0.0)
    return null;
```

Connect to the GIS server

This Web service makes use of objects in the GIS server, so you will add code to connect to the GIS server and get the *IServerObjectManager* interface. In this example, the GIS server is running on a machine called "melange".

This example uses the ServerConnection and ServerObjectManager objects to connect to the server and get references to the geocode server object. As an alternative, this Web service could make use of the geocodeConnection Web control in the ESRI.ArcGIS.Server.WebControls assembly. The Web control would manage the connection and the geocode server objects' server context as well as provide convenient methods for working with the geocode server object using .NET.

1. Add the following lines of code:

```
ESRI.ArcGIS.Server.WebControls.ServerConnection connection = new
ESRI.ArcGIS.Server.WebControls.ServerConnection();
connection.Host = "melange";
connection.Connect();

IServerObjectManager som = connection.ServerObjectManager;
```

The Web service will make use of the *PortlandGC* server object that you created with ArcCatalog. You get a server object by asking the server for a server context containing the object.

2. Add the following line of code to get the server object's context:

```
IServerContext sc =
som.CreateServerContext("PortlandGeocode","GeocodeServer");
```

In order to successfully connect to the GIS server, your Web service must impersonate a user who is a member of the ArcGIS Server users group on the GIS server. There are a number of strategies for implementing impersonation in .NET. For more details on impersonation, refer to Chapter 5 and your .NET documentation.

It's the responsibility of the developer to release the server object's context when finished using it. It's important to ensure that your method calls *ReleaseContext* on the server context when the method no longer needs the server object or any other objects running in the context. It's also important that you ensure the context is released in the event of an error. So, the remainder of the code will run within a try/catch block. If an error occurs in the code in the try block, the code in the catch block will be executed. So, you will add the call to release the context at the end of the try block and in the catch block.

3. Add the following code to your Web service:

```
try
{
  sc.ReleaseContext();
}
catch
{
  sc.ReleaseContext();
}
```

Your *FindToxicLocations* method should now look like the following:

```
[WebMethod]
 public ToxicSite[] FindToxicLocations(string Address, string ZipCode,
double Distance)
{
  using (WebObject webObj = new WebObject())
  {
    if (Address == null || ZipCode == null || Distance == 0.0)
      return null;

    ESRI.ArcGIS.Server.WebControls.ServerConnection connection = new
ESRI.ArcGIS.Server.WebControls.ServerConnection();
    connection.Host = "melange";
    connection.Connect();

    IServerObjectManager som = connection.ServerObjectManager;
    IServerContext sc =
som.CreateServerContext("PortlandGeocode","GeocodeServer");

    try
    {
     sc.ReleaseContext();
    }
    catch
    {
     sc.ReleaseContext();
    }

    return null;
  }
}
```

Geocode the input address

Now that you have connected to the GIS server and have a context containing the geocode server object, you will add the code to use the server object to geocode the input address and store the resulting point as *gcPoint*.

Add the following lines of code to your try block:

```
IServerObject so = sc.ServerObject;
IGeocodeServer gc = so as IGeocodeServer;

IPropertySet ps = sc.CreateObject("esriSystem.PropertySet") as
IPropertySet;
ps.SetProperty("street",Address);
ps.SetProperty("Zone",ZipCode);

IPropertySet res = gc.GeocodeAddress(ps,null);
IPoint gcPoint = res.GetProperty("Shape") as IPoint;
```

A PropertySet is a generic class that is used to hold a set of any properties. A PropertySet's properties are stored as name/value pairs. Examples for the use of a property set are to hold the properties required for opening an SDE workspace or geocoding an address. To learn more about PropertySet objects, see the online developer documentation.

Buffer the result and query the toxic sites

You will buffer this point and use the resulting geometry to query the toxic sites from the *ToxicSites* feature class. Since the *ToxicSites* feature class is in the same workspace as the geocode server object's locator's reference data, you do not have to create a new connection to the geodatabase but can use the geocode server's connection.

1. Add the following code to your try block to buffer the point, open the feature class, and query it:

```
ISegmentCollection segc = sc.CreateObject("esriGeometry.Polygon") as
ISegmentCollection;
segc.SetCircle(gcPoint, Distance);
IGeometry geom = segc as IGeometry;

IGeocodeServerObjects gcso = gc as IGeocodeServerObjects;
IReferenceDataTables reftabs = gcso.AddressLocator as
IReferenceDataTables;
IEnumReferenceDataTable enumreftabs = reftabs.Tables;
enumreftabs.Reset();
IReferenceDataTable reftab = enumreftabs.Next();
IDatasetName dsname = reftab.Name as IDatasetName;
IName wsnm = dsname.WorkspaceName as IName;
IFeatureWorkspace fws = wsnm.Open()  as IFeatureWorkspace;

IFeatureClass fc = fws.OpenFeatureClass("ToxicSites");
ISpatialFilter sf = sc.CreateObject("esriGeoDatabase.SpatialFilter") as
ISpatialFilter;
sf.Geometry = geom;
sf.GeometryField = fc.ShapeFieldName;
sf.SpatialRel = esriSpatialRelEnum.esriSpatialRelIntersects;
IFeatureCursor fcursor = fc.Search(sf, true);
```

Since you can get the workspace from the geocode server object, the connection is pooled by the geocode server. The code as written still has to open the ToxicSites feature class. This application could be further optimized by using a pooled map server object that has a single layer whose source data is the ToxicSites feature class. The application would then get the feature class from the map server object, effectively using the map server to pool the feature class.

If the ToxicSites feature class was not in the same workspace as the locator, this method would be the recommended approach for pooling both the workspace connection and the feature class.

You'll use the *ManageLifetime* method on the *WebObject* to add your feature cursor to the set of objects that the *WebObject* will explicitly release at the end of the using block.

2. Add the following code to your try block:

```
webObj.ManageLifetime(fcursor);
```

You will now add code to loop through the features returned by the query and use the *Name* and *SiteType* field values and the *X* and *Y* properties of the feature's geometry as the arguments for the *ToxicSite* class constructor to create a new *ToxicSite* object for each feature. Because the number of features returned by the query is unknown until the cursor has been exhausted, you can't declare your ToxicSite array, as its size is unknown. First, you will store these *ToxicSite* objects in an *ArrayList* collection, then copy the contents of that *ArrayList* to an array of *ToxicSite* objects that will be returned by the method.

3. Add the following code to your try block:

```
int lName;
int lType;
IFields flds = fc.Fields;
lName = flds.FindField("NAME");
lType = flds.FindField("SITETYPE");

ArrayList toxicList = new ArrayList();
IFeature f;
while ((f = fcursor.NextFeature()) != null)
{
  IPoint pt = f.Shape as IPoint;
  toxicList.Add (new ToxicSite((string) f.get_Value(lName),
(string)f.get_Value(lType),pt.X,pt.Y));
}
ToxicSite[] toxicArray = new ToxicSite[toxicList.Count];
toxicList.CopyTo(toxicArray);
```

4. To complete the function add the following line to the end of the try block to return the array of toxic sites:

```
return toxicArray;
```

Your *FindToxicLocations* method should now look like the following:

```
[WebMethod]
public ToxicSite[] FindToxicLocations(string Address, string ZipCode,
double Distance)
{
  using (WebObject webObj = new WebObject())
   {
    if (Address == null || ZipCode == null || Distance == 0.0)
      return null;

    ESRI.ArcGIS.Server.WebControls.ServerConnection connection = new
ESRI.ArcGIS.Server.WebControls.ServerConnection();
    connection.Host = "padisha";
    connection.Connect();

    IServerObjectManager som = connection.ServerObjectManager;

    IServerContext sc =
som.CreateServerContext("PortlandGeocode","GeocodeServer");
```

The objects and interfaces used for creating and working with geometries can be found in the Geometry object library. To learn more about geometry objects, see the online developer documentation.

The Locator object is the fine-grained ArcObjects component associated with a GeocodeServer object. Once you have a reference to the Locator (which you can get from the GeocodeServer via the IGeocodeServerObjects interface), you can access other objects associated with the Locator, such as the Locator's reference data.

The objects and interfaces used for performing spatial queries and for working with the results of those queries can be found in the GeoDatabase object library. To learn more about geodatabase objects, see the online developer documentation.

```
try
{
 IServerObject so = sc.ServerObject;
 IGeocodeServer gc = so as IGeocodeServer;

 IPropertySet ps = sc.CreateObject("esriSystem.PropertySet") as
IPropertySet;
 ps.SetProperty("street",Address);
 ps.SetProperty("Zone",ZipCode);

 IPropertySet res = gc.GeocodeAddress(ps,null);
 IPoint gcPoint = res.GetProperty("Shape") as IPoint;

 ISegmentCollection segc = sc.CreateObject("esriGeometry.Polygon") as
ISegmentCollection;
 segc.SetCircle(gcPoint, Distance);
 IGeometry geom = segc as IGeometry;

 IGeocodeServerObjects gcso = gc as IGeocodeServerObjects;
 IReferenceDataTables reftabs = gcso.AddressLocator as
IReferenceDataTables;
 IEnumReferenceDataTable enumreftabs = reftabs.Tables;
 enumreftabs.Reset();
 IReferenceDataTable reftab = enumreftabs.Next();
 IDatasetName dsname = reftab.Name as IDatasetName;
 IName wsnm = dsname.WorkspaceName as IName;
 IFeatureWorkspace fws = wsnm.Open() as IFeatureWorkspace;

 IFeatureClass fc = fws.OpenFeatureClass("ToxicSites");
 ISpatialFilter sf = sc.CreateObject("esriGeoDatabase.SpatialFilter")
as ISpatialFilter;
 sf.Geometry = geom;
 sf.GeometryField = fc.ShapeFieldName;
 sf.SpatialRel = esriSpatialRelEnum.esriSpatialRelIntersects;

 IFeatureCursor fcursor = fc.Search(sf, true);
 webObj.ManageLifetime(fcursor);

 int lName;
 int lType;
 IFields flds = fc.Fields;
 lName = flds.FindField("NAME");
 lType = flds.FindField("SITETYPE");

 ArrayList toxicList = new ArrayList();
 IFeature f;
 while ((f = fcursor.NextFeature()) != null)
 {
```

```
        IPoint pt = f.Shape as IPoint;
        toxicList.Add (new ToxicSite((string) f.get_Value(lName),
(string)f.get_Value(lType),pt.X,pt.Y));
        }
      ToxicSite[] toxicArray = new ToxicSite[toxicList.Count];
      toxicList.CopyTo(toxicArray);

      return toxicArray;

      sc.ReleaseContext();
     }
     catch
     {
       sc.ReleaseContext();
     }
     return null;
    }
  }
```

Your Web service is now ready to be tested. Compile the project (*Build/Build Solution*), and fix any errors.

Test the Web service

If you run the Web service from within Visual Studio, it will open a browser and list the *FindToxicLocations* method, which you can invoke from within the browser.

1. Click *Debug*, then click *Start*.

2. On the browser that opens, click the *FindToxicLocations* link.

3. Type the following values for the Web service parameters:

Address: 2111 Division St

ZipCode: 97202

Distance: 10000

4. Click *Evoke*. A new browser will open; confirm the following results from the Web service:

When you run your Web service within Visual Studio, it will open a browser listing the Web service's methods. You can evoke the methods by clicking the links on the browser and typing the method's inputs. You can also see the WSDL for the Web service by clicking the ServiceDescription link.

```
<?xml version="1.0" encoding="utf-8" ?>
- <ArrayOfAnyType xmlns:xsd="http://www.w3.org/2001/XMLSchema"
xmlns:xsi="http://www.w3.org/2001/XMLSchema-instance" xmlns="http://
tempuri.org/">
- <anyType xsi:type="ToxicSite">
  <Name>TRI MET CENTER STREET GARAGE</Name>
  <Type>Hazardous waste generator</Type>
  <X>7651285.4405499762</X>
  <Y>672416.22979136126</Y>
  </anyType>
- <anyType xsi:type="ToxicSite">
  <Name>EAST SIDE PLATING INC PLANT 4</Name>
  <Type>Hazardous waste generator</Type>
  <X>7647860.94568513</X>
```

```
<Y>679162.18254093162</Y>
</anyType>
- <anyType xsi:type="ToxicSite">
<Name>Portland Office of Transportation</Name>
<Type>Brownfield Pilot</Type>
<X>7646057.6159455236</X>
<Y>684318.6635023664</Y>
</anyType>
- <anyType xsi:type="ToxicSite">
<Name>Portland Office of Transportation</Name>
<Type>Brownfield Pilot</Type>
<X>7646057.6159455236</X>
<Y>684318.6635023664</Y>
</anyType>
</ArrayOfAnyType>
```

This indicates that four toxic sites were found within 10,000 feet of the given address.

When you evoke the Web method, a new browser will open, displaying the results returned from the method in XML.

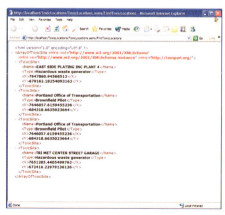

Create a client application

Since your Web service exposes a language-neutral interface that can be called using HTTP, your Web service can be called from any language that understands HTTP and WSDL. An elaborate client application (which itself could be a Web application, a desktop application, or even another Web service) is not demonstrated here, but the following is an example of how such an application would call the Web service method.

The example uses C#. Rather than describe such an application in detail, assume that this is a Windows desktop application that contains a button called *btnCallWS* whose *Click* event calls into your Web service. The code for this might look like the following, assuming you have added your Web service as a Web reference called *ToxicLocation*:

```
private void btnCallWS_Click(object sender, System.EventArgs e)
{
  ToxicLocation.ToxicSiteLocator toxloc = new
ToxicLocation.ToxicSiteLocator();
  object[] objs = toxloc.FindToxicLocations("2111 Division
St","97202",10000);

  for (int i = 0; i < objs.Length; i++)
  {
   ToxicLocation.ToxicSite toxsite = objs[i] as ToxicLocation.ToxicSite;
   // do something with the toxic site object
  }
}
```

DEPLOYMENT

Presumably you developed this Web service using your development Web server. To deploy this Web service on your production Web server, you can use the built-in Visual Studio .NET tools to copy the project.

1. In the Solution Explorer, click *ToxicLocations*.

2. Click Project, then click *Copy Project*.

3. In the Copy Project dialog box, specify the location on your Web server to copy the project to.

4. Click OK.

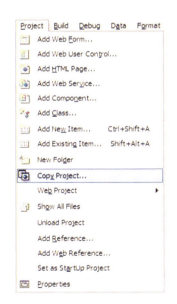

ADDITIONAL RESOURCES

This scenario includes functionality and programming techniques covering a number of different aspects of ArcObjects, the ArcGIS Server API, and .NET Web controls.

You are encouraged to read Chapter 4 of this book to get a better understanding of core ArcGIS Server programming concepts and programming guidelines for working with server contexts and ArcObjects running within those contexts.

This scenario makes use of the ArcGIS Server's .NET ADF Web controls to provide the GIS server connection object for this Web service. To learn more about the .NET Web controls, see Chapter 5. If you are unfamiliar with

ASP.NET Web development, it's also recommended that you refer to your .NET developer documentation to become more familiar with Web application development.

ArcGIS Server applications exploit the rich GIS functionality of ArcObjects. This Web service is no exception. It includes the use of ArcObjects to work with the components of a GeocodeServer to locate an address, manipulate geometries, and perform spatial queries against feature classes in a geodatabase. To learn more about these aspects of ArcObjects, refer to the online developer documentation on the Location, GeoDatabase, and Geometry object libraries.

Rather than walk through this scenario, you can get the completed Web service from the samples installation location. The sample is installed as part of the ArcGIS developer samples.

This walk-through is for developers who need to build and deploy a Java application Web service incorporating geocoding and spatial query functionality using the ArcGIS Server API. It describes the process of building, deploying, and consuming the *Application_web_service* sample, which is part of the ArcGIS developer samples.

You can find this sample in:
<install_location>\DeveloperKit\Samples\Developer_Guide_Scenarios\ ArcGIS_Server\Application_web_service

PROJECT DESCRIPTION

The purpose of this scenario is to create a Java Web service that uses ArcObjects to locate all of the toxic waste sites within a specified distance of a specified address. The Web service returns an array of application-defined toxic waste site objects.

This Web service is intended to be called by other programs, and an example of such a client program is a Java class. This scenario will also provide an example of how a client application would consume this Web service.

CONCEPTS

A Web service is a set of related application functions that can be programmatically invoked over the Internet. The function can be one that solves a particular application problem, as in this example; a Web service that finds all of the toxic waste sites within a certain distance of an address; or performs some other type of GIS function. Web services can be implemented using the native Web service framework of your Web server, such as ASP.NET Web service (WebMethod), or Java Web service (Axis).

When using native frameworks, such as ASP.NET and J2EE, to create and consume your application Web services, you need to use native or application-defined types as both arguments and return values from your Web methods. Clients of the Web service will not be ArcObjects applications, and as such, your Web service should not expect ArcObjects types as arguments and should not directly return ArcObjects types.

To learn more about WSDL, refer to http://www.w3.org.

Any development language that can use standard HTTP to invoke methods can consume this Web service. The Web service consumer can get the methods and types exposed by the Web service through its Web Service Description Language. As you walk through this scenario, you will see where special attributes need to be added to your methods and classes such that they can be expressed in WSDL and serialized as XML.

One key aspect of designing your application is whether it is stateful or stateless. You can make either stateful or stateless use of a server object running within the GIS server. A stateless application makes read-only use of a server object, meaning the application does not make changes to the server object or any of its associated objects. A stateful application makes read–write use of a server object where the application does make changes to the server object or its related objects.

Web services are, by definition, stateless applications.

DESIGN

This Web service is designed to make stateless use of the GIS server. It uses ArcObjects on the server to locate an address and query a feature class. In order to support this application, you need to add a pooled geocode server object to your ArcGIS Server using ArcCatalog.

Both coarse-grained calls to remote ArcObjects, such as the methods on the MapServer and GeocodeServer, and fine-grained calls to remote ArcObjects, such as creating new geometries, are exposed through the ArcGIS Server API and can be used in your Web service. The Web service will connect to the GIS and can be used in your Web service. The Web service will connect to the GIS server and use an instance of the geocode server object to locate the address supplied to the Web method from the calling application. To then buffer the resultant point and use that buffered geometry to query for toxic waste sites, you will use the geocode server's server context to work in. Since the geodatabase has been designed such that the address locator is stored in the same geodatabase as the feature class containing the toxic waste sites, you can use the fine-grained objects associated with the geocode server to get a reference to that workspace.

REQUIREMENTS

The requirements for working through this scenario are that you have ArcGIS Server and ArcGIS Desktop installed and running. The machine on which you develop this Web server must have the ArcGIS Server Java Application Developer Framework installed.

You must have a geocode server object configured and running on your ArcGIS Server that uses the Portland.loc locator installed with the ArcGIS developer samples. In ArcCatalog, create a connection to your GIS server and use the Add Server Object command to create a new server object with the following properties:

Name: PortlandGC

Type: GeocodeServer

Description: Geocode server object for metropolitan Portland

Locator:
<install_location>\DeveloperKit\Samples\Data\ServerData\Toxic\Portland.loc

The Add Server Object wizard

Batch size: 10 (default)

Pooling: The Web service makes stateless use of the server object. Accept the defaults for the pooling model (pooled server object with minimum instances = 2, max instances = 4).

Accept the defaults for the remainder of the configuration properties.

After creating the server object, start it and right-click it to verify that it is correctly configured and that the geocoding properties are displayed.

You can refer to Chapter 3 for more information on using ArcCatalog to connect to your server and create a new server object. Once the server object is configured and running, you can begin to code your Web service.

The following ArcObjects—Java packages will be used

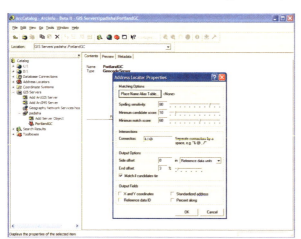

ArcCatalog is used for managing your spatial data holdings, defining your geographic data schemas, and managing your ArcGIS Server. Once you have created your PortlandGC server object, open its properties using ArcCatalog to verify it is correctly configured.

in this example:

- com.esri.arcgis.geodatabase
- com.esri.arcgis.geometry

- com.esri.arcgis.location
- com.esri.arcgis.server
- com.esri.arcgis.system

The development environment does not require any ArcGIS licensing; however, connecting to a server and using a geocode server object does require that the GIS server be licensed to run ArcObjects in the server. None of the packages used require an extension license.

The IDE used in this example is JBuilder 9 Enterprise Edition, and all IDE specific steps will assume this is the IDE you are using. This Web service can be implemented with other Java IDEs.

IMPLEMENTATION

To begin, you must create a new JBuilder project and a working directory that will be used for this project.

Create a new project

The Project wizard

1. Create a new folder called "webservice_projects". This is where your Web service project files will be created.

2. Start JBuilder.

3. Click File, then click New Project to open the Project Wizard window.

4. In the Project Wizard dialog box, "type toxiclocation_project" for the Name, then type the path to the webservice_projects you created above for the Directory.

Add references to the ArcObjects—Java libraries

A proxy object is a local representation of a remote object. The proxy object controls access to the remote object by forcing all interaction with the remote object to be via the proxy object. The supported interfaces and methods on a proxy object are the same as those supported by the remote object. You can make method calls on, and get and set properties of, a proxy object as if you were working directly with the remote object.

In order to program using ArcGIS Server, references to the ArcObjects—Java libraries must be added in your development environment. These libraries contain proxies to the ArcObjects components the Web service will use. These libraries are installed when you install the ArcGIS Server Java ADE. To add these references, you will create a new library and add it to your project.

1. Click Tools, then click Configure Libraries.

2. In the Configure Libraries dialog box, click New to create a new library.

3. In the New Library wizard, type "arcgis_lib" for the Name.

4. Click Add and browse to the location and select <install_location>/ArcGIS/java/jintegra.jar, then click OK.

5. Repeat step 4 to add <install_location>/ArcGIS/java/opt/arcobjects.jar.

6. Click OK to close the Configure Libraries dialog box.

Now you will add the new library to your project.

7. Click Project, then click Project Properties.

8. In the Project Properties dialog box, click the Required Libraries tab.

The Configure Libraries dialog box

The New Library wizard

The Configure Libraries dialog box

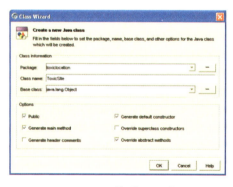

The Class wizard

Classes that implement Serializable can be serialized into XML. Any custom type that is returned by a Web service must implement Serializable.

9. Click Add.

10. Under User Home, click arcgis_lib and click OK.

11. Click OK to close the Project Properties dialog box.

Create the ToxicSite class

The point of this Web service is to expose a method that is accessible via HTTP-based SOAP requests and that returns all of the toxic site locations within a specified distance of a specified address. Before you create your new method, first define the toxic waste class that will be returned when this Web service is invoked.

1. Click File, then click New Class.

2. In the Class Wizard, type "toxiclocation" for the *Package*, and type "ToxicSite" for the *Class Name*.

3. Click OK.

This will add a new class to your project and will open the code for the class with some autogenerated code. The autogenerated code will include the name of the package (toxiclocation) and a stubbed out class definition for ToxicSite and a default constructor:

```
package toxiclocation;

public class ToxicSite {
  public ToxicSite () {
  }
```

Results returned from Web services must be serializable in XML. The ToxicSite type must be marked as serializable by implementing Serializable. Because Serializable is defined in the java.io.Serializable Java package, you need to add an import statement for that package.

4. Add the following import statement to your code:

```
import java.io.Serializable;
```

5. Change the following line:

```
public class ToxicSite {
```

to:

```
public class ToxicSite implements Serializable{
```

Your code should now look like this:

```
package toxiclocation;

import java.io.Serializable;

public class ToxicSite implements Serializable{
  public ToxicSite () {
  }
```

You are interested in returning the type of toxic site, the name of the organization associated with the toxic site, and the x and y coordinates of the toxic site feature. So, you will add four private fields to your *ToxicSite* class, an overloaded constructor to set the data, and Get and Set methods for each field.

6. Add the following lines of code to your *ToxicSite* class:

```
private String name;
private String type;
private double X;
private double Y;
```

7. To add the overloaded constructor, type the following in your class definition:

```
public ToxicSite(String name, String type, double x, double y) {
    this.name = name;
    this.type = type;
    this.X = x;
    this.Y = y;
}
```

8. To add the Get and Set methods, type the following in your class definition:

```
public void setName(String name){
    this.name= name;
}

public String getName(){
    return name;
}

public void setType(String type){
    this.type = type;
}

public String getType(){
    return type;
}

public void setX(double x){
    this.X = x;
}

public double getX(){
    return X;
}
public void setY(double y){
    this.Y = y;
}

public double getY(){
    return Y;
}
```

The code for your *ToxicSite* class should now look like the following:

```java
package toxiclocation;
import java.io.Serializable;

public class ToxicSite implements Serializable{
  private String name;
  private String type;
  private double X;
  private double Y;

  public ToxicSite() {
  }

  public ToxicSite(String name, String type, double x, double y) {
    this.name = name;
    this.type = type;
    this.X = x;
    this.Y = y;
  }

  public void setName(String name){
      this.name= name;
  }

  public String getName(){
      return name;
  }

  public void setType(String type){
      this.type = type;
  }

  public String getType(){
      return type;
  }

  public void setX(double x){
      this.X = x;
  }

  public double getX(){
      return X;
  }
  public void setY(double y){
      this.Y = y;
  }

  public double getY(){
      return Y;
  }
}
```

Create the ToxicSiteLocator class

Now that you have defined your *ToxicSite* class, you will implement your Web service method. As described above, the point of this Web service is to expose a method that is accessible via HTTP-based SOAP requests and that returns all of the toxic site locations within a specified distance of a specified address. This class will have the method that will be exposed as a Web service.

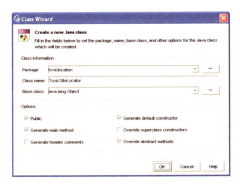

The Class wizard

1. Click File, then click New Class.

2. In the Class Wizard, type "toxiclocation" for the *Package*, and type "ToxicSiteLocator" for the *Class Name*.

3. Click OK.

This will add a new class to your project and will open the code for the class with some autogenerated code. The autogenerated code will include the name of the package (toxiclocation) and a stubbed out class definition for ToxicSite and a default constructor:

```
package toxiclocation;

public class ToxicSiteLocator {
  public ToxicSiteLocator () {
  }
```

In order to program using ArcGIS Server, you need to import ArcObjects—Java API packages that contain the proxies for the ArcObjects that the Web service will use. In addition, you need to import some standard Java packages.

4. Add the following import statement to your class:

```
import java.util.ArrayList;
import com.esri.arcgis.datasourcesGDB.*;
import com.esri.arcgis.geodatabase.*;
import com.esri.arcgis.geometry.*;
import com.esri.arcgis.location.*;
import com.esri.arcgis.server.*;
import com.esri.arcgis.system.*;
```

You will create a method called *FindToxicLocations* that takes as arguments an address, ZIP Code, and search distance and returns an array of ToxicSite objects.

The Project Properties dialog box

5. Add the following lines of code to your *ToxicSiteLocator* class:

```
public ToxicSite[] FindToxicLocations(String address, String
zipCode,double distance) throws Exception {

}
```

The FindToxicLocations *method returns an array of* ToxicSite *objects, rather than a generic collection, such as an* ArrayList. *If the method did not have a return type of* ToxicSite, *then the custom class wouldn't be expressed in the Web service's WSDL.*

Your code should now look like the following:

```
package toxiclocation;

import java.util.ArrayList;
import com.esri.arcgis.datasourcesGDB.*;
import com.esri.arcgis.geodatabase.*;
```

```
import com.esri.arcgis.geometry.*;
import com.esri.arcgis.location.*;
import com.esri.arcgis.server.*;
import com.esri.arcgis.system.*;

public class ToxicSiteLocator {

  public ToxicSite[] FindToxicLocations(String address, String
zipCode,double distance) throws Exception {
  }
}
```

Validate input parameters

The first thing the Web service will do is verify the provided parameters are valid and, if not, return null.

1. Type the following in your FindToxicLocations method (note, all code additions from this step on are made to this method):

```
if (address == null || zipCode == null || distance == 0.0 )
   return null;
```

You will now create some local variables to hold the server context, server connection, and an *ArrayList* of resultant *ToxicSite* objects.

2. Add the following lines of code to your method.

```
IServerContext sc = null;
IServerConnection con = new ServerConnection();
ArrayList toxicSiteList = new ArrayList();
```

It's the responsibility of the developer to release the server object's context when it's finished using it. It's important to ensure that your method calls *ReleaseContext* on the server context when the method no longer needs the server object or any other objects running in the context. It's also important that you ensure the context is released in the event of an error. If an error occurs in the code in the try block, the code in the catch block will be executed. So, you will add the call to release the context at the end of the try block and in the catch block.

3. Add the following lines of code to your method.

```
try {
  sc.releaseContext();
}
catch (Exception e) {
  sc.releaseContext();
  e.printStackTrace();
  return null;
}
```

Connect to the GIS server

This Web service makes use of objects in the GIS server, so you will add code to connect to the GIS server and get the *IServerObjectManager* interface. In this example, the GIS server is running on a machine called "melange" as user "amelie" with a password of "xyz".

1. Add the following lines of code to your try block.

```
ServerInitializer si = new ServerInitializer();
si.initializeServer("melange", "amelie", "xyz");
con.connect("melange");
IServerObjectManager som = con.getServerObjectManager();
```

The Web server will make use of the *PortlandGC* server object that you created as a requirement to this Web service. You get a server object by asking the server for a server context containing the object.

2. Add the following line of code to get the server object's context.

```
sc = som.createServerContext("PortlandGC", "GeocodeServer");
```

Geocode the input address

In order to successfully connect to the GIS server, your Web service must impersonate a user who is a member of the ArcGIS Server users group on the GIS server. There are a number of strategies for implementing impersonation in Java. For more details on impersonation, refer to Chapter 6 and your Java documentation.

Now that you have connected to the GIS server and have a context containing the geocode server object, you will add the code to use the server object to geocode the input address and store the resulting point as gcPoint. Add the following lines of code to your try block:

```
IServerObject so = new IServerObjectProxy(sc.getServerObject());
IGeocodeServer gc = new IGeocodeServerProxy(so);


IPropertySet ps = new
IPropertySetProxy(sc.createObject(PropertySet.getClsid()));
ps.setProperty("Street", address);
ps.setProperty("Zone", zipCode);


IPropertySet res = gc.geocodeAddress(ps, null);
IPoint gcPoint = new IPointProxy(res.getProperty("Shape"));
```

A PropertySet is a generic class that is used to hold any set of properties. A PropertySet's properties are stored as name/value pairs. Examples for the use of a property set are to hold the properties required for opening an SDE workspace or geocoding an address. To learn more about PropertySet objects, see the online developer documentation.

Buffer the result and query the toxic sites

You will buffer this point and use the resulting geometry to query the toxic sites from the *ToxicSites* feature class. Since the *ToxicSites* feature class is in the same workspace as the geocode server object's locator, you do not have to create a new connection to the geodatabase but can use the geocode server's connection.

1. Add the following code to your try block to buffer the point, open the feature class, and query it.

Since you can get the workspace from the geocode server object, the connection is pooled by the geocode server. The code as written still has to open the ToxicSites feature class. This application could be further optimized by using a pooled map server object that has a single layer whose source data is the ToxicSites feature class. The application would then get the feature class from the map server object, effectively using the map server to pool the feature class.

If the ToxicSites feature class was not in the same workspace as the locator, this method would be the recommended approach for pooling both the workspace connection and the feature class.

```
ISegmentCollection segc = new ISegmentCollectionProxy(sc.createObject(
Polygon.getClsid()));
segc.setCircle(gcPoint, distance);
IGeometry geom = new IGeometryProxy(segc);


IGeocodeServerObjects gcso = new IGeocodeServerObjectsProxy(gc);
IReferenceDataTables reftabs = new
IReferenceDataTablesProxy(gcso.getAddressLocator());
IEnumReferenceDataTable enumreftabs = new
IEnumReferenceDataTableProxy(reftabs.getTables());
enumreftabs.reset();
IReferenceDataTable reftab = new
IReferenceDataTableProxy(enumreftabs.next());
IDatasetName dsname = new IDatasetNameProxy(reftab.getName());
IName wsnm = new INameProxy(dsname.getWorkspaceName());
IFeatureWorkspace fws = new IFeatureWorkspaceProxy(wsnm.open());
```

```
IFeatureClass fc = fws.openFeatureClass("ToxicSites2");
ISpatialFilter sf = new
ISpatialFilterProxy(sc.createObject(SpatialFilter.getClsid()));
sf.setGeometryByRef(geom);
sf.setGeometryField(fc.getShapeFieldName());
sf.setSpatialRel(esriSpatialRelEnum.esriSpatialRelIntersects);

IFeatureCursor fCursor = fc.search(sf, true);
```

You will now loop through the features returned by the query and use the Name and SiteType field values and the x and y properties of the feature's geometry as the arguments for the ToxicSite class constructor to create a new *ToxicSite* object for each feature. Because the number of features returned by the query is unknown until the cursor has been exhausted, you can't declare your ToxicSite array, as its size is unknown. First, you will store these *ToxicSite* objects in an ArrayList collection, then copy the contents of that ArrayList to an array of *ToxicSite* objects that will be returned by the method.

2. Add the following code to your try block:

```
int lName;
int lType;
IFields flds = fc.getFields();
lName = flds.findField("NAME");
lType = flds.findField("SITETYPE");

IFeature f;
while ( (f = fCursor.nextFeature()) != null) {
  IPoint pt = new IPointProxy(f.getShape());
  ToxicSite toxicSite = new ToxicSite((String) f.getValue(lName),(String)
f.getValue(lType),pt.getX(),pt.getY());
  toxicSiteList.add(toxicSite);
}
ToxicSite[] sites = new ToxicSite[toxicSiteList.size()];
for (int i = 0; i < sites.length; i++) {
  sites[i] = (ToxicSite) toxicSiteList.get(i);
}
sc.releaseContext();
return sites;
```

The class is now ready to be exposed as a Web service. Your *ToxicSiteLocator* class should now look like the following:

```
package toxiclocation;

import java.util.ArrayList;
import com.esri.arcgis.datasourcesGDB.*;
import com.esri.arcgis.geodatabase.*;
import com.esri.arcgis.geometry.*;
import com.esri.arcgis.location.*;
import com.esri.arcgis.server.*;
import com.esri.arcgis.system.*;

public class ToxicSiteLocator {
```

```
public ToxicSite[] FindToxicLocations(String address, String
zipCode,double distance) throws Exception {

  if (address == null || zipCode == null || distance == 0.0) {
    return null;
  }

  IServerContext sc = null;
  IServerConnection con = new ServerConnection();
  ArrayList toxicSiteList = new ArrayList();
  try {
    ServerInitializer si = new ServerInitializer();
si.initializeServer("melange", "amelie", "xyz");
con.connect("melange");
    IServerObjectManager som = con.getServerObjectManager();
    sc = som.createServerContext("PortlandGC", "GeocodeServer");
    IServerObject so = new IServerObjectProxy(sc.getServerObject());
    IGeocodeServer gc = new IGeocodeServerProxy(so);

    IPropertySet ps = new
IPropertySetProxy(sc.createObject(PropertySet.getClsid()));
    ps.setProperty("Street", address);
    ps.setProperty("Zone", zipCode);

    IPropertySet res = gc.geocodeAddress(ps, null);
    IPoint gcPoint = new IPointProxy(res.getProperty("Shape"));

    ISegmentCollection segc = new
ISegmentCollectionProxy(sc.createObject(Polygon.getClsid()));
    segc.setCircle(gcPoint, distance);
    IGeometry geom = new IGeometryProxy(segc);

    IGeocodeServerObjects gcso = new IGeocodeServerObjectsProxy(gc);
    IReferenceDataTables reftabs = new
IReferenceDataTablesProxy(gcso.getAddressLocator());
    IEnumReferenceDataTable enumreftabs = new
IEnumReferenceDataTableProxy(reftabs.getTables());
    enumreftabs.reset();
    IReferenceDataTable reftab = new
IReferenceDataTableProxy(enumreftabs.next());
    IDatasetName dsname = new IDatasetNameProxy(reftab.getName());
    IName wsnm = new INameProxy(dsname.getWorkspaceName());
    IFeatureWorkspace fws = new IFeatureWorkspaceProxy(wsnm.open());

    IFeatureClass fc = fws.openFeatureClass("ToxicSites");
    ISpatialFilter sf = new
ISpatialFilterProxy(sc.createObject(SpatialFilter.getClsid()));
    sf.setGeometryByRef(geom);
    sf.setGeometryField(fc.getShapeFieldName());
    sf.setSpatialRel(esriSpatialRelEnum.esriSpatialRelIntersects);
```

The objects and interfaces used for creating and working with geometries can be found in the Geometry object library. To learn more about geometry objects, see the online developer documentation.

The Locator object is the fine-grained ArcObjects component associated with a GeocodeServer object. Once you have a reference to the Locator, which you can get from the GeocodeServer via the IGeocodeServerObjects interface, you can access other objects associated with the Locator, such as the Locator's reference data.

The objects and interfaces used for performing spatial queries and for working with the results of those queries can be found in the GeoDatabase object library. To learn more about geodatabase objects, see the online developer documentation.

```
IFeatureCursor fCursor = fc.search(sf, true);

int lName;
int lType;
IFields flds = fc.getFields();
lName = flds.findField("NAME");
lType = flds.findField("SITETYPE");

IFeature f;

while ( (f = fCursor.nextFeature()) != null) {
  IPoint pt = new IPointProxy(f.getShape());
  ToxicSite toxicSite = new ToxicSite((String)f.getValue(lName),
(String)f.getValue(lType), pt.getX(),pt.getY());
    toxicSiteList.add(toxicSite);
  }
  ToxicSite[] sites = new ToxicSite[toxicSiteList.size()];
  for (int i = 0; i < sites.length; i++) {
    sites[i] = (ToxicSite) toxicSiteList.get(i);
  }
  sc.releaseContext();
  return sites;
}
catch (Exception e) {
  sc.releaseContext();
  e.printStackTrace();
  return null;
}
}
```

Create the Web service

JBuilder has a version of Apache Axis toolkit built into it, and you will use this to expose the Java classes as a Web service.

1. In the Project pane, right-click *ToxicSiteLocator.java* and click *Export as Web Service*.

2. In the Web Services Configuration Wizard, click *New*.

3. In the Web Application wizard, type "toxicloc" for the *Name* and click OK.

4. Click Finish.

5. Accept the defaults in the Export as Axis Web Service dialog box, click Finish.

The Export As Axis Web Service wizard

The toxicloc Web application is created. JBuilder has Tomcat Servlet engine built in; you can run the Web application and host your Web service by running this Web application.

1. In the Project pane, click toxicloc, then click Root Directory.

2. Right-click *index.html* and from the context menu click Web Run using "Web Services Server".

By default JBuilder will start Tomcat on port 8080, and you will see the JBuilder Apache Axis Admin Console. You can view the WSDL generated for the Web service you created.

1. Click the link to View the list of deployed Web services.

2. Click the link to ToxicSiteLocator (wsdl).

When you run your Web service within JBuilder, it will display a browser with links to view the list of deployed services and the WSDL for your Web service.

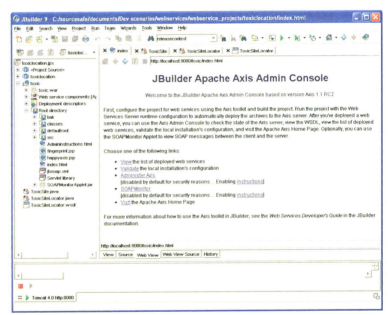

The WSDL for the Web service will be displayed. You can also use an external Web browser to view the WSDL by typing the URL "http://localhost:8080/toxicloc/services/ToxicSiteLocator?wsdl".

Create a client application

Since your Web service exposes a language-neutral interface that can be called using HTTP, your Web service can be called from any language that understands HTTP and WSDL. An elaborate client application (which itself could be a Web application, a desktop application, or even another Web service) is not demonstrated here, but the following is an example of how such an application would call the Web service method. This will be a simple Java class that will call the Web service in its main method.

The Project wizard

The Object Gallery dialog box

The Import a Web Service With Axis wizard

Create a new JBuilder project for the client application. JBuilder will generate stub classes for the client by using the WSDL2Java utility of Apache Axis.

1. Click File, then click *New Project*.

2. In the Project Wizard dialog box, type "toxicLocationClient" for the *Name*, then type the path to your webservice_projects folder for the *Directory*.

3. Click Finish.

You will now create the stub classes from the Web service's WSDL.

1. Click File, then click New.

2. In the Object Gallery dialog box, click the *Web Services* tab.

3. Click *Import a Web Service* and click OK.

4. In the Import a Web Service With Axis dialog box, click the Browse button beside WSDL URL.

5. In the File or URL text box enter "http://localhost:8080/toxicloc/services/ToxicSiteLocator?wsdl", then click OK.

6. Click Finish.

Rebuild the project (Project/Rebuild Project "toxicLocationClient.jpx"). The project should rebuild without any errors. Now that the stub classes for the client have been successfully compiled, you will create the client class.

1. Click File, then click New Class.

2. In the Class wizard, type "client" for the *Package*, and type "ToxicSitelocatorClient" for the *Class Name*.

3. Click OK.

Open the file *ToxicSitelocatorClient* by double-clicking on it in the Project pane.

Add the following import statements for referencing the generated stub classes.

```
import localhost.toxicloc.services.ToxicSiteLocator.*;
import toxiclocation.*;
```

Next, you will create a main function that will be executed when you run this project.

Add the following lines of code to your class:

```
public static void main(String[] args) {
}
```

Next, you will create a service from the generated stub classes.

Add the following lines of code to your main function:

```
// Make a service using wasdl2java generated client proxy class
ToxicSiteLocatorService service = new ToxicSiteLocatorServiceLocator();
```

Use the findToxicLocations() method to locate the address "2111 Division St" with ZIP Code 97202 and find all the toxic locations within a distance of 10,000 units.

Add the following lines of code to your main function.

```
try {
  // Now use the service to get the stub class
  ToxicSiteLocator port = service.getToxicSiteLocator();

  // Make the actual call; use stub class for remote service just like a
  local object
  ToxicSite[] sites = (ToxicSite[]) port.findToxicLocations("2111 Division
St", "97202", 10000);

  for (int i = 0; i < sites.length; i++) {
    ToxicSite site = sites[i];
    System.out.println("-------- Site : "+ (i+1) + " ---------");
    System.out.println("Name : "+ site.getName());
    System.out.println("Type : "+ site.getType());
    System.out.println("X : "+ site.getX());
    System.out.println("Y : "+ site.getY());
  }
}
catch (Exception ex) {
  System.out.println(ex.getStackTrace());
}
```

The code for your *ToxicSitelocatorClient* should look like the following:

```
package client;

import localhost.toxicloc.services.ToxicSiteLocator.*;
import toxiclocation.*;

public class ToxicSiteLocatorClient {
  public static void main(String[] args) {
    // Make a service using wasdl2java generated client proxy class
    ToxicSiteLocatorService service = new
ToxicSiteLocatorServiceLocator();

    try {
      // Now use the service to get the stub class
      ToxicSiteLocator port = service.getToxicSiteLocator();

      // Make the actual call; use stub class for remote service just like
    a local object
      ToxicSite[] sites = (ToxicSite[]) port.findToxicLocations(
          "2111 Division St", "97202", 10000);

      for (int i = 0; i < sites.length; i++) {
        ToxicSite site = sites[i];
        System.out.println("-------- Site : "+ (i+1) + " ---------");
        System.out.println("Name : "+ site.getName());
```

```
        System.out.println("Type : "+ site.getType());
        System.out.println("X : "+ site.getX());
        System.out.println("Y : "+ site.getY());
        }
      }
    catch (Exception ex) {
     System.out.println(ex.getStackTrace());
     }
   }
 }
```

Test the Web service

Execute the client to work with the Web service.

In the Project pane, right-click the node ToxicSiteLocatorClient.java and click Run using defaults.

The results should resemble the following:

```
-------- Site : 1 ---------
Name : Portland Office of Transportation
Type : Brownfield Pilot
X : 7646057.614036874
Y : 684318.6770439692
-------- Site : 2 ---------
Name : Tri-County Metropolitan Transportation District of
Type : Brownfield Pilot
X : 7646057.614036874
Y : 684318.6770439692
-------- Site : 3 ---------
Name : EAST SIDE PLATING INC PLANT 4
Type : Hazardous waste generator
X : 7647860.949718554
Y : 679162.1745111668
-------- Site : 4 ---------
Name : TRI MET CENTER STREET GARAGE
Type : Hazardous waste generator
X : 7651285.4340737425
Y : 672416.2253208841
```

DEPLOYMENT

You can deploy the Web service you created on a production server by deploying the Web application archive file "toxicloc.war" in any J2EE-compliant servlet engine or application server. Depending on the final URL for the WSDL file of the Web service, you may need to generate the client stubs again to recompile your client application.

ADDITIONAL RESOURCES

This scenario includes functionality and programming techniques covering a number of different aspects of ArcObjects and the ArcGIS Server API.

You are encouraged to read Chapter 4 of this book to get a better understanding of core ArcGIS Server programming concepts and programming guidelines for working with server contexts and ArcObjects running within those contexts.

This scenario makes use of the ArcGIS Server Java ADF to provide the GIS server connection object for this Web service. To learn more about the Java Web controls, see Chapter 6. If you are unfamiliar with Java Web development, it's also recommended that you refer to your Java developer documentation to become more familiar with Web application development.

ArcGIS Server applications exploit the rich GIS functionality of ArcObjects. This Web service is no exception. It includes the use of ArcObjects to work with the components of a GeocodeServer to locate an address, manipulate geometries, and perform spatial queries against feature classes in a geodatabase. To learn more about these aspects of ArcObjects, refer to the online developer documentation on the Location, GeoDatabase, and Geometry object libraries.

Rather than walk through this scenario, you can get the completed application from the samples installation location. The sample is installed as part of the ArcGIS developer samples.

This walk-through is for developers who need to build and deploy a .NET windows desktop application incorporating mapping functionality using ArcGIS Server Web services directly. It describes the process of building, deploying, and consuming the *ArcGIS_web_service_client* sample, which is part of the ArcGIS Developer Samples.

You can find this sample in:
<install_location>\DeveloperKit\Samples\Developer_Guide_Scenarios\ArcGIS_Server\ArcGIS_web_service_client

PROJECT DESCRIPTION

The purpose of this scenario is to create a windows application using Visual Studio .NET that uses MapServer Web services published in Web service catalogs to navigate the maps served by those Web services.

The following is how the user will interact with the application:

1. Specify the URL of a Web service catalog and click *Get Web Services*.

2. Click the *MapServer Web service* list and click the desired MapServer to browse.

3. Click the *Data frame* dropdown to choose the data frame of interest from the MapServer.

4. Navigate the map using the *Bookmark* list and the *Zoom In*, *Zoom Out*, *Full Extent*, and *Pan* tools on the toolbar.

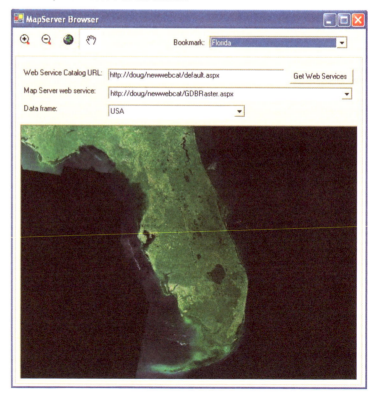

CONCEPTS

A Web service is a set of related application functions that can be programmatically invoked over the Internet. ArcGIS Server Web services can be accessed by any development language that can submit SOAP-based requests to a Web service and process SOAP-based responses. The Web service consumer can get the methods and types exposed by the Web service through its Web Service Description Language. ArcGIS Web Services are based on the SOAP doc/literal format and are interoperable across Java and .NET.

To learn more about WSDL, refer to http://www.w3.org.

On the Web server, the Web service catalog templates can be used by developers to publish any server object as a Web service over HTTP. For any published server object, the template creates an HTTP endpoint (URL) to which SOAP requests can be submitted using standard HTTP POST. The endpoint also supports returning the WSDL for the Web service using a standard HTTP GET with "wsdl" as the query string. The implementation of the HTTP endpoint is thin, while the actual processing of the SOAP request and the generation of the SOAP response take place within the GIS server. The WSDLs that describe the definitions of the SOAP requests and responses are also part of the GIS server and are installed as part of the ArcGIS Server install under <install directory>\XMLSchema.

If you are building an application using ArcGIS Desktop or ArcGIS Engine, you can use the objects in the GISClient object library to access ArcGIS Server Web services. This is an example of using those Web services directly from an application that does not require any ArcObjects components to run.

For more information on using the objects in GISClient, see Chapter 4 and the developer help.

ArcGIS Web services can be consumed from .NET or Java. As a consumer of an ArcGIS Web service, you can use the methods exposed by the Web service by including a reference to the Web service in your .NET or Java project. Web services implemented on a Java Web server can be consumed from a .NET client and vice versa.

DESIGN

This application makes use of the methods and objects defined by the ArcGIS Server Web service catalog and MapServer Web service. These methods and objects correspond to the set of ArcObjects necessary to call the stateless methods exposed by the MapServer object.

If you have already worked with the stateless methods on the MapServer using ArcObjects, you'll notice that there are differences working with those methods on the MapServer Web service.

In order to support this application, you need to have access to a Web service catalog that contains at least one MapServer Web service. Note: The Web service catalog itself can be either .NET or Java. For purposes of this example, the assumption is that the Web service is .NET.

The application will connect to a Web service catalog and present to the user a list of the available MapServer Web services in the Web service catalog. The application will list the available data frames in the map and any bookmarks associated with the data frame that the user can pick from. In addition, the user can navigate the map using navigation tools in a toolbar.

REQUIREMENTS

The requirement for working through this scenario is that you have access to an ArcGIS Server Web service catalog that contains at least one MapServer Web service. The machine on which you develop this application does not need to have any ArcGIS software or licensing installed on it.

For the purposes of this example, assume you have access to a Web service

While the examples given are .NET Web service catalogs, if you have access to Java Web services catalogs, you can use those in your .NET application.

catalog created with .NET with the following URL:

`http://padisha/MyWebServiceCatalog/default.aspx`

The Web service catalog contains a MapServer Web service for a MapServer object called "Map" whose URL is:

`http://padisha/MyWebServceCatalog/Map.aspx`

It will be easiest to follow this walk-through if you create a Web service catalog with the same name as above (MyWebServiceCatalog) and a MapServer server object and Web service (Map), although this is not necessary. For more information about creating a map server object, see Chapter 3. For information on creating a Web service catalog and exposing your MapServer as a Web service using .NET, see Chapter 5.

If you have your own Web service catalog and MapServer Web services with different names, the points at which the difference in names will impact the code for this application will be pointed out.

The IDE used in this example is Visual Studio .NET 2003. This Web application can be implemented with other .NET IDEs.

IMPLEMENTATION

In this scenario, you will use the Windows Application template project that is installed as part of Visual Studio .NET that you will add your functionality to. The code for this scenario will be written in C#; however, you can also write this application using VB.NET.

The first step is to create the new project.

Create a new project

1. Start Visual Studio .NET.

2. Click *File*, click *New*, then click *Project*.

3. In the New Project dialog box, under Project Types, click the *Visual C# Projects* category. Under *Templates*, click *Windows Application*.

4. For the application name, type "MapServerBrowser".

5. Click *OK*. This will create a new project that contains a single Windows form.

Add references to the Web services

For your application to have access to the methods and objects exposed by Web service catalog and MapServer Web services, you need to add Web references to a Web service catalog and a reference to a MapServer Web service to your application. A Web reference enables you to use objects and methods provided by a Web service in your code. After adding a Web reference to your project, you can use any element or functionality provided by that Web service within your application.

The New Project dialog box

The Solution Explorer dialog box

The Add Web Reference dialog box

The Class View dialog box

1. In the Solution Explorer, right-click References and click *Add Web Reference*.

2. For URL, type the URL of your Web service catalog with "wsdl" as the query string. In this example, the URL would be:

   `http://padisha/MyWebServiceCatalog/default.aspx?wsdl`

3. Click Go.

4. Once the Web service is found, type "WebCatalog" for the *Web reference name*, then click *Add Reference*.

5. In the Solution Explorer, right-click References and click Add Web Reference.

6. For URL, type the URL of your MapServer Web service with "wsdl" as the query string. In this example, the URL would be:

   `http://padisha/MyWebServiceCatalog/Map.aspx?wsdl`

7. Click Go.

8. Once the Web service is found, type MapServerWS for the Web reference name, then click *Add Reference*.

In the project's class view, expand MapServerBrowser, then expand MapServerWS and WebCatalog to see the classes that have been added to your project by referencing the Web services. Now that these references have been added to the project, you will start programming your Windows application to make use of the classes and methods provided by these Web service references to consume ArcGIS Server Web services.

Set the properties of the form

To accommodate the functionality for this application, you will add a number of user interface controls to the windows form. Before doing that, you need to set some properties on the form itself, such as its size and text.

1. In the Solution Explorer, double-click "Form1.cs". This will open the form in design mode.

2. In the Properties for the form, type "584, 596" for the *Size* property and type "MapServer Browser" for the *Text* property.

Add controls to the form

This application utilizes a number of user controls that you need to add to and arrange on the form.

The first control you'll add is a picture box that will display the map images returned by the MapServer Web service.

1. In the Microsoft Visual Studio .NET toolbox, click the Windows Forms tab to display the Windows Forms tools.

2. In the toolbox, click *PictureBox* and drag a picture box onto the form.

3. In the picture box's properties, type "552, 400" for the *Size* property and type "12, 152" for the Location property.

4. Click *Fixed3D* for the *BorderStyle* property.

The next set of controls you'll add is to handle the user input for the URL of the Web service catalog that the user of the application wants to connect to.

While this example includes exact control placement and size on the form, you can also arrange these controls interactively by dragging them and sizing them with the mouse.

1. In the Microsoft Visual Studio .NET toolbox, click the Windows Forms tab to display the Windows Forms tools.

2. In the toolbox, click *Label* and drag a label onto the form.

3. In the label's properties, type "Web Service Catalog URL:" for the *Text* property, "140, 20" for the *Size* property, and "12, 64" for the *Location* property.

4. In the Windows Forms toolbox, click TextBox and drag a text box onto the form.

5. In the text box's properties, type "txtServer" for the *(Name)* property, "288, 20" for the *Size* property, and type "156, 64" for the *Location* property.

6. In the Windows Forms toolbox, click Button and drag a button onto the form.

7. In the button's properties, type "btnConnect" for the *(Name)* property, "Get Web Services" for the *Text* property, "104, 23" for the *Size* property, and "452, 64" for the *Location* property.

The next controls you'll add are the controls to list the MapServer Web services that are in the Web service catalog specified by the user.

1. In the Microsoft Visual Studio .NET toolbox, click the Windows Forms tab to display the Windows Forms tools.

2. In the toolbox, click *Label* and drag a label onto the form.

3. In the label's properties, type "Map Server Web service:" for the *Text* property, "128, 20" for the Size property, and "12, 92" for the *Location* property.

4. In the Microsoft Visual Studio .NET toolbox, click the Windows Forms tab to display the Windows Forms tools.

5. In the toolbox, click *ComboBox* and drag a combo box onto the form.

6. In the combo box's properties, type "cboMapServer" for the *(Name)*, "400, 21" for the *Size* property, and "156, 92" for the *Location* property.

7. Click *DropDownList* for the *DropDownStyle* property.

8. Click False for the Enabled property.

The next set of controls you'll add is the controls to list the data frames in the MapServer Web service selected by the user.

1. In the Microsoft Visual Studio .NET toolbox, click the Windows Forms tab to display the Windows Forms tools.

2. In the toolbox, click *Label* and drag a label onto the form.

3. In the label's properties, type "Data frame:" for the *Text* property, "64, 20" for the *Size* property and type "12, 120" for the *Location* property.

4. In the Microsoft Visual Studio .NET toolbox, click the Windows Forms tab to display the Windows Forms tools.

5. In the toolbox, click *ComboBox* and drag a combo box onto the form.

6. In the combo box's properties, type "cboDataFrame" for the *(Name)*, "224, 21" for the *Size* property, and "156, 120" for the *Location* property.

7. Click *DropDownList* for the *DropDownStyle* property.

8. Click False for the Enabled property.

The next set of controls you'll add is the toolbar and its buttons for navigating the map.

1. In the Microsoft Visual Studio .NET toolbox, click the Windows Forms tab to display the Windows Forms tools.

2. In the toolbox, click *ToolBar* and drag a toolbar onto the form. The toolbar will automatically size and position itself along the top of the form.

3. Click Flat for the Appearance property.

Before adding the tools to the toolbar, you will add an image list to the form that will contain the images for the commands on the toolbar.

1. In the Microsoft Visual Studio .NET toolbox, click the Windows Forms tab to display the Windows Forms tools.

2. In the toolbox, click *ImageList* and drag an image list onto the form. The image list is a nonvisual control, so it will appear in the IDE below the form.

3. Click the *Images* property and click the Browse button. This will open the Image Collection Editor.

The Image Collection Editor dialog box

4. Click Add and click <install_location>\DeveloperKit\Samples\Data\ServerData\zoomin.gif

5. Repeat step 4 to add the following images to the image collection:

<install_location>\DeveloperKit\Samples\Data\ServerData\zoomout.gif

<install_location>\DeveloperKit\Samples\Data\ServerData\fullext.gif

<install_location>\DeveloperKit\Samples\Data\ServerData\pan.gif

6. Type "28,28" for the *ImageSize* property.

7. Click OK to close the Image Collection Editor.

8. On the form, click the toolbar control.

9. In the toolbar control's properties, click imageList1 for the ImageList property.

Now that you have added the image collection, you will add the commands to the toolbar.

1. In the toolbar control's properties, click the Buttons property and click the Browse button. This will open the ToolBarButton Collection Editor.

2. Click Add. This will add a new toolbar button.

3. Type "tbZoomIn" for the *Name*.

4. Click *False* for the *Enabled* property.

5. Click *0* for the *ImageIndex* property.

6. Click *PushButton* for the *Style* property.

7. Type "Zoom in" for the *ToolTipText* property.

8. Repeat steps 2 to 7 to add a button with the following properties:

    Name: tbZoomOut

    Enabled: False

    ImageIndex: 1

    Style: PushButton

    ToolTipText: Zoom out

9. Repeat steps 2 to 7 to add a button with the following properties:

    Name: tbFullExt

    Enabled: False

The ToolBarButton Collection Editor dialog box

    ImageIndex: 2

    Style: PushButton

    ToolTipText: Full Extent

10. Repeat steps 2 to 7 to add a button with the following properties:

    Name: tbPan

    Enabled: False

    ImageIndex: 3

    Style: ToggleButton

    ToolTipText: Pan

The last controls you'll add are the controls to list the bookmarks in the MapServer Web service's data frame selected by the user.

1. In the Microsoft Visual Studio .NET toolbox, click the Windows Forms tab to display the Windows Forms tools.

2. In the toolbox, click *Label* and drag a label onto the form.

3. In the label's properties, type "Bookmark:" for the *Text* property, "60, 16" for the *Size* property, and "260, 16" for the *Location* property.

4. In the Microsoft Visual Studio .NET toolbox, click the Windows Forms tab to display the Windows Forms tools.

5. In the toolbox, click *ComboBox* and drag a combo box onto the form.

6. In the combo box's properties, type "cboBookMark" for the *(Name)*, "228, 21" for the *Size* property, and "320, 12" for the *Location* property.

7. Click *DropDownList* for the *DropDownStyle* property.

8. Click *False* for the *Enabled* property.

Now you have added all the controls necessary for the application to the form. Your form should look like the following in design mode:

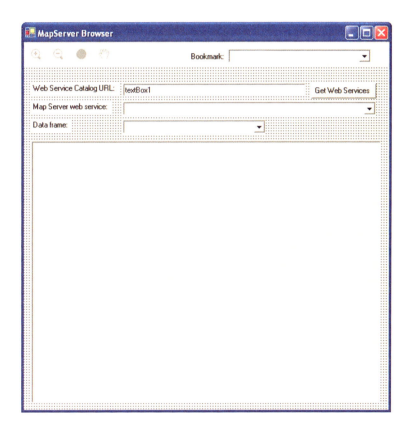

Now you'll add the code to the events for the controls in the application.

Add member variables to the application

This application requires a number of private member variables. Each variable will not be explained here, but as you add code to the various control events, you will use these variables.

1. Right-click the form and click ViewCode. This will open the code window for the form.

You will see that there are already a number of private member variables that were added by Visual Studio .NET for the controls you added to the form.

2. Below these member variables, add the following lines of code:

```
private string m_sSelectedMap;
private MapServerWS.MapDescription m_sMapDesc;
private string m_sDataFrame;
private double startX;
private double startY;
private int startDragX;
private int startDragY;
private int deltaDragX;
private int deltaDragY;
private MapServerWS.ImageDisplay idisp;
private System.Drawing.Image pImage;
```

Add code to get the list of Web services in the catalog

When using this application, the user will type the URL of a Web service catalog, then click the Get Web Services button. You will add code to the click event of the button to get the MapServer Web services in the Web service catalog and add them to the Web service combo box (cboMapServer).

1. In the Solution Explorer, double-click Form1.cs to open the form in design mode or click the Form1.cd [Design] tab.

2. Double-click the Get Web Services button you added to the form. This will open the code window and place the cursor in the default event for the button, which is the click event. The code for the click event should look like the following:

```
private void btnConnect_Click(object sender, System.EventArgs e)
{

}
```

You will add code to the click event to connect to the Web service catalog and get the list of MapServer Web services to add to the Web service combo box.

First, since the operation may take a few seconds to execute, you'll want to indicate to the user that the application is busy. To do this, you will set the cursor to a wait cursor. You'll also add code to clear the contents of the Web services combo box.

3. Add the following lines of code to the click event:

```
this.Cursor = Cursors.WaitCursor;
cboMapServer.Items.Clear();
```

Double-click the GetWeb Services button to open the code window in the button's click event.

The remainder of the code will run within a try/catch block. If an error occurs in the code in the try block, the code in the catch block will be executed. The catch block will change the cursor back to a normal cursor and display a message box containing an error message.

4. Add the following lines of code to the click event:

```
try
{
  this.Cursor = Cursors.Default;
}
catch (Exception exception)
{
  this.Cursor = Cursors.Default;
  MessageBox.Show(exception.Message  ,"An error has occurred");
}
```

If you named your Web service catalog Web reference something other than WebCatalog, then you'll have to modify this code to reflect the naming difference.

For example, if you named your Web reference MyWebRef, then the object would be MyWebRef.Default.

Next, you need to add code to create an instance of a WebCatalog.Default object. WebCatalog is the name of the Web service catalog Web reference you added earlier in this scenario, and Default is the name of the actual Web service. Once you create a WebService.Default, you'll set the url property to be the URL specified by the use in the txtServer text box.

5. Add the following lines of code to your try block:

```
WebCatalog.Default sc = new WebCatalog.Default();
sc.Url = txtServer.Text;
```

Next you'll call the GetServiceDescriptions method on the Web service catalog to get an array of ServiceDescription objects. You'll loop through this array and get the URLs for the MapServer Web services and add their URLs to the Web service combo box (cboMapServer). Finally you'll add code to enable the cboMapServer combo box.

6. Add the following lines of code to the try block:

```
WebCatalog.ServiceDescription[] wsdesc = sc.GetServiceDescriptions();
WebCatalog.ServiceDescription sd = null;

for (int i = 0;i < wsdesc.Length;i++)
{
  sd = wsdesc[i];
  if (sd.Type == "MapServer")
  {
    cboMapServer.Items.Add(sd.Url);
  }
}
cboMapServer.Enabled = true;
```

The completed code for the click event of your button should look like the following:

```
private void btnConnect_Click(object sender, System.EventArgs e)
{
```

```
this.Cursor = Cursors.WaitCursor;
cboMapServer.Items.Clear();

try
{
 WebCatalog.Default sc = new WebCatalog.Default();
 sc.Url = txtServer.Text;
 WebCatalog.ServiceDescription[] wsdesc = sc.GetServiceDescriptions();
 WebCatalog.ServiceDescription sd = null;

 for (int i = 0;i < wsdesc.Length;i++)
  {
   sd = wsdesc[i];
   if (sd.Type == "MapServer")
    {
     cboMapServer.Items.Add(sd.Url);
    }
  }

 cboMapServer.Enabled = true;
 this.Cursor = Cursors.Default;
}
catch (Exception exception)
{
 this.Cursor = Cursors.Default;
 MessageBox.Show(exception.Message  ,"An error has occurred");
}
}
```

Add code to get the data frames from the MapServer

Once connected to the Web service catalog, the user will pick a MapServer Web service to work with. You will add code to the selected index changed event of the Web service combo box (cboMapServer) to get the list of data frames in the MapServer and add them to the data frames combo box (cboDataFrame). You'll also set some of the member variables added earlier.

1. In the Solution Explorer, double-click Form1.cs to open the form in design mode, or click the Form1.cd [Design] tab.

2. Double-click the MapServer Web service combo box you added to the form. This will open the code window and place the cursor in the default event for the combo box, which is the selected index changed event. The code for the selected index changed event should look like the following:

```
private void cboMapServer_SelectedIndexChanged(object sender,
System.EventArgs e)
{

}
```

Double-click the Web service combo box to open the code window in the combo box's selected index changed event.

One of the member variables you added to the form was a string called m_sSelectedMap. You will use this variable to remember what the URL of the currently selected MapServer Web service is. This will be the value of the Text property of the map server combo box.

3. Add the following line of code to your event:

```
m_sSelectedMap = cboMapServer.Text;
```

The rest of the code for this event will be in a try/catch block.

4. Add the following lines of code to the event.

```
try
{

}
catch(Exception exception)
{
  MessageBox.Show(exception.Message  ,"An error has occurred");
}
```

Next you need to add code to create an instance of a MapServerWS.Map object. MapServerWS is the name of the MapServer Web service Web reference you added earlier in this scenario, and Map is the name of the actual Web service. Once you create a MapServerWS.Map, you'll set the url property to be the value of the m_sSelectedMap member variable.

If the MapServerWeb service that you created your Web reference from was not called Map, then the name of the object will not be MapServerWS.Map, but will be the name of your MapServer.

For example, if you referenced a MapServerWeb service called "World", then the object's name would be MapServerWS.World.

5. Add the following lines of code to your try block:

```
MapServerWS.Map map = new MapServerWS.Map();
map.Url = m_sSelectedMap;
```

m_sDataFrame is another string member variable that you added to the form. This member variable will store the name of the currently selected data frame. By default, when a new MapServer Web service is selected, the default data frame is picked for the user. So, you'll set the value of m_sDataFrame to be the default data frame of the MapServer.

6. Add the following line of code to your try block:

```
m_sDataFrame = map.GetDefaultMapName();
```

Next, you'll add code to loop through all the data frames in the Web service and add them to the data frame combo box (cboDataFrame). Once the combo box is populated, set its text property to be the value of m_sDataFrame.

7. Add the following code to your try block:

```
// get dataframes, populate drop down
cboDataFrame.Items.Clear();
for (int i=0;i<map.GetMapCount();i++)
{
  cboDataFrame.Items.Add(map.GetMapName(i));
}
cboDataFrame.Text = m_sDataFrame;
```

When you set the Text property on the cboDataFrame combo box, it will trigger its selected index changed event. You'll add code to that event next.

8. Finally, add the following code to your try block to enable the other controls and toolbar buttons on the form.

```
// enable controls
cboDataFrame.Enabled = true;
cboBookMark.Enabled = true;
IEnumerator benum = toolBar1.Buttons.GetEnumerator();
```

```
ToolBarButton btn = null;
while (benum.MoveNext())
{
  btn = benum.Current as ToolBarButton;
  btn.Enabled = true;
}
```

The completed code for the selected index changed event should look like the following:

```
private void cboMapServer_SelectedIndexChanged(object sender,
System.EventArgs e)
{
  m_sSelectedMap = cboMapServer.Text;
  try
  {
    MapServerWS.Map map = new MapServerWS.Map();
    map.Url = m_sSelectedMap;

    m_sDataFrame = map.GetDefaultMapName();

    // get dataframes, populate drop down
    cboDataFrame.Items.Clear();
    for (int i=0;i<map.GetMapCount();i++)
     {
      cboDataFrame.Items.Add(map.GetMapName(i));
     }
    cboDataFrame.Text = m_sDataFrame;

    // enable controls
    cboDataFrame.Enabled = true;
    cboBookMark.Enabled = true;
    IEnumerator benum = toolBar1.Buttons.GetEnumerator();
    ToolBarButton btn = null;
    while (benum.MoveNext())
     {
      btn = benum.Current as ToolBarButton;
      btn.Enabled = true;
     }
  }
  catch(Exception exception)
  {
    MessageBox.Show(exception.Message ,"An error has occurred");
  }
}
```

Add code to get the bookmarks for the selected data frame

You will add code to the selected index changed event of the data frame combo box (cboDataFrame) to get the list of spatial bookmarks for the selected data frame and add them to the bookmarks combo box (cboBookMark). You'll also set some of the member variables added earlier.

1. In the Solution Explorer, double-click Form1.cs to open the form in design mode, or click the Form1.cd [Design] tab.

2. Double-click the data frame combo box you added to the form. This will open the code window and place the cursor in the default event for the combo box, which is the selected index changed event. The code for the selected index changed event should look like the following:

Double-click the data frame combo box to open the code window in the combo box's selected index changed event.

```
private void cboDataFrame_SelectedIndexChanged(object sender,
System.EventArgs e)
{

}
```

Since the data frame has changed, you'll set the data frame member variable (m_sDataFrame) to be the Text value of the data frame combo box.

Add the following line of code to the event.

```
m_sDataFrame = cboDataFrame.Text;
```

The rest of the code for this event will be in a try/catch block.

3. Add the following lines of code to the event.

```
try
{

}
catch(Exception exception)
{
  MessageBox.Show(exception.Message  ,"An error has occurred");
}
```

If the MapServer Web service that you created your Web reference from was not called Map, then the name of the object will not be MapServerWS.Map but will be the name of your MapServer.

For example, if you referenced a MapServer Web service called "World", then the object's name would be MapServerWS.World.

Next, you need to add code to create an instance of a MapServerWS.Map object and set the url property to be the value of the m_sSelectedMap member variable.

4. Add the following lines of code to your try block:

```
MapServerWS.Map map = new MapServerWS.Map();
map.Url = m_sSelectedMap;
```

In order to get a list of the spatial bookmarks in the map, you'll add code to get a reference to the MapServer's MapServerWS.MapServerInfo object for the selected data frame, then get an array of bookmarks from MapServerInfo.

Once you have the array of bookmarks, you'll loop through the array and add the names of the bookmarks to the bookmarks combo box (cboBookMark). Notice the code also adds a "<Default Extent>" item to the combo box. This allows the user to return to the default extent of the data frame.

5. Add the following lines of code to your try block.

```
// get bookmarks, populate bookmarks drop down
MapServerWS.MapServerInfo mapi = map.GetServerInfo(m_sDataFrame);
MapServerWS.MapServerBookmark[] pMSBookMarks = mapi.Bookmarks;

cboBookMark.Items.Clear();
cboBookMark.Items.Add("<Default Extent>");
MapServerWS.MapServerBookmark pMDBook;
for (int j = 0;j<pMSBookMarks.Length;j++)
{
  pMDBook = pMSBookMarks[j];
  cboBookMark.Items.Add(pMDBook.Name);
}
cboBookMark.SelectedItem = "<Default Extent>";
```

When you set the Text property on the cboBookMark combo box, it will trigger its selected index changed event. You'll add code to that event later.

The completed code for the selected index changed event should look like the following:

```
private void cboDataFrame_SelectedIndexChanged(object sender,
System.EventArgs e)
{
  m_sDataFrame = cboDataFrame.Text;

  try
  {
   // find the info about the selected data frame from the map server
   MapServerWS.Map map = new MapServerWS.Map();
   map.Url = m_sSelectedMap;

   // get bookmarks, populate bookmarks drop down
   MapServerWS.MapServerInfo mapi = map.GetServerInfo(m_sDataFrame);
   MapServerWS.MapServerBookmark[] pMSBookMarks = mapi.Bookmarks;

   cboBookMark.Items.Clear();
   cboBookMark.Items.Add("<Default Extent>");
   MapServerWS.MapServerBookmark pMDBook;
   for (int j = 0;j<pMSBookMarks.Length;j++)
    {
     pMDBook = pMSBookMarks[j];
     cboBookMark.Items.Add(pMDBook.Name);
    }
   cboBookMark.SelectedItem = "<Default Extent>";
  }
  catch(Exception exception)
  {
   MessageBox.Show(exception.Message ,"An error has occurred");
  }
}
```

Add helper function to draw the map

To this point, the code you have added to the application has used methods on the Web service catalog and MapServer Web services to get information about a particular Web service catalog or MapServer Web service and populate controls on the form.

The next set of controls whose events you'll add code to will actually draw the map and display the result in the picture box. In order to draw the map, you'll add a helper function that these events will call to do the map drawing.

1. Add the following function to your form.

```
private void drawMap(ref MapServerWS.MapDescription pMapDescription)
{

}
```

By definition, Web services are stateless. For more information about programming ArcGIS Server and how it relates to managing state in an application that makes stateless use of server objects, see Chapter 4.

As you can see, the drawMap function takes a single argument of type MapServerWS.MapDescription. The map description object is used by a MapServer when drawing maps to allow, at draw time, various aspects of the map to be changed, without changing the running MapServer object. These properties include the extent to draw, the layers to turn on or off, and so on. In this respect, the map description serves the purpose of allowing stateless use of a MapServer while allowing these aspects of state to be saved in the application by saving and modifying a local copy of the map description.

This function doesn't do anything to the map description except use it to draw the map. You'll see later how the code that calls the drawMap function modifies and uses the map description to allow the user to navigate around the map.

If the MapServer Web service that you created your Web reference from was not called Map, then the name of the object will not be MapServerWS.Map, but will be the name of your MapServer.

For example, if you referenced a MapServer Web service called "World", then the object's name would be MapServerWS.World.

The first code you'll add to this function creates an instance of a *MapServerWS.Map* object and sets the url property to be the value of the m_sSelectedMap member variable.

2. Add the following lines of code to the function block:

```
MapServerWS.Map map = new MapServerWS.Map();
map.Url = m_sSelectedMap;
```

3. Next, add the following lines of code to create an image description object that you'll use when drawing the map. Notice that the *ImageDisplay* object (idisp) is one of the member variables you added to the form. You will see later that this *ImageDisplay* object will be used to query the map in the events for the Pan command.

```
// set up the image description for the output
MapServerWS.ImageType it = new MapServerWS.ImageType();
it.ImageFormat  = MapServerWS.esriImageFormat.esriImageJPG;
it.ImageReturnType =
MapServerWS.esriImageReturnType.esriImageReturnMimeData;

idisp = new MapServerWS.ImageDisplay();
idisp.ImageHeight = 400;
idisp.ImageWidth  = 552;
idisp.ImageDPI  = 150;
```

The size of the ImageDisplay is the same as the size of the picture box control. This will produce an image that fits exactly in the picture box on the form.

```
MapServerWS.ImageDescription pID = new MapServerWS.ImageDescription();
pID.ImageDisplay = idisp;
pID.ImageType = it;
```

Finally, add the following lines of code to call the ExportMapImage method on the MapServer to draw the map, then convert the result into a .NET Image object and set it as the Image property of the picture box, then return. Notice that the *Image* object (pImage) is a member variable of the form. This will be used again later in the events for the Pan button.

```
MapServerWS.MapImage pMI = map.ExportMapImage(pMapDescription, pID);
System.IO.Stream pStream = new
System.IO.MemoryStream((byte[])pMI.ImageData);
pImage = Image.FromStream(pStream);

pictureBox1.Image = pImage;
pictureBox1.Refresh();

return;
```

The completed code for the drawMap function should look like the following:

```
private void drawMap(ref MapServerWS.MapDescription pMapDescription)
{
  MapServerWS.Map map = new MapServerWS.Map();
  map.Url = m_sSelectedMap;

  // set up the image description for the output
  MapServerWS.ImageType it = new MapServerWS.ImageType();
  it.ImageFormat = MapServerWS.esriImageFormat.esriImageJPG;
  it.ImageReturnType =
MapServerWS.esriImageReturnType.esriImageReturnMimeData;

  idisp = new MapServerWS.ImageDisplay();
  idisp.ImageHeight = 400;
  idisp.ImageWidth = 552;
  idisp.ImageDPI = 150;

  MapServerWS.ImageDescription pID = new MapServerWS.ImageDescription();
  pID.ImageDisplay = idisp;
  pID.ImageType = it;

  MapServerWS.MapImage pMI = map.ExportMapImage(pMapDescription, pID);
  System.IO.Stream pStream = new
System.IO.MemoryStream((byte[])pMI.ImageData);
  pImage = Image.FromStream(pStream);

  pictureBox1.Image = pImage;
  pictureBox1.Refresh();

  return;
}
```

Add code to draw the extent of the selected bookmark

When the user clicks the bookmarks combo box and picks a bookmark, the picture box will display the map for the extent of the selected bookmark. To do this you'll add code to the selected index changed event of the bookmarks combo box (cboBookMark).

1. In the Solution Explorer, double-click Form1.cs to open the form in design mode, or click the Form1.cd [Design] tab.

2. Double-click the bookmarks combo box you added to the form. This will open the code window and place the cursor in the default event for the combo box which is the selected index changed event. The code for the selected index changed event should look like the following:

```
private void cboBookMark_SelectedIndexChanged(object sender,
System.EventArgs e)
{

}
```

Since the operation may take a few seconds to execute, you'll want to indicate to the user that the application is busy. To do this, you will set the cursor to a wait cursor.

3. Add the following line of code to the click event:

```
this.Cursor = Cursors.WaitCursor;
```

The rest of the code for this event will be in a try/catch block.

4. Add the following lines of code to the event.

```
try
{

}
catch(Exception exception)
{
  this.Cursor = Cursors.Default;
  MessageBox.Show(exception.Message  ,"An error has occurred");
}
```

If the MapServer Web service that you created your Web reference from was not called Map, then the name of the object will not be MapServerWS.Map but will be the name of your MapServer.

For example, if you referenced a MapServer Web service called "World", then the object's name would be MapServerWS.World.

Next you need to add code to create an instance of a *MapServerWS.Map* object. and set the url property to be the value of the m_sSelectedMap member variable. Then you'll get a *MapServerInfo* object for the selected data frame (m_sDataFrame),

5. Add the following lines of code to your try block:

```
MapServerWS.Map map = new MapServerWS.Map();
map.Url = m_sSelectedMap;
MapServerWS.MapServerInfo mapi = map.GetServerInfo(m_sDataFrame);
```

6. Next, add code to create a MapServerWS.MapDescription variable. If the Text value of the bookmarks combo box is "<Default Extent>", then set the map description to be the default map description for the data frame, update the map description member variable m_sMapDesc, draw the map, and return. Otherwise, set the map description to be the map description member variable.

```
MapServerWS.MapDescription pMapDescription;

// if they chose the default extent, get the map description from the map
// server, then exit
if(cboBookMark.Text == "<Default Extent>")
{
  pMapDescription = mapi.DefaultMapDescription;
  m_sMapDesc = pMapDescription;
  drawMap (ref pMapDescription);
  this.Cursor = Cursors.Default;
  return;
}

pMapDescription = m_sMapDesc;
```

By definition, Web services are stateless. For more information about programming ArcGIS Server and how it relates to managing state in an application that makes stateless use of server objects, see Chapter 4.

A copy of the map description is being stored in a member variable, because as you implement other commands, such as the zoom and pan commands, you'll want to keep track of the user's current extent to know what to zoom or pan from. These commands will modify the extent of the local copy of the map description, then use it as input to the drawMap function.

7. Next, add the following code to find the bookmark that corresponds to the bookmark picked by the user, get its extent, set the extent into the map description, then draw the map. Notice that after the map description is modified, the member variable m_sMapDesc is updated so the user's current extent is remembered.

```
// find the chosen bookmark
MapServerWS.MapServerBookmark[] pMSBookMarks = mapi.Bookmarks;
MapServerWS.MapServerBookmark pMDBook = null;

for (int i = 0;i < pMSBookMarks.Length;i++)
{
    pMDBook = pMSBookMarks[i];
    if (pMDBook.Name == cboBookMark.Text)
      break;
}

// set the extent of the map description to the bookmark's extent
MapServerWS.MapArea pMA = pMDBook;
pMapDescription.MapArea = pMA;

// save the map description
m_sMapDesc = pMapDescription;
```

```
    drawMap (ref pMapDescription);

  this.Cursor = Cursors.Default;
```

The completed code for the selected index changed event should look like the following:

```
private void cboBookMark_SelectedIndexChanged(object sender,
System.EventArgs e)
{
  this.Cursor = Cursors.WaitCursor;

  try
  {
   MapServerWS.Map map = new MapServerWS.Map();
   map.Url = m_sSelectedMap;

   MapServerWS.MapServerInfo mapi = map.GetServerInfo(m_sDataFrame);
   MapServerWS.MapDescription pMapDescription;

   // if they chose the default extent, get the map description from the
   // map server, then exit
   if(cboBookMark.Text == "<Default Extent>")
    {
     pMapDescription = mapi.DefaultMapDescription;
     m_sMapDesc = pMapDescription;
     drawMap (ref pMapDescription);
     this.Cursor = Cursors.Default;
      return;
    }

   pMapDescription = m_sMapDesc;

   // find the chosen bookmark
   MapServerWS.MapServerBookmark[] pMSBookMarks = mapi.Bookmarks;
   MapServerWS.MapServerBookmark pMDBook = null;

   for (int i = 0;i < pMSBookMarks.Length;i++)
    {
        pMDBook = pMSBookMarks[i];
      if (pMDBook.Name == cboBookMark.Text)
         break;
    }

   // set the extent of the map description to the bookmark's extent
```

```
MapServerWS.MapArea pMA = pMDBook;
pMapDescription.MapArea = pMA;

// save the map description
m_sMapDesc = pMapDescription;
drawMap (ref pMapDescription);

this.Cursor = Cursors.Default;
}
catch (Exception exception)
{
this.Cursor = Cursors.Default;
MessageBox.Show(exception.Message ,"An error has occurred");
}
}
```

Add code for the Zoom, Full Extent buttons

Some of the map navigation functionality in this application includes fixed zoom in, fixed zoom out, and zoom to full extent commands. You added these commands as buttons to the toolbar on the form. You'll add the code to execute when these commands are clicked to the button-click event of the toolbar (toolBar1).

1. In the Solution Explorer, double-click Form1.cs to open the form in design mode, or click the Form1.cd [Design] tab.

2. Double-click the toolbar you added to the form. This will open the code window and place the cursor in the default event for the toolbar, which is the button-click event. The code for the button-click event should look like the following:

```
private void toolBar1_ButtonClick(object sender,
System.Windows.Forms.ToolBarButtonClickEventArgs e)
{

}
```

3. Add the following code to determine which button was clicked.

```
// Evaluate the Button property to determine which button was clicked.
switch(toolBar1.Buttons.IndexOf(e.Button))
{
  case 0: // zoom in
    break;
  case 1: // zoom out
    break;
  case 2: // full extent
    break;
  case 3: // pan
    break;
}
```

If the MapServer Web service that you created your Web reference from was not called Map, then the name of the object will not be MapServerWS.Map but will be the name of your MapServer.

For example, if you referenced a MapServer Web service called "World", then the object's name would be MapServerWS.World.

For the zoom in case, you'll add code to get the *MapServerWS.Map* object, set its url, get the map description saved as m_sMapDesc, shrink its extent, draw the map, then update m_sMapDesc with the new extent.

4. Add the following code to case 0.

```
this.Cursor = Cursors.WaitCursor;
try
{
  MapServerWS.Map map = new MapServerWS.Map();
  map.Url = m_sSelectedMap;

  MapServerWS.MapDescription pMapDescription = m_sMapDesc;

  // get the current extent and shrink it, then set the new extent into
  // the map description
  MapServerWS.EnvelopeN pEnvelope = pMapDescription.MapArea.Extent as
MapServerWS.EnvelopeN;

  double eWidth = Math.Abs(pEnvelope.XMax - pEnvelope.XMin);
  double eHeight = Math.Abs(pEnvelope.YMax - pEnvelope.YMin);
  double xFactor = (eWidth - (eWidth * 0.75))/2;
  double yFactor = (eHeight - (eHeight * 0.75))/2;
  pEnvelope.XMax = pEnvelope.XMax - xFactor;
  pEnvelope.XMin = pEnvelope.XMin + xFactor;
  pEnvelope.YMax = pEnvelope.YMax - yFactor;
  pEnvelope.YMin = pEnvelope.YMin + yFactor;

  MapServerWS.MapExtent pMapExtext = new MapServerWS.MapExtent();
  pMapExtext.Extent = pEnvelope;
  pMapDescription.MapArea = pMapExtext;

  // save the map description and draw the map
  m_sMapDesc = pMapDescription;
  drawMap (ref pMapDescription);

  this.Cursor = Cursors.Default;
}
catch (Exception exception)
{
  this.Cursor = Cursors.Default;
  MessageBox.Show(exception.Message  ,"An error has occurred");
}
```

For the zoom out case, you'll add similar code as that for zoom in, except you'll expand the map description's extent.

5. Add the following code to case 1.

```
this.Cursor = Cursors.WaitCursor;
try
{
  MapServerWS.Map map = new MapServerWS.Map();
  map.Url = m_sSelectedMap;

  MapServerWS.MapDescription pMapDescription = m_sMapDesc;
```

```
// get the current extent and shrink it, then set the new extent into
// the map description
MapServerWS.EnvelopeN pEnvelope = pMapDescription.MapArea.Extent as
MapServerWS.EnvelopeN;

double eWidth = Math.Abs(pEnvelope.XMax - pEnvelope.XMin);
double eHeight = Math.Abs(pEnvelope.YMax  - pEnvelope.YMin);
double xFactor = ((eWidth * 1.25) - eWidth)/2;
double yFactor = ((eHeight * 1.25) - eHeight)/2;
pEnvelope.XMax = pEnvelope.XMax + xFactor;
pEnvelope.XMin = pEnvelope.XMin - xFactor;
pEnvelope.YMax = pEnvelope.YMax + yFactor;
pEnvelope.YMin = pEnvelope.YMin - yFactor;

MapServerWS.MapExtent pMapExtext = new MapServerWS.MapExtent();
pMapExtext.Extent = pEnvelope;
pMapDescription.MapArea = pMapExtext;

// save the map description and draw the map
m_sMapDesc = pMapDescription;
drawMap (ref pMapDescription);

this.Cursor = Cursors.Default;
}
catch (Exception exception)
{
  this.Cursor = Cursors.Default;
  MessageBox.Show(exception.Message  ,"An error has occurred");
}
```

For the full extent case, the code is similar, except you get the full extent from the map server's *MapServerInfo* object and set it into the map description.

6. Add the following code to case 2.

```
this.Cursor = Cursors.WaitCursor;
try
{
  MapServerWS.Map map = new MapServerWS.Map();
  map.Url = m_sSelectedMap;
  MapServerWS.MapServerInfo mapi = map.GetServerInfo(m_sDataFrame);

  MapServerWS.MapDescription pMapDescription = m_sMapDesc;

  // get the full extent of the map and set it as the map description's extent
  MapServerWS.Envelope pEnvelope = mapi.FullExtent;

  MapServerWS.MapExtent pMapExtext = new MapServerWS.MapExtent();
  pMapExtext.Extent = pEnvelope;
  pMapDescription.MapArea = pMapExtext;
```

```
    // save the map description and draw the map
    m_sMapDesc = pMapDescription;
    drawMap(ref pMapDescription);

    this.Cursor = Cursors.Default;
  }
  catch (Exception exception)
  {
    this.Cursor = Cursors.Default;
    MessageBox.Show(exception.Message  ,"An error has occurred");
  }
```

The code for the button-click event should look like the following:

```
private void toolBar1_ButtonClick(object sender,
System.Windows.Forms.ToolBarButtonClickEventArgs e)
{
  // Evaluate the Button property to determine which button was clicked.
  switch(toolBar1.Buttons.IndexOf(e.Button))
  {
    case 0:
      this.Cursor = Cursors.WaitCursor;
      // zoom in
      try
        {
        MapServerWS.Map map = new MapServerWS.Map();
        map.Url = m_sSelectedMap;

        MapServerWS.MapDescription pMapDescription = m_sMapDesc;

        // get the current extent and shrink it, then set the new extent
        // into the map description
        MapServerWS.EnvelopeN pEnvelope = pMapDescription.MapArea.Extent
as MapServerWS.EnvelopeN;

        double eWidth = Math.Abs(pEnvelope.XMax - pEnvelope.XMin);
        double eHeight = Math.Abs(pEnvelope.YMax  - pEnvelope.YMin);
        double xFactor = (eWidth - (eWidth * 0.75))/2;
        double yFactor = (eHeight - (eHeight * 0.75))/2;
        pEnvelope.XMax = pEnvelope.XMax - xFactor;
        pEnvelope.XMin = pEnvelope.XMin + xFactor;
        pEnvelope.YMax = pEnvelope.YMax - yFactor;
        pEnvelope.YMin = pEnvelope.YMin + yFactor;

        MapServerWS.MapExtent pMapExtext = new MapServerWS.MapExtent();
        pMapExtext.Extent = pEnvelope;
        pMapDescription.MapArea = pMapExtext;

        // save the map description and draw the map
        m_sMapDesc = pMapDescription;
        drawMap (ref pMapDescription);
```

```
        this.Cursor = Cursors.Default;
        }
    catch (Exception exception)
        {
        this.Cursor = Cursors.Default;
        MessageBox.Show(exception.Message  ,"An error has occurred");
        }

    break;
case 1: // zoom out

    this.Cursor = Cursors.WaitCursor;

    try
        {
        MapServerWS.Map map = new MapServerWS.Map();
        map.Url = m_sSelectedMap;

        MapServerWS.MapDescription pMapDescription = m_sMapDesc;

        // get the current extent and shrink it, then set the new extent
        // into the map description
        MapServerWS.EnvelopeN pEnvelope = pMapDescription.MapArea.Extent
    as MapServerWS.EnvelopeN;

        double eWidth = Math.Abs(pEnvelope.XMax - pEnvelope.XMin);
        double eHeight = Math.Abs(pEnvelope.YMax  - pEnvelope.YMin);
        double xFactor = ((eWidth * 1.25) - eWidth)/2;
        double yFactor = ((eHeight * 1.25) - eHeight)/2;
        pEnvelope.XMax = pEnvelope.XMax + xFactor;
        pEnvelope.XMin = pEnvelope.XMin - xFactor;
        pEnvelope.YMax = pEnvelope.YMax + yFactor;
        pEnvelope.YMin = pEnvelope.YMin - yFactor;

        MapServerWS.MapExtent pMapExtext = new MapServerWS.MapExtent();
        pMapExtext.Extent = pEnvelope;
        pMapDescription.MapArea = pMapExtext;

        // save the map description and draw the map
        m_sMapDesc = pMapDescription;
        drawMap (ref pMapDescription);

        this.Cursor = Cursors.Default;
        }
    catch (Exception exception)
        {
        this.Cursor = Cursors.Default;
        MessageBox.Show(exception.Message  ,"An error has occurred");
        }
    break;
```

```
      case 2: // full extent

        this.Cursor = Cursors.WaitCursor;
         try
         {
         MapServerWS.Map map = new MapServerWS.Map();
         map.Url = m_sSelectedMap;
         MapServerWS.MapServerInfo mapi = map.GetServerInfo(m_sDataFrame);

         MapServerWS.MapDescription pMapDescription = m_sMapDesc;

         // get the full extent of the map and set it as the map
         // description's extent
         MapServerWS.Envelope pEnvelope = mapi.FullExtent;

         MapServerWS.MapExtent pMapExtext = new MapServerWS.MapExtent();
         pMapExtext.Extent = pEnvelope;
         pMapDescription.MapArea = pMapExtext;

         // save the map description and draw the map
         m_sMapDesc = pMapDescription;
         drawMap(ref pMapDescription);

         this.Cursor = Cursors.Default;
         }
        catch (Exception exception)
         {
         this.Cursor = Cursors.Default;
         MessageBox.Show(exception.Message  ,"An error has occurred");
         }
        break;
      case 3: //pan
        break;
       }
   }
```

You will implement case 3 (pan) next.

Add code for the Pan button

The final piece of functionality you'll add to the application is the ability for the user to interactively pan the map. To do this, you'll add code to a number of events on the picture box control, specifically, the mouse down, mouse up, mouse move, and paint events. Before adding code to those events, you'll add code to the button-click event of the toolbar to change the cursor for the picture box when the Pan button is pushed in.

1. Add the following lines of code to case 3 in the toolbar-click event.

```
ToolBarButton btn = e.Button;
if (btn.Pushed)
  pictureBox1.Cursor = Cursors.Hand;
else
  pictureBox1.Cursor = Cursors.Default;
```

The picture box control's Properties dialog box

If the MapServer Web service that you created your Web reference from was not called Map, then the name of the object will not be MapServerWS.Map but will be the name of your MapServer.

For example, if you referenced a MapServer Web service called "World", then the object's name would be MapServerWS.World.

Next, you will add code to the mouse down event for the picture box. This code will check to see if the Pan button is pushed and, if so, record the point on the map (for example, picture box) that is clicked in both screen and map coordinates, then save those coordinates in two of the member variables you added to the form.

1. In the Solution Explorer, double-click Form1.cs to open the form in design mode, or click the Form1.cd [Design] tab.

2. Click the picture box control on the form.

3. In the picture box's properties, click the Event button, then double-click the MouseDown event. This will open the code window and place the cursor in the mouse down event. The code for the mouse event should look like the following:

```
private void pictureBox1_MouseDown(object sender,
System.Windows.Forms.MouseEventArgs e)
{

}
```

4. Add the following code to the event to verify if the left mouse button is clicked and, if not, return.

```
if(e.Button != MouseButtons.Left)
    return;
```

Next, verify the Pan button is pushed and, if it is, create an instance of the *MapServerWS.Map* object, and use the map description and image display variables (m_sMapDesc and idisp) with the ToMapPoints method on the MapServer to convert the screen coordinates of the point clicked to map coordinates. Then save the map and screen coordinates as the member variables startX, startY, startdragX, startdragY.

5. Add the following code to the event.

```
// is pan enabled?
IEnumerator benum = toolBar1.Buttons.GetEnumerator();
ToolBarButton btn = null;
while (benum.MoveNext())
{
  btn = benum.Current as ToolBarButton;
  if (toolBar1.Buttons.IndexOf(btn) == 3 && btn.Pushed == true)
  {
    MapServerWS.Map map = new MapServerWS.Map();
    map.Url = m_sSelectedMap;

    MapServerWS.MapDescription pMapDescription = m_sMapDesc;
    int[] Xs – {e.X};
    int[] Ys = {e.Y};
    MapServerWS.MultipointN mpnt = map.ToMapPoints(m_sMapDesc,idisp,Xs,Ys)
as MapServerWS.MultipointN;
    MapServerWS.Point[] pnta = mpnt.PointArray;
    MapServerWS.PointN pnt = pnta[0] as MapServerWS.PointN;
    startX = pnt.X;
```

```
    startY = pnt.Y;
    startDragX = e.X;
    startDragY = e.Y;
  }
}
```

The completed code for the mouse down event should look like the following:

```
private void pictureBox1_MouseDown(object sender,
System.Windows.Forms.MouseEventArgs e)
{
  if(e.Button != MouseButtons.Left)
    return;

  // is pan enabled?
  IEnumerator benum = toolBar1.Buttons.GetEnumerator();
  ToolBarButton btn = null;
  while (benum.MoveNext())
   {
    btn = benum.Current as ToolBarButton;
    if (toolBar1.Buttons.IndexOf(btn) == 3 && btn.Pushed == true)
     {
      MapServerWS.Map map = new MapServerWS.Map();
      map.Url = m_sSelectedMap;

      MapServerWS.MapDescription pMapDescription = m_sMapDesc;
      int[] Xs = {e.X};
      int[] Ys = {e.Y};
      MapServerWS.MultipointN mpnt =
map.ToMapPoints(m_sMapDesc,idisp,Xs,Ys) as MapServerWS.MultipointN;
      MapServerWS.Point[] pnta = mpnt.PointArray;
      MapServerWS.PointN pnt = pnta[0] as MapServerWS.PointN;
      startX = pnt.X;
      startY = pnt.Y;
      startDragX = e.X;
      startDragY = e.Y;
     }
   }
}
```

The next event you'll implement is the mouse move. In the mouse move event, you'll add code similar to the code you added in mouse down event to verify the Pan button is pushed and the mouse button is the left mouse button. The code that executes when these conditions are met calculates the amount the mouse has moved from the original point that was clicked on the picture box (startX, startY) and stores the difference as member variables deltaDragX and deltaDragY. It then forces a redraw of the picture box by calling its Invalidate method.

6. In the Solution Explorer, double-click Form1.cs to open the form in design mode, or click the Form1.cd [Design] tab.

7. Click the picture box control on the form.

8. In the picture box's properties, click the Event button, then double-click the MouseMove event. This will open the code window and place the cursor in the mouse move event. The code for the mouse move event should look like the following:

```
private void pictureBox1_MouseMove(object sender,
System.Windows.Forms.MouseEventArgs e)
{

}
```

9. Add code to your mouse move event, so it looks like the following:

```
private void pictureBox1_MouseMove(object sender,
System.Windows.Forms.MouseEventArgs e)
{
  if (e.Button != MouseButtons.Left)
    return;

  IEnumerator benum = toolBar1.Buttons.GetEnumerator();
  ToolBarButton btn = null;
  while (benum.MoveNext())
  {
    btn = benum.Current as ToolBarButton;
    if (toolBar1.Buttons.IndexOf(btn) == 3 && btn.Pushed == true)
    {
      // drag the image
      pictureBox1.Image = null;
      deltaDragX = startDragX - e.X;
      deltaDragY = startDragY - e.Y;
      pictureBox1.Invalidate();
    }
  }
}
```

The Invalidate method triggers the picture box's Paint event. Next you'll add code to the paint event to draw the image offset from the picture box as the mouse is dragged.

10. In the Solution Explorer, double-click Form1.cs to open the form in design mode, or click the Form1.cd [Design] tab.

11. Click the picture box control on the form.

12. In the picture box's properties, click the Event button, then double-click the Paint event. This will open the code window and place the cursor in the paint event. The code for the paint event should look like the following:

```
private void pictureBox1_MouseMove(object sender,
System.Windows.Forms.MouseEventArgs e)
{

}
```

The code you add to this event gets the image from the member variable pImage, then applies the offset of the upper left corner of the image based on the values of deltaDragX and deltaDragY and creates a new rectangle the same width as the picture, but offset, and draws the image in that offset rectangle. The effect is that as the user drags the mouse, it will appear as though the map is being dragged. This feedback makes it easier for the user to effectively pan the map.

13. Add code to your paint event, such that it looks like the following:

```
private void pictureBox1_Paint(object sender,
System.Windows.Forms.PaintEventArgs e)
{
  // get the image.
  Image newImage = pImage;

  if (newImage != null)
  {
    // Create rectangle for displaying image.
    Point loc = pictureBox1.Location;
    Rectangle destRect = new Rectangle(pictureBox1.Left - loc.X -
deltaDragX, pictureBox1.Top - loc.Y - deltaDragY, pictureBox1.Width,
pictureBox1.Height);
    // Draw image to screen.
    e.Graphics.DrawImage(newImage, destRect);
  }
}
```

The last event to implement is the mouse up event. The code in this event will execute when the user has completed the pan and releases the button.

14. In the Solution Explorer, double-click Form1.cs to open the form in design mode, or click the Form1.cd [Design] tab.

15. Click the picture box control on the form.

16. In the picture box's properties, click the Event button, then double-click the MouseUp event. This will open the code window and place the cursor in the mouse up event. The code for the mouse up should look like the following:

```
private void pictureBox1_MouseUp(object sender,
System.Windows.Forms.MouseEventArgs e)
{

}
```

If the MapServer Web service that you created your Web reference from was not called Map, then the name of the object will not be MapServerWS.Map but will be the name of your MapServer.

For example, if you referenced a MapServer Web service called "World", then the object's name would be MapServerWS.World.

Like the mouse down and mouse move events, your code will check to verify the left mouse button is clicked and that the Pan button is pushed. Once that is verified, your code will get the *MapServerWS.Map* object, get the user's current extent from the map description (m_sMapDesc), apply the deltaX and deltaY offset to its coordinates, update the map description (m_sMapDesc), and redraw the map. It also resets the deltaDragX and deltaDragY member variables to 0 for the next pan operation.

17. Add code to your mouse up event, so it looks like the following.

```
private void pictureBox1_MouseUp(object sender,
System.Windows.Forms.MouseEventArgs e)
{
  if(e.Button != MouseButtons.Left)
    return;

  // is pan enabled?
  IEnumerator benum = toolBar1.Buttons.GetEnumerator();
  ToolBarButton btn = null;
  while (benum.MoveNext())
  {
    btn = benum.Current as ToolBarButton;
    if (toolBar1.Buttons.IndexOf(btn) == 3 && btn.Pushed == true)
    {
      this.Cursor = Cursors.WaitCursor;
      MapServerWS.Map map = new MapServerWS.Map();
      map.Url = m_sSelectedMap;

      MapServerWS.MapDescription pMapDescription = m_sMapDesc;
      int[] Xs = {e.X};
      int[] Ys = {e.Y};
      MapServerWS.MultipointN mpnt =
map.ToMapPoints(m_sMapDesc,idisp,Xs,Ys) as MapServerWS.MultipointN;
      MapServerWS.Point[] pnta = mpnt.PointArray;
      MapServerWS.PointN pnt = pnta[0] as MapServerWS.PointN;

      double deltaX = pnt.X - startX;
      double deltaY = pnt.Y - startY;

      //change the extent and draw
      MapServerWS.EnvelopeN pEnvelope = pMapDescription.MapArea.Extent as
MapServerWS.EnvelopeN;

      pEnvelope.XMax = pEnvelope.XMax - deltaX;
      pEnvelope.XMin = pEnvelope.XMin - deltaX;
      pEnvelope.YMax = pEnvelope.YMax - deltaY;
      pEnvelope.YMin = pEnvelope.YMin - deltaY;

      MapServerWS.MapExtent pMapExtext = new MapServerWS.MapExtent();
      pMapExtext.Extent = pEnvelope;
      pMapDescription.MapArea = pMapExtext;

      // save the map description and draw the map
      m_sMapDesc = pMapDescription;
      drawMap (ref pMapDescription);

      deltaDragX = 0;
      deltaDragY = 0;
```

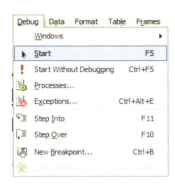

```
        pictureBox1.Invalidate();
        this.Cursor = Cursors.Default;
      }
    }
  }
```

Your application is now ready to be tested. Compile the project (*Build/Build Solution*) and fix any errors.

Test the application

If you run the application from within Visual Studio (Debug/Start), it will open the form.

1. Type the URL of a Web service catalog.

2. Choose one of the MapServer Web services.

3. Click on the Bookmark combo box and choose a bookmark to zoom to.

4. Use the Zoom In, Zoom Out, Full Extent, and Pan buttons to navigate around the map.

5. Try choosing different MapServer Web services and data frames.

You can use this application to connect to any ArcGIS Server Web service catalog, whether that Web service catalog was written in .NET or in Java.

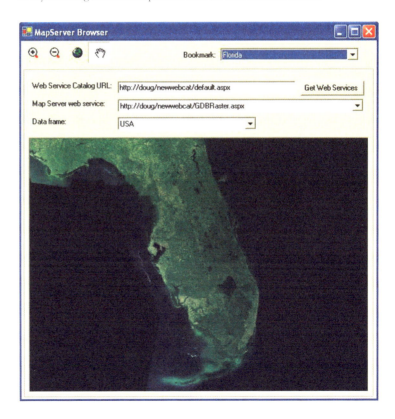

DEPLOYMENT

Because this application requires no ArcGIS software or licensing to run, you can simply build an executable and copy it to any machine that has the .NET framework installed. If a user has access and knows the URL to an ArcGIS Server Web service catalog, the application's functionality can be utilized.

ADDITIONAL RESOURCES

This scenario includes functionality and programming techniques for consuming ArcGIS Server Web services over the Internet. You are encouraged to read Chapter 4 of this book to get a better understanding of ArcGIS Server Web service concepts and programming guidelines.

If you are unfamiliar with developing applications that make use of Web services with .NET, it's also recommended that you refer to your .NET developer documentation to become more familiar with .NET application development. If you want to learn more about Web services in general, visit www.w3.org.

Rather than walk through this scenario, you can get the completed application from the samples installation location. The sample is installed as part of the ArcGIS developer samples.

This walk-through is for developers who need to build and deploy a Java desktop application incorporating mapping functionality using ArcGIS Server Web services directly. It describes the process of building, deploying, and consuming the *ArcGIS_web_service_client* sample, which is part of the ArcGIS developer samples.

You can find this sample in:
<install_location>\DeveloperKit\Samples\Developer_Guide_Scenarios\ArcGIS_Server\ArcGIS_web_service_client

PROJECT DESCRIPTION

The purpose of this scenario is to create a standalone application using JBuilder that uses MapServer Web services published in Web service catalogs to navigate the maps served by those Web services.

The following is how the user will interact with the application:

1. Specify the URL of a Web service catalog and click *Get Web Services*.

2. Click the *MapServer Web service* list and click the desired MapServer.

3. Click the *Data frame* list to choose the data frame of interest from the MapServer.

4. Navigate the map using the *Bookmark* list and the *Zoom In*, *Zoom Out*, *Full Extent*, and *Pan* tools on the toolbar.

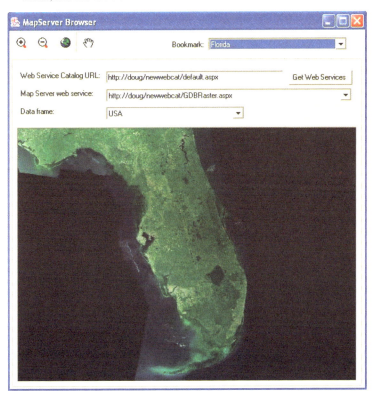

CONCEPTS

A Web service is a set of related application functions that can be programmatically invoked over the Internet. ArcGIS Server Web services can be accessed by any development language that can submit SOAP-based requests to a Web service and process SOAP-based responses. The Web service consumer can get the methods and types exposed by the Web service through its Web Service Description Language. ArcGIS Web Services are based on the SOAP doc/literal format and are interoperable across Java and .NET.

To learn more about WSDL, refer to http://www.w3.org.

On the Web server, the Web service catalog templates can be used by developers to publish any server object as a Web service over HTTP. For any published server object, the template creates an HTTP endpoint (URL) to which SOAP requests can be submitted using standard HTTP POST. The endpoint also supports returning the WSDL for the Web service using a standard HTTP GET with "wsdl" as the query string. The implementation of the HTTP endpoint is thin, while the actual processing of the SOAP request and the generation of the SOAP response take place within the GIS server. The WSDLs that describe the definitions of the SOAP requests and responses are also part of the GIS server and are installed as part of the ArcGIS Server install under <install directory>\XMLSchema.

If you are building an application using ArcGIS Desktop or ArcGIS Engine, you can use the objects in the GISClient object library to access ArcGIS Server Web services. This is an example of using those Web services directly from an application that does not require any ArcObjects components to run.

For more information on using the objects in GISClient, see Chapter 4 and the developer help.

ArcGIS Web services can be consumed from .NET or Java. As a consumer of an ArcGIS Web service, you can use the methods exposed by the Web service by including a reference to the Web service in your .NET or Java project. Web services implemented on a Java Web server can be consumed from a .NET client and vice versa.

DESIGN

If you have already worked with the stateless methods on the MapServer using ArcObjects, you'll notice that there are differences working with those methods on MapServer Web service.

This application makes use of the methods and objects defined by the ArcGIS Server Web service catalog and MapServer Web service. These methods and objects correspond to the set of ArcObjects necessary to call the stateless methods exposed by the MapServer server object.

In order to support this application, you need to have access to a Web service catalog that contains at least one MapServer Web service. Note: The Web service catalog itself can be either .NET or Java. For purposes of this example, the assumption is that the Web service is Java.

The application will connect to a Web service catalog and present to the user a list of the available MapServer Web services in the Web service catalog. The application will list the available data frames in the map and any bookmarks associated with the data frame that the user can pick from. In addition, the user can navigate the map using navigation tools in a toolbar.

REQUIREMENTS

The requirements for working through this scenario are that you have access to an ArcGIS Server Web service catalog that contains at least one MapServer Web service. The machine on which you develop this application does not need to have any ArcGIS software or licensing installed on it.

While the examples given are Java Web service catalogs, if you have access to .NET Web services catalogs, you can use those in your Java application.

For the purposes of this example, assume you have access to a Web service catalog created with Java with the following URL:

```
http://doug/MyWebServiceCatalog/default.jsp
```

The Web service catalog contains a MapServer Web service for a MapServer object called "Map" whose URL is:

```
http://doug/MyWebServceCatalog/Map_Mapserver.jsp
```

It will be easiest to follow this walk-through if you create a Web service catalog with the same name as above (MyWebServiceCatalog) and a MapServer server object and Web service (Map), although this is not necessary. For more information about creating a map server object, see Chapter 3. For information on creating a Web service catalog and exposing your MapServer as a Web service using Java, see Chapter 6.

If you have your own Web service catalog and MapServer Web services with different names, the points at which the difference in names will impact the code for this application will be pointed out.

The IDE used in this example is JBuilder™ 9 Enterprise. This application can be implemented with other Java IDEs.

IMPLEMENTATION

The first step is to create the new project.

To begin, you must create a new JBuilder project and a working directory that will be used for this project.

Create a new project

1. Create a new folder called *webservice_projects*. This is where your Web service project files will be created.

2. Start JBuilder.

3. Click *File*, then click *New Project* to open the Project Wizard window.

4. In the Project Wizard dialog box, type "MapServerBrowser" for the *Name*, then type the path to the webservice_projects you created above for the *Directory*.

5. Click *Finish*.

The Project wizard

Add references to the Web services

For your application to have access to the methods and objects exposed by the Web service catalog and MapServer Web services, you need to add references to a Web service catalog and a reference to a MapServer Web service to your application. A reference enables you to use objects and methods provided by a Web service in your code. After adding a reference to your project, you can use any element or functionality provided by that Web service within your application.

You will now create a reference to the Web services and create the stub classes from the Web service's WSDL.

The Object Gallery dialog box

1. Click *File*, then click *New*.

2. In the Object Gallery dialog box, click the *Web Services* tab.

3. Click *Import A Web Service* and click *OK*.

4. In the Import A Web Service With Axis dialog box, click the Browse button beside *WSDL URL*. In the File or URL text box type "http://padisha/ MyWebServiceCatalog/default.jsp?wsdl", then click *OK*.

5. Click *Finish*.

6. Click *File*, then click *New*.

7. In the Object Gallery dialog box, click the *Web Services* tab.

8. Click *Import A Web Service* and click OK.

9. In the Import A Web Service With Axis dialog box, Click the Browse button beside WSDL URL. In the File or URL text box type "http://padisha/ MyWebServiceCatalog/ Map_MapServer.jsp?wsdl", then click OK.

10. Click *Finish*.

The Import A Web Service With Axis wizard

Rebuild the project (*Project|Rebuild Project "MapServerBrowser.jpx"*). The project should rebuild without any errors. In the project browser you can see the classes that have been added to your project by referencing the Web services. Now that these references have been added to the project, you will start programming your application to make use of the classes and methods provided by these Web service references to consume ArcGIS Server Web services.

The Object Gallery dialog box

The Application wizard

Create a Swing application

1. Click *File*, then click *New*.

2. In the Object Gallery dialog box, click the *General* tab.

3. Click *Application* and click *OK*.

4. In the Application Wizard dialog box, type "mapserverbrowser" for the *Package*, then type "MapServerBrowser" for *Class name*.

5. Click *Finish*.

Two new classes will be created as a result: *MapServerBrowser.java* and *Frame1.java*.

Create a new class *MapPanel* as the subclass of *JPanel*. This is will be used to display the map image returned by the ArcGIS Server.

1. Click *File*, then click *New Class*.

2. In the Class Wizard dialog box, type "mapserverbrowser" for the *Package*, then type "MapPanel" for *Class name* and "javax.swing.JPanel" for *Base Class*.

3. Click *OK*.

This will add a new class to your project and will open the code for the class with some autogenerated code. The autogenerated code will include the name of the package (*mapserverbrowser*) and a stubbed out class definition for *MapPanel*.

The Class wizard

4. Add import statements for the additional Java packages you will use in this project. At the top of the code window, add the following import statements:

```
import java.awt.Image;
import java.awt.Graphics;
```

5. Add the following member variable and functions to the *MapPanel* class.

```
Image mapimage = null;
public void drawMap(Image image) {
   mapimage = image;
   repaint();
}


public void paint(Graphics g) {
   super.paint(g);
   if (mapimage != null) {
     g.drawImage(mapimage, 0, 0, this);
   }
}
```

The code for your *MapPanel* class should now look like the following:

```
package mapserverbrowser;

import javax.swing.JPanel;
import java.awt.Image;
import java.awt.Graphics;

public class MapPanel extends JPanel {
  Image mapimage = null;

   public void drawMap(Image image) {
      mapimage = image;
      repaint();
   }

   public void paint(Graphics g) {
      super.paint(g);
      if (mapimage != null) {
        g.drawImage(mapimage, 0, 0, this);
      }
   }
  public MapPanel() {
  }
}
```

6. Next, copy the following images from *<install_location>\DeveloperKit\Samples\Data\ServerData* to *mapserverbrowser/ src/mapserverbrowser/images*:

- fullext.gif

- pan.gif

- zoomin.gif

- zoomout.gif

7. Click *Project/Refresh*.

These images will be used for icons that you will add to your application later. To use these images as icons, you'll create member variables of type ImageIcons to be used as image icons for the buttons.

8. In the Project pane, double-click *Frame1.java*. This will open the code window for Frame1.

9. Add the following lines of code to your *Frame1* class.

```
private ImageIcon icon1 = new
ImageIcon(this.getClass().getResource("images/zoomin.gif"));
private ImageIcon icon2 = new
ImageIcon(this.getClass().getResource("images/zoomout.gif"));
private ImageIcon icon3 = new
ImageIcon(this.getClass().getResource("images/fullext.gif"));
private ImageIcon icon4 = new
ImageIcon(this.getClass().getResource("images/pan.gif"));
```

10. Rebuild the project (*Project|Rebuild Project "MapServerBrowser.jpx"*).

Add controls to the frame

To accommodate the functionality for this application, you will add a number of user interface controls to the frame. This application utilizes a number of user controls that you need to add to and arrange on the form.

1. Open Frame1.java by double-clicking *Frame1.java* in the Project pane.

2. Switch to the design mode by clicking *View/Switch Viewer to "Design"*.

3. Right-click the properties browser and set the *Property Exposure Level* to *Hidden*.

The JBuilder IDE

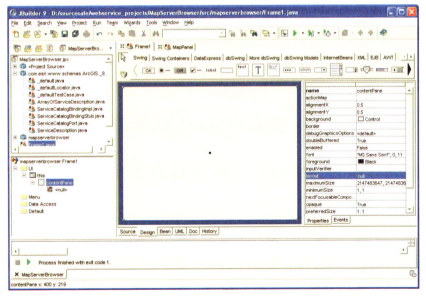

4. In the component tree window, click *this*. Type "Map Server Browser" for the *Title* property and "600,600" for the *Size* property.

5. Click *contentPane* in the component tree window. Click *null* for the *Layout* property.

While this example includes exact control placement and size on the form, you can also arrange these controls interactively by dragging them and sizing them with the mouse.

The next set of controls you'll add is the toolbar and its buttons for navigating the map.

1. Click the *Swing Containers* tab in the Component palette and click *JToolbar*. Drag a new toolbar on the content pane.

2. In the toolbar's properties, type "5,5,160,27" for the *Bounds* property.

3. Click the *Swing* tab in the Component palette and click *JButton*. Drag a new button onto the toolbar you added to the Content pane.

4. In the button's properties, type "zoominButton" for the *Name* property, "9,1,37,25" for the Bounds property, and "1,1,1,1" for the *Margin* property and click *icon1* for the *Icon* property.

5. Click *JButton* in the Component palette. Drag a new button onto the toolbar you added to the Content pane.

6. In the button's properties, type "zoomoutButton" for the *Name* property, "46,1,37,25" for the *Bounds* property, and "1,1,1,1" for the *Margin* property and click *icon2* for the *Icon* property.

7. Click *JButton* in the Component palette. Drag a new button onto the toolbar you added to the Content pane.

8. In the button's properties, type "fullextButton" for the *Name* property, "87,1,37,25" for the Bounds property, and "1,1,1,1" for the *Margin* property and click *icon3* for the *Icon* property.

9. Click *JToggleButton* in the Component palette. Drag a new toggle button onto the toolbar you added to the Content pane.

10. In the toggle button's properties, type "panButton" for the *Name* property and "1,1,1,1" for the *Margin* property and click *icon4* for the *Icon* property.

The next set of controls you'll add includes the controls to list the bookmarks in the MapServer Web service's data frame selected by the user.

1. Click *JLabel* in the Component palette. Drag a new label onto the Content pane.

2. In the label's properties, type "255,8,65,25" for the *Bounds* property and "Bookmark:" for the *Text* property.

3. Click *JComboBox* in the Component palette. Drag a new combo box onto the Content pane.

4. In the combo box's properties, type "bookmarkComboBox" for the *Name* property and type "320,8,250,25" for the *Bounds* property.

The next set of controls you'll add is to handle the user input for the URL of the Web service catalog that the user of the application wants to connect to.

1. Click *JLabel* in the Component palette. Drag a new label onto the Content pane.

2. In the label's properties, type "10,50,162,24" for the *Bounds* property and "Web Service Catalog URL:" for the *Text* property.

3. Click *JTextField* in the Component palette. Drag a new text field onto the Content pane.

4. In the text field's properties, type "urlTextField" for the *Name* property and type "170,50,267,25" for the *Bounds* property.

5. Click *JButton* in the Component palette. Drag a new button onto the toolbar you added to the Content pane.

6. In the button's properties, type "getWebServices" for the *Name* property, "445, 50,130,26" for the *Bounds* property, and "Get Web Services" for the *Text* property.

The next set of controls you'll add includes the controls to list the MapServer Web services that are in the Web service catalog specified by the user.

1. Click *JLabel* in the Component palette. Drag a new label onto the content pane.

2. In the label's properties, type "10,85,149,20" for the *Bounds* property and "Map Server Web Service:" for the *Text* property.

3. Click *JComboBox* in the Component palette. Drag a new combo box onto the Content pane.

4. In the combo box's properties, type "mapserverComboBox" for the *Name* property and type "170,85,405,25" for the *Bounds* property.

The next set of controls you'll add includes the controls to list the data frames in the MapServer Web service selected by the user.

1. Click *JLabel* in the Component palette. Drag a new label onto the content pane.

2. In the label's properties, type "10,120,140,20" for the *Bounds* property and "Data Frame:" for the *Text* property.

3. Click *JComboBox* in the Component palette. Drag a new combo box onto the Content pane.

4. In the combo box's properties, type "dfComboBox" for the *Name* property and "170,120,269,25" for the *Bounds* property.

The next set of controls you'll add is the MapPanel that will display the map images returned by the MapServer Web service.

1. Click the *Bean Chooser* (left-most icon on the component pallette), then click *Select*.

2. In the Bean Chooser dialog box, type "mapserverbrowser.MapPanel" for the *Class name*.

3. Click *OK*.

The Bean Chooser dialog box

4. Add the map panel by drawing a box on the Content pane.

5. In the properties for the map panel, type "mapPanel" for the *Name* property and "10,160,572,418" for the *Bounds* property. Click *LoweredBevel* for the *Border* property.

Now you have added all the controls necessary for the application. Your application should look like the following in design mode:

The JBuilder IDE

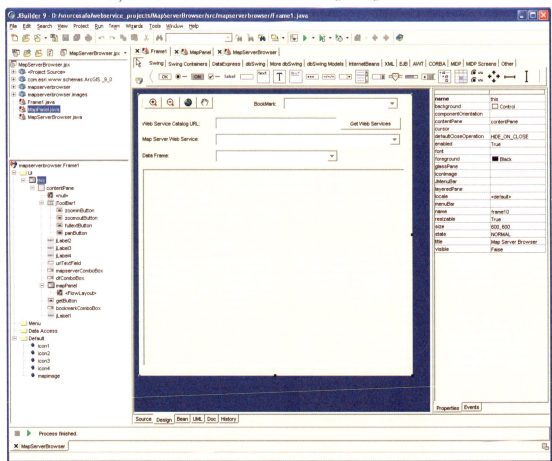

Now you'll add the code to the events for the controls in the application.

Add import statements and member variables to the application

This application requires a number of import statements and private member variables. Each variable will not be explained here, but as you add code for the various events, you will use these variables.

Switch to the source code view of *Frame1.java* by clicking the *Source* tab at the bottom of the Content pane. You will see that there are already a number of private member variables that were added by JBuilder for the controls you added to the application.

1. Add the following import statements:

```
import java.net.*;
import java.awt.Color;
import com.esri.www.schemas.ArcGIS._9_0.*;
import com.esri.www.schemas.ArcGIS._9_0.Point;
```

2. Add the following member variables to your *Frame1* class:

```
private Image mapimage = null;
private int startX;
private int startY;
private MapDescription mapDescription = null;
private ImageDisplay idisp = null;
```

Next, you'll need to add an *init* method to your *Frame1* class that will hold the event listeners for all the controls that you added to your application. Add the following method to your *Frame1* class:

```
private void init() {
}
```

Add code to get the list of Web services in the catalog

Upon using this application, the user will type the URL of a Web service catalog, then click the *Get Web Services* button. You will add code to add an action listener to a button that calls a function to get the MapServer Web services in the Web service catalog and add them to the Web service combo box (*mapserverComboBox*).

1. Add the following lines of code inside the *init()* method.

```
// add event listener to the "Get Web Services" button
getWebServices.addActionListener(new ActionListener() {
  public void actionPerformed(ActionEvent e) {
    getMapServer();
  }
});
```

Now create the *getMapserver()* function to get all the MapServer Web services and add their URLs to the map server combo box (*mapserverComboBox*).

2. Add the following method to your *Frame1* class.

```
private void getMapServer() {
}
```

You will add code to the function *getMapServer()* to connect to the Web service catalog and get the list of MapServer Web services to add to the Web service combo box.

First, make sure the *urlTextField* is not empty. Since the operation may take a few seconds to execute, you'll want to indicate to the user that the application is busy. To do this, you will set the cursor to a wait cursor. You'll also add code to clear the contents of the Web services combo box. Add the code within a try/catch block. If an error occurs in the code in the try block, the code in the catch block, then the code in the finally block will be executed. The catch block will print the stack trace of the error, and the finally block will change the cursor back to a normal cursor.

3. Add the following lines of code to the *getMapServer* function:

```
try {
  // make sure input text field is not empty
  if (urlTextField.getText().length() == 0) {
    return;
  }
  this.setCursor(new Cursor(Cursor.WAIT_CURSOR));
  // clear old items in the MapServer JComboBox
  mapserverComboBox.removeAllItems();
}
catch (Exception ex) {
  ex.printStackTrace();
}
finally {
  this.setCursor(new Cursor(Cursor.DEFAULT_CURSOR));
}
```

Next, you will use the stub classes created by Apache Axis from the WSDL to create a Web service catalog. Use the URL specified by the user in the *urlTextField* to discover the services in the catalog

4. Add the following lines of code to your try block:

```
 // get MapServers
URL url = new URL(urlTextField.getText());
ServiceCatalogBindingStub servicecatalog = (ServiceCatalogBindingStub)new
        _defaultLocator().getServiceCatalogPort(url);
```

Next you'll call the *getServiceDescriptions* method on the Web service catalog to get an array of *serviceDescription* objects. You'll loop through this array and get the URL for the MapServer Web services and add their URLs to the Web service combo box (*mapserverComboBox*).

5. Add the following lines of code to your try block:

```
ArrayOfServiceDescription sdArray =
servicecatalog.getServiceDescriptions();
ServiceDescription sd[] = sdArray.getServiceDescription();
// add MapServers to the JComboBox
for (int i = 0; i < sd.length; i++) {
  if (sd[i].getType().equals("MapServer")) {
    mapserverComboBox.addItem(sd[i].getUrl());
  }
}
```

The code for the *init* method and the completed code for the *getMapServer* method should look like the following:

```java
private void init() {
  // add event listener to the "Get Web Services" button
  getWebServices.addActionListener(new ActionListener() {
    public void actionPerformed(ActionEvent e) {
      getMapServer();
    }
  });
}
private void getMapServer() {
    try {
     // make sure input text field is not empty
     if (urlTextField.getText().length() == 0) {
       return;
      }
     this.setCursor(new Cursor(Cursor.WAIT_CURSOR));
     // clear old items in the MapServer JComboBox
     mapserverComboBox.removeAllItems();

     // get MapServers
     URL url = new URL(urlTextField.getText());
     ServiceCatalogBindingStub servicecatalog =
(ServiceCatalogBindingStub)new
       _defaultLocator().getServiceCatalogPort(url);
     ArrayOfServiceDescription sdArray =
servicecatalog.getServiceDescriptions();
     ServiceDescription sd[] = sdArray.getServiceDescription();

     // add MapServers to the JComboBox
     for (int i = 0; i < sd.length; i++) {
      if (sd[i].getType().equals("MapServer")) {
        mapserverComboBox.addItem(sd[i].getUrl());
        }
      }
    }
    catch (Exception ex) {
     ex.printStackTrace();
    }
    finally {
     this.setCursor(new Cursor(Cursor.DEFAULT_CURSOR));
    }
}
```

Add code to get the data frames from the MapServer

Once connected to the Web service catalog, the user will pick a MapServer Web service to work with. You will add code to add an event listener to the Web service combo box (*mapserverComboBox*) that calls a function to get the list of data frames in the MapServer and add them to the data frames combo box (*dfComboBox*).

1. Add the following code inside the *init()* method.

```
// add event listener to the mapserver JComboBox
mapserverComboBox.addActionListener(new ActionListener() {
  public void actionPerformed(ActionEvent e) {
    getDataFrame( ( (JComboBox) e.getSource()).getSelectedItem());
  }
});
```

Next, create the *getDataFrame()* method to get all the data frames of the selected MapServer.

2. Add the following method to your *Frame1* class.

```
private void getDataFrame(Object obj) {
}
```

The rest of the code will be in a try/catch block. Next you need to add code to create an instance of the selected map server Web service.

3. Add the following lines of code to your try block:

```
try {
  URL url = new URL(obj.toString());
  MapServerBindingStub mapserver = (MapServerBindingStub)new
MapLocator().getMapServerPort(url);
}
catch (Exception ex) {
  ex.printStackTrace();
}
```

If the MapServer Web service that you created your Web reference from was not called Map, then the name of the object will not be MapLocator, but will be the name of your MapServer<Locator>. For example, if you referenced a MapServer Web service called "World", then the object's name would be WorldLocator.

Next, you'll add code to loop through all the data frames in the Web service and add them to the data frame combo box (*dfComboBox*). Remove items from the list before populating it.

4. Add the following code to your try block.

```
dfComboBox.removeAllItems();
for (int i = 0; i < mapserver.getMapCount(); i++) {
  dfComboBox.addItem(mapserver.getMapName(i));
}
```

The code for the *init* method and the completed code for the *getDataFrame* method should look like the following:

```
private void init() {
  // add event listener to the "Get Web Services" button
  getWebServices.addActionListener(new ActionListener() {
  public void actionPerformed(ActionEvent e) {
    getMapServer();
    }
  });

  // add event listener to the mapserver JComboBox
  mapserverComboBox.addActionListener(new ActionListener() {
  public void actionPerformed(ActionEvent e) {
    getDataFrame( ( (JComboBox) e.getSource()).getSelectedItem());
    }
  });
}
private void getDataFrame(Object obj) {
  try {
    URL url = new URL(obj.toString());
    MapServerBindingStub mapserver = (MapServerBindingStub)new
MapLocator().getMapServerPort(url);
    dfComboBox.removeAllItems();
    for (int i = 0; i < mapserver.getMapCount(); i++) {
     dfComboBox.addItem(mapserver.getMapName(i));
    }
  }
  catch (Exception ex) {
    ex.printStackTrace();
  }
}
```

Add code to get the bookmarks for the selected data frame

You will add code to add an action listener to the combo box that calls a function to get the list of spatial bookmarks for the selected data frame and add them to the bookmarks combo box (*bookmarksComboBox*).

1. Add the following line of code to the *init()* method:

```
// add event listener to the bookmark JComboBox
dfComboBox.addActionListener(new ActionListener() {
public void actionPerformed(ActionEvent e) {
  getBookmark( ( (JComboBox) e.getSource()).getSelectedItem());
  }
});
```

Next create the *getBookmark()* function to get all the spatial bookmarks and add them to the bookmarks combo box (*bookmarksComboBox*).

2. Add the following method to your *Frame1* class.

```
private void getBookmark(Object obj) {
}
```

The rest of the code will be in a try/catch block. Next you need to add code to create an instance of the selected map server Web service.

3. Add the following lines of code to your try block:

```
try{
  URL url = new URL(mapserverComboBox.getSelectedItem().toString());
    MapServerBindingStub mapserver = (MapServerBindingStub)new
MapLocator().
    getMapServerPort(url);
}
 catch (Exception ex) {
    ex.printStackTrace();
}
```

If the MapServer Web service that you created your Web reference from was not called Map, then the name of the object will not be MapLocator but will be the name of your MapServer<Locator>. For example, if you referenced a MapServer Web service called "World", then the object's name would be WorldLocator.

In order to get a list of the spatial bookmarks in the map, you'll add code to get a reference to the MapServer's *MapServerInfo* object for the selected data frame, then get an array of bookmarks from *MapServerInfo*. Once you have the array of bookmarks, you'll loop through the array and add the names of the bookmarks to the bookmarks combo box (*bookmarksComboBox*). Notice the code also adds a "<Default Extent>" item to the combo box. This allows the user to return to the default extent of the data frame.

4. Add the following lines of code to your try block:

```
String name = mapserver.getMapName(dfComboBox.getSelectedIndex());
MapServerInfo mapInfo = mapserver.getServerInfo(name);
ArrayOfMapServerBookmark bookmarkArray = mapInfo.getBookmarks();
MapServerBookmark[] bookmarks = bookmarkArray.getMapServerBookmark();
bookmarkComboBox.removeAllItems();
bookmarkComboBox.addItem("<Default Extent>");
if (bookmarks != null) {
  for (int i = 0; i < bookmarks.length; i++) {
    bookmarkComboBox.addItem(bookmarks[i].getName());
  }
}
 bookmarkComboBox.setSelectedItem("<Default Extent>");
```

The code for the *init* method and the completed code for the *getBookmark* method should look like the following:

```
private void init() {
  // add event listener to the "Get Web Services" button
  getWebServices.addActionListener(new ActionListener() {
  public void actionPerformed(ActionEvent e) {
    getMapServer();
    }
  });
  // add event listener to the mapserver JComboBox
  mapserverComboBox.addActionListener(new ActionListener() {
  public void actionPerformed(ActionEvent e) {
    getDataFrame( ( (JComboBox) e.getSource()).getSelectedItem());
```

```
    }
  });

  // add event listener to the bookmark JComboBox
  dfComboBox.addActionListener(new ActionListener() {
  public void actionPerformed(ActionEvent e) {
    getBookmark( ( (JComboBox) e.getSource()).getSelectedItem());
    }
  });

}
private void getBookmark(Object obj) {
  try {

    URL url = new URL(mapserverComboBox.getSelectedItem().toString());
    MapServerBindingStub mapserver = (MapServerBindingStub)new
MapLocator().
        getMapServerPort(url);

    String name = mapserver.getMapName(dfComboBox.getSelectedIndex());
    MapServerInfo mapInfo = mapserver.getServerInfo(name);
    ArrayOfMapServerBookmark bookmarkArray = mapInfo.getBookmarks();
    MapServerBookmark[] bookmarks = bookmarkArray.getMapServerBookmark();
    bookmarkComboBox.removeAllItems();
    bookmarkComboBox.addItem("<Default Extent>");
    for (int i = 0; i < bookmarks.length; i++) {
      bookmarkComboBox.addItem(bookmarks[i].getName());
      }
      bookmarkComboBox.setSelectedItem("<Default Extent>");
  }
  catch (Exception ex) {
    ex.printStackTrace();
  }
}
```

Add helper function to draw the map

To this point the code you have added to the application has used methods on the Web service catalog and MapServer Web services to get information about a particular Web service catalog or MapServer Web service and populate controls on the application.

The next set of controls whose events you'll add code to will actually draw the map and display the result in the picture box. In order to draw the map, you'll add a helper function that these events will call to do the map drawing.

1. Add the following function to your *Frame1* class:

```
private void drawMap(MapDescription mapDesc) {
  try {
  }
  catch (Exception ex) {
    ex.printStackTrace();
```

```
        }
    }
```

As you can see, the drawMap function takes a single argument of type *MapDescription*. The map description object is used by a MapServer when drawing maps to allow, at draw time, various aspects of the map to be changed, without changing the running MapServer object. These properties include the extent to draw, the layers to turn on or off, and so on. In this respect, the map description serves the purpose of allowing stateless use of a MapServer, while allowing these aspects of state to be saved in the application by saving and modifying a local copy of the map description.

This function doesn't do anything to the map description except use it to draw the map. You'll see later how the code that calls the *drawMap* function modifies and uses the map description to allow the user to navigate around the map.

The first code you'll add to this function is to create an instance of the selected map server Web service.

2. Add the following lines of code to the try block of the function:

```
URL url = new URL(mapserverComboBox.getSelectedItem().toString());
MapServerBindingStub mapserver = (MapServerBindingStub) new
MapLocator().getMapServerPort(url);
```

3. Next, add the following lines of code to create an image description object that you'll use when drawing the map.

```
ImageType it = new ImageType();
it.setImageFormat(EsriImageFormat.esriImageJPG);
        it.setImageReturnType(EsriImageReturnType.esriImageReturnMimeData);
ImageDisplay idisp = new ImageDisplay();
idisp.setImageHeight(400);
idisp.setImageWidth(552);
idisp.setImageDPI(150);
ImageDescription iDesc = new ImageDescription();
iDesc.setImageDisplay(idisp);
iDesc.setImageType(it);
```

Finally, add the following lines of code to call the *exportMapImage* method on the MapServer to draw the map. Get the bytes from the resulting image and create an *Image* to be drawn on the *mapPanel*.

```
MapImage mi = mapserver.exportMapImage(mapDesc, iDesc);
byte[] data = mi.getImageData();
mapimage = java.awt.Toolkit.getDefaultToolkit().createImage(data);

//wait for the image to load
MediaTracker tracker = new MediaTracker(this);
tracker.addImage(mapimage,1);
tracker.waitForID(1);
```

By definition, Web services are stateless. For more information about programming ArcGIS Server and how it relates to managing state in an application that makes stateless use of server objects, see Chapter 4.

The size of the ImageDisplay is the same as the size of the picture box control. This will produce an image that fits exactly in the picture box on the form.

```
  // draw the image
  mapPanel.drawMap(mapimage);
```

The completed code for the *drawMap* function should look like the following:

```java
private void drawMap(MapDescription mapDesc) {
  try {
    URL url = new URL(mapserverComboBox.getSelectedItem().toString());
    MapServerBindingStub mapserver = (MapServerBindingStub) new
MapLocator().getMapServerPort(url);
    ImageType it = new ImageType();
    it.setImageFormat(EsriImageFormat.esriImageJPG);
    it.setImageReturnType(EsriImageReturnType.esriImageReturnMimeData);

    idisp = new ImageDisplay();
    idisp.setImageHeight(400);
    idisp.setImageWidth(552);
    idisp.setImageDPI(150);

    ImageDescription iDesc = new ImageDescription();
    iDesc.setImageDisplay(idisp);
    iDesc.setImageType(it);

    MapImage mi = mapserver.exportMapImage(mapDesc, iDesc);
    byte[] data = mi.getImageData();
    mapimage = java.awt.Toolkit.getDefaultToolkit().createImage(data);

    // wait for the image to load
    MediaTracker tracker = new MediaTracker(this);
    tracker.addImage(mapimage,1);
    tracker.waitForID(1);

    // draw the image
    mapPanel.drawMap(mapimage);
  }
  catch (Exception ex) {
    ex.printStackTrace();
  }
}
```

Add code to draw the extent of the selected bookmark

When the user clicks the Bookmarks combo box and picks a bookmark, the panel will display the map for the extent of the selected bookmark. To do this you'll add code to the selected index changed event of the bookmarks combo box (*cboBookMark*).

1. Add the following lines of code to the *init()* method.

```java
// add event listener to the bookmark JComboBox
bookmarkComboBox.addActionListener(new ActionListener() {
public void actionPerformed(ActionEvent e) {
```

```
  bookmarkSelected(((JComboBox)e.getSource()).getSelectedItem());
 }
});
```

Next, create the *bookmarkSelected()* method to update the map with the corresponding extent.

2. Add the following method to your *Frame1* class.

```
private void bookmarkSelected(Object obj) {
  try {
  }
  catch (Exception ex) {
   ex.printStackTrace();
  }
}
```

The rest of the code will be in a try/catch block. Next you need to add code to create an instance of the selected map server Web service. Then you'll get a *MapServerInfo* object for the selected data frame.

If the MapServer Web service that you created your Web reference from was not called Map, then the name of the object will not be MapLocator but will be the name of your MapServer<Locator>. For example, if you referenced a MapServer Web service called "World", then the object's name would be WorldLocator.

3. Add the following lines of code to your try block:

```
URL url = new URL(mapserverComboBox.getSelectedItem().toString());
MapServerBindingStub mapserver = (MapServerBindingStub)new
MapLocator().getMapServerPort(url);
String name = mapserver.getMapName(dfComboBox.getSelectedIndex());
MapServerInfo mapInfo = mapserver.getServerInfo(name);
```

Next, add code to create a *MapDescription* variable. Add a condition statement to check if the user picked a bookmark or if it is a default extent and execute the corresponding code. If the text value of the bookmarks combo box is "<Default Extent>", then set the map description to be the default map description for the data frame and draw the map. If the user picked a bookmark, find the bookmark that corresponds to the bookmark picked by the user, get its extent, set the extent into the map description, then draw the map.

4. Add the following lines of code to the try block:

```
MapDescription mapDesc = null;
if (dfComboBox.getSelectedItem().toString().equals("<Default Extent>")) {
  mapDesc = mapInfo.getDefaultMapDescription();
} else {
  mapDesc = mapDescription;
  ArrayOfMapServerBookmark bookmarkArray = mapInfo.getBookmarks();
  MapServerBookmark[] bookmarks = bookmarkArray.getMapServerBookmark();
  if (bookmarks != null) {
    for (int i = 0; i < bookmarks.length; i++) {
       if (
bookmarks[i].getName().equals(dfComboBox.getSelectedItem().toString()) ) {
         mapDesc.setMapArea(bookmarks[i]);
       }
    }
  }
}
```

By definition, Web services are stateless. For more information about programming ArcGIS Server and how it relates to managing state in an application that makes stateless use of server objects, see Chapter 4.

```
  mapDescription = mapDesc;
  drawMap(mapDescription);
```

A copy of the map description is being stored in a member variable, because as you implement other commands, such as the Zoom and Pan commands, you'll want to keep track of the user's current extent to know what to zoom or pan from. These commands will modify the extent of the local copy of the map description, then use it as input to the *drawMap* function.

The code for the *init* method and the completed code for the *bookmarkSelected()* method should look like the following:

```
private void init() {
  // add event listener to the "Get Web Services" button
  getWebServices.addActionListener(new ActionListener() {
  public void actionPerformed(ActionEvent e) {
    getMapServer();
    }
  });
  // add event listener to the mapserver JComboBox
  mapserverComboBox.addActionListener(new ActionListener() {
  public void actionPerformed(ActionEvent e) {
    getDataFrame( ( (JComboBox) e.getSource()).getSelectedItem());
    }
  });

  // add event listener to the bookmark JComboBox
  dfComboBox.addActionListener(new ActionListener() {
  public void actionPerformed(ActionEvent e) {
    getBookmark( ( (JComboBox) e.getSource()).getSelectedItem());
    }
  });
  // add event listener to the bookmark JComboBox
  bookmarkComboBox.addActionListener(new ActionListener() {
  public void actionPerformed(ActionEvent e) {

    bookmarkSelected(((JComboBox)e.getSource()).getSelectedItem());
   }
  });
  }
private void bookmarkSelected(Object obj) {
  try {
   URL url = new URL(mapserverComboBox.getSelectedItem().toString());
   MapServerBindingStub mapserver = (MapServerBindingStub)new
MapLocator().getMapServerPort(url);

   String name = mapserver.getMapName(dfComboBox.getSelectedIndex());
   MapServerInfo mapInfo = mapserver.getServerInfo(name);
   MapDescription mapDesc = null;
   if (dfComboBox.getSelectedItem().toString().equals("<Default
Extent>")) {
    mapDesc = mapInfo.getDefaultMapDescription();
```

```
      } else {
        ArrayOfMapServerBookmark bookmarkArray = mapInfo.getBookmarks();
        MapServerBookmark[] bookmarks =
bookmarkArray.getMapServerBookmark();
        for (int i = 0; i < bookmarks.length; i++) {
          if (
bookmarks[i].getName().equals(dfComboBox.getSelectedItem().toString()) ) {
            mapDesc.setMapArea(bookmarks[i]);
          }
        }
      }
      mapDescription = mapDesc;
      drawMap(mapDescription);
    }
    catch (Exception ex) {
      ex.printStackTrace();
    }
  }
```

Add code for the Zoom, Full Extent buttons

Some of the map navigation functionality in this application includes fixed zoom
in, fixed zoom out, and zoom to full extent commands. You added these commands as buttons to the toolbar on the frame. You'll add action listeners to these
buttons to execute the code when these buttons are clicked.

1. Add the following lines of code to the *init()* method.

```
// add event listener the "Zoom In" button
zoominButton.addActionListener(new ActionListener() {
public void actionPerformed(ActionEvent e) {
  zoomin();
  }
});


// add event listener to the "Zoom Out" button
zoomoutButton.addActionListener(new ActionListener() {
public void actionPerformed(ActionEvent e) {
  zoomout();
  }
});


// add event listener to the "Full Ext" button
fullextButton.addActionListener(new ActionListener() {
public void actionPerformed(ActionEvent e) {
  fullext();
  }
```

```
});
```

Now you will add code to the functions *zoomin()*, *zoomout()* and *fullext()*.

Add code to create the *zoomin()* function to get the envelope from the *mapDescription* member variable, shrink its extent, draw the map, then update the variable *mapDescription* with the new extent.

2. Add the following method to your *Frame1* class.

```
private void zoomin() {
  try {
    EnvelopeN env = (EnvelopeN) mapDescription.getMapArea().getExtent();
    double w = Math.abs(env.getXMax().doubleValue() -
env.getXMin().doubleValue());
    double h = Math.abs(env.getYMax().doubleValue() -
env.getYMin().doubleValue());
    double xFactor = (w -(w * 0.75))/2;
    double yFactor = (h -(h * 0.75))/2;
    env.setXMax(new Double(env.getXMax().doubleValue() - xFactor));
    env.setXMin(new Double(env.getXMin().doubleValue() + xFactor));
    env.setYMax(new Double(env.getYMax().doubleValue() - yFactor));
    env.setYMin(new Double(env.getYMin().doubleValue() + yFactor));

    MapExtent mapExt = new MapExtent();
    mapExt.setExtent(env);
    mapDescription.setMapArea(mapExt);

    drawMap(mapDescription);
  }
  catch (Exception ex) {
    ex.printStackTrace();
  }
}
```

Add code to create the *zoomout()* function to get the envelope from the *mapDescription* member variable, expand its extent, draw the map, then update the variable *mapDescription* with the new extent.

3. Add the following method to your *Frame1* class.

```
private void zoomout() {
  try {
    EnvelopeN env = (EnvelopeN) mapDescription.getMapArea().getExtent();
    double w = Math.abs(env.getXMax().doubleValue() -
env.getXMin().doubleValue());
    double h = Math.abs(env.getYMax().doubleValue() -
env.getYMin().doubleValue());
    double xFactor = ((w * 1.25) - w)/2;
    double yFactor = ((h * 1.25) - h)/2;
    env.setXMax(new Double(env.getXMax().doubleValue() + xFactor));
    env.setXMin(new Double(env.getXMin().doubleValue() - xFactor));
    env.setYMax(new Double(env.getYMax().doubleValue() + yFactor));
    env.setYMin(new Double(env.getYMin().doubleValue() - yFactor));
```

```
    MapExtent mapExt = new MapExtent();
    mapExt.setExtent(env);
    mapDescription.setMapArea(mapExt);

    drawMap(mapDescription);
  }
  catch (Exception ex) {
    ex.printStackTrace();
  }
}
```

Add code to create the *fullext()* function to get the full extent from the map server's *MapServerInfo* object and set it into the map description.

4. Add the following method to your *Frame1* class.

If the MapServerWeb service that you created your Web reference from was not called Map, then the name of the object will not be MapLocator, but will be the name of your MapServer<Locator>. For example, if you referenced a MapServerWeb service called "World", then the object's name would be WorldLocator.

```
private void fullext() {
  try {
    URL url = new URL(mapserverComboBox.getSelectedItem().toString());
    MapServerBindingStub mapserver = (MapServerBindingStub)new
MapLocator().getMapServerPort(url);

    String name = mapserver.getMapName(dfComboBox.getSelectedIndex());
    MapServerInfo mapInfo = mapserver.getServerInfo(name);
    EnvelopeN env = (EnvelopeN) mapInfo.getFullExtent();
    MapExtent mapExt = new MapExtent();
    mapExt.setExtent(env);
    mapDescription.setMapArea(mapExt);

    drawMap(mapDescription);
  }
  catch (Exception ex) {
    ex.printStackTrace();
  }
}
```

Add code for the Pan button

The final piece of functionality you'll add to the application is the ability for the user to interactively pan the map. To do this, you'll add code to a number of events on the *mapPanel* control, specifically, the mouse pressed, mouse released, mouse dragged events. Before adding code to those events, you'll add an actionListener to the *panButton*. Add code to change the cursor for the *mapPanel* when the Pan button is pushed in.

1. Add the following lines of code to the *init()* method.

```
// add event listener to the Pan button
panButton.addActionListener(new ActionListener() {
  public void actionPerformed(ActionEvent e) {
    if ( ((JToggleButton) (e.getSource())).isSelected() ) {
      mapPanel.setCursor(new Cursor(Cursor.HAND_CURSOR));
    } else {
      mapPanel.setCursor(new Cursor(Cursor.DEFAULT_CURSOR));
    }
```

```
  }
});
```

Next you will add code to add a mouse listener to the *mapPanel*. You will also add code to implement the *mousePressed()* and the *mouseReleased()* methods.

2. Add the following lines of code to the *init()* method.

```
mapPanel.addMouseListener(new MouseAdapter() {
  public void mousePressed(MouseEvent e) {
  }

  public void mouseReleased(MouseEvent e) {
  }
});
```

Next, you will add code to implement the *mousePressed()* event for the *mapPanel*. This code will check to see if the Pan button is pushed and, if so, save the screen coordinates in two of the member variables you added to the frame.

3. Add the following method to your *Frame1* class.

```
public void mousePressed(MouseEvent e) {
  if (panButton.isSelected()) {
    startX = e.getX();
    startY = e.getY();
  }
}
```

If the MapServer Web service that you created your Web reference from was not called Map, then the name of the object will not be MapLocator, but will be the name of your MapServer<Locator>. For example, if you referenced a MapServer Web service called "World", then the object's name would be WorldLocator.

Next, you will add code to implement the *mouseReleased()* event for the mapPanel. This code will check to see if the Pan button is pushed and, if it is, create an instance of the *MapLocator* object and use the map description and image display variables (*mapDescription* and *idisp*) with the *toMapPoint* method on the MapServer to convert the screen coordinates of the point clicked to map coordinates. Record the point on the map that is clicked in both screen and map coordinates.

4. Add the following method to your *Frame1* class:

```
public void mouseReleased(MouseEvent e) {
  if (! panButton.isSelected())
    return;

  try {
    // get selected map server
    URL url = new URL(mapserverComboBox.getSelectedItem().toString());
    MapServerBindingStub mapserver = (MapServerBindingStub)new
MapLocator().getMapServerPort(url);

    // calculate mouse move distance in map unit
    ArrayOfInt arraystartX = new ArrayOfInt();
    arraystartX.set_int(new int[] {startX});
    ArrayOfInt arraystartY = new ArrayOfInt();
    arraystartY.set_int(new int[] {startY});
    ArrayOfInt arrayendX = new ArrayOfInt();
```

```
    arrayendX.set_int(new int[] {e.getX()});
    ArrayOfInt arrayendY = new ArrayOfInt();
    arrayendY.set_int(new int[] {e.getY()});

    MultipointN mpStart = (MultiPointN)
mapserver.toMapPoints(mapDescription,idisp,arraystartX,arraystartY);
    MultipointN mpEnd =
(MultiPointN)mapserver.toMapPoints(mapDescription,idisp,arrayendX,arrayendY);
    PointN pStart = ((PointN) mpStart).getPointArray().getPoint(0);
    PointN pEnd = ((PointN) mpEnd).getPointArray().getPoint(0);
    double deltaX = pEnd.getX() - pStart.getX();
    double deltaY = pEnd.getY() - pStart.getY();

    //change map extent and redraw
    EnvelopeN env = (EnvelopeN) mapDescription.getMapArea().getExtent();

    env.setXMax(new Double(env.getXMax().doubleValue() - deltaX));
    env.setXMin(new Double(env.getXMin().doubleValue() - deltaX));
    env.setYMax(new Double(env.getYMax().doubleValue() - deltaY));
    env.setYMin(new Double(env.getYMin().doubleValue() - deltaY));

    MapExtent mapExt = new MapExtent();
    mapExt.setExtent(env);
    mapDescription.setMapArea(mapExt);

    drawMap(mapDescription);
  }
  catch (Exception ex) {
    ex.printStackTrace();
  }
}
```

Next you will add code to add a mouse motion listener to the *mapPanel*. You will also add code to implement the *mouseDragged()* method.

5. Add the following lines of code to the *init()* method.

```
// add mouse motion listener to the MapPanel
mapPanel.addMouseMotionListener(new MouseMotionAdapter() {
  public void mouseDragged(MouseEvent e) {
  }
});
```

Next you will add code to implement the *mouseDragged()* event. In the *mouseDragged* event, you'll add code similar to the code you added in *mousePressed* to verify the Pan button is pushed. The code that executes when these conditions are met calculates the amount the mouse has moved from the original point that was clicked on the picture box. It then forces a redraw of the image.

6. Add the following method to your *Frame1* class:

```
public void mouseDragged(MouseEvent e) {
  if (panButton.isSelected()) {
    Graphics g = mapPanel.getGraphics();
    int w = (int) mapPanel.getSize().getWidth();
    int h = (int) mapPanel.getSize().getHeight();
    g.clearRect(0, 0, w, h);
    g.drawImage(mapimage, e.getX() - startX, e.getY() - startY, mapPanel);
  }
}
```

Add init() method to the Frame constructor

Add the *init()* method you created to the constructor of the frame. This will enable all the action listeners associated with the controls to be initialized.

Add the following line of code to the *Frame1* constructor.

```
init();
```

The complete code for the *Frame1* constructor should look like the following:

```
public Frame1() {
  enableEvents(AWTEvent.WINDOW_EVENT_MASK);
  try {
    jbInit();
    init();
  }
  catch (Exception e) {
    e.printStackTrace();
  }
}
```

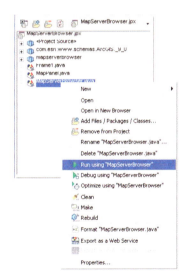

Your application is now ready to be tested. Compile the project (*Project/Rebuild project "MapServerBrowser_project.jpx"*).

Test the application

Run the application from within JBuilder.

1. Right-click on *MapServerBrowser.java* from the project explorer.

2. Click *Run using "MapServerBrowser"*; it will open the frame.

3. Type the URL of a Web service catalog.

4. Choose one of the MapServer Web services.

5. Click the Bookmark combo box and choose a bookmark to zoom to.

6. Use the Zoom In, Zoom Out, Full Extent, and Pan buttons to navigate around the map.

7. Try choosing different MapServer Web services and data frames.

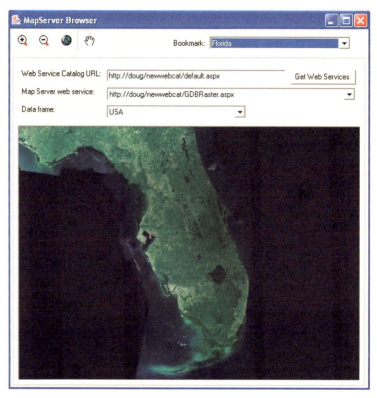

DEPLOYMENT

Because this application requires no ArcGIS software or licensing to run, you can simply build an executable and copy it to any machine that has the Java runtime installed. If users have access and know the URL to an ArcGIS Server Web service catalog, they can make use of the application's functionality.

You can use this application to connect to any ArcGIS Server Web service catalog, whether that Web service catalog was written in .NET or Java.

ADDITIONAL RESOURCES

This scenario includes functionality and programming techniques for consuming ArcGIS Server Web services over the Internet. You are encouraged to read Chapter 4 of this book to get a better understanding of ArcGIS Server Web service concepts and programming guidelines.

If you are unfamiliar with developing applications that make use of Web services with Java, it's also recommended that you refer to your Java developer documentation to become more familiar with Java application development. If you want to learn more about Web services in general, visit www.w3.org.

Rather than walk through this scenario, you can get the completed Web application from the samples installation location. The sample is installed as part of the ArcGIS developer samples.

This walk-through is for developers who need to build and deploy a .NET Web application, incorporating GIS functionality using the ArcGIS Server API that makes heavy use of fine-grained ArcObjects method calls. It describes the process of building, deploying, and consuming the *Extending_the_server* developer scenario sample, which is part of the ArcGIS developer samples.

You can find this sample in:
<install_location>\DeveloperKit\Samples\Developer_Guide_Scenarios\
ArcGIS_Server\Extending_the_server

PROJECT DESCRIPTION

While this example makes use of extending the server to support functionality in a Web application, you can use the same technique when creating other types of applications, such as Web services.

The purpose of this scenario is to create an ASP.NET Web application using Visual Studio .NET to extend the MapViewer ArcGIS Server Web application project template. The application uses ArcObjects to clip the geometries of vegetation polygons to a buffer of user-specified size around a user-specified point. It will then report the total area of each different vegetation type within the buffered area.

The following is how the user will interact with the application:

1. Open a Web browser and specify the URL of the Web application.

2. Type the buffer distance (in meters).

3. Click the *Summarize Vegetation* tool and click the point on the map to buffer.

A graphic displaying the clipped vegetation polygons is displayed on the map, and a table summarizing the total area of each vegetation type within the buffer is displayed in the browser.

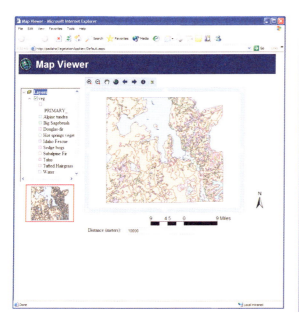

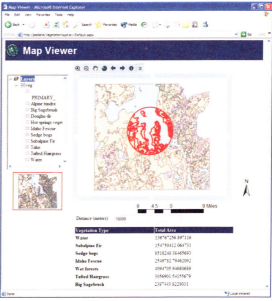

CONCEPTS

Both coarse-grained calls to remote ArcObjects, such as the methods on the MapServer and GeocodeServer, as well as fine-grained calls to remote ArcObjects, such as looping through all the vertices of a polygon, are exposed through the ArcGIS Server API and can be used in your application. However, it's important to note that when making a call against an object running in the server from your Web application, you are making that call across processes. The Web server is running in one process, while the object is running in another process.

Calls to objects across processes are significantly slower than calls to objects in the same process. It's also likely that your Web application is running on a Web server that is actually a different machine from the one the object is running on, so the calls are not only cross process but also cross machine.

This application includes functionality that requires making a large number of fine-grained ArcObjects calls to loop through features, get their geometry, clip the geometry, summarize the areas for the different vegetation types, create a graphic for each feature, and so on. Since the user of the application is free to specify a buffer distance that may include a large number of features, the number of features that would be analyzed is indeterminate, which could easily result in 1,000s of fine-grained ArcObjects calls.

To minimize this large number of remote calls, you can create a simple utility COM object in VB, C++, or .NET that does the majority of the fine-grained ArcObjects work but exposes a single coarse-grained method that the Web application calls. Using this technique, the application has the same functionality, but the cross process/cross machine cost is minimized by running the bulk of the code in the GIS server. In this example, this technique will help to maximize the performance of the Web application.

DESIGN

This Web application is designed to make stateless use of the GIS server. It uses events on the map Web control to get a point from the user, then uses that point and ArcObjects to perform analysis on the vegetation polygons. In order to support this application, you need to add a pooled map server object to your ArcGIS Server using ArcCatalog.

The Web application will use the Web controls to manage the connection to the GIS server, and the MapViewer Web application template will provide the basic mapping functionality that is required for this application. You will add a new tool to the Toolbar control that allows the user to click the map as input to the analysis. The results are displayed on the map as a set of graphics and summarized in a table on the Web page. The bulk of the ArcObjects processing will be implemented as a separate utility COM object that is installed on the server and created in the server by the Web application.

REQUIREMENTS

The requirements for working through this scenario are that you have ArcGIS Server and ArcGIS Desktop installed and running. The machine on which you develop this Web server must have the ArcGIS Server .NET Application Developer Framework SDK installed.

One key aspect of designing your application is whether it is stateful or stateless. You can make either stateful or stateless use of a server object running within the GIS server. A stateless application makes read-only use of a server object, meaning the application does not make changes to the server object or any of its associated objects. A stateful application makes read–write use of a server object where the application does make changes to the server object or its related objects.

Web services are, by definition, stateless applications.

You must have a map server object configured and running on your ArcGIS Server that uses the Yellowstone.mxd map document, which is installed with the ArcGIS developer samples. In ArcCatalog, create a connection to your GIS server and use the Add Server Object command to create a new server object with the following properties:

Name: Yellowstone

Type: MapServer

Description: Map server containing vegetation data for Yellowstone National Park

Map document:
<install_location>\DeveloperKit\Samples\Data\ServerData\Yellowstone\Yellowstone.mxd

Output directory: Choose from the output directories defined by your server.

Pooling: The Web service makes stateless use of the server object. Accept the defaults for the pooling model (pooled server object with minimum instances = 2, max instances = 4).

Accept the defaults for the remainder of the configuration properties.

After creating the server object, start it and click the Preview tab to verify that it is correctly configured and that the map draws.

You can refer to Chapter 3 for more information on using ArcCatalog to connect to your server and create a new server object. Once the server object is configured and running, you can begin to code your Web service.

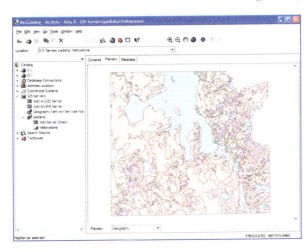

The Add Server Object wizard

ArcCatalog is used for managing your spatial data holdings, defining your geographic data schemas, and managing your ArcGIS server. Once you have created your Yellowstone server object, preview it using ArcCatalog to verify it is correctly configured.

The following ArcObjects .NET assemblies will be used in this example:

- ESRI.ArcGIS.Carto
- ESRI.ArcGIS.Geodatabase
- ESRI.ArcGIS.Geometry
- ESRI.ArcGIS.Server
- ESRI.ArcGIS.Server.WebControls
- ESRI.ArcGIS.esriSystem

To create your utility COM object, the following COM object libraries will be used:

- esriCarto
- esriDisplay
- esriGeoDatabase
- esriGeometry
- esriSystem

The development environment does not require any ArcGIS licensing; however, connecting to a server and using a map server object does require that the GIS server is licensed to run ArcObjects in the server. None of the assemblies or object libraries used require an extension license.

The IDE used in this example is Visual Studio .NET 2003 for the Web application, and all IDE specific steps will assume that it is the IDE you are using. This Web service can be implemented with other .NET IDEs. The utility COM object will be written using Visual Basic 6, but it can be implemented using any COM language/IDE.

IMPLEMENTATION

The code for this scenario is divided into two parts. The first is the implementation of the utility COM object, which will run within the GIS server, and exposes the coarse-grained method that the Web application will call. This will be written in VB using VB6.

The second is the Web application itself. The Web application code will be written in C#; however, you can also write this Web application using VB.NET. The Web application executes within the Web server.

Part 1: Creating the utility COM object

To begin, you must create a new project in VB6:

Create a new project

1. Start Visual Basic 6.

2. Click *File*, then click *New Project*.

3. In the New Project dialog box, click *ActiveX DLL* as the project type.

4. Click OK. This will create a blank application.

5. Click *Project*, then click *Project1 Properties*. In the Properties dialog box, for the Project name, type "VegUtilities".

6. Click OK.

7. In the project browser, click *Class1 (Class1)*, then in the Properties for Class1, type "clsVegUtils" for the name.

8. Click *File*, then click *Save Project* to save the project.

Add references to ESRI object libraries

1. Click *Project*, then click *References*.

2. In the References dialog box, check the following object libraries:

- ESRI Carto Object Library

- ESRI GeoDatabase Object Library

- ESRI Geometry Object Library

- ESRI System Object Library

- ESRI Display Object Library

This COM object will also make use of the *Scripting.Dictionary* object.

To learn more about COM and developing with ArcObjects and COM, refer to Appendix D and the online developer documentation.

The New Project dialog box

The Project Properties dialog box

The Project browser

The References dialog box

The Project browser box

The Add Class dialog box

3. To have access to this object, in the References dialog box, check the following:

Microsoft Scripting Runtime

4. Click OK.

Create the results class

The Web application you are going to build requires two things from the utility function: an array of graphics to draw on the map and a record set of the vegetation types and their total area. To do this, you will implement a new object that contains these two aspects of the results of the function.

1. In the project browser, right-click *VegUtilities*, click *Add*, then click *Class Module*.

2. In the Add Class Module dialog box, click *Class Module*, then click *Open*. This will add the new class module to your project and open it.

3. In the project browser, click *Class1 (Class1)*, then in the Properties for Class1, type "clsVegResults" for the name.

The *clsVegResults* class will have two read–write properties:

ResGraphics: a collection of graphic elements representing the geometry of the vegetation polygons intersecting the buffer.

Stats: a record set with a record for each type of vegetation and the total area of that vegetation type contained within the buffer.

The class will hold the graphics array and the record set in two private variables.

4. Add the following code to your *clsVegResults* class:

```
Option Explicit
Private m_resGraphics As esriCarto.IGraphicElements
Private m_resStats As esriGeoDatabase.IRecordSet
```

For your function to set these properties and for your Web application to get these properties, you will add public *Get* and *Set* properties for each.

5. Add the following code to your clsVegResults class:

```
Public Property Get ResGraphics() As esriCarto.IGraphicElements
    Set ResGraphics = m_resGraphics
End Property

Public Property Set ResGraphics(pResGraphics As
esriCarto.IGraphicElements)
    Set m_resGraphics = pResGraphics
End Property

Public Property Get Stats() As esriGeoDatabase.IRecordSet
    Set Stats = m_resStats
End Property

Public Property Set Stats(pStats As esriGeoDatabase.IRecordSet)
    Set m_resStats = pStats
End Property
```

The *clsVegResults* class is complete. Your code for the *clsVegResults* class should look like the following:

```
Option Explicit
Private m_resGraphics As esriCarto.IGraphicElements
Private m_resStats As esriGeoDatabase.IRecordSet

Public Property Get ResGraphics() As esriCarto.IGraphicElements
    Set ResGraphics = m_resGraphics
End Property

Public Property Set ResGraphics(pResGraphics As
esriCarto.IGraphicElements)
    Set m_resGraphics = pResGraphics
End Property

Public Property Get Stats() As esriGeoDatabase.IRecordSet
    Set Stats = m_resStats
End Property

Public Property Set Stats(pStats As esriGeoDatabase.IRecordSet)
    Set m_resStats = pStats
End Property
```

The results class includes a ResGraphics property, which is a GraphicElements collection. As you will see later, the MapServer's MapDescription object allows you to add graphics to a map at draw time. The MapDescription takes those graphics as a GraphicElements collection. This is an example of how you can write your utility COM object to best satisfy the requirements of an ArcGIS Server application.

Implement the analysis function

You will now add the public function called *sumVegetationType* that will do the analysis, create and populate the *clsVegResults* object, and pass it back to the calling application (your Web application in this case) to the *clsVegUtils* class. In addition, you will implement some private helper functions.

The *sumVegetationType* function will take the following parameters:

pVegClass: the feature class containing the polygon features to analyze.

pPoint: the point to buffer.

dDistance: a double that represents the distance to buffer the point.

sSummaryField: the name of the field whose unique values will be summarized by the area of the polygons in the buffer.

The function returns a *clsVegResults* object.

1. To define your public function, add the following lines of code to the *clsVegUtils* class:

```
Option Explicit
Public Function sumVegetationType(pVegClass As IFeatureClass, pPoint As
IPoint, dDistance As Double, sSummaryFld As String) As clsVegResults

End Function
```

The first step this function needs to take is to buffer the input (*pPoint*) the distance specified by *dDistance* to create a new geometry called *pGeom*.

The objects and interfaces used for creating and working with geometries can be found in the Geometry object library. To learn more about geometry objects, see the online developer documentation.

2. Add the following lines of code to your *sumVegetationType* function:

```
' buffer the point
Dim pTopoOp As ITopologicalOperator
Set pTopoOp = pPoint
Dim pGeom As IGeometry
Set pGeom = pTopoOp.Buffer(dDistance)
```

Next, you will use the buffer as the geometry for a spatial query to query the feature class for all of the polygons that intersect the buffer.

3. Add the following lines of code to your *sumVegetationType* function:

```
' query the feature class
Dim pSFilter As ISpatialFilter
Set pSFilter = New SpatialFilter
Set pSFilter.Geometry = pGeom
pSFilter.SpatialRel = esriSpatialRelIntersects
pSFilter.GeometryField = pVegClass.ShapeFieldName

Dim pFCursor As IFeatureCursor
Set pFCursor = pVegClass.Search(pSFilter, True)
```

The objects and interfaces used for performing spatial queries and for working with the results of those queries can be found in the GeoDatabase object library. To learn more about geodatabase objects, see the online developer documentation.

Before looping through the features, you need to create a *GraphicElements* collection to hold the graphics, a simple fill symbol to apply to each graphic element, a dictionary object that you will use to categorize the different vegetation types, and some other needed variables. Note that the fill symbol is created using a helper function called *newFillS*. You will create this function later.

4. Add the following lines of code to your *sumVegetationType* function:

```
' loop through the features, clip each geometry to the buffer
' and total areas by attribute value
Set pTopoOp = pGeom
Dim pNewGeom As IGeometry

Dim lPrim As Long
lPrim = pVegClass.FindField(sSummaryFld)

Dim pFeature As IFeature
Dim pArea As IArea
Dim pFE As IFillShapeElement
Dim pElement As IElement
Dim sType As String

Dim dict As Scripting.Dictionary
Set dict = New Scripting.Dictionary

' create the symbol and collection for the graphics
Dim pSFS As ISimpleFillSymbol
Set pSFS = newFillS
Dim pGraphics As IGraphicElements
Set pGraphics = New GraphicElements
```

The next step is to loop through the features in the *pVegClass* feature class that intersect the buffer geometry and clip each vegetation polygon to the buffer. The resulting clipped geometry is then used to create a graphic that is added to the graphics collection. The area of the clipped geometry is added to the total area of the feature's type (as defined by the value of *sSummaryField*) in the dictionary object.

5. Add the following lines of code to your *sumVegetationType* function:

```
Set pFeature = pFCursor.NextFeature
Do Until pFeature Is Nothing
  ' create the graphic
  Set pFE = New PolygonElement
  Set pElement = pFE

  ' clip the geometry
  Set pNewGeom = pTopoOp.Intersect(pFeature.Shape, esriGeometry2Dimension)
  pElement.Geometry = pNewGeom
  pFE.Symbol = pSFS
  pGraphics.Add pFE

  ' add to dictionary
  Set pArea = pNewGeom
  sType = pFeature.Value(lPrim)
  If dict.Exists(sType) Then
    dict.Item(sType) = dict.Item(sType) + pArea.Area
  Else
    dict.Item(sType) = pArea.Area
  End If

  Set pFeature = pFCursor.NextFeature
Loop
```

The objects and interfaces used for creating and working with graphic elements can be found in the Carto object library. To learn more about carto objects, see the online developer documentation.

After this code executes, the dictionary object will have a key for each unique value of the *sSummaryField* field whose item is the total area for that unique value within the buffer. The next step is to create a record set object and copy the keys and items from the dictionary into rows and fields in the record set. This is accomplished using the *sumRS* helper function. You will implement the *sumRS* helper function later.

6. Add the following lines to your *sumVegetationType* function:

```
' create the summary recordset
Dim psumRS As IRecordSet
Set psumRS = sumRS(dict)
```

A RecordSet object is a collection of rows that are not mapped to a physical table in a database. You can use record sets to create in-memory tables and rows without writing them to a physical table. The objects and interfaces associated with record sets can be found in the GeoDatabase object library. To learn more about geodatabase objects, see the online developer documentation.

Finally, since the *sumVegetationType* function returns a *clsVegResults* object, the last part of the function creates a new *clsVegResults* object, sets the graphics collection and summary record set in the object, and returns the object to the caller.

7. Add the following lines of code to complete your *sumVegetationType* function:

```
' create the results object
Dim pResClass As clsVegResults
Set pResClass = New clsVegResults
Set pResClass.ResGraphics = pGraphics
```

```
Set pResClass.Stats = psumRS

Set sumVegetationType = pResClass
```

The code for your *sumVegetationType* function is now complete. The next section will define the helper functions that the *sumVegetationType* calls.

Implement helper functions

As described above, the *sumVegetationType* function makes use of two helper functions to create a fill symbol (*newFillS*) and to copy the contents of a dictionary object to a record set (*sumRS*). You will now implement these helper functions in your *clsVegUtils* class.

The newFillS function creates and returns a new *SimpleFillSymbol* object. This fill symbol is a hollow fill symbol with a red outline.

1. To define the *newFillS* function, add the following code to your *clsVegUtils* class:

The objects and interfaces used for creating symbols and colors can be found in the Display object library. To learn more about display objects, see the online developer documentation.

```
Private Function newFillS() As ISimpleFillSymbol
  Dim pSLS As ISimpleLineSymbol
  Set pSLS = New SimpleLineSymbol
  Dim pcolor As IRgbColor
  Set pcolor = New RgbColor
  pcolor.Red = 255
  pcolor.Green = 0
  pcolor.Blue = 0
  pSLS.Color = pcolor
  pSLS.Style = esriSLSSolid
  pSLS.Width = 2

  Dim pSFS As ISimpleFillSymbol
  Set pSFS = New SimpleFillSymbol
  pSFS.Outline = pSLS
  pSFS.Style = esriSFSHollow

  Set newFillS = pSFS
End Function
```

The *sumRS* function is a little more involved. This function takes a dictionary object as an argument, and returns a record set. The function creates a new record set with a field for the *sSummaryField* unique values and a field for the total area. It then loops through the keys and items in the dictionary and creates a row in the record set for each key/item pair.

2. Add the following code to your *clsVegUtils* class:

```
Private Function sumRS(dict As Scripting.Dictionary) As IRecordSet
  ' create the new record set
  Dim pNewRs As IRecordSet
  Set pNewRs = New Recordset
  Dim prsInit As IRecordSetInit
  Set prsInit = pNewRs

  Dim pFields As IFields
```

The objects and interfaces associated with record sets, fields, and cursors can be found in the GeoDatabase object library. To learn more about geodatabase objects, see the online developer documentation.

```
Dim pFieldsEdit As IFieldsEdit
Dim pField As IField
Dim pFieldEdit As IFieldEdit
Set pFields = New Fields
Set pFieldsEdit = pFields
pFieldsEdit.FieldCount = 2

Set pField = New Field
Set pFieldEdit = pField

With pFieldFdit
  .Name = "Type"
  .Type = esriFieldTypeString
  .Length = 50
End With
Set pFieldsEdit.Field(0) = pField

Set pField = New Field
Set pFieldEdit = pField

With pFieldEdit
  .Name = "Area"
  .Type = esriFieldTypeDouble
End With
Set pFieldsEdit.Field(1) = pField

prsInit.CreateTable pFields

' add all the area/type pairs
Dim pIC As ICursor
Set pIC = prsInit.Insert
Dim pRowBuf As IRowBuffer
Set pRowBuf = prsInit.CreateRowBuffer

Dim keys As Variant
Dim items As Variant
keys = dict.keys
items = dict.items
Dim i As Long
For i = LBound(keys) To UBound(keys)
  pRowBuf.Value(0) = keys(i)
  pRowBuf.Value(1) = items(i)
  pIC.InsertRow pRowBuf
Next i

Set sumRS = pNewRs
End Function
```

Compile and register the utility COM DLL

Now that the coding for the utility COM object is complete, you need to compile it into a DLL.

1. Click *File*, then click *Make VegUtilities.dll*.

2. In the Make Project dialog box, click OK.

3. Fix any errors.

Set the binary compatibility for the component:

4. Click *Project* then click *VegUtilities Properties*.

5. Click the *Component* tab.

6. Click the *Binary Compatibility* option and browse for *VegUtilities.dll*.

7. Click OK.

8. Recompile the dll by following steps 1 to 3.

Rather than copying the actual COM object onto the Web server machine, you can generate a remote server file when compiling the Visual Basic project. Because the object actually runs within the GIS server, the Web server needs only the object's proxies to work with it.

Both your Web server and your GIS server's server object container machines need to have this COM DLL copied and registered on their system. The dll must be registered on the Web servers that the Web application will run on so that it has the correct types to work with the utility COM object's proxies. If the machine on which you developed this utility COM object is also the machine that hosts your server object and is the machine on which you will develop the Web application, then the above will automatically register it.

Part 2: Creating the Web application

Now that you have developed your COM object that includes your coarse-grained method call, you can build your Web application to make use of it. In this example, you will use the MapViewer template project that is installed as part of the ArcGIS Server Application Developer Framework's SDK to provide some basic mapping functionality, and you will extend this template with your own functionality.

The first step is to create the new project.

Create a new project

1. Start Visual Studio .NET.

2. Click *File*, click *New*, then click *Project*.

3. In the New Project dialog box, under *Project Types*, click the *ArcGIS Server Projects* category, then the *Visual C#* category. Under *Templates*, click *MapViewer Web Application*.

4. For the Web application name, type "http://localhost/VegetationWebApp".

5. Click OK. This will create a new project that contains all of the functionality included in the MapViewer template.

The New Project dialog box

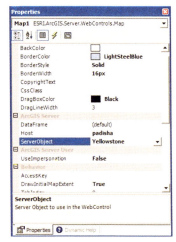

The Map control's properties

If the Yellowstone map server object is not listed, verify the server object is started.

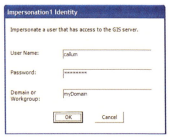

The Impersonation control's identity dialog box

For your Web application to successfully connect to the GIS server, the account you specify in the impersonation control's properties must be a member of the ArcGIS Server users group on the GIS server. Since the Impersonation control sets impersonation at the Web page level, there is an impersonation control on both the Default.aspx page and the Identify.aspx page.

Set the necessary properties on the Web controls

The template includes a *Map* control, an *OverviewMap* control, and an *Impersonation* control. All require properties to be set, specifically the GIS server name, the MapServer object that the map and overview map controls will use, and the user account that the Web application will run as for the impersonation control.

1. In the Solution Explorer, double-click *Default.aspx*. This will open the Web form in design mode.

2. On the Web form, click the *Map1* map control.

3. In the Properties for the map control, type the name of the GIS server for the *host* property, then click the *ServerObject* dropdown and click *Yellowstone* for the server object.

4. On the Web form, click the *OverviewMap1* control.

The overview map can use the same or a different server object to display the current extent of the map in the map control. In this case, you will use the same map as the map control.

5. In the Properties for the overview map control, type the name of the GIS server for the *host* property, then click the *ServerObject* dropdown and click *Yellowstone* for the server object.

6. On the Web form, click the *Impersonation1* control.

7. In the Properties for the impersonation control, click the *Identity* property and click the Browse button. This will open the identity dialog box.

8. Type the username, password, and domain for the account that your Web application will run as, then click OK.

Add the SumVeg tool

This application allows the user to identify a point to buffer by clicking the map on the map control. To allow the user to do this, you will add a new tool to the toolbar control's tool collection and a text box for the user to specify the buffer distance.

The first step is to add to your Web form the text box that the user will type the buffer distance into.

1. Click the *Design* tab to return to the Web form design mode.

2. In the Microsoft Visual Studio .NET toolbox, click the Web Forms tab to display the Web Forms tools.

3. In the toolbox, click *Label* and drag a label onto the form next to the button.

4. In the label's properties, type *Distance (meters):* for the *Text* property.

5. In the toolbox, click *TextBox* and drag a text box onto the form to the right of the label.

6. In the TextBox's properties, type "10000" for the *Text* property and "txtDistance" for the *ID* property.

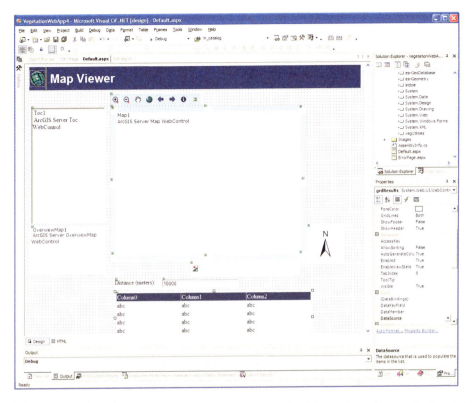

The VegetationWebApp application in the Visual Studio .NET IDE

7. In the toolbox, click *DataGrid* and drag a data grid onto the form below the other controls.

8. In the DataGrid's properties, type "grdResults" for the *ID* property.

Now you will add the tool to your toolbar that will actually let users click a point on the map. Before adding the tool to the toolbar, you must create a new class that implements the *IMapServerToolAction* interface and defines the functionality for the tool.

Create the SumVeg class

The first step is to add the new class to the project.

1. In the Solution Explorer, right-click the *VegetationApp* project, click *Add*, then click *Add New Item*.

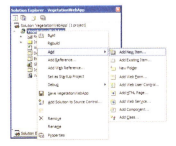

The Solution Explorer

The Add New Item dialog box

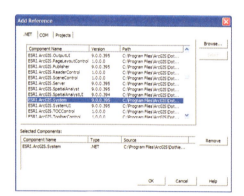

The Add Reference dialog box

2. In the Add New Item dialog box, under *Templates*, click *Class*.

3. For the name, type "SumVeg.cs".

4. Click Open.

This will add a new class (*SumVeg*) to your project and will open the code for the class with some autogenerated code.

Add additional references for the SumVeg class

The Web application template includes references to a collection of ESRI assemblies. However, the functionality that you will add to the *SumVeg* class will require additional ESRI assemblies. These assemblies were installed when you install the ArcGIS Server .NET Application Developer Framework.

In addition to these ESRI assemblies, you also need to add a reference to your custom COM DLL that you created in Part 1. To add references to these assemblies and the COM DLL, do the following:

1. In the Solution Explorer, right-click *References* and click *Add Reference*.

2. In the Add Reference dialog box, double-click the following assembly:

   ESRI.ArcGIS.System

3. Click the *COM* tab.

4. Click Browse and click the *VegUtilities.dll*.

5. Click OK.

Notice, this will automatically add the following additional references to ESRI COM Object Libraries:

- esriCarto
- esriGeodatabase
- esriGeometry
- esriSystem

6. Add using statements for the assemblies you will use in this class. At the top of the code window, add the following using statements:

```
using ESRI.ArcGIS.Server;
using ESRI.ArcGIS.Server.WebControls;
using ESRI.ArcGIS.Server.WebControls.Tools;
using ESRI.ArcGIS.Geometry;
using ESRI.ArcGIS.Geodatabase;
using ESRI.ArcGIS.Carto;
using ESRI.ArcGIS.esriSystem;
using System.Web.UI.WebControls;
using VegUtilities;
```

Implement the SumVeg class

Now you are ready to implement the *SumVeg* class that contains the code to execute when the map is clicked by the user. Since this class is a map server tool action, it must implement the *IMapServerToolAction* interface.

1. Change the following line:

```
public class SumVeg
```

to:

```
public class SumVeg : IMapServerToolAction
```

2. In the Class View window, expand the class list to the *Bases and Interfaces* of the *SumVeg* class.

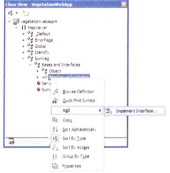

The class view

3. Right-click the *IMapServerToolAction* interface, click *Add*, then click *Implement Interface*.

Visual Studio stubs out the members of *IMapServerToolAction* in the code window automatically, bracketing the stubs within a region named *IMapServerToolAction Members*.

The *IMapServerToolAction* has a single method to implement called *ServerAction*. This method is where you will put the code to execute when the user clicks the map. The following code will be added to your class:

```
#region IMapServerToolAction Members

public void ServerAction(ToolEventArgs args)
{
  // TODO: Add SumVeg.ServerAction implementation
}

#endregion
```

The remainder of the code will be added to this method.

In order to create the utility COM object in the server and execute its function, you need to get a server context from the server. You will be getting a point from the map server and will be ultimately adding graphics to the map. Since objects work better together when they are in the same context, you will get the map server's context from the map control's WebMap object and create the ArcObjects you will use within that context. Since the Yellowstone map server object is a pooled server object, the WebMap will take care of releasing the context for you when it goes out of scope. You will scope the use of the WebMap within a *Using* block.

The *args* object passed into the *ServerAction* method includes a reference to the map control.

4. Add the following code to your *ServerAction* method:

```
if (args.Control is ESRI.ArcGIS.Server.WebControls.Map)
{
  ESRI.ArcGIS.Server.WebControls.Map mapctl = args.Control as
ESRI.ArcGIS.Server.WebControls.Map;
  using (WebMap webMap = mapctl.CreateWebMap() )
  {
  }
}
```

You need to get a reference to the WebMap's server context and map server object. You will then use the *IMapServerObjects* interface to access the map server's fine-grained ArcObjects to get a reference to the first layer in the map (this is the layer whose feature class you will query in this application).

5. Add the following lines of code to your using block:

```
IServerContext sc = webMap.ServerContext;
IMapServer map = webMap.MapServer;

IMapServerObjects mapobj = map as IMapServerObjects;
IMap fgmap= mapobj.get_Map(map.DefaultMapName);
IFeatureLayer fl = fgmap.get_Layer(0) as IFeatureLayer;
```

The objects and interfaces associated with working with a map can be found in the Carto object library. To learn more about map objects, see the online developer documentation.

The first argument to the *sumVegetation* function in your *clsVegResults* class is the feature class to perform the query against. In this case, you will use the feature class from the first layer in the map, obtained from the code above.

6. Add the following line of code to your using block:

```
IFeatureClass fc = fl.FeatureClass;
```

The second argument to the function is the point. You will get the screen coordinates of where on the map control the user clicked from the *ServerAction* method's *args* object. The WebMap includes a method to convert the screen coordinates to an ESRI Point object. You will use this point object when you call the COM function.

7. Add the following line of code to your Using block:

```
PointEventArgs pargs = args as PointEventArgs;
IPoint pt = webMap.ToMapPoint(pargs.ScreenPoint.X, pargs.ScreenPoint.Y);
```

The last two arguments to the function are the distance to buffer the point (double) and the name of the field on the feature class on which to summarize the areas of the polygons. The distance was input by the user into the *txtDistance* text box on the Web form. You will use the *Page* property of the map control to get a reference to the page and text box controls and to convert the *Text* property of the control into a double.

The field name will be hard coded to be "PRIMARY_".

8. Add the following lines of code to your using block:

```
System.Web.UI.WebControls.TextBox txb =
mapctl.Page.FindControl("txtDistance") as
System.Web.UI.WebControls.TextBox;
double dDist = Convert.ToDouble(txb.Text);
string fldName = "PRIMARY_";
```

Use the CreateObject method when you need to create an object for use in your application. ArcGIS Server applications should not use New to create ArcObjects but should always create objects by calling CreateObject on IServerContext. CreateObject will return a proxy to the object that is in the server context. Your application can make use of the proxy as if the object were created locally within its process. To learn more about working with objects in server contexts, see Chapter 4 and the online developer documentation.

Now you have all of the arguments needed to call the *sumVegetation* function you created in Part 1 of this scenario: the input feature class (*fc*), the point (*pt*), the distance to buffer (*dDist*), and the field to total the areas on (*fldName*). To call this function, you need to create an instance of the utility COM object in the map server's context.

You will do this by using the *CreateObject* method and specifying the progID of the COM object. Once you have created the object, you can call the method that returns a *clsVegResults* object.

9. Add the following lines of code to your using block:

```
VegUtilities.clsVegUtils veg = sc.CreateObject("VegUtilities.clsVegUtils")
as VegUtilities.clsVegUtils;
VegUtilities.clsVegResults vegres = veg.sumVegetationType(ref fc,ref
pt,ref dDist, ref fldName);
```

As described in Part 1, the *clsVegResults* class has two properties:

resGraphics: a collection of graphics of the clipped polygons to draw on the map.

Stats: a record set of vegetation type, total area pairs.

The next section of code takes these results and makes use of them by adding the graphics collection to the WebMap's map description and displaying the contents of the record set in a data grid on the Web form.

Because the map server object you are using is pooled and the application is stateless, you do not want to add the result graphics directly to the map using its *GraphicsContainer* object, as this would change the state of the map server object. To draw the graphics on the map, you will add them to the WebMap's copy of the map description. When you do this, the graphics will be drawn and saved by the WebMap as part of the map description in session state for you. Once you add the graphics to the map description, you will ask the WebMap to redraw itself using its *Refresh* method, which will reflect the changes you have made to the map description.

The results class's ResGraphics property returns a GraphicElements collection. The MapServer's MapDescription object allows you to add graphics to a map at draw time. The MapDescription takes those graphics as a GraphicElements collection. This is an example of how you can write your utility COM object to best satisfy the requirements of an ArcGIS Server application.

10. Add the following lines of code to your using block:

```
IGraphicElements ge = vegres.get_ResGraphics() as IGraphicElements;
IMapDescription md = webMap.MapDescription;
md.CustomGraphics = ge;
webMap.Refresh();
```

Finally, you will display the contents of the *Stats* record set from the *clsVegResults* object in the data grid you added to the Web form (*grdResults*). To do this, you could get the record set object from the *clsVegResults* object, then open a cursor on the record set. However, this would mean that you would be making a number of remote calls to loop through the record set and get the properties for each row. If the record set has a large number of rows, these remote calls can cause the same performance problem that you tried to avoid by creating the utility COM object.

The Converter convenience class also has a ToDataset method for converting record sets to .NET datasets, independent of a Web control. The Converter object is in the ESRI.ArcGIS.Server.WebControls assembly.

The WebMap has a convenience method called ConvertRecordSetToDataSet that converts the record set to a .NET DataSet object using a single remote call. Once you have a .NET dataset, you can bind the grdResults grid control to the dataset.

11. Add the following lines of code to your using block:

```
IRecordSet rs = vegres.get_Stats() as IRecordSet;
System.Data.DataSet rsDataset = webMap.ConvertRecordSetToDataSet(rs,
false, false);

System.Web.UI.WebControls.DataGrid resGrid =
mapctl.Page.FindControl("grdResults") as
System.Web.UI.WebControls.DataGrid;
resGrid.DataSource = rsDataset;
resGrid.DataBind();
```

By compiling the project, this allows you to pick MapView.NewConservationPlan for the ServerToolActionClass in the Toolbar CollectionEditor.

The MapViewer template's toolbar already contains tools and commands for navigating the map (Zoom In, Zoom Out, Pan) and for identifying features in the map.

The Toolbar control's properties

The ClientToolAction specifies what code is executed in the client (the Web browser). In this case, JavaScript for drawing a polygon on the map control is the client tool action.

The ServerToolAction is the code in the server that is executed when the client tool action has completed. In this case, the server tool action is defined by the NewConservationPlan class.

The ToolbarItem Collection Editor dialog box

Now that the class is defined, you must compile your project to add your new *SumVeg* class to the .NET assembly list.

1. Click *Build*, then click *Build Solution*.

2. Fix any errors.

Add the Summarize Vegetation tool to the toolbar

Now that you have implemented your class, you will add the tool to your toolbar. The template includes a toolbar control that already has a number of tools (*Zoom In*, *Zoom Out*, *Pan*, *Identify*, etc.); you will add a tool that will allow the user to click a point on the map and execute the code in the *SumVeg* class.

Before adding the tool, you will need to copy a set of images that you will use for the tool in the toolbar. Copy the following image files from `<install_location>\DeveloperKit\Samples\Data\ServerData\Yellowstone` to your application's Images folder (this will be `c:\inetpub\wwwroot\VegetationWebApp\Images`):

• click_point.gif

• click_pointU.gif

• click_pointD.gif

1. In the Solution Explorer, click *Default.aspx* to open the Web form in design mode.

2. Click the toolbar control.

3. In the properties for the toolbar control, click the *ToolbarItemsCollection* property and click the Browse button. This will open the ToolbarItem Collection Editor.

4. Click the *Add* dropdown, then click *Tool*. This will add a new tool to the toolbar collection.

5. Click the *Name* property for the new tool and type "tbSummarizeVegetation" for the name.

6. Click the *Text* property and type "Summarize vegetation" for the text.

7. Click the *ToolTip* property and type "Summarizes vegetation types within a buffer" for the tooltip.

8. Click the *ClientToolAction* property drop down list and click *Point*.

9. Click the *ServerToolActionAssembly* property and type "VegetationWebApp" for the assembly name.

10. Click the *ServerToolActionClass* property dropdown list and click *VegetationWebApp.SumVeg* for the class (this is the class you just created).

11. Click the *DefaultImage* property and type "Images\click_point.gif" for the default image.

The new tool that you add through the ToolbarItem Collection Editor will appear on the toolbar.

12. Click the *HoverImage* property and type "Images\click_pointU.gif" for the hover image.

13. Click the *SelectedImage* property and type "Images\click_pointD.gif" for the selected image.

14. Click OK. This will add the new tool to your toolbar.

Your Web application is now ready to be tested. Compile the project (*Build/Build Solution*)and fix any errors.

Test the Web application

If you run the Web application from within Visual Studio, it will open a browser and connect to the application's startup page (*Default.aspx*).

1. Click *Debug*, then click *Start*.

2. On the browser that opens, click the *Summarize vegetation* button.

3. Click on the map.

All the polygons within 10,000 meters of your point (clipped to the buffer) will draw in red on the map, and the total area of each vegetation type within that buffer will be displayed in the grid.

You can experiment by changing the buffer distance and clicking other locations on the map.

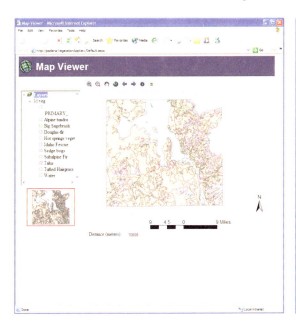

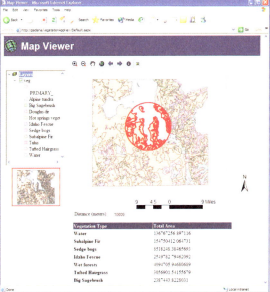

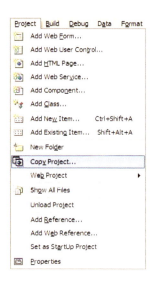

DEPLOYMENT

Presumably you developed this Web application using your development Web server. To deploy this Web application on your production Web server, you can use the built-in Visual Studio .NET tools to copy the project.

1. In the Solution Explorer, click the *VegetationWebApp* project.

2. Click *Project*, then click *Copy Project*.

3. In the Copy Project dialog box, specify the location on your Web server to copy the project to.

4. Click OK.

In addition to copying the project, you must copy and register your *VegUtilities.dll* COM object on both your Web server and the GIS server's server object container machines.

ADDITIONAL RESOURCES

This scenario includes functionality and programming techniques covering a number of different aspects of ArcObjects, the ArcGIS Server API, .NET application templates, and Web controls.

You are encouraged to read Chapter 4 of this book to get a better understanding of core ArcGIS Server programming concepts such as stateful versus stateless server application development. Chapter 4 also covers concepts and programming guidelines for working with server contexts and ArcObjects running within those contexts, as well as discussion on extending the GIS server as demonstrated in this scenario.

This scenario makes use of a Web application template and the ArcGIS Server's .NET ADF Web controls to provide the majority of the user interface for this Web application. To learn more about this Web application template and other template applications that are included with the .NET ADF, see Chapter 5 of this book. Chapter 5 also includes detailed descriptions and examples of using the .NET Web controls, including the map and toolbar Web controls that you made use of while programming this Web application. If you are unfamiliar with ASP.NET Web development, it's also recommended that you refer to your .NET developer documentation to become more familiar with Web application development.

ArcGIS Server applications exploit the rich GIS functionality of ArcObjects. This application is no exception. It includes the use of ArcObjects to work with the components of a MapServer, buffer and clip geometries, query a geodatabase, and create graphics. To learn more about these aspects of ArcObjects, refer to the online developer documentation on the on the Carto, Display, GeoDatabase, and Geometry object libraries.

Server library

ArcGIS Server is fundamentally an object server that manages a set of GIS server objects. These server objects are software objects that serve a GIS resource such as a map or a locator. Applications and application developers make use of server objects in their applications. Server objects are ArcObjects. ArcObjects is a collection of software objects that make up the foundation of ArcGIS. The Server library contains a set of ArcObjects that you use to create, manage, and work with these objects. The Server library also contains ArcObjects for administering the GIS server itself.

This chapter details using ArcObjects to perform ArcGIS Server-related tasks. Topics covered include:
• connecting to the GIS server • getting and creating objects in the server • working with and managing objects in the server • getting and administering the properties of the GIS server and its collection of server objects

IGISServerConnection

GISServer-
Connection

The GISServerConnection class provides connections to a GIS server over a LAN and hands out references to the server's ServerObjectManager and ServerObjectAdmin.

The *GISServerConnection* class provides connections to a GIS server over a LAN.

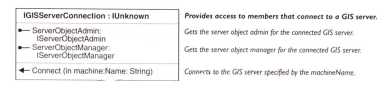

IGISServerConnection : IUnknown	*Provides access to members that connect to a GIS server.*
►— ServerObjectAdmin: IServerObjectAdmin	*Gets the server object admin for the connected GIS server.*
►— ServerObjectManager: IServerObjectManager	*Gets the server object manager for the connected GIS server.*
◄— Connect (in machine:Name: String)	*Connects to the GIS server specified by the machineName.*

IGISServerConnection contains one method for connecting to the GIS server. You call the *Connect* method and provide the name or IP address of the machine on which the server object manager is running.

Once connected to the GIS server, *IGISServerConnection* has properties that hand out references to the *ServerObjectManager* and the *ServerObjectAdmin* for making use of server objects and administering server objects, respectively.

In order to successfully connect to the GIS server using *GISServerConnection*, the user account running the application must be a member of the agsusers group on the GIS server. If the account running the application is not a member of this group, the *Connect* method on *IGISServerConnection* will return an error.

If the user account running the application is a member of the agsadmin users group on the GIS server, the *ServerObjectAdmin* property on *IGISServerConnection* can be used to get a reference on the *ServerObjectAdmin*. If the user account running the application is not in the agsadmin users group, the *ServerObjectAdmin* property will return an error.

The following code is an example of how to use the *GISServerConnection* to connect to a GIS server running on a machine "melange" and print out the names and types of the server object configurations configured on the server:

Application running as **Amelie** can connect with access to ServerObjectManager and ServerObjectAdmin.

Application running as **Fred**; connection is refused.

Application running as **Cal** can connect with access to ServerObjectManager.

GIS Server

agsusers
  Cal
  Liz

agsadmin
  Amelie

When applications make connections to the GIS server, they are authenticated against the agsusers and agsadmin users groups on the GIS server.

```
Dim IGISServerConnection As IGISServerConnection
Set IGISServerConnection = New GISServerConnection
IGISServerConnection.Connect "melange"
```

```
Dim pServerObjectManager As IServerObjectManager
Set pServerObjectManager = pGISServerConnection.ServerObjectManager
```

```
Dim pEnumConfigInfo As IEnumServerObjectConfigurationInfo
Set pEnumConfigInfo = pServerObjectManager.GetConfigurationInfos
```

```
Dim pConfigInfo As IServerObjectConfigurationInfo
Set pConfigInfo = pEnumConfigInfo.Next
Do Until pConfigInfo Is Nothing
  Debug.Print pConfigInfo.Name & ": " & pConfigInfo.TypeName
  Set pConfigInfo = pEnumConfigInfo.Next
Loop
```

Server Consumer Objects

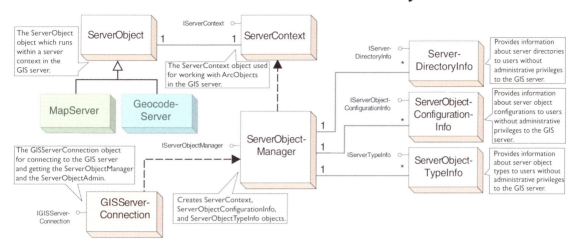

The ServerObject object which runs within a server context in the GIS server.

ServerObject

ServerContext

IServerContext

1 | 1

The ServerContext object used for working with ArcObjects in the GIS server.

MapServer

Geocode-Server

Server-DirectoryInfo

IServer-DirectoryInfo

*

Server-DirectoryInfo

Provides information about server directories to users without administrative privileges to the GIS server.

IServerObject-ConfigurationInfo

ServerObject-Configuration-Info

*

Provides information about server object configurations to users without administrative privileges to the GIS server.

The GISServerConnection object for connecting to the GIS server and getting the ServerObjectManager and the ServerObjectAdmin.

IServerObjectManager

ServerObject-Manager

1

1

IServerTypeInfo

ServerObject-TypeInfo

*

Provides information about server object types to users without administrative privileges to the GIS server.

IGISServer-Connection

GISServer-Connection

Creates ServerContext, ServerObjectConfigurationInfo, and ServerObjectTypeInfo objects.

When developing applications that connect to a GIS server, you need to be able to access objects running in the server and create objects in the server for your application's use. If the user your application runs as is a member of the agsusers user group on the GIS server, then that application will be able to connect to the GIS server using the *GISServerConnection* object and can get a reference to the set of objects that provide the ability for the application to make use of ArcObjects running in the server.

IScrverObjectManager o——

ServerObject-
Manager

The ServerObjectManager class provides methods for getting information about the GIS server and for creating server contexts for use by an application.

The *ServerObjectManager* class provides access to information about the GIS server to nonadministrators, and creates *ServerContexts* for use by applications. Any application that runs as a user account in the agsusers user group on the GIS server can use the *IGISServerConnection* interface to connect to the GIS server and to get a reference to the *ServerObjectManager*.

ISeverObjectManager : IUnknown	Provides access to properties of, and members to work with, a GIS server's server object manager.
← CreateServerContext (in configName: String, in TypeName: String) : IServerContext	Gets a reference to a server context. The server context can be based on a specified server object configuration or can be an empty server context if no server object configuration is specified.
← GetConfigurationInfo (in Name: String, in TypeName: String) : IServerObjectConfigurationInfo	Gets the information for server object configuration with the specified Name and TypeName.
← GetConfigurationInfos: IEnumServerObjectConfigurationInfo	An enumerator over all the GIS server's configuration infos.
← GetServerDirectoryInfos: IEnumServerDirectoryInfo	An enumerator over all the GIS server's directory infos.
← GetTypeInfos: IEnumServerObjectTypeInfo	An enumerator over all the GIS server's type infos.

Use the *IServerObjectManager* interface when your application connects to the GIS server to use and create server objects. The *IServerObjectManager* interface has the necessary methods for an application to get the collection of server object configurations, server object types, and server directories configured in the server as *ServerObjectConfigurationInfo*, *ServerObjectTypeInfo*, and *ServerDirectoryInfo* objects, respectively.

The *CreateServerContext* method on *IServerObjectManager* is used to get a reference to a context on the server. A context is a process managed by the server within which a server object runs. You can use *CreateServerContext* to create a context based on a server object configuration, or you can create empty contexts solely for the purpose of creating ArcObjects on the fly within the server.

When using *CreateServerContext* to create a context based on a server object configuration, if the server object configuration is pooled, you may get a reference to a context that is already created and running in the server. When you have completed using that context, it is important to release it explicitly by calling the *ReleaseContext* method on *IServerContext* to return it to the pool. When using *CreateServerContext* to create a context based on a non-pooled server object configuration, or when creating an empty context, a new context is created on the server. You still need to call *ReleaseContext* when you are finished using it, and the context is destroyed on the server.

The following code is an example of how to connect to the GIS server "melange", create a server context based on the RedlandsMap server object configuration, get a reference to the RedlandsMap *MapServer* object running in the context, and print its *DefaultMapName* property. Note the call to *ReleaseContext* when the use of the context is complete:

```
Dim pGISServerConnection As IGISServerConnection
Set pGISServerConnection = New GISServerConnection
pGISServerConnection.Connect "melange"

Dim pServerObjectManager As IServerObjectManager
Set pServerObjectManager = pGISServerConnection.ServerObjectManager
```

```
Dim pServerContext As IServerContext
Set pServerContext =
pServerObjectManager.CreateServerContext("RedlandsMap", "MapServer")

Dim pServerObject As IServerObject
Set pServerObject = pServerContext.ServerObject
Dim pMapServer As IMapServer
Set pMapServer = pServerObject

Debug.Print pMapServer.DefaultMapName

pServerContext.ReleaseContext
```

The following code is an example of how to create an empty server context and, within that context, create a new polygon and print its area. Note the call to *ReleaseContext* when the use of the context is complete:

```
Dim pGISServerConnection As IGISServerConnection
Set pGISServerConnection = New GISServerConnection
pGISServerConnection.Connect "melange"

Dim pServerObjectManager As IServerObjectManager
Set pServerObjectManager = pGISServerConnection.ServerObjectManager

Dim pServerContext As IServerContext
Set pServerContext = pServerObjectManager.CreateServerContext("", "")

' create a new polygon in the server context
Dim pGonColl As IPointCollection
Set pGonColl = pServerContext.CreateObject("esriGeometry.Polygon")

' create the points in the server context
Dim pPoint(4) As IPoint
Dim i As Long
For i = 0 To 3
 Set pPoint(i) = pServerContext.CreateObject("esriGeometry.Point")
Next

pPoint(0).PutCoords 0, 0
pPoint(1).PutCoords 10, 0
pPoint(2).PutCoords 10, 10
pPoint(3).PutCoords 0, 10
'Add the points to the polygon
pGonColl.AddPoints 4, pPoint(0)

Dim pArea As IArea
Set pArea = pGonColl

Debug.Print pArea.Area

pServerContext.ReleaseContext
```

The ServerContext class provides access to a context in the GIS server and provides methods for creating and managing objects within that context.

A *ServerContext* is a reserved space within the server dedicated to a set of running objects. GIS server objects also reside in a server context. When developing applications with ArcGIS Server, all ArcObjects that your application creates and uses reside within a server context. To obtain a server object, you actually get a reference to its context, then get the server object from the context.

You get a *ServerContext* by calling the *CreateServerContext* method on *IServerObjectManager*.

IServerContext : IUnknown	Provides access to members for managing a server context, and the objects running within that server context.
ServerObject: IServerObject	The map or geocode server object running in the server context.
CreateObject (in CLSID: String) : IUnknown Pointer	Create an object in the server context whose type is specified by the CLSID.
GetObject (in Name: String) : IUnknown Pointer	Get a reference to an object in the server context's object dictionary by its Name.
LoadObject (in str: String) : IUnknown Pointer	Create an object in the server context from a string that was created by saving an object using SaveObject.
ReleaseContext	Release the server context back to the server so it can be used by another client (if pooled), or so it can be destroyed (if non-pooled).
Remove (in Name: String)	Remove an object from the server context's object dictionary.
RemoveAll	Remove all objects from the server context's object dictionary.
SaveObject (in obj: IUnknown Pointer) : String	Save an object in the server context to a string.
SetObject (in Name: String, in obj: IUnknown Pointer)	Add an object running in the server context to the context's object dictionary.

IServerContext contains methods for creating and managing objects running within the *ServerContext*. You can get at the server object running within a server context using the *ServerObject* property on *IServerContext*.

The following code is an example of creating a server context and getting a reference to the *MapServer* server object running in the server context.

```
Dim IServerContext as IServerContext
Set IServerContext =
pServerObjectManager.CreateServerContext("RedlandsMap","MapServer")
Dim pMapServer as IMapServer
Set pMapServer = pServerContext.ServerObject
```

You can also create empty server contexts. You can use an empty context to create ArcObjects on the fly within the server to do ad hoc GIS processing. The following code is an example of creating an empty server context.

```
Dim IServerContext as IServerContext
Set IServerContext = pServerObjectManager.CreateServerContext("","")
Dim pWorkspaceFactory as IWorkspaceFactory
Set pWorkspaceFactory =
pServerContext.CreateObject("esriDataSourcesGDB.SdeWorkspaceFactory")
```

All ArcObjects that your application uses should be created within a server context using the *CreateObject* method on *IServerContext*. Also, objects that are used together should be in the same context. For example, if you create a *Point* object to use in a spatial selection to query features in a feature class, the point should be in the same context as the feature class.

ArcGIS Server applications should not use New to create ArcObjects but should always create objects by calling *CreateObject* on *IServerContext*:

Incorrect:

```
Dim pPoint as IPoint
Set pPoint = New Point
```

Correct:

```
Dim pPoint as IPoint
Set pPoint = pServerContext.CreateObject("esriGeometry.Point")
```

Garbage collection is the process by which .NET and Java reclaim memory from objects that are created by applications. Garbage collection will happen based on memory allocations being made. When garbage collection occurs is when objects that are not referenced are actually cleaned up, which may be some time after they go out of the scope of your application.

When your application is finished working with a server context, it must release it back to the server by calling the *ReleaseContext* method. If you allow the context to go out of scope without explicitly releasing it, it will remain in use and be unavailable to other applications until it is garbage collected. Once a context is released, the application can no longer make use of any objects in that context. This includes both objects that you may have obtained from the context or objects that you created in the context.

```
Dim IServerContext as IServerContext
Set IServerContext =
pServerObjectManager.CreateServerContext("RedlandsMap","MapServer")
Dim pMapServer as IMapServer
Set pMapServer = pServerContext.ServerObject
' Do something with the object
pContext.ReleaseContext
```

The *IServerContext* interface has a number of methods for helping you manage the objects you create within server contexts. The following is a description of how and when you would use these methods.

Use the *CreateObject* method when you need to create an object for use in your application.

```
Dim pPointCollection as IPointCollection
Set pPointCollection = pServerContext.CreateObject("esriGeometry.Polygon")
```

A proxy object is a local representation of a remote object. The proxy object controls access to the remote object by forcing all interaction with the remote object to be via the proxy object. The supported interfaces and methods on a proxy object are the same as those supported by the remote object. You can make method calls on, and get and set properties of, a proxy object as if you were working directly with the remote object.

CreateObject will return a proxy to the object that is in the server context. Your application can make use of the proxy as if the object were created locally within its process. If you call a method on the proxy that hands back another object, that object will actually be in the server context and your application will be handed back a proxy to that object. In the above example, if you get a point from the point collection using *IPointCollection::Point()*, the point returned will be in the same context as the point collection.

If you add a point to the point collection using *IPointCollection::AddPoint()*, the point should be in the same context as the point collection.

```
Dim pPointCollection as IPointCollection
Set pPointCollection = pServerContext.CreateObject("esriGeometry.Polygon")

Dim pPoint as IPoint
Set pPoint = pServerContext.CreateObject("esriGeometry.Point")
pPoint.X = 1
pPoint.Y = 1

pPointCollection.AddPoint pPoint
```

The objects and interfaces used for creating and working with geometries can be found in the Geometry object library. To learn more about geometry objects, see the online developer documentation.

Also, you should not directly use objects in a server context with local objects in your application and vice versa. You can indirectly use objects or make copies of them. For example, if you have a *Point* object in a server context, you can get its x, y properties and use them with local objects or use them to create a new local point. Don't directly use the point in the server context as, for example, the geometry of a local graphic element object.

Consider the following examples. In each example, assume that objects with Remote in their names are objects in a server context as in:

```
Dim pRemotePoint as IPoint
Set pRemotePoint = pServerContext.CreateObject("esriGeometry.Point")
```

while objects with Local in their name are objects created locally as in:

```
Dim pLocalPoint as IPoint
Set pLocalPoint = New Point
```

You can't set a local object to a remote object:

```
' this is incorrect
Set pLocalPoint = pRemotePoint


' this is also incorrect
Set pLocalElement.Geometry = pRemotePoint
```

Do not set a local object, or a property of a local object, to be an object obtained from a remote object:

```
' this is incorrect
Set pLocalPoint = pRemotePointCollection.Point(0)
```

When calling a method on a remote object, don't pass in local objects as parameters:

```
' this is incorrect
Set pRemoteWorkspace = pRemoteWorkspaceFactory.Open(pLocalPropertySet,0)
```

You can get simple data types (double, long, string, etc.) that are passed by value from a remote object and use them as properties of a local object as in:

```
' this is OK
pLocalPoint.X = pRemotePoint.X
pLocalPoint.Y = pRemotePoint.Y
```

SetObject and *GetObject* allow you to store references to objects in the server context. A context contains a dictionary that you can use as a convenient place to store objects that you create within the context. Note that this dictionary is itself valid only as long as you hold on to the server context, and it is emptied when you release the context. You can use this dictionary to share objects created within a context between different parts of your application that have access to the context.

myWorkspace	
	esriGeodatabase.Workspace
myPoint	
	esriGeometry.Point
myProperties	
	esriSystem.PropertySet
(other)	

A server context contains an object dictionary that serves as a convenient place for you to store references to commonly used objects. Use the SetObject and GetObject methods on IServerContext to work with the object dictionary.

SetObject adds objects to the dictionary, and *GetObject* retrieves them. An object that is set in the context will be available until it is removed (by calling *Remove* or *RemoveAll*) or until the context is released.

```
Dim pPointCollection as IPointCollection
Set pPointCollection = pServerContext.CreateObject("esriGeometry.Polygon")

pServerContext.SetObject "myPoly", pPointCollection
Dim pPoly as IPolygon
set pPoly = pServerContext.GetObject("myPoly")
```

Use the *Remove* and *RemoveAll* methods to remove objects from a context that have been set using *SetObject*. Once an object is removed, a reference can no longer be made to it using *GetObject*. Note that if you do not explicitly call *Remove* or *RemoveAll*, you can still not get references to objects set in the context after the context has been released.

```
pServerContext.Remove "myPoly"
```

The *SaveObject* and *LoadObject* methods allow you to serialize objects in the server context to strings, then deserialize them back into objects. Any object that supports *IPersistStream* can be saved and loaded using these methods. These methods allow you to copy objects between contexts. For example, if you use a *GeocodeServer* object to locate an address, then you want to draw the point that *GeocodeAddress* returns on your map, you need to copy the point into your *MapServer's* context.

```
Dim pServerContext As IServerContext
Set pServerContext = pSOM.CreateServerContext("RedlandsMap", "MapServer")

Dim pServerContext2 As IServerContext
Set pServerContext2 = pSOM.CreateServerContext("RedlandsGeocode",
"GeocodeServer")

Dim pGCServer As IGeocodeServer Set pGCServer =
pServerContext2.ServerObject
Dim pPropertySet As IPropertySet
Set pPropertySet = pServerContext2.CreateObject("esriSystem.PropertySet")
pPropertySet.SetProperty "Street", "380 New York St"

Dim pResults As IPropertySet
Set pResults = pGCServer.GeocodeAddress(pPropertySet, Nothing)
Dim pPoint As IPoint  Set pPoint = pResults.GetProperty("Shape")

' copy the Point to the Map's server context
Dim sPoint As String sPoint = pServerContext2.SaveObject(pPoint)
Dim pPointCopy As IPoint Set pPointCopy =
pServerContext.LoadObject(sPoint)
```

A PropertySet is a generic class that is used to hold any set of properties. A PropertySet's properties are stored as name/value pairs. Examples for the use of a property set are to hold the properties required for opening an SDE workspace or geocoding an address. To learn more about PropertySet objects, see the online developer documentation.

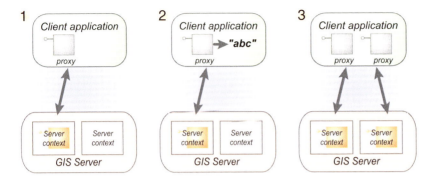

This diagram illustrates the use of SaveObject and LoadObject to copy objects between server contexts:

1. The client application gets or creates an object within a server context.

2. The application uses the SaveObject method on the object's context to serialize the object as a string that is held in the application's session state.

3. The client application gets a reference to another server context and calls the LoadObject method, passing in the string created by SaveObject. LoadObject creates a new instance of the object in the new server context.

Another important use of these methods is to manage state in your application while making stateless use of a pooled server object. A good example of this is in a mapping application. The initial session state for all users is the same and is equal to the map description for the map server object. Each user can then change map description properties such as the extent and layer visibility, which need to be maintained in the user's session state. The application does this by saving a serialized map description as part of each user's session state. Using the serialized string representation allows the application to take advantage of the standard session state management facilities of the Web server. The application uses the *LoadObject* and *SaveObject* methods to reconstitute the session's map description whenever it needs to make edits to it in response to user changes or whenever it needs to pass the map descriptor to the map server object for drawing the map according to the user's specifications.

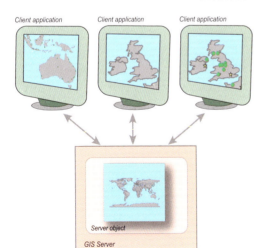

It is possible to write a stateful application that makes stateless use of server objects in the GIS server by maintaining aspects of application state such as the extent of the map, layer visibility, and application-added graphics using the application's state.

A ServerObject is a coarse-grained ArcObjects component that runs in a server context.

A *ServerObject* is a coarse-grained ArcObjects component, that is a high-level object that simplifies the programming model for doing certain operations and hides the fine-grained ArcObjects that do the work. Server objects support coarse-grained interfaces that support methods that do large units of work, such as "draw a map", or "geocode a set of addresses". *ServerObjects* also have SOAP interfaces, which makes it possible to expose server objects as Web services that can be consumed by clients across the Internet.

ArcGIS Server has two *ServerObjects*: *MapServer* and *GeocodeServer*.

IServerObject : IUnknown	Provides access to properties of a map or geocode server object.
■— ConfigurationName: String	Name of the server object configuration that defines the server object.
■— TypeName: String	Type of the server object (MapServer or GeocodeServer).

IServerObject is an interface supported by all server objects such as the *MapServer* and *GeocodeServer*. The *IServerObject* interface is returned as the *ServerObject* property on *IServerContext*.

The *IServerObject* interface has properties to indicate the name and type of the server object configuration that created the server object. You can query interface for interfaces supported by the server object type, such as *IMapServer* for a *MapServer* object or *IGeocodeServer* for a *GeocodeServer* object.

The following code shows how to connect to a GIS server, create a server context based on a server object configuration, and use the *ServerObject* property to get the *IServerObject* interface on the server context's server object:

```
Dim pGISServerConnection As IGISServerConnection
Set pGISServerConnection = New GISServerConnection
pGISServerConnection.Connect "padisha"

Dim pSOM As IServerObjectManager
Set pSOM = pGISServerConnection.ServerObjectManager

Dim pContext As IServerContext
Set pContext = pSOM.CreateServerContext("MyMapServer", "MapServer")

Dim pServerObject As IServerObject
Set pServerObject = pContext.ServerObject

Dim pMapServer As IMapServer
Set pMapServer = pServerObject

' Do something with the mapserver

pContext.ReleaseContext
```

The MapServer object is a coarse-grained server object that provides access to the contents of a map document and methods for querying and drawing the map.

The GeocodeServer object is a coarse-grained server object that provides access to an address locator and methods for single address and batch geocoding.

The ServerObjectConfigurationInfo *class provides read-only access to some of the properties of a server object configuration.*

The Server object library's Info classes provide read-only access to users and developers who are not administrators to the properties of server directories, server types, and server object configurations. These properties are necessary for developing applications using the GIS server.

Each Info class has a corresponding class that is only accessible to administrators (users in the agsadmin users group), which exposes the properties of the Info object with read/write access as well as additional properties.

A *ServerDirectoryInfo* object is an Info object that describes the properties of a server directory.

The ServerObjectTypeInfo *class provides read-only access to some of the properties of a server object type.*

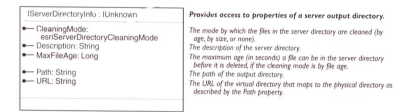

The GIS server manages a set of server directories. A server directory is a location on a file system that the GIS server is configured to clean up files it writes. The *ServerDirectoryInfo* class gives users and developers who are not administrators access to the list of server directories and the set of their properties that are necessary for programming applications that use them as locations to write output. You can get information about server directories using the *GetServerDirectoryInfos* method on *IServerObjectManager* to get the *IServerDirectoryInfo* interface.

Files in a server directory can be cleaned based on file age or based on when the file was last accessed. The maximum file age or time since last accessed is a property of a server directory. If the *CleaningMode* is *esriDCAbsolute*, then all files created by the GIS server that are older than the maximum age are automatically cleaned up by the GIS server. If the *CleaningMode* is *esriDCSliding*, then all files created by the GIS server that have not been accessed for a duration defined by maximum age are automatically cleaned up by the GIS server.

Note that when creating files in a server directory, they must be prefixed with "_ags_" to be cleaned up by the GIS server. Any files in a server directory not prefixed with "_ags_" will not be cleaned up.

IServerDirectoryInfo provides read-only access to a subset of the server directory's properties. These properties include:

Path: the physical path of the directory in disk

URL: the URL of the virtual directory corresponding to the physical directory

Description: the description of the server directory

CleaningMode: indicates whether the directory is cleaned by file age, by last accessed, or if its contents are not cleaned up

MaxFileAge: indicates the maximum age or the maximum time since last accessed that files can be in the server directory before they are cleaned up

The properties listed above are those necessary for developers of server applications to make use of the various GIS servers' server directories.

The following code shows how to list the server directories of a GIS server using the *ServerDirectoryInfo* class:

```
Dim pSC As IGISServerConnection
Set pSC = New GISServerConnection
pSC.Connect "padisha"

Dim pSOM As IServerObjectManager
Set pSOM = pSC.ServerObjectManager

Dim pEnumSDirInfo As IEnumServerDirectoryInfo
Set pEnumSDirInfo = pSOM.GetServerDirectoryInfos

Dim pSDirInfo As IServerDirectoryInfo
Set pSDirInfo = pEnumSDirInfo.Next
Do Until pSDirInfo Is Nothing
  Debug.Print pSDirInfo.Path
  Set pSDirInfo = pEnumSDirInfo.Next
Loop
```

A *ServerObjectConfigurationInfo* object is an Info object that describes the properties of a server object configuration.

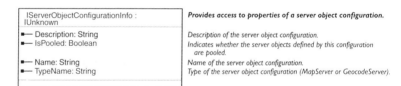

IServerObjectConfigurationInfo : IUnknown	**Provides access to properties of a server object configuration.**
▪— Description: String	*Description of the server object configuration.*
▪— IsPooled: Boolean	*Indicates whether the server objects defined by this configuration are pooled.*
▪— Name: String	*Name of the server object configuration.*
▪— TypeName: String	*Type of the server object configuration (MapServer or GeocodeServer).*

The GIS server manages a set of server objects running across one or more host (container) machines. How those server objects are configured and run is defined by a set of server object configurations. The *ServerObjectConfigurationInfo* class gives users and developers who are not administrators access to the list of server object configurations and the set of their properties that are necessary for programming applications with them. You can get information about server object configurations using the *GetConfigurationInfos* method on *IServerObjectManager* to get the *IServerObjectConfigurationInfo* interface. Note that the *GetConfigurationInfos* method returns only server object configurations that are started.

IServerObjectConfigurationInfo provides read-only access to a subset of the server object configuration's properties. These properties include:

Name: the name of the server object configuration

TypeName: the type of server object configuration (for example, MapServer, GeocodeServer)

Description: the description of the server object configuration

IsPooled: indicates whether the server objects described by this configuration are pooled or non-pooled

The properties listed above are those necessary for developers of server applications to make use of the various server objects configured on the GIS server.

The following example shows how to connect to a GIS server and use the *IServerObjectConfigurationInfo* interface to print out the name and type of all the server object configurations:

```
Dim pGISServerConnection As IGISServerConnection
Set pGISServerConnection = New GISServerConnection
pGISServerConnection.Connect "melange"

Dim pSOM As IServerObjectManager
Set pSOM = pGISServerConnection.ServerObjectManager

Dim pEnumSOCInfo As IEnumServerObjectConfigurationInfo
Set pEnumSOCInfo = pSOM.GetConfigurationInfos

Dim pSOCInfo As IServerObjectConfigurationInfo
Set pSOCInfo = pEnumSOCInfo.Next
Do Until pSOCInfo Is Nothing
  Debug.Print pSOCInfo.Name & ": " & pSOCInfo.TypeName
  Set pSOCInfo = pEnumSOCInfo.Next
Loop
```

A *ServerObjectTypeInfo* object is an Info object that describes the properties of a server object type.

The *ServerObjectTypeInfo* class gives users and developers who are not administrators access to the list of server object types and the set of their properties that are necessary for programming applications with them. You can get information about server object types using the *GetTypeInfos* method on *IServerObjectManager* to get the *IServerObjectTypeInfo* interface.

IServerObjectTypeInfo : IUnknown	Provides access to properties of a server object type.
■— Description: String ■— Name: String	Description of the server object type. Name of the server object type.

IServerObjectTypeInfo provides read-only access to a subset of the server object type's properties. These properties include:

Name: the name of the server object type (for example, MapServer, GeocodeServer)

Description: the description of the server object type

Server Administration Objects

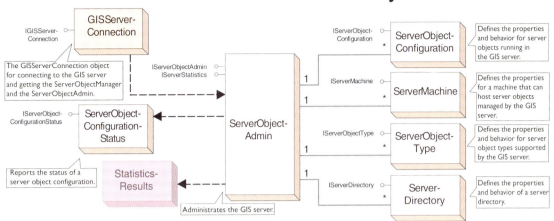

When developing applications that connect to a GIS server for the purposes of administering the GIS server and its server objects, you need to have access to the objects for administering these aspects of the GIS server. If the user your application runs as is a member of the agsadmin users group on the GIS server, then that application will be able to connect to the GIS server using the *GISServerConnection* object and can get a reference to the set of objects that provide the ability for the application to administrate the GIS server and its server objects.

Applications that make use of ArcObjects running in the server do not require access to the administration objects.

IServerObjectAdmin
IServerStatistics

ServerObject-
Admin

The ServerObjectAdmin class provides methods for administrating the GIS server and its server objects.

The *ServerObjectAdmin* class administrates a GIS server. Any application that runs as a user account in the agsadmin users group on the GIS server can use the *IGISServerConnection* interface to connect to the GIS server and to get a reference to the *ServerObjectAdmin*. If the user account is not part of the agsadmin users group, the *ServerObjectAdmin* property on *IGISServerConnection* will return an error. Applications that are running as accounts that can connect to the server but are not part of the agsadmin users group can use the *ServerObjectManager* property on *IGISServerConnection* to get a reference on the *ServerObjectManager*.

Use *ServerObjectAdmin* to administrate both the set of server object configurations and types associated with the server as well as to administer aspects of the server itself. The following administration functionality of the GIS Server is provided by *ServerObjectAdmin*:

Administer server object configurations:

- Add and delete server object configurations.

- Update a server object configuration's properties.

- Start, stop, and pause server object configurations.

- Report the status of a server object configuration.

- Get all server object configurations and their properties.

- Get all server object types and their properties.

Administer aspects of the server itself:

- Add and remove server container machines.

- Get all server container machines.

- Add and remove server directories.

- Get all server directories.

- Configure the server's logging properties.

- Get statistics about events in the server.

ISeverObjectAdmin : IUnknown	Provide access to members that administrate the GIS server.
■—■ Properties: IPropertySet	The logging properties for the GIS server.
◄— AddConfiguration (in config: IServerObjectConfiguration)	Adds a server object configuration (created with CreateConfiguration) to the GIS server.
◄— AddMachine (in machine: IServerMachine)	Adds a host machine (created with CreateMachine) to the GIS server.
◄— AddServerDirectory (in pSD: IServerDirectory)	Adds a server directory (created with CreateServerDirectory) to the GIS server.
◄— CreateConfiguration: IServerObjectConfiguration	Creates a new server object configuration.
◄— CreateMachine: IServerMachine	Creates a new host machine.
◄— CreateServerDirectory: IServerDirectory	Creates a new server directory.
◄— DeleteConfiguration (in Name: String, in TypeName: String)	Deletes a server object configuration from the GIS server.
◄— DeleteMachine (in machineName: String)	Deletes a host machine from the GIS server, making it unavailable to host server objects.
◄— DeleteServerDirectory (in Path: String)	Deletes a server directory such that its cleanup is no longer managed by the GIS server. It does not delete the physical directory from disk.
◄— GetConfiguration (in Name: String, in TypeName: String) : IServerObjectConfiguration	Get the server object configuration with the specified Name and TypeName.
◄— GetConfigurations: IEnumServerObjectConfiguration	An enumerator over all the server object configurations.
◄— GetConfigurationStatus (in Name: String, in TypeName: String) : IServerObjectConfigurationStatus	Get the configuration status for a server object configuration with the specified Name and TypeName.
◄— GetMachine (in Name: String) : IServerMachine	Get the host machine with the specified Name.
◄— GetMachines: IEnumServerMachine	An enumerator over all the GIS server's host machines.
◄— GetServerDirectories: IEnumServerDirectory	An enumerator over the GIS server's output directories.
◄— GetServerDirectory (in Path: String) : IServerDirectory	Get the server directory with the specified Path.
◄— GetTypes: IEnumServerObjectType	An enumerator over all the server object types.
◄— PauseConfiguration (in Name: String, in TypeName: String)	Makes the configuration unavailable to clients for processing requests, but does not shut down running instances of server objects, or interrupt requests in progress.
◄— StartConfiguration (in Name: String, in TypeName: String)	Starts a server object configuration and makes it available to clients for processing requests.
◄— StopConfiguration (in Name: String, in TypeName: String)	Stops a server object configuration and shuts down any running instances of server objects defined by the configuration.
◄— UpdateConfiguration (in config: IServerObjectConfiguration)	Updates the properties of a server object configuration.
◄— UpdateMachine (in machine: IServerMachine)	Updates the properties of a host machine.
◄— UpdateServerDirectory (in pSD: IServerDirectory)	Updates the properties of a server directory.

You can use *IServerObjectAdmin* to administrate either the server's set of server object configurations and server object types and to administrate aspects of the server itself, such as the list of machines that may host server objects.

If your application is connecting to the server to make use of objects in the server, use the *IServerObjectManager* interface.

ADMINISTER SERVER OBJECT CONFIGURATIONS

The *AddConfiguration* method will add a *ServerObjectConfiguration* to your GIS server. A new *ServerObjectConfiguration* can be created using the *CreateConfiguration* method. Use the *IServerObjectConfiguration* interface to set the various properties of the configuration, then use the *AddConfiguration* method on *IServerObjectAdmin* to add the new configuration to the GIS server.

Once a configuration is added to the server, you can use *StartConfiguration* to make it available for applications to use.

The following code shows how to connect to the GIS server called "melange" and use the *CreateConfiguration*, *AddConfiguration*, and *StartConfiguration* methods to create a new geocode server object configuration, add it to the server, and make it available for use:

```
Dim pGISServerConnection As IGISServerConnection
Set pGISServerConnection = New GISServerConnection
pGISServerConnection.Connect "melange"

Dim pServerObjectAdmin As IServerObjectAdmin
Set pServerObjectAdmin = pGISServerConnection.ServerObjectAdmin

' create the new configuration
Dim pConfiguration As IServerObjectConfiguration
Set pConfiguration = pServerObjectAdmin.CreateConfiguration

pConfiguration.Name = "California"
pConfiguration.TypeName = "GeocodeServer"

Dim pProps As IPropertySet
Set pProps = pConfiguration.Properties
pProps.SetProperty "LocatorWorkspacePath",
"\\melange\Geocoding\California"
pProps.SetProperty "Locator", "California"
pProps.SetProperty "SuggestedBatchSize", "500"

pConfiguration.IsPooled = True
pConfiguration.MinInstances = 1
pConfiguration.MaxInstances = 1

pConfiguration.WaitTimeout = 10
pConfiguration.UsageTimeout = 120

' add the configuration to the server
pServerObjectAdmin.AddConfiguration pConfiguration
```

The *UpdateConfiguration* method will update the *ServerObjectConfiguration* that is specified when the method is called. You can use the *GetConfiguration* or *GetConfigurations* methods on *IServerObjectAdmin* to get a reference to the *ServerObjectConfiguration* you want to update.

Note that the server object configuration must be stopped before you call *UpdateConfiguration*. You can use *StopConfiguration* to stop the server object configuration.

The following code shows how to connect to the GIS server called "melange" and use *GetConfiguration* to get a *ServerObjectConfiguration*, change its *MinInstances* property, then update the configuration using the *UpdateConfiguration* method:

```
Dim pGISServerConnection As IGISServerConnection
Set pGISServerConnection = New GISServerConnection
pGISServerConnection.Connect "melange"

Dim pServerObjectAdmin As IServerObjectAdmin
Set pServerObjectAdmin = pGISServerConnection.ServerObjectAdmin
Dim pConfig as IServerObjectConfiguration
Set pConfig = pServerObjectAdmin.GetConfiguration ("RedlandsMap",
"MapServer")
```

```
pConfig.MinInstances = 3
pServerObjectAdmin.UpdateConfiguration pConfig
```

Use *DeleteConfiguration* to delete a server object configuration from your GIS server. Note: In order to call *DeleteConfiguration*, the server object configuration must be stopped. If it is not stopped, then *DeleteConfiguration* will return an error.

The following code shows how to stop, then delete a server object configuration called RedlandsMap:

```
Dim pGISServerConnection As IGISServerConnection
Set pGISServerConnection = New GISServerConnection
pGISServerConnection.Connect "melange"

Dim pServerObjectAdmin As IServerObjectAdmin
Set pServerObjectAdmin = pGISServerConnection.ServerObjectAdmin


pServerObjectAdmin.StopConfiguration "RedlandsMap", "MapServer"
pServerObjectAdmin.DeleteConfiguration "RedlandsMap", "MapServer"
```

The *GetConfigurations* method returns an enumeration of all the server object configurations that are configured in the GIS server.

The following code shows how to connect to the GIS server named "melange" and print the name and type of all its server object configurations.

```
Dim pGISServerConnection As IGISServerConnection
Set pGISServerConnection = New GISServerConnection
pGISServerConnection.Connect "melange"

Dim pServerObjectAdmin As IServerObjectAdmin
Set pServerObjectAdmin = pGISServerConnection.ServerObjectAdmin

Dim pEnumSOC As IEnumServerObjectConfiguration
Set pEnumSOC = pServerObjectAdmin.GetConfigurations

Dim pSOC As IServerObjectConfiguration
Set pSOC = pEnumSOC.Next
Do Until pSOC Is Nothing
  Debug.Print pSOC.Name & ": " & pSOC.TypeName
  Set pSOC = pEnumSOC.Next
Loop
```

The *GetConfigurationStatus* method will return a *ServerObjectConfigurationStatus* object for the specified server object configuration. You can use *IServerObjectConfigurationStatus* to get information such as the number of instances of that configuration that are running and the number in use.

The ServerObjectConfigurationStatus class provides information about the status of a server object configuration.

IServerObjectConfigurationStatus also provides information as to the startup status of the configuration.

You can use *IServerObjectConfigurationStatus* to monitor server object configuration usage. For example, you can use the *InstanceInUseCount* to determine if any instances of the configuration are in use before stopping the configuration.

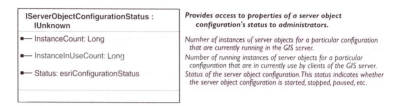

The following code demonstrates using *GetConfigurationStatus* and the *InstancesInUse* property to verify no instances of a configuration are in use before stopping it:

```
Dim pGISServerConnection As IGISServerConnection
Set pGISServerConnection = New GISServerConnection
pGISServerConnection.Connect "melange"

Dim pServerObjectAdmin As IServerObjectAdmin
Set pServerObjectAdmin = pGISServerConnection.ServerObjectAdmin

Dim pSOCStatus As IServerObjectConfigurationStatus
Set pSOCStatus = pServerObjectAdmin.GetConfigurationStatus("RedlandsMap",
"MapServer")

If pSOCStatus.InstanceInUseCount = 0 Then
  pServerObjectAdmin.StopConfiguration "MyMapServer", "RedlandsMap"
End If
```

The *GetTypes* method returns an enumeration of the server object types that are installed in the GIS server. By default, the *MapServer* and *GeocodeServer* object types are installed in the server.

The *StartConfiguration* method will start the server object configuration and make it available for use by clients. When a server object configuration is started, if it is a pooled configuration, the minimum number of server objects as described by the configuration will be preloaded. The *StartConfiguration* method will complete only when all the server object instances have been created to satisfy the minimum.

The *StopConfiguration* method stops a server object configuration. Stopping a server object configuration both makes the server object configuration unusable by clients (calls to *CreateServerObject* referencing the configuration will fail) andshuts down any of the configurations running server objects. If any of those objects are in use and executing requests, they will be interrupted and shut down.

You should stop a configuration when you need to change its properties, such as the pooling model or recycling model. Use the *StartConfiguration* method to restart the configuration.

If you want to stop the configuration without interrupting clients that are making use of server objects, you can pause the configuration using *PauseConfiguration* and wait until such a time that all server object usage has ended, then call *StopConfiguration* to stop the configuration.

The *PauseConfiguration* method will pause the server object configuration. Pausing the configuration does not interrupt any requests that are being executed by instances of server objects managed by that configuration. A paused server object, however, will refuse future requests.

You should pause server objects if you want to lock requests from being processed, but you want requests in process to complete. Once all objects are no longer in use, you can perform necessary operations (database maintenance, stopping the server object configuration to changes properties, and so forth). Use *GetConfigurationStatus* and *IServerObjectConfigurationStatus* to determine if any server objects are in use.

Use the *StartConfiguration* method to cancel pause on the server object configuration.

ADMINISTER SERVER MACHINES

ArcGIS Server is a distributed system. Server objects managed by the GIS server can run on one or more host machines. A machine that can host server objects must have the Server Object Container installed on it, and the machine must be added to the list of host machines managed by the server object manager.

Use the *AddMachine* method to add new host machines to your GIS server. Once a machine has been added to the GIS server, as new server object instances are created, the server object manager will make use of the new machine.

Use the *CreateMachine* method to create a new server machine that you can pass as an argument to the *AddMachine* method to add new host machines to your GIS server.

The following code shows how to use the *CreateMachine* and *AddMachine* methods to add a machine to your GIS server:

```
Dim pGISServerConnection As IGISServerConnection
Set pGISServerConnection = New GISServerConnection
pGISServerConnection.Connect "melange"

Dim pServerObjectAdmin As IServerObjectAdmin
Set pServerObjectAdmin = pGISServerConnection.ServerObjectAdmin

Dim pServerMachine As IServerMachine
Set pServerMachine = pServerObjectAdmin.CreateMachine
pServerMachine.Name = "callum"

pServerObjectAdmin.AddMachine pServerMachine
```

Use the *GetMachines* method to retrieve the names of the machines that have been added to the server to host server objects.

The *DeleteMachine* method removes a machine from the machines that can host server objects for the GIS server. When you delete a machine, any instances of server objects that are running on that machine will be shut down and replaced with instances running on the GIS server's other host machines.

ADMINISTER SERVER DIRECTORIES

You can use *IServerObjectAdmin* to add and remove server directories from the GIS server. Both server objects and server applications typically need to write either temporary data or result data to some location for it to be delivered or presented to the end user. For example, a map server object's *ExportMapImage* method can create an image file that is then displayed on a Web application. These files are typically transient and temporary by nature. For example, when a map server writes an image to satisfy a request from a Web application, that image is needed only for the time it takes to display it on the Web application. An application that creates check-out personal geodatabases for download would provide a finite amount of time during which that geodatabase is created and when it can be downloaded.

Because server applications support many user sessions, these output files can accumulate and need to be periodically cleaned up. The server provides the capability to automatically clean up these output files if they are written to one of the server's output directories.

Use the *CreateServerDirectory* method to create a new server directory that you can pass as an argument to the *AddServerDirectory* method to add new server directories to your GIS server. Once you have added the server directory, you can configure your server objects and server applications to make use of the server directory.

Note: Server directories must be accessible by all host machines configured in the GIS server.

The following code shows how to use the *CreateServerDirectory* and *AddServerDirectory* methods to add a directory to your GIS server:

```
Dim pGISServerConnection As IGISServerConnection
Set pGISServerConnection = New GISServerConnection
pGISServerConnection.Connect "melange"

Dim pServerObjectAdmin As IServerObjectAdmin
Set pServerObjectAdmin = pGISServerConnection.ServerObjectAdmin

Dim pSDir As IServerDirectory
Set pSDir = pServerObjectAdmin.CreateServerDirectory

pSDir.CleaningMode = esriSDCByFileAge
pSDir.Description = "Default output directory"
pSDir.Path = "\\melange\serveroutput"

pSDir.URL = http://melange/serveroutput
```

```
pSDir.MaxFileAge = 100

pSOM.AddServerDirectory pSDir
```

The *UpdateServerDirectory* method will update the *ServerDirectory* that is specified when the method is called. You can use the *GetServerDirectory* or *GetServerDirectories* methods on *IServerObjectAdmin* to get a reference to the *ServerDirectory* you want to update.

The *UpdateServerDirectory* is useful for modifying the cleanup mode (*CleaningMode*) and cleanup schedule.

The following code shows how to connect to the GIS server called "melange" and use the *GetServerDirectory* method to get a *ServerDirectory*, change its *MaxFileAge* property, then update the directory using the *UpdateServerDirectory* method:

```
Dim pGISServerConnection As IGISServerConnection
Set pGISServerConnection = New GISServerConnection
pGISServerConnection.Connect "padisha"

Dim pServerObjectAdmin As IServerObjectAdmin
Set pServerObjectAdmin = pGISServerConnection.ServerObjectAdmin

Dim pSDir As IServerDirectory
Set pSDir =
pServerObjectAdmin.GetServerDirectory("\\melange\serveroutput")
pSDir.MaxFileAge = 200
pServerObjectAdmin.UpdateServerDirectory pSDir
```

The *DeleteServerDirectory* method removes a directory from the set of directories managed by the GIS server. The *DeleteServerDirectory* method will not affect the physical directory.

When a server directory is removed with this method, the GIS server will no longer manage the cleanup of output files written to that directory. Applications or server objects that are configured to write their output to the physical directory that is referenced by the server directory will continue to work, but the files they write will not be cleaned up by the server.

ADMINISTER SERVER LOGGING AND TIME-OUT PROPERTIES

The *Properties* property on *IServerObjectAdmin* returns the properties for the GIS server. The properties are for the GIS server's logging and for server object creation time-out.

The GIS server logs its activity, including server object configuration startup, shutdown, server context creation and shutdown, and errors generated through any failed operation or request in the GIS server.

You can control the logging properties through the *PropertySet* returned by *Properties*. The following is a description of the logging properties:

LogPath: this is the path to the location on disk to which log files are written. By default, the LogPath is <install location>\log.

LogSize: this is the size to which a single log file can grow (in MB) before a new log file is created. By default, the LogSize is 10.

LogLevel: this is a number between 0 and 5 that indicates the level of detail that the server logs. By default, the LogLevel is 3. The following is a description of the each log level:

0 (None): No logging

1 (Error): Serious problems that require immediate attention

2 (Warning): Problems that require attention

3 (Normal): Common administrative messages of the server

4 (Detailed): Common messages from user use of the server, including server objects

5 (Debug): Verbose messages to aid in troubleshooting

All aspects of logging can be changed when the GIS server is running. When they are changed, the server will immediately use the new logging settings.

The following example shows how to use the *Properties* property on *IServerObjectAdmin* to modify the logging properties of the GIS server:

A PropertySet is a generic class that is used to hold any set of properties. A PropertySet's properties are stored as name/value pairs. Examples for the use of a property set are to hold the properties required for opening an SDE workspace or geocoding an address. To learn more about PropertySet objects, see the online developer documentation.

```
Dim pGISServerConnection As IGISServerConnection
Set pGISServerConnection = New GISServerConnection
pGISServerConnection.Connect "melange"


Dim pServerObjectAdmin As IServerObjectAdmin
Set pServerObjectAdmin = pGISServerConnection.ServerObjectAdmin


Dim pLogProps As IPropertySet
Set pLogProps = pServerObjectAdmin.Properties


pLogProps.SetProperty "LogPath", "c:\ServerLogs"
pLogProps.SetProperty "LogLevel", 5


pServerObjectAdmin.Properties = pLogProps
```

Server object creation may hang for a variety of reasons. To prevent this from adversely affecting the GIS server, it has a *ConfigurationStartTimeout* property that defines the maximum time in seconds a server object instance has to initialize itself before its creation is cancelled.

GET STATISTICS ABOUT EVENTS IN THE SERVER

As the GIS server creates and destroys server objects, handles client requests, and so on, statistics about these events are logged in the GIS server's logs. In addition to the log, statistics are also kept in memory and can be queried using the *IServerStatistics* interface.

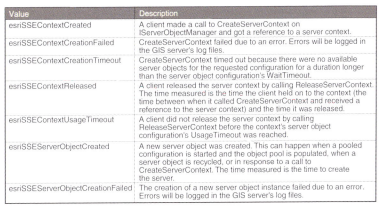

IServerStatistics : IUnknown	Provides access to members that report statistics for a GIS server to administrators.
← GetAllStatisticsForTimeInterval (in event: esriServerStatEvent, in period: esriServerTimePeriod, in index: Long, in length: Long, in Name: String, in Type: String, in machine: String) : IStatisticsResults	Gets a set of statistics, such as count, min, max, mean for an event (such as context usage time) for a specified time period.
← GetSpecificStatisticForTimeIntervals (in event: esriServerStatEvent, in function: esriServerStatFunction, in period: esriServerTimePeriod, in index: Long, in length: Long, in Name: String, in Type: String, in machine: String) : IDoubleArray	Gets a specific statistic (such as total count of server contexts created) for a specified time period.
← Reset	Clears out the currently gathered statistics.

You can query the GIS server for statistics on the following events described by *esriServerStatEvent*:

ArcCatalog provides administrators with a user interface for querying the GIS server's statistics.

Value	Description
esriSSEContextCreated	A client made a call to CreateServerContext on IServerObjectManager and got a reference to a server context.
esriSSEContextCreationFailed	CreateServerContext failed due to an error. Errors will be logged in the GIS server's log files.
esriSSEContextCreationTimeout	CreateServerContext timed out because there were no available server objects for the requested configuration for a duration longer than the server object configuration's WaitTimeout.
esriSSEContextReleased	A client released the server context by calling ReleaseServerContext. The time measured is the time the client held on to the context (the time between when it called CreateServerContext and received a reference to the server context) and the time it was released.
esriSSEContextUsageTimeout	A client did not release the server context by calling ReleaseServerContext before the context's server object configuration's UsageTimeout was reached.
esriSSEServerObjectCreated	A new server object was created. This can happen when a pooled configuration is started and the object pool is populated, when a server object is recycled, or in response to a call to CreateServerContext. The time measured is the time to create the server.
esriSSEServerObjectCreationFailed	The creation of a new server object instance failed due to an error. Errors will be logged in the GIS server's log files.

You can query these events using the statistical functions described by *esriServerStatFunction*:

- *esriSSFCount*
- *esriSSFMinimum*
- *esriSSFMaximum*
- *esriSSFSum*
- *esriSSFSumSquares*
- *esriSSFMean*
- *esriSSFStandardDeviation*

Note: For *esriSSEContextCreationFailed*, *esriSSEContextCreationTimeout*, *esriSSEContextUsageTimeout*, and *esriSSEServerObjectCreationFailed*, the only relevant statistical function is *esriSSFCount*, as these events do not have time associated with them. The other functions reflect the statistics of the elapsed time associated with the event.

While the GIS server's logs maintain a record of all events in the server, the set of statistics that are in memory and that can be queried are accumulated summaries of time slices since the GIS server was started. The granularity of these time slices is more coarse the further back in time you go. These statistics can be queried for the following time intervals:

- By second for the current minute

- By minute for the current hour

- By hour for the current day

- By day for events that happened previous to the current day

Each time period is an accumulated total of the statistics for that time period. For example, if you query the total number of requests to create server contexts for the last 30 days, you would get statistics from now to the beginning of the day 30 days ago (not to the current time on that day). This is because the in-memory statistics have been combined for that entire day.

This means that you may actually get statistics for a longer period that you specified in your query. When you query the GIS server for statistics, you can use the *IServerTimeRange* interface to get a report of the actual time period that your queries results reflect.

The *IServerStatistics* interface has methods for querying a specific statistical function for an event or for querying all statistical functions for an event.

Use the *GetSpecificStatisticForTimeIntervals* method to query the GIS server for a specific statistic for an event at discrete time intervals. For example, you can use this method to get the count of all server contexts that were created for each minute of the last hour.

Use the *GetAllStatisticsForTimeInterval* to query the GIS server for all statistics for an event. For example, you can use this method to get the sum, mean, and so forth, of server contexts usage time for the last two days.

These methods can be used to query based on the events occurring in the server as a whole (that is, across all machines) or for those occurring on a specific machine. In addition, these methods can be used to query based on the events using all server objects or for events on a particular server object.

You specify the time interval for which you want to query using an index of time periods relative to the current time based on the time period described by *esriServerTimePeriod*. The index argument to the *GetSpecificStatisticForTimeIntervals* and *GetAllStatisticsForTimeInterval* methods describe the index of the time period to start from, and the length argument describes the number of time periods to query.

For example, to query for all statistics in the last two hours, specify a time period of *esriSTPHour*, an index of 0, and a length of 2.

To query for all statistics since the server started, specify a time period of *esriSTPNone*, an index of 0, and a length of 1.

If you are unsure of the actual time period that the results of your query reflect, use the *IServerTimeRange* interface to get a report of the actual time period that your query results reflect.

IServerTimeRange : IUnknown	Provides access to members that report the actual time range for GIS server statistics reported by IServerStatistics to administrators.
■— EndTime: Date ■— StartTime: Date	*The end time for the period that the statistics represent.* *The start time for the period that the statistics represent.*

Use the *Reset* method to clear the statistics in memory.

The following code shows how to query the GIS server for statistics on the create context event for all host machines and all configurations since the GIS server was started. It also demonstrates the usage of the *IServerTimeRange* interface to report the actual time the statistics represent:

```
Dim pGISServerConnection As IGISServerConnection
Set pGISServerConnection = New GISServerConnection
pGISServerConnection.Connect "melange"

Dim pServerStats As IServerStatistics
Set pServerStats = pGISServerConnection.ServerObjectAdmin

Dim pStatsRes As IStatisticsResults
Dim pSTR As IServerTimeRange

Set pStatsRes =
pServerStats.GetAllStatisticsForTimeInterval(esriSSEContextCreated,
esriSTPNone, 0, 1, "", "", "")
Set pSTR = pStatsRes

Debug.Print pStatsRes.Count & ": " & pSTR.StartTime & " to " &
pSTR.EndTime
```

The following code shows the same query but for statistics for only the last two hours:

```
Dim pGISServerConnection As IGISServerConnection
Set pGISServerConnection = New GISServerConnection
pGISServerConnection.Connect "melange"

Dim pServerStats As IServerStatistics
Set pServerStats = pGISServerConnection.ServerObjectAdmin

Dim pStatsRes As IStatisticsResults
Dim pSTR As IServerTimeRange
```

```
Set pStatsRes =
pServerStats.GetAllStatisticsForTimeInterval(esriSSEContextCreated,
esriSTPHour, 0, 2, "", "", "")
Set pSTR = pStatsRes

Debug.Print pStatsRes.Count & ": " & pSTR.StartTime & " to " &
pSTR.EndTime
```

The following code shows the query for statistics since the server started for only the Yellowstone map server object:

```
Dim pGISServerConnection As IGISServerConnection
Set pGISServerConnection = New GISServerConnection
pGISServerConnection.Connect "melange"

Dim pServerStats As IServerStatistics
Set pServerStats = pGISServerConnection.ServerObjectAdmin

Dim pStatsRes As IStatisticsResults
Dim pSTR As IServerTimeRange

Set pStatsRes =
pServerStats.GetAllStatisticsForTimeInterval(esriSSEContextCreated,
esriSTPNone, 0, 1, "Yellowstone", "MapServer", "")
Set pSTR = pStatsRes

Debug.Print pStatsRes.Count & ": " & pSTR.StartTime & " to " &
pSTR.EndTime
```

The following code shows the query for the count of server contexts created for each minute in the last 60 minutes for all host machines and all server objects:

```
Dim pGISServerConnection As IGISServerConnection
Set pGISServerConnection = New GISServerConnection
pGISServerConnection.Connect "melange"

Dim pServerStats As IServerStatistics
Set pServerStats = pGISServerConnection.ServerObjectAdmin

Dim dr As IDoubleArray

Set dr =
pServerStats.GetSpecificStatisticForTimeIntervals(esriSSEContextCreated,
esriSSFCount, esriSTPMinute, 0, 60, "", "", "")

For i = 0 To dr.Count - 1
  Debug.Print dr.Element(i)
Next i
```

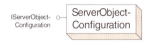

The ServerObjectConfiguration class describes the configuration for a server object that is managed by the GIS server.

The *ServerObjectConfiguration* class describes the configuration for a server object that is managed by the GIS server. *ServerObjectConfigurations* can be added, removed, and modified by users or developers who are members of the agsadmin users group and, therefore, have administrator privileges on the GIS server.

The administrator-level properties of *ServerObjectConfiguration* are:

- *Properties*
- *RecycleProperties*
- *MinInstances*
- *MaxInstances*
- *IsolationLevel*
- *StartupType*
- *WaitTimeout*
- *UsageTimeout*

A read-only subset of properties of a *ServerObjectConfiguration* are available to nonadministrators via the *GISServerConnectionInfo* object. These nonadministrator level properties are:

- *Name*
- *TypeName*
- *Description*
- *IsPooled*

ISecverObjectConfiguration : IUnknown	*Provides access to members that control the behavior and properties of a server object configuration to administrators.*
■—■ Description: String	*Description of the server object configuration.*
■—■ IsolationLevel: esriServerIsolationLevel	*The isolation level of the server objects defined by the server object configuration.*
■—■ IsPooled: Boolean	*Indicates whether the server objects defined by this configuration are pooled.*
■—■ MaxInstances: Long	*The maximum number of server object instances for a server object configuration.*
■—■ MinInstances: Long	*The minimum number of server object instances for a pooled server object configuration.*
■—■ Name: String	*Name of the server object configuration.*
■—◻ Properties: IPropertySet	*Initialization parameters and properties for the server objects created by the server object configuration.*
■—◻ RecycleProperties: IPropertySet	*The recycling properties for the server object configuration.*
■—■ StartupType: esriStartupType	*The startup type for this server object configuration. Startup type describes whether the server object configuration is started when the server object manager service is started for the GIS server.*
■—■ TypeName: String	*Type of the server object configuration (MapServer or GeocodeServer).*
■—■ UsageTimeout: Long	*Maximum time (in seconds) a client can hold on to an instance of a server object for this server object configuration before releasing it back to the server. It is the maximum time allowed between calling CreateServerContext and ReleaseServerContext.*
■—■ WaitTimeout: Long	*Maximum time (in seconds) a client will wait for an instance of a server object for this server object configuration using the CreateServerContext method on IServerObjectManager before timing out.*

The *IServerObjectConfiguration* interface is a read/write interface on a server object configuration that allows administrators to configure new server object configurations to add to the server, update existing server object configurations, and view the configuration properties of a server object configuration.

If you use *IServerObjectConfiguration* to modify any of a configuration's properties, you must call *UpdateConfiguration* on *IServerObjectAdmin* for those changes to be reflected in the server.

The following code shows how to connect to the GIS server "melange" and use the *IServerObjectConfiguration* interface to set the properties of a new server object configuration created and added to the server with the *CreateConfiguration* and *AddConfiguration* methods on *IServerObjectAdmin* to create a new geocode server object configuration and add it to the server:

```
Dim pGISServerConnection As IGISServerConnection
Set pGISServerConnection = New GISServerConnection
pGISServerConnection.Connect "melange"

Dim pServerObjectAdmin As IServerObjectAdmin
Set pServerObjectAdmin = pGISServerConnection.ServerObjectAdmin

' create the new configuration
Dim pConfiguration As IServerObjectConfiguration
Set pConfiguration = pServerObjectAdmin.CreateConfiguration

pConfiguration.Name = "California"
pConfiguration.TypeName = "GeocodeServer"

Dim pProps As IPropertySet
Set pProps = pConfiguration.Properties
pProps.SetProperty "LocatorWorkspacePath",
"\\melange\Geocoding\California"
pProps.SetProperty "Locator", "California"
pProps.SetProperty "SuggestedBatchSize", "500"

pConfiguration.IsPooled = True
pConfiguration.MinInstances = 1
pConfiguration.MaxInstances = 1
pConfiguration.WaitTimeout = 10
pConfiguration.UsageTimeout = 120

' add the configuration to the server
pServerObjectAdmin.AddConfiguration pConfiguration
```

Use the *IsolationLevel* property to get the server object isolation, or set it for a new configuration or to update an existing configuration.

Server objects can have either high isolation (*esriServerIsolationLevelHigh*) or low isolation (*esriIsolationLevelLow*). Each instance of a server object with high isolation runs in a dedicated process on the server that it does not share with other server objects. Instances of server objects with low isolation may share the same process with other server object instances of the same configuration.

Use the *IsPooled* property to indicate if the server objects created by this server object configuration are pooled or non-pooled.

Server objects can be either pooled or non-pooled. Pooled server objects can be shared across multiple sessions and applications and are held on to by an applica-

```
' add the configuration to the server
pServerObjectAdmin.AddConfiguration pConfiguration

' start the configuration
pServerObjectAdmin.StartConfiguration pConfiguration.Name,
pConfiguration.TypeName
```

Recycling allows for server objects that have become unusable to be destroyed and replaced with fresh server objects and to reclaim resources taken up by stale server objects.

Pooled server objects are typically shared between multiple applications and users of those applications. Through reuse, a number of things can happen to a server object to make them unavailable for use by applications. For example, an application may incorrectly modify a server object's state, or an application may incorrectly hold a reference to a server object, making it unavailable to other applications or sessions. In some cases, a server object may become corrupted and unusable.

Recycling allows you to keep the pool of server objects fresh and cycle out stale or unusable server objects.

You can get the recycling properties and change them using the *RecyclingProperties* property on the server object configuration. The *RecyclingProperties* property returns an *IPropertySet*. Use *GetProperty* and *SetProperty* on *IPropertySet* to get and set these properties. If you change these properties, you must call *UpdateConfiguration* to change them in the server object configuration.

The properties associated with recycling are:

- *StartTime*: the time at which the recycling interval is initialized. The time specified is in 24-hour notation. For example, to set the start time at 2:00 p.m., the StartTime property would be 14:00.

- *Interval*: the time between recycling operations in seconds. For example, to recycle the configuration every hour, this property would be set to 3600.

The *StartupType* indicates if the configuration is automatically started (*esriSTAutomatic*) when the server object manager windows service is started. Server object configurations that are not configured to start up automatically (*esriSTManual*) must be started manually using ArcCatalog or by calling the *StartConfiguration* method on *IServerObjectAdmin*.

The amount of time it takes between a client requesting a server object (using the *CreateServerContext* method on *IServerObjectManager*) and getting a server object is called the wait time. A server object can be configured to have a maximum wait time by specifying the *WaitTimeout* property on *IServerObjectConfiguration*. If a client's wait time exceeds the maximum wait time for a server object, then the request will time out.

The *WaitTimeout* property is in seconds.

Once a client gets a reference to a server object, it can hold on to that server object as long as desired before releasing it. The amount of time between when a client gets a reference to a server object and when it is released is the usage time.

A PropertySet is a generic class that is used to hold any set of properties. A PropertySet's properties are stored as name/value pairs. Examples for the use of a property set are to hold the properties required for opening an SDE workspace or geocoding an address. To learn more about PropertySet objects, see the online developer documentation.

To ensure that clients don't hold references to server objects for too long (for example., they don't correctly release server objects), a server object can be configured to have a maximum usage time by specifying the *UsageTimeout* property on *IServerObjectConfiguration*. If a client holds on to a server object longer than the maximum usage time, then the server object is automatically released and the client will lose the reference to the server object.

The *UsageTimeout* is in seconds.

The ServerMachine class is used to define a machine that can host server objects managed by the GIS server.

The *ServerMachine* class is used to define a machine that can host server objects managed by the GIS server.

ArcGIS Server is a distributed system. Server objects managed by the GIS server can run on one or more host machines. A machine that can host server objects must have the Server Object Container installed on it, and the machine must be added to the list of host machines managed by the Server Object Manager.

IServerMachine : IUnknown	Provides access to members that control the behavior and properties of a server host machine to administrators.
■—■ Description: String	*The description of the host machine.*
■—■ Name: String	*The name of the machine that can host server objects for the GIS server.*

IServerMachine allows you to configure the properties of a machine to add it to the GIS server. You must set the *Name* property for the machine, which will be the name of the machine on the network. The description is optional.

Use the *AddMachine* method to add new machines to your GIS server. All server objects configured in the GIS server can run on any of the host machines, so all host machines must have access to the necessary data and output directories used by all the server objects.

The following code shows how to use the *IServerMachine* interface and *AddMachine* method to add a machine to your GIS server:

```
Dim pGISServerConnection As IGISServerConnection
Set pGISServerConnection = New GISServerConnection
pGISServerConnection.Connect "melange"

Dim pServerObjectAdmin As IServerObjectAdmin
Set pServerObjectAdmin = pGISServerConnection.ServerObjectAdmin

Dim pServerMachine As IServerMachine
Set pServerMachine = pServerObjectAdmin.CreateMachine
pServerMachine.Name = "callum"

pServerObjectAdmin.AddMachine pServerMachine
```

The ServerObjectType class defines the type of a server object.

The *ServerObjectType* class defines the type of a server object. The type of server object configuration and, therefore, server objects that can be created on a GIS server, is one of a defined set of server object types that a GIS server can support. By default, the supported types are *MapServer* and *GeocodeServer*. The server object configuration type defines the types of server object instances that a particular server object configuration starts up and provides to applications to make use of.

The *ServerObjectType* also defines the set of properties associated with a particular server object configuration that must be specified when creating the configuration. For example, a *MapServer* requires a map document, a *GeocodeServer* requires a locator.

You must be connected to the GIS server as an administrator to access *ServerType* objects. A read-only subset of properties of a *ServerObjectType* is available to nonadministrators via the *ServerObjectTypeInfo* object.

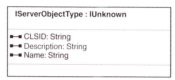

IServerObjectType : IUnknown	Provides access to members that control the behavior and properties of a server object type to administrators.
■—■ CLSID: String ■—■ Description: String ■—■ Name: String	The GUID of the COM class corresponding to the server object type. Description of the server object type. Name of the server object type.

The *IServerObjectType* interface is a read/write interface on a server object type that allows administrators to configure new server object types to add to the server, update existing server object types, and view the properties of a server object type.

The ServerDirectory *class defines the properties of a server directory.*

The *ServerDirectory* class defines the properties of a server directory. A server directory is a location on a file system that the GIS server is configured to clean up files that it writes. By definition, a server directory can be written to by all container machines.

The GIS server hosts and manages server objects and other ArcObjects for use in applications. In many cases, the use of those objects requires writing output to files. For example, when a map server object draws a map, it writes images to disk on the server machine. Other applications may write their own data; for example, an application that checks out data from a geodatabase may write the check-out personal geodatabase to disk on the server.

Typically, these files are transient and need only be available to the application for a short time, for example, the time for the application to draw the map or the time required to download the check-out database. As applications do their work and write out data, these files can accumulate quickly. The GIS server will automatically clean up its output if that output is written to a server directory.

Files in a server directory can be cleaned based on file age or based on when the file was last accessed. The maximum file age or time since last accessed is a property of a server directory. If the *CleaningMode* is *esriDCAbsolute*, then all files created by the GIS server that are older than the maximum age are automatically cleaned up by the GIS server. If the *CleaningMode* is *esriDCSliding*, then all files created by the GIS server that have not been accessed for a duration defined by maximum age are automatically cleaned up by the GIS server.

Note that when creating files in a server directory, they must be prefixed with "_ags_" in order to be cleaned up by the GIS server. Any files in a server directory not prefixed with "_ags_" will not be cleaned up.

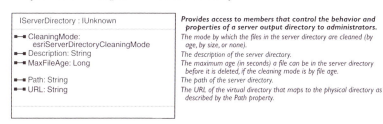

The *IServerDirectory* interface allows you to configure the properties of a server directory to add it to the GIS server. You must set the *Path*, *CleaningMode*, and *MaxFileAge* (if cleaning mode is absolute or sliding) properties for the server directory, which will be the directories' path on disk. The *Description* and *URL* properties are optional.

The *URL* property is the virtual directory that corresponds to the physical directory specified by the *Name* property. Server objects, such as a map server object, can use the *Name* property to write their output files to a directory where they will be cleaned up and can pass back to clients the *URL* for the location of the files they write. Clients (for example, Web applications) will then not require direct access to the physical directory.

Use the *AddServerDirectory* method on *IServerObjectAdmin* to add the new server directory to your GIS server.

Configuration and log files

The ArcGIS Server manages a set of files that are written by the server object manager. These files can be grouped into two categories: log files and configuration files.

Understanding the structure of these files and how to use them will be critical for an ArcGIS Server administrator.

This appendix gives a description of the structure and content for both the log and configuration files, including details of how to interpret and use them.

The ArcGIS Server logs events that occur in the server, and any errors associated with those events, to log files. Events such as when server objects are started, when server contexts are created, and when machines are added to the server are some of the examples of events logged by the server.

The server object manager (SOM) is the centralized logging mechanism for ArcGIS server. Through the SOM, all events that occur within the SOM, events that occur in server object containers (SOCs), and their contained objects are logged by the SOM.

The logs are actually a pair of files: an XML file and a .dat file. The XML file contains the <log> tags, and the .dat file contains the messages that are constantly being appended. You can open the XML file in any XML aware tool to view the messages in the .dat file.

By default, these log files are written to <install_location>\log on the SOM machine. Each time the SOM service starts, a new log file is created, and the server will continue to write messages to that log until it reaches the maximum log size. Once the log file exceeds the maximum size, it is retired and a new log file is created. By default the maximum log size is 10 megabytes.

Log messages can vary in their level of severity from "error", which indicates a problem that requires immediate attention, to "detailed", which is a common message generated through regular use of the server. The messages that are logged are defined by the log level that is set in the server. The following are the ArcGIS Server's logging levels:

0   (None): No logging

1   (Error): Serious problems that require immediate attention are logged.

2   (Warning): Problems that require attention and errors are logged.

3   (Normal): Common administrative messages of the server, warnings, and errors are logged.

ArcCatalog Server Properties page General tab

4   (Detailed): Common messages from user use of the server, including server objects, normal messages, warnings, and errors are logged.

5   (Debug): Verbose messages to aid in troubleshooting; detailed messages, normal messages, warnings, and errors are logged.

By default, the log level of the server is set to level 3 (Normal), meaning messages whose severity is Error, Warning, or Normal will be logged. All messages whose level is Detailed or Debug are not logged.

The log location, maximum log size, and logging level can be changed at any time using either ArcCatalog or the server API. Changes made will be reflected immediately in the server. The log properties can also be modified by editing the Server.cfg configuration file. In that case, the changes will not be reflected until the SOM is stopped and restarted.

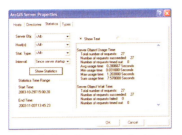

ArcCatalog Server Properties page Statistics tab

USING THE LOG FILES

Log files are an important tool for monitoring and troubleshooting problems with your GIS server. GIS server administrators will monitor the server's statistics and use the statistics to help determine when they need to consult the log files. The GIS server's statistics give general information about the state of the GIS server and whether errors have been occurring. The statistics are available to server administrators through ArcCatalog.

Through both the GIS server's statistics and through reports from users of the system, administrators will be confronted with errors and other problems occurring in the system. The log files provide the information to identify what the errors are and, through the information in the log, infer what to investigate to solve the problem.

For example, an administrator may view the server statistics and note that there were a number of errors associated with creating new server objects in the server. By further examining the statistics, the problem area can be narrowed down to a particular server object container machine. The information in the log file can then be used to determine what errors have been logged and to discover, for example, that the server object container machine on which the errors were occurring did not have access to the output directory. Using this information, the administrator can rectify the directory access problem, then use statistics and the log files to determine if the error occurs again.

The log files also serve as a history of the statistics and events that occur over time in the server. The server's statistics are in memory and are accumulated summaries of time slices since the GIS server started. The granularity of these time slices is more coarse the further back in time you go. Statistics are also cleared when the server is stopped. The GIS server's logs maintain a record of all events in the server and are not deleted when the server is stopped.

READING THE LOG FILES

The information contained in the log file messages has a consistent structure. Understanding this structure is important when interpreting the information in the log files.

Log message structure

Loggable messages are generated by many subsystems of ArcGIS Server. Messages are generated as a result of SOM startup and administrative and user usage. Each message has a target that can be either "Server" or a server object. Server-targeted messages log events associated with the core functionality of the SOM, while server object messages log events associated with a specific server object configuration and instances of that server object configuration.

Independent of target, all loggable messages have six explicit properties: time, level, code, target, thread, and message. The time is the time at which the logged event occurred. The level is the level of detail of the message in relation to other messages (as described above, the levels are: Error, Warning, Normal, Detailed, and Debug). The code is the result code associated with the message. The target is either "Server" or the name of a server object configuration associated with the message. Thread is the SOM process thread that generated the message. Finally, message is the human readable description of the logged event that includes the process and thread IDs of the container processes where the object that generated the message is running and the server object container machine on which that process is running (if applicable).

The message may contain any error description that comes from the server object itself, such as an error indicating that it cannot write its output.

Additional properties that may be included with certain messages include machine and elapsed. Machine indicates the server object container machine for which the event occurred. For example, the sever may log an error that a server context for a particular server object configuration could not be created on a particular machine. Only those messages that apply to statistics that are recorded for a specific machine will include a machine property.

The elapsed property is the time it took for the event that is being logged to complete. For example, the create server object event has an elapsed time to indicate the amount of time it took to create the server object instance.

The following is an example of a typical log message. This log message indicates that an instance of the Yellowstone map server object was created on the server object container machine "padisha", and it took 2.443 seconds to create the object.

```
<Msg time="2003-10-31T14:36:05"
       level="Detailed"
       code="4004"
       target="Yellowstone.MapServer"
       machine="padisha"
       thread="2936"
       elapsed="2.443">
       Server Object instance is successfully created on machine padisha.
</Msg>
```

Message targets

As described above, the targets for the log messages can be Server or a server object. Messages associated with server objects have the name and type of the server object as the target property. For example, the Yellowstone map server object will appear as Yellowstone.MapServer.

There are two additional targets that will appear in the log that are internal server object configurations. While the administrator does not manage these configurations directly, errors may still occur with them that need to be dealt with.

The internal server objects are SDM.ServerDirectoryManager and Engine. SDM.ServerDirectoryManager is the object that cleans files from the GIS server's server directories. An instance of the SDM.ServerDirectoryManager is created when the SOM starts. Each time a directory is cleaned, this instance of SDM.ServerDirectoryManager does the cleaning for all server directories. Any errors it encounters are reported. Errors that are typically reported by SDM.ServerDirectoryManager include the inability to access a directory it needs to clean.

Each time a server directory is created or removed, the SDM.ServerDirectoryManager is stopped and restarted.

The Engine.Engine server object represents the empty server context configuration. When a client asks to create an empty context, the Engine.Engine configuration creates one for the client.

Code	Meaning	Target	Comments
1000	Add machine failed	Server	Error logged if you attempt to add a new server container machine that does not exist or cannot be found by the SOM.
1001	Remove machine failed	Server	Error logged if removing a machine fails.
1002	Illegal configuration name	Server	Error logged when you attempt to add a server object configuration whose name contains illegal characters.
1003	Load server object configuration failed	Server	Error logged when the server cannot load a server object configuration, for example, if the server object configuration file is corrupt or contains invalid values.
1004	Add server object configuration failed	Server	Error logged when adding a new server object configuration fails, for example, if you attempt to add a server object configuration with the same name and type as an existing server object configuration.
1005	Delete server object configuration failed	Server	Error logged when deleting a server object configuration fails, for example, if you attempt to delete a server object configuration that does not exist.
1006	Stopped server object configuration failed	Server object	Error logged when stopping a server object configuration fails, for example, if you attempt to stop a server object configuration that is already stopped.
1007	Pause server object configuration failed	Server object	Error logged when pausing a server object configuration fails, for example, if you attempt to pause a stopped or already paused server object configuration.
1008	Start server object configuration failed	Server object	Error logged when starting a server object configuration fails, for example, if you attempt to start an already started configuration, or none of the minimum instances could be started.
1010	Server object type not found	Server	Error logged when a server object configuration is trying to use a server object type that is not registered with the server. The registered types are MapServer and GeocodeServer.
1012	Failed to load the server configuration file	Server	Error logged when the Server.cfg cannot be read on SOM startup, for example, if the SOM startup encounters a corrupted Server.cfg or if Server.cfg contains invalid values.
1013	Failed to create a server context	Server object	Errors logged if the server fails to create a server context for a client, for example, if the client asks the server to create a context for a server object configuration that does not exist.
1014	Failed to create a server container process	Server object	Error logged if the server failed to create a new container process on one of the server object container machines.
1015	Failed to create a thread in a container process	Server object	Error logged if the server failed to create a new thread in a container process on one of the server object container machines.
1016	Failed to create a server object in a container process thread	Server object	Error logged if an instance of a server object failed to create on a server object container machine. Look at the message for more specific error information.
1017	Container process crashed	Server object	Error logged if a container process containing server objects for the target server object configuration crashes. If this is a pooled server object, the server will create a new process and server objects to repopulate the pool.
1018	Server directory has invalid path	Server	Error logged if a server directory has an invalid path.
1019	Add server directory failed	Server	Error logged if the addition of a server directory failed. This can happen if you attempt to add a server directory if one already exists.
1020	Delete server directory failed	Server	Error logged if the deletion of a server directory failed. This can happen if you attempt to delete a server directory that does not exist.
1021	Update server directory failed	Server	Error logged if the update of a server directory failed. This can happen if you attempt to update a server directory that does not exist.
1022	Server object types file not found	Server	Error logged if the server object types file cannot be found when the SOM starts. The server object types file is <install_location>\bin\ServerTypes.txt
1023	Server object type not registered	Server	Error logged if the server object types file is empty. The server objects types file is <install_location>\bin\ServerTypes.txt

Log codes

The codes are numbered based on the following:

0–5999: Messages that are generated and written by the SOM.

6000 and greater: Messages that are generated by the server object components themselves (MapServer, GeocodeServer, SDM.ServerDirectoryManager, Engine.Engine).

Code	Meaning	Target	Comments
2000	Server object container machine not found	Server	Message logged if the SOM attempts to create a new container process or thread on a machine that it can't find. This error will typically occur if the machine becomes unavailable (due to a network problem) while the SOM is running.
2001	Server object container machine exists	Server	Message logged when you attempt to add a server object container machine that has already been added to the GIS server.
2002	Server object configuration exists	Server	Message logged when you attempt to add a new server object configuration with the same name and type as an existing server object configuration.
2003	Server object configuration is not found	Server	Message logged when a client requests a server object configuration that does not exist.
2006	The minimum and maximum instance values are invalid	Server	When adding or updating server object configuration, minimum or maximum instances specified are invalid, for example, a minimum that is larger than the maximum.
2007	A server context's usage time-out was exceeded	Server object	A client held onto a server object for longer than its usage time-out, and the server automatically released it.
2008	A server object's creation time-out was exceeded	Server object	An instance of a server object took longer to create than the creation time-out, and the server cancelled the creation.
2009	Server configuration not found	Server	Message logged if while the SOM is running, it cannot read the Server.cfg file.
2010	Server configuration contains an invalid parameter	Server	Message logged if the SOM encounters invalid tags or values when reading the Server.cfg file.
2011	A server directory was not found.	Server	Message logged if you attempt to find or delete a server directory that does not exist.
2012	Server directory exists	Server	Message logged if you attempt to add a new server directory that already exists.

Code	Meaning	Target	Comments
4000	Container process created	Server object	This message is logged each time a container process for the target server object configuration is started on a server object container machine.
4001	Thread created in a container process	Server object	This message is logged each time a thread is created in a container process for the target server object configuration.
4002	Container process removed	Server object	This message is logged each time a container process for the target server object configuration is removed from a server object container machine.
4003	Thread removed from a container process	Server object	This message is logged each time a thread is removed from a container process for the target server object configuration.
4004	Server object created in a server container thread	Server object	This message is logged each time an instance of a server object is created in a thread within a container process for the target server object configuration. The elapsed time logged is the amount of time taken to create the server object.
4006	Server context created	Server object	This message is logged any time a client calls CreateServerContext for the target server object configuration. The elapsed time is the wait time between the call to CreateServerContext and when the client receives the server context.
4007	Server context released	Server object	This message is logged any time a client calls ReleaseContext for a context created for the target server object configuration. The elapsed time is the usage time between when the client's call to CreateServerContext is completed and when the client is released.
4008	Recycling started	Server object	Message logged when the target server object configuration recycling is started.
4009	Recycle next object	Server object	Message logged when an instance of the target server object configuration is recycled.

Some codes can apply to messages of different levels, for example, a message with a code may be an error for one event, while it may be a warning for another event. These tables summarize the log codes, their meaning, and the applicable targets.

Code	Meaning	Target	Comments
3000	Server object container machine added	Server	Message logged when the user adds a new server object container machine to the GIS server.
3001	Server object container machine deleted	Server	Message logged when the user removes a server object container machine from the GIS server.
3002	Server object configuration load begin	Server	Message logged for each server object configuration when the SOM starts.
3003	Server object configuration add begin	Server	Message logged when the server attempts to add a server object configuration. This event occurs when you either add a new or update an existing server object configuration.
3004	Server object configuration delete begin	Server	Message logged when the server attempts to delete a server object configuration. This event occurs when you either delete or update a server object configuration.
3005	Server object configuration stop begin	Server	Message logged when the server attempts to stop a server object configuration.
3006	Server object configuration pause begin	Server	Message logged when the server attempts to pause a server object configuration.
3007	Server object configuration start begin	Server	Message logged when the server attempts to start a server object configuration. This message is not logged when server object configurations are started during SOM startup.
3008	Server object configuration loaded	Server	This message is logged for each server object configuration that was successfully loaded on starting the SOM. This message is logged after a message with code 3002. If the load failed, an error with code 1003 is logged.
3009	Server object configuration added	Server	Message logged when the SOM starts and successfully adds the SDM (ServerDirectoryManager) and internal Engine server object configurations.
3010	Server object configuration deleted	Server	Message logged when a server object configuration is deleted.
3011	Server object configuration stopped	Server object	Message logged when a server object configuration is stopped.
3012	Server object configuration paused	Server object	Message logged when a server object configuration is paused.
3013	Server object configuration started	Server object	Message logged when a server object configuration is started. Note that a successful start may occur after errors have been logged.
3014	Server object manager started	Server	Message logged when the SOM has successfully started. Note that a successful start may occur after errors have been logged.
3015	Server object manager stopped	Server	Message logged when the SOM is stopped. No additional messages will be added to a log file after this message.
3016	Server configuration loaded	Server	Message logged when the SOM starts and the information in Server.cfg is successfully loaded.
3017	Server directory added	Server	Message logged when a new server directory is added to the server.
3018	Server directory updated	Server	This message is logged when the properties of a server directory are updated.
3019	Server directory deleted	Server	Message logged when a server directory is deleted from the server.

The ArcGIS administrator manages the various configuration properties of both the GIS server and its collection of server objects. All of these properties are maintained by the server in a collection of configuration files. These configuration files are read by the SOM when its started, and they define such things as the set of server object container machines, the output directories, the server object configurations, their pooling model, and so on.

The properties of the server itself are maintained in a file called Server.cfg, while the properties of each of the server's server object configurations are maintained in files called <server object configuration name>.<server object type>.cfg. For example, the configuration file for the map server object configuration called Yellowstone would be Yellowstone.MapServer.cfg.

These configuration files are in XML and are located in the <install_location>\cfg folder on the SOM machine. When properties of the server and its configurations are set or modified using ArcCatalog, these changes are reflected in both the GIS server and the appropriate configuration file. Developers can also change these properties using the server API to update the server's properties and to create, update, and delete the server's configurations.

The server's configuration files can also be modified manually using a text editor. Unlike using ArcCatalog or the server API, any changes made to the server's configuration files manually are not reflected in the server until it is restarted. While this appendix documents the structure of these files, it is recommended that you use ArcCatalog or the server API to create and modify their contents, which is done indirectly though modifying the server's properties.

The remainder of this appendix details the content and use of the server configuration files.

THE SERVER.CFG CONFIGURATION FILE

The server's properties are maintained in the Server.cfg configuration file. The contents of this file are read when the SOM is started. The server will report a successful start once this file has been successfully read and any initialization detailed in it completed. If there are errors in the file, the SOM will log an error and attempt to start using default values for the missing or invalid properties.

When the SOM is installed on a machine, the Server.cfg file does not exist. Server.cfg is created after the SOM is started, and either a server object container machine or a server directory is added to the GIS server.

The following is an example of a Server.cfg file for a GIS server with a single container machine (padisha) and a single output directory (\\padisha\images), whose logging level is 3.

```
<Server>
  <ServerMachines>
    <Machine>
      <Name>padisha</Name>
      <Description>Server container machine 1</Description>
    </Machine>
  </ServerMachines>
  <ServerDirectories>
```

```
<Directory>
  <Path>\\padisha\images</Path>
  <URL>http://padisha/images</URL>
  <Description>default output location</Description>
  <Cleaning>sliding</Cleaning>
  <MaxFileAge>600</MaxFileAge>
</Directory>
</ServerDirectories>
<Properties>
  <LogPath>C:\Program Files\ArcGIS\log\</LogPath>
  <LogSize>10</LogSize>
  <LogLevel>3</LogLevel>
  <ConfigurationStartTimeout>300</ConfigurationStartTimeout>
</Properties>
</Server>
```

The IServerObjectAdmin interface provides the methods and properties for updating the GIS server's properties and, therefore, the contents of Server.cfg.

IServerObjectAdmin : IUnknown	**Provide access to members that administrate the GIS server.**
Properties: IPropertySet	*The logging properties for the GIS server.*
AddConfiguration (in config: IServerObjectConfiguration)	*Adds a server object configuration (created with CreateConfiguration) to the GIS server.*
AddMachine (in machine: IServerMachine)	*Adds a host machine (created with CreateMachine) to the GIS server.*
AddServerDirectory (in pSD: IServerDirectory)	*Adds a server directory (created with CreateServerDirectory) to the GIS server.*
CreateConfiguration: IServerObjectConfiguration	*Creates a new server object configuration.*
CreateMachine: IServerMachine	*Creates a new host machine.*
CreateServerDirectory: IServerDirectory	*Creates a new server directory.*
DeleteConfiguration (in Name: String, in TypeName: String)	*Deletes a server object configuration from the GIS server.*
DeleteMachine (in machineName: String)	*Deletes a host machine from the GIS server, making it unavailable to host server objects.*
DeleteServerDirectory (in Path: String)	*Deletes a server directory such that its cleanup is no longer managed by the GIS server. It does not delete the physical directory from disk.*
GetConfiguration (in Name: String, in TypeName: String) : IServerObjectConfiguration	*Get the server object configuration with the specified Name and TypeName.*
GetConfigurations: IEnumServerObjectConfiguration	*An enumerator over all the server object configurations.*
GetConfigurationStatus (in Name: String, in TypeName: String) : IServerObjectConfigurationStatus	*Get the configuration status for a server object configuration with the specified Name and TypeName.*
GetMachine (in Name: String) : IServerMachine	*Get the host machine with the specified Name.*
GetMachines: IEnumServerMachine	*An enumerator over all the GIS server's host machines.*
GetServerDirectories: IEnumServerDirectory	*An enumerator over the GIS server's output directories.*
GetServerDirectory (in Path: String) : IServerDirectory	*Get the server directory with the specified Path.*
GetTypes: IEnumServerObjectType	*An enumerator over all the server object types.*
PauseConfiguration (in Name: String, in TypeName: String)	*Makes the configuration unavailable to clients for processing requests, but does not shut down running instances of server objects, or interrupt requests in progress.*
StartConfiguration (in Name: String, in TypeName: String)	*Starts a server object configuration and makes it available to clients for processing requests.*
StopConfiguration (in Name: String, in TypeName: String)	*Stops a server object configuration and shuts down any running instances of server objects defined by the configuration.*
UpdateConfiguration (in config: IServerObjectConfiguration)	*Updates the properties of a server object configuration.*
UpdateMachine (in machine: IServerMachine)	*Updates the properties of a host machine.*
UpdateServerDirectory (in pSD: IServerDirectory)	*Updates the properties of a server directory.*

The ArcCatalog ArcGIS Server Properties page Hosts tab

Server.cfg tags

The following are the tags, their meanings and example values in a Server.cfg file.

<ServerMachines>

The list of server object container machines. This tag contains <Machine> subtags for each server machine.

<Machine>

A server object container machine. This tag contains two subtags: <Name> and <Description>.

<Name>

A string that represents the name of the server object container machine. If this tag is missing, the rest of the <Machine> tag is ignored. On start, the SOM does not validate that the value of this tag is a valid server object container machine. If it's invalid, errors will be logged as the SOM attempts to create server objects on it.

The machine names must be unique. Duplicate machines will be ignored.

<Description>

An optional string that describes the server object container machine.

The following is an example of the <ServerMachine> tag, <Machine> tag, and its subtags.

```
<ServerMachines>
  <Machine>
    <Name>padisha</Name>
    <Description>Server container machine 1</Description>
  </Machine>
  <Machine>
    <Name>melange</Name>
    <Description>Server container machine 2</Description>
  </Machine>
</ServerMachines>
```

<ServerDirectories>

The list of server directories. This tag contains <Directory> subtags for each server directory.

<Directory>

A server directory. This tag contains the required subtag <Path> and a number of optional subtags.

<Path>

The ArcCatalog ArcGIS Server Properties page Directories tab

A string that represents the path of the server directory. This property is required and must be unique per server. Directories with duplicate <Path> tags will be ignored. Note, however, that if a single location has multiple paths—for example, two shares with different names—the SOM will not recognize those as being the same directory.

<URL>

An optional string that represents the URL of a virtual directory that points to the physical location specified in the <Path> tag. The URL will be in URL form as in http://padisha/images.

<Description>

An optional string that is a description of the server directory.

<Cleaning>

An optional string that specifies the server directory's cleaning mode. Valid values are "off", "sliding", or "absolute". If <Cleaning> is "off", then the server will not clean up its files in the directory. If <Cleaning> is "sliding", then the server will delete files for which the time specified by the <MaxFileAge> tag has elapsed since they were last accessed. If <Cleaning> is "absolute", then the server will delete files for which the time specified by the <MaxFileAge> tag has elapsed since they were created. If this tag is missing, the default value is "sliding".

<MaxFileAge>

An optional integer (greater than 0) that represents the amount of time in seconds that needs to elapse since files were last accessed (sliding) or were created (absolute) before they are deleted. If this tag is missing, the default is 10.

The following is an example of the <ServerDirectory> tag, <Directory> tag, and its subtags.

```
<ServerDirectories>
  <Directory>
    <Path>\\padisha\images</Path>
    <URL>http://padisha/images</URL>
    <Description>default output location</Description>
    <Cleaning>sliding</Cleaning>
    <MaxFileAge>600</MaxFileAge>
  </Directory>
  <Directory>
    <Path>\\melange\ServerOutput</Path>
    <URL>http://melange/ServerOutput</URL>
    <Description>large file location</Description>
    <Cleaning>absolute</Cleaning>
    <MaxFileAge>6000</MaxFileAge>
  </Directory>
</ServerDirectories>
```

<Properties>

The list of properties of the GIS server, including the logging properties and server object creation time-out. All the subtags of Properties are optional.

<LogPath>

An optional string representing the path to the location on disk that log files are written. The default is <install_location>\log. Note, the GIS server account must have write access to this location.

ArcCatalog Server Properties page General tab

\<LogSize\>

An optional integer representing the size to which a single log file can grow (in MB) before a new log file is created. The default is 10.

\<LogLevel\>

An optional integer that indicates the level of detail that the server logs. The levels are:

0   (None): No logging

1   (Error): Serious problems that require immediate attention are logged.

2   (Warning): Problems that require attention and errors are logged.

3   (Normal): Common administrative messages of the server, warnings, and errors are logged.

4   (Detailed): Common messages from user use of the server, including server objects, normal messages, warnings, and errors are logged.

5   (Debug): Verbose messages to aid in troubleshooting, detailed messages, normal messages, warnings, and errors are logged.

The default log level is 3.

\<ConfigurationTimeout\>

An optional integer that represents the time in seconds that the GIS server will wait for a server object instance to start. If a server object takes longer to start than \<ConfigurationTimeout\>, then it will time out and an error will be logged. The default time-out is 300.

The following is an example of the \<Properties\> tag and its subtags.

```
<Properties>
  <LogPath>C:\Program Files\ArcGIS\log\</LogPath>
  <LogSize>10</LogSize>
  <LogLevel>3</LogLevel>
  <ConfigurationStartTimeout>300</ConfigurationStartTimeout>
</Properties>
```

THE SERVER OBJECT CONFIGURATION FILES

The properties of server object configurations are maintained in a file for each server object configuration. A GIS server's set of configurations is defined by the list of configuration files in the cfg directory. When a new configuration is added to the GIS server, a new file is created. When a configuration is deleted, its file is deleted from the cfg directory.

A MapServer server object configuration will have a file called \<configuration_name\>.MapServer.cfg, and a GeocodeServer server object configuration will have a file named \<configuration_name\>.GeocodeServer.cfg.

For example, a GIS server that has map server objects called USA, Redlands, and Yellowstone and geocode server objects named Portland and USAStreets would have the following files in its cfg directory:

```
Portland.GeocodeServer.cfg
Redlands.MapServer.cfg
```

```
Server.cfg
USA.MapServer.cfg
USAStreets.GeocodeServer.cfg
Yellowstone.MapServer.cfg
```

It's possible to add a configuration to the GIS server by manually creating a configuration file in the cfg directory, and it's possible to delete a configuration by deleting its file from the cfg directory. In both cases, the new or deleted configuration will not be recognized by the server until the SOM is restarted. If the SOM encounters a corrupted configuration file, the SOM will log a warning and ignore the configuration.

The following is an example of a server object configuration file for a MapServer:

```
<ServerObjectConfiguration>
  <Description>Map server containing vegetation data for Yellowstone
National Park</Description>

  <Properties>
    <FilePath>D:\ArcGIS_Server_Data\DocDevScenarioData\Yellowstone\Yellowstone.mxd</
FilePath>
    <OutputDir>\\padisha\images</OutputDir>
    <VirtualOutputDir>http://padisha/images</VirtualOutputDir>
    <MaxRecordCount>500</MaxRecordCount>
    <MaxBufferFeatures>100</MaxBufferFeatures>
    <MaxImageWidth>2048</MaxImageWidth>
    <MaxImageHeight>2048</MaxImageHeight>
  </Properties>

  <Recycling>
    <StartTime>00:00</StartTime>
    <Interval>36000</Interval>
  </Recycling>

  <IsPooled>true</IsPooled>
  <MinInstances>2</MinInstances>
  <MaxInstances>4</MaxInstances>
  <WaitTimeout>60</WaitTimeout>
  <UsageTimeout>600</UsageTimeout>
  <Isolation>high</Isolation>
  <StartupType>automatic</StartupType>
</ServerObjectConfiguration>
```

The IServerObjectConfiguration interface provides the methods and properties for setting a server object configuration's properties and, therefore, the contents of the configuration files.

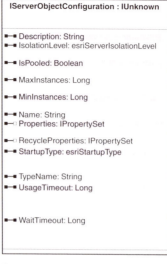

ISeverObjectConfiguration : IUnknown	Provides access to members that control the behavior and properties of a server object configuration to administrators.
■―■ Description: String	Description of the server object configuration.
■―■ IsolationLevel: esriServerIsolationLevel	The isolation level of the server objects defined by the server object configuration.
■―■ IsPooled: Boolean	Indicates whether the server objects defined by this configuration are pooled.
■―■ MaxInstances: Long	The maximum number of server object instances for a server object configuration.
■―■ MinInstances: Long	The minimum number of server object instances for a pooled server object configuration.
■―■ Name: String	Name of the server object configuration.
■―◻ Properties: IPropertySet	Initialization parameters and properties for the server objects created by the server object configuration.
■―◻ RecycleProperties: IPropertySet	The recycling properties for the server object configuration.
■―■ StartupType: esriStartupType	The startup type for this server object configuration. Startup type describes whether the server object configuration is started when the server object manager service is started for the GIS server.
■―■ TypeName: String	Type of the server object configuration (MapServer or GeocodeServer).
■―■ UsageTimeout: Long	Maximum time (in seconds) a client can hold on to an instance of a server object for this server object configuration before releasing it back to the server. It is the maximum time allowed between calling CreateServerContext and ReleaseServerContext.
■―■ WaitTimeout: Long	Maximum time (in seconds) a client will wait for an instance of a server object for this server object configuration using the CreateServerContext method on IServerObjectManager before timing out.

The ArcCatalog Server Object Properties page General tab

Server object configuration tags

The following are the tags, their meanings, and example values in a server object configuration file.

<Description>

An optional string that is the description of the server object configuration.

<Properties>

The list of properties of the server object configuration. The subtags are properties specific to the server object configuration type. For MapServer configurations, the subtags are <FilePath>, <OutputDir>, <VirtualOutputDir>. For GeocodeServer configurations, the subtags are <Locator>, <LocatorWorkspacePath>, <LocatorWorkspaceConnectionString>, and <SuggestedBatchSize>.

<FilePath>

A string representing the path to the map document (.mxd) or published map document (.pmf) that the MapServer will serve.

<OutputDir>

A string that represents the path to a location on the file system to which the MapServer will write its output. When ArcCatalog creates a new MapServer configuration, this property is copied from the server directory's path that the user specifies. If you want the MapServer's output to be cleaned by the GIS server, this path should be a path to a server directory.

<VirtualOutputDir>

A string that represents the URL of the virtual directory that points to the physical location specified in the <OutputDir> tag. When ArcCatalog creates a new MapServer configuration, this property is copied from the server directory's URL that the user specifies.

The ArcCatalog Server Object Properties page Parameters tab for a MapServer object

The ArcCatalog Server Object Properties page Parameters tab for a GeocodeServer object

\<MaxRecordCount\>

An integer that represents the maximum number of result records returned by the following methods on the map server:

- QueryFeatureData
- Find
- Identify

\<MaxBufferCount\>

An integer that represents the maximum number of features that can be buffered by the map server at draw time per layer.

\<MaxImageWidth\>

An integer representing the maximum width (in pixels) of images the map server will export.

\<MaxImageHeight\>

An integer representing the maximum height (in pixels) of images the map server will export.

\<ConnectionCheckInterval\>

An integer representing the number of seconds the MapServer or GeocodeServer will wait before checking if ArcSDE servers that have become unavailable are once again available. Once the ArcSDE server is available, the MapServer or GeocodeServer instance will be replaced to repair the connection to the ArcSDE server. By default, this property is not included in the .cfg file and this time is 300 seconds. You can add this tag to the .cfg file to modify this time. You can disable this behavior by specifying a value of 0.

The following is an example of the \<Properties\> tag and its subtags for a MapServer configuration.

```
<Properties>
  <FilePath>\\padisha\ArcGIS_Server_Data\Yellowstone.mxd</FilePath>
  <OutputDir>\\padisha\images</OutputDir>
  <VirtualOutputDir>http://padisha/images</VirtualOutputDir>
  <MaxRecordCount>500</MaxRecordCount>
  <MaxBufferCount>100</MaxBufferCount>
  <MaxImageWidth>2048</MaxImageWidth>
  <MaxImageHeight>2048</MaxImageHeight>
</Properties>
```

\<Locator\>

A string that represents the name of the locator for the GeocodeServer.

\<LocatorWorkspacePath\>

A required string for file-based locators that represents the path to the location on disk where the locator file is stored.

\<LocatorWorkspaceConnectionString\>

A required string for ArcSDE software-based locators that represents the connection string to the ArcSDE database.

\<SuggestedBatchSize\>

An integer that represents the number of records that will be located at a time for batch geocoding.

\<MaxResultSize\>

An integer that represents the maximum number of result records returned by the *FindAddressCandidates* method on the geocode server:

\<MaxBatchSize\>

An integer that represents the maximum number of records that can be input into the geocode server's *GeocodeAddresses* method.

The following is an example of the \<Properties\> tag and its subtags for a GeocodeServer configuration whose locator is an ArcSDE locator.

```
<Properties>
 <Locator>GDB.Portland</Locator>
  <LocatorWorkspaceConnectionString>
  ENCRYPTED_PASSWORD=0002c06e3bc49d6412c06c1baa554d00;
  SERVER=doug;
  INSTANCE=5151;
  USER=gdb;
  VERSION=SDE.DEFAULT
  </LocatorWorkspaceConnectionString>
  <SuggestedBatchSize>1000</SuggestedBatchSize>
  <MaxResultSize>500</MaxResultSize>
  <MaxBatchSize>1000</MaxBatchSize>
 </Properties>
```

The ArcCatalog Server Object Properties page Pooling tab

\<Recycling\>

The list of recycling properties of the server object configuration. This tag contains subtags \<Start\> and \<Interval\>. Note: If the \<Recycling\> tag is missing, or any of its subtags are invalid, recycling will be switched off for the configuration.

\<Start\>

A required string that represents the recycling start time, which is the time at which recycling is initialized. The time specified is in 24-hour notation. For example, to set the start time at 2:00 p.m., the StartTime property would be 14:00.

\<Interval\>

A required integer that defines the time between recycling operations in seconds. For example, to recycle the configuration every hour, this property would be set to 3600.

The following is an example of the \<Recycling\> tag and its subtags.

The ArcCatalog Server Object Properties page Processes tab

```
<Recycling>
 <StartTime>00:00</StartTime>
 <Interval>36000</Interval>
</Recycling>
```

<MinInstances>

An integer specifying the minimum number of instances for the server object's object pool. The default is 0.

<MaxInstances>

An integer specifying the maximum number of server object instances that can be running at any time. The default is 0.

<WaitTimeout>

An optional integer specifying the maximum amount of time in seconds allowed between a client requesting a server object and getting a server object. The default is 60.

<UsageTimeout>

An optional integer specifying the maximum amount of time in seconds a client can hold onto a server object before it is automatically released. The default is 600.

<IsPooled>

A required string indicating whether the server objects created by this configuration are pooled (true) or not pooled (false).

<Isolation>

A required string indicating if the configuration's server object has high isolation (high) or low isolation (low).

<StartupType>

An optional string that specifies if the configuration is started by the SOM when the SOM starts, or if it needs to be manually started by an administrator. The valid values are "automatic" or "manual". The default is "automatic".

The following are examples of these tags:

```
<IsPooled>true</IsPooled>
<MinInstances>2</MinInstances>
<MaxInstances>4</MaxInstances>
<WaitTimeout>60</WaitTimeout>
<UsageTimeout>600</UsageTimeout>
<Isolation>high</Isolation>
<StartupType>automatic</StartupType>
```

C

Developing applications with EJBs

This chapter includes a discussion and exercise for developers who wish to build Enterprise JavaBeans (EJBs) that make use of the ArcGIS Server API. It will take a close look at some key concepts concerning EJBs and present a simple development scenario demonstrating the integration of this important enterprise technology with powerful GIS capabilities, found within the ArcGIS Server.

Enterprise JavaBeans are Java components that live and operate within a Java 2 Platform, Enterprise Edition (J2EE) application server, a server that provides the EJBs with all of the essential services that allow them to perform and scale to appropriate standards for mission critical, industrial-strength, and highly secure enterprise infrastructures.

EJBs are pure Java components, designed and written to the J2EE EJB specification. This specification ensures that EJBs abide by a standard set of rules, enabling them to be portable across all J2EE-based application servers and across all operating systems that support the Java platform.

As part of the specification, EJBs must not make native calls into native libraries, such as Windows DLLs or UNIX shared objects. This is especially critical in the case of the ArcGIS Server architecture, since ArcObjects, the library of components that make up ArcGIS, is highly "native" in nature, whether it be native to Windows or to UNIX.

Another equally important restriction for EJBs is that they are not allowed to create or manage threads. Because thread management can often be complex and platform-specific, the J2EE container better handles this implementation detail. This becomes a key issue when properly programming EJB applications to work with ArcGIS Server.

So why are EJBs important? While there are many advantages provided by the use of EJBs and J2EE for hosting enterprise systems, the focus here is on two of the main benefits.

First, EJBs and J2EE are important because they adhere to tested and proven computing standards. This is key for interoperability and portability of enterprise applications. No matter what the specific business process the EJB is performing, we can be sure that other EJBs from various providers, running either locally or remotely, will work together in an enterprise system because they adhere to the same rules and standards. They are built in pure Java, which, in itself, is a standard computing platform and language. The application servers that host these various EJBs and EJB containers are designed and built to the J2EE standard as well. This ensures that EJBs can be ported from one application server environment to another without too much overhead and without any rewriting or recompiling of source code.

Second, EJB technology enables development of business objects and system-level services as separate and distinct entities among the various members of an enterprise maintenance staff. The J2EE architecture decouples the many tiers of its application configuration in such a way that all the members of an enterprise development team are free to focus on their specific roles and tasks, without any dependency on the system architecture. An EJB programmer can focus solely on the business logic, or main process, that the EJB will be responsible for performing. The EJB programmer will not need to be concerned with writing code to handle transaction management, threading, object pooling, security, or scaling of that EJB. In fact, the EJB specification prohibits this. These concerns are the responsibility of the application server designers and vendors who build these services, using standard APIs for transaction management, database connectivity, security, scaling, and so on.

For a comprehensive listing of J2EE rules and restrictions for EJBs, go to http://java.sun.com/blueprints/qanda/ejb_tier/restrictions.html.

So how do ArcGIS Java programmers build EJBs that utilize the ArcGIS platform? They do so through the use of a very important J2EE specification feature: the J2EE Connector architecture (JCA).

THE J2EE CONNECTOR ARCHITECTURE

Since EJBs are pure Java objects and ArcObjects components are not, a bridge is required that allows Java objects to make calls into nonnative libraries, such as those that compose ArcObjects.

J2EE provides a connector architecture specification for this kind of integration between heterogeneous enterprise information systems (EIS). The JCA specification allows EIS vendors—for instance, ESRI—to provide a standard resource adapter to handle communications with their EIS.

For more information regarding the Java Connector architecture, go to http://java.sun.com/j2ee/connector/index.jsp.

JAVA ADF RESOURCE ADAPTER

The Java Application Developer Framework (ADF) includes a resource adapter specifically implemented to connect any J2EE-compliant application server to ArcObjects. Because this resource adapter conforms to the JCA specification, it can be plugged in to any compliant application server and used to connect to ArcGIS Server. EJBs can work with instances of the ArcGIS Server Object Manager (SOM) through a connection to the resource adapter. Through the SOM, EJBs can obtain references to other ArcGIS Server objects.

The diagram at left illustrates how the adapter handles the object "brokering" between EJBs and ArcObjects components. In the diagram, "J-Integra JCACOM" represents the Common Client Interface (CCI), or the classes and interfaces that produce connection factory objects and connection objects. They connect to the J-Integra runtime libraries, which handle all of the plumbing required for accessing and "bridging" server objects from the ArcGIS Server to the EJBs running in the application server. Some of the key system-level processes that are handled by the J-Integra runtime for the EJBs are the creation and management of threads, as previously mentioned, as well as opening and managing socket connections. The explicit use of socket objects for network communication is also prohibited from within EJBs.

The model diagram below shows the J-Integra CCI implementation classes and their relationships. The important pieces here are the ConnectionFactory and Connection interfaces. The factory produces connection objects that can be obtained both locally and remotely.

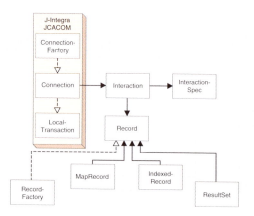

The intrinsycjca.rar file

The resource adapter included with the ADF is bundled into a single resource archive (RAR), intrinsycjca.rar. This archive contains all of the classes and interfaces that conform to the JCA specification, enabling connections to ArcObjects from the application server.

It also contains the J-Integra runtime libraries and the ArcObjects Java proxy classes that represent and marshal requests to and from ArcObjects native objects. The contents and structure of the intrinsycjca.rar are:

- **intrinsycjca.jar**—Contains the implementing classes of the CCI, enabling connections and connection pooling to take place between the J2EE application server and ArcGIS Server containers

- **jintegra.jar**—Contains all J-Integra runtime objects

- META-INF/**ra.xml**—A configuration file, also known as a deployment descriptor, which provides information to the J2EE application server describing how to deploy the resource adapter

The intrinsycjca.rar file contains everything that an ArcGIS EJB application will need for deployment and runtime. The deployment procedures for the adapter itself will depend on the specifics of the application server on which it is being deployed. Each application server vendor will provide procedures for deploying resource adapters. In the next section, the developer scenario will walk you through the deployment of a resource adapter.

The scenario in this section presents some basic mapping functionality, managed by an EJB, that is exposed through a simple Web application. In addition, the scenario illustrates EJBs consuming a MapService in a "back-end" application server environment. Again, the JCA resource adapter is the critical piece making this EJB to ArcGIS Server relationship possible.

PROJECT DESCRIPTION

In this scenario, you will create a Web application for your client that presents a map with some basic tools for zooming in and out. Requests made from a JavaServer Pages (JSP) Web application to a stateless EJB result in a map image being returned from a MapService, running in ArcGIS Server.

You can find the sample in:

```
<install_location>\DeveloperKit\Samples\Developer_Guide_Scenarios\
ArcGIS_Server\Application_EJB
```

DESIGN

The core of this application is a stateless EJB that performs map retrievals from a MapServer object, running in ArcGIS Server. Implementing the EJB, called StatelessMapImageBean, as a stateless session will provide better overall performance and flexibility. Stateless session beans are not tied to any one client and can be shared between many. The EJB container creates a pool of identical stateless session beans and distributes them to handle incoming client requests. As soon as a business method on an EJB is finished, the bean is placed back in the EJB containers pool, becoming available for reuse by another client. The reuse of stateless session beans enables a small number of beans to handle a large number of concurrent clients, allowing the bean to scale better. Stateless session beans also enable a looser coupling between the client and the bean. Since a client is not tied to one specific bean containing state, the client can be of many types, including Web services.

In the scenario, StatelessMapImageBean will include methods for zooming and panning a map. Since the bean is stateless, the client must maintain a certain level of map state that can be passed back to the StatelessMapImageBean for processing. The following example uses the serialized map and image description strings returned from the map's associated MapServer object to provide that state. Each method of the StatelessMapImageBean returns a java.util.Hashtable, which contains properties representing the current state of the map.

The java.util.Hashtable holds the URL of the newly generated map image in a property called *mapurl*. In addition, java.util.Hashtable contains *MapDescription* and *ImageDescription* properties that hold serialized string values returned from the assigned MapServer object. The MapDescription and ImageDescription strings are passed so that current map state is maintained within the client and *not* within the EJB.

All method calls on the EJB sample return java.util.Hashtables; this reduces the amount of network traffic from the levels that would occur if individual method calls were used to return a single property.

REQUIREMENTS

In order to work through this scenario, you must have ArcGIS Server, ArcGIS Server Java ADF, and ArcGIS Desktop installed and running. The Java ADF installation provides two critical archive files that are required for building and deploying an EJB for ArcGIS Server: *intrinsycjca.jar* and *intrinsycjca.rar*. These must be present on the same machine that the application server and EJB application are running on.

All EJBs must live and operate within a J2EE-compliant application server. Such application servers vary widely in levels of complexity, availability, scalability, performance, and cost. This simple exercise deploys your EJB application to IBM's WebSphere Application Server v5.0. However, you can choose any application server available to you, including the free J2EE 1.3.1 reference implementation, which can be downloaded from http://java.sun.com/j2ee/sdk_1.3/index.html.

You must have a MapServer object configured and running on your ArcGIS Server. It can use any map document (.mxd) available to you.

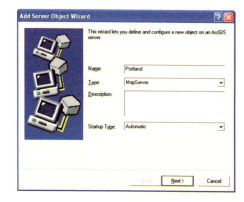

The Add Server Object wizard

In ArcCatalog, create a connection to your GIS server and use the Add Server Object command to create a new server object. For example:

Name: Portland.

Type: MapServer.

Description: This is a service that will produce a map of my favorite part of the world.

Map document: < path to your .mxd document >.

Output directory: Choose from the output directories configured in your server.

Pooling: The EJB application makes stateless use of the server object.

Accept the defaults for the pooling model (pooled server object with minimum instances = 2, maximum instances = 4).

Accept the defaults for the remainder of the configuration properties.

Server object properties dialog box

After creating the server object, start it and right-click it to verify that it is correctly configured and that the map properties are displayed. Refer to Chapter 3, 'Administering an ArcGIS Server', for more information on how to use ArcCatalog to connect to your server and create new server objects.

Once your server object is configured and running, you can begin to code your EJB and client application.

The following ArcObjects Java packages are used in this example:

- com.esri.arcgis.carto

- com.esri.arcgis.server

In order to obtain GIS server objects from within an EJB, the implementation classes provided in the JCA-compliant resource adapter are needed. These classes are found in the com.intrinsyc.jca package, archived in intrinsycjca.jar

The development environment does not require any ArcGIS licensing; however, connecting to a server and using a MapServer object does require that the GIS server be licensed to run ArcObjects in the server. None of the packages used require an extension license.

In this scenario, you can use the development tool of your choice. The steps assume that you are using your favorite text editor or IDE to write your source code but won't reference any specific vender product for tools and/or wizards. The building and deployment of the end product depends on the various tools and application servers you choose to host your EJB implementation. For demonstration purposes, the scenario uses ANT as the build tool and IBM's WebSphere Application Server v5.0 for deployment.

IMPLEMENTATION

The first step in this walk-through is for you to write a relatively simple EJB that will produce a map image from ArcGIS Server.

Creating the StatelessMapImageBean EJB

As mentioned earlier, this EJB will provide a simple zooming function for a map. Since it is a type of session bean, the StatelessMapImageBean must implement the javax.ejb.SessionBean interface.

Use your favorite text editor or IDE to write your source code.

1. Use the skeleton code below, in a text editor or IDE, to create the StatelessMapImageBean class.

```
package com.esri.arcgis.samples.ejb;
import javax.ejb.SessionBean;
import javax.ejb.EJBException;
import javax.ejb.SessionContext;
import com.esri.arcgis.samples.ejb.value.RgbColor;
import com.esri.arcgis.samples.ejb.value.Envelope;
import com.esri.arcgis.samples.ejb.LocalTransaction;
import com.esri.arcgis.samples.ejb.XAResource;
import com.esri.arcgis.geometry.IEnvelope;
import com.esri.arcgis.geometry.IPoint;
public class StatelessMapImageBean implements javax.ejb.SessionBean {
private javax.ejb.SessionContext m_context;
private com.intrinsyc.jca.ConnectionFactory m_connectionFactory;
private com.intrinsyc.jca.Connection m_connection;
private com.intrinsyc.jca.JintConnectionRequestInfo
m_connectionRequestInfo;
private java.lang.Object m_serverConnectionObject;
private com.esri.arcgis.server.ServerConnection m_serverConnection;
private com.esri.arcgis.server.IServerContext m_serverContext;
private com.esri.arcgis.carto.IMapServer m_mapServer;
private com.esri.arcgis.carto.IMapServerInfo m_mapServerInfo;
private com.esri.arcgis.carto.IMapDescription m_mapDescription;
private com.esri.arcgis.carto.IGraphicElements m_graphicElements;
private com.esri.arcgis.carto.IImageType m_imageType;
private com.esri.arcgis.carto.IImageDisplay m_imageDisplay;
```

```
private com.esri.arcgis.carto.IImageDescription m_imageDescription;
private com.esri.arcgis.carto.IMapImage m_mapImage;
private String m_mapDescriptionString;
private String m_imageDescriptionString;
private String m_arcServerHostName;
private String m_arcServerHostDomain;
private String m_arcServerHostUserName;
private String m_arcServerHostPassword;
private String m_mapServerName;

public void ejbActivate() {
}
public void ejbPassivate() {
}
public void ejbRemove() {
}
public void ejbCreate() {
}
public void setSessionContext(SessionContext sessionContext) throws
EJBException{
}
```

2. Before constructing the zooming function, you need to create a connection to the resource adapter. In the body of the ejbCreate method, create a JCA *ConnectionFactory* object.

```
javax.naming.Context initialContext, paramEnv;
try{
      initialContext = new javax.naming.InitialContext();
      paramEnv = (javax.naming.Context)initialContext.lookup("java:comp/
env");
      this.m_arcServerHostName =
(String)paramEnv.lookup("ArcServerHostName");
      this.m_arcServerHostDomain =
(String)paramEnv.lookup("ArcServerDomainName");
      this.m_arcServerHostUserName =
(String)paramEnv.lookup("ArcServerHostUserName");
      this.m_arcServerHostPassword =
(String)paramEnv.lookup("ArcServerHostPassword");
        this.m_mapServerName = (String)paramEnv.lookup("MapServerName");

        m_connectionFactory =
(com.intrinsyc.jca.ConnectionFactory)initialContext.lookup("intrinsycjca");

}catch(Exception exception){
      m_connectionFactory = null;
      throw new Exception("JCA Lookup Failed: " + exception);
}
```

The paramEnv context provides the EJB with the ArcGIS Server specific parameters needed to connect to a specific ArcGIS Server. The parameter strings are located in the ejb-jar.xml deployment descriptor for this bean, which will be discussed and defined later in the exercise.

3. Continuing in this method, create an instance of *JintConnectionRequestInfo*. This contains the necessary parameters for connecting to the ArcGIS Server.

```
XAResource xaResource = new XAResource();
LocalTransaction localTransaction = new LocalTransaction();
```

tion for the duration of a single request. Pooled server objects are meant for applications that make stateless use of those objects.

Non-pooled server objects are dedicated to a single application session and are held on to for the duration of an application session. Non-pooled server objects are not shared between application sessions and are meant for applications that make stateful use of those objects.

When *StartConfiguration* is called on a server object configuration whose *IsPooled* property is true, a set of server objects will be preloaded based on the *MinInstances* property of the server object configuration.

When *StartConfiguration* is called on a server object configuration whose *IsPooled* property is false, no server objects are preloaded. Server objects are loaded and initialized when an application gets one from the server using *CreateServerContext*.

The *MaxInstances* property indicates the maximum number of server objects that can be running and handle requests at any one time. If the maximum number of server objects are running and busy, additional requests will be queued until a server object becomes free.

For a pooled server object, the *MaxInstances* represents the maximum simultaneous requests that can be processed by the server object configuration. For a non-pooled server object, the *MaxInstances* represents the maximum number of simultaneous application users of that particular server object configuration.

The *MaxInstances* property must be greater than 0 and must be greater than the *MinInstances* property.

The *MinInstances* property applies to only pooled server object configurations. It represents the number of server object instances that are preloaded when the server object configuration is started. The GIS server will ensure that the minimum number of instances are always running within the server for a given configuration.

When there are more simultaneous requests than server object instances running, additional server object instances will be started until *MaxInstances* is reached.

Non-pooled server object configurations always have a *MinInstances* property of 0.

The *MinInstances* property must be less than the *MaxInstances* property.

The *Name* property in combination with the *TypeName* property is used to identify a server object configuration in methods such as *GetConfiguration*, *UpdateConfiguration*, *StartConfiguration*, and so on.

Name is case sensitive and can have a maximum of 120 characters. Names can contain only the following characters:

A–Z

a–z

0–9

_ (underscore)

- (minus)

The *TypeName* property indicates the type of server object that this configuration creates and runs. Examples are *MapServer* and *GeocodeServer*.

Server objects that are defined by server object configurations have a collection of initialization parameters and properties associated with them. An example of an initialization parameter is the map document associated with a *MapServer* object. An example of a property is the batch geocode size for a *GeocodeServer* object.

You can get these properties and change them using the *Properties* property on the server object configuration. The *Properties* property returns an *IPropertySet*. Use *GetProperty* and *SetProperty* on *IPropertySet* to get and set these properties. If you change these properties, you must call *UpdateConfiguration* to change them in the server object configuration.

You also use the *Properties* property to get a reference on the *PropertySet* for a new server object configuration to set its properties before adding it to the server by calling *AddConfiguration*.

The following code shows how to connect to the GIS server "melange" and use the *CreateConfiguration*, *AddConfiguration*, and *StartConfiguration* methods to create a new geocode server object configuration, add it to the server, and make it available for use. Note how the *Properties* property is called to get a reference to the server object configuration's properties.

```
Dim pGISServerConnection As IGISServerConnection
Set pGISServerConnection = New GISServerConnection
pGISServerConnection.Connect "melange"

Dim pServerObjectAdmin As IServerObjectAdmin
Set pServerObjectAdmin = pGISServerConnection.ServerObjectAdmin

' create the new configuration
Dim pConfiguration As IServerObjectConfiguration
Set pConfiguration = pServerObjectAdmin.CreateConfiguration

pConfiguration.Name = "California"
pConfiguration.TypeName = "GeocodeServer"

Dim pProps As IPropertySet
Set pProps = pConfiguration.Properties
pProps.SetProperty "LocatorWorkspacePath",
  "\\melange\Geocoding\California"
pProps.SetProperty "Locator", "California"
pProps.SetProperty "SuggestedBatchSize", "500"

pConfiguration.IsPooled = True
pConfiguration.MinInstances = 1
pConfiguration.MaxInstances = 1
Dim pRecProps As IPropertySet
Set pRecProps = pConfiguration.RecycleProperties
pRecProps.SetProperty "StartTime", "00:00"
pRecProps.SetProperty "Interval", "3600"
```

is a generic class that is used to set of properties. A PropertySet's ties are stored as name/value pairs. for the use of a property set are to operties required for opening an SDE pace or geocoding an address. To learn ut PropertySet objects, see the online developer documentation.

The references to the classes xaResource and localTransaction are required. These classes implement standard J2EE interfaces for transaction management. As the scope of this scenario is limited, this will not discussed in more detail. However, these implementation classes are required for this sample. If desired, you can find their source files in the sample's JAR file, ejbsample.jar, located in <install_location>\DeveloperKit\Samples\ Developer_Guide_Scenarios\ArcGIS_Server\ Application_EJB.

```
m_connectionRequestInfo = new com.intrinsyc.jca.
        JintConnectionRequestInfo("localhost",
        this.m_arcServerHostDomain, this.m_arcServerHostUserName,
        this.m_arcServerHostPassword, "", xaResource,
        localTransaction);
```

4. Obtain a JCA connection from the *ConnectionFactory*, passing along the JintConnectionRequestInfo object.

```
try{
    m_connection = (com.intrinsyc.jca.Connection)m_connection
Factory.getConnection(m_connectionRequestInfo);

    java.util.Vector params = new java.util.Vector();
    params.add(this.m_arcServerHostName);
```

5. Once successfully connected to the JCA, use the connection to obtain the *ServerConnection* object from ArcGIS Server itself. Once this server object is created, initialize the server by creating an instance of *ServerInitializer*.

```
    if(m_connection != null){

        m_serverConnectionObject =
m_connection.getObject("com.esri.arcgis.server.ServerConnection",
            params, this.m_arcServerHostDomain,
this.m_arcServerHostUserName, this.m_arcServerHostPassword);

        if((m_serverConnectionObject != null) &&
(m_serverConnectionObject instanceof
com.esri.arcgis.server.ServerConnection)){

            com.esri.arcgis.system.ServerInitializer
serverInitializer = new com.esri.arcgis.system.ServerInitializer();
            serverInitializer.setDefault(this.m_arcServerHostDomain,
this.m_arcServerHostUserName,
                            this.m_arcServerHostPassword);
            serverInitializer.setThreadDefault(
this.m_arcServerHostDomain, this.m_arcServerHostUserName,
  this.m_arcServerHostPassword);
            serverInitializer.trackObjectsInCurrentThread();

            m_serverConnection =
(com.esri.arcgis.server.ServerConnection)m_serverConnectionObject;
                                }
                        }
                }
        }
    catch(Exception exception){
        System.out.println("MapBean initialization failed: " +
exception.getMessage());
        }
    }
```

All of the connection initialization will happen upon creation of the EJB instance.

The body of your ejbCreate() method should now appear as follows:

```
public void ejbCreate(){
        javax.naming.Context initialContext, paramEnv;
            try{
                initialContext = new javax.naming.InitialContext();
                 paramEnv =
(javax.naming.Context)initialContext.lookup("java:comp/env");
                this.m_arcServerHostName =
(String)paramEnv.lookup("ArcServerHostName");
                this.m_arcServerHostDomain =
(String)paramEnv.lookup("ArcServerDomainName");
                this.m_arcServerHostUserName =
(String)paramEnv.lookup("ArcServerHostUserName");
                this.m_arcServerHostPassword =
(String)paramEnv.lookup("ArcServerHostPassword");
                            this.m_mapServerName =
(String)paramEnv.lookup("MapServerName");

                m_connectionFactory =
(com.intrinsyc.jca.ConnectionFactory)initialContext.lookup("intrinsycjca");
             }
            catch(Exception exception){
                m_connectionFactory = null;
                throw new Exception("JCA Lookup Failed: " + exception);
             }

        XAResource xaResource = new XAResource();
        LocalTransaction localTransaction = new LocalTransaction();

        m_connectionRequestInfo = new
com.intrinsyc.jca.JintConnectionRequestInfo(
        "localhost", this.m_arcServerHostDomain,
this.m_arcServerHostUserName, this.m_arcServerHostPassword,
            "", xaResource, localTransaction);
            try{
        m_connection =
(com.intrinsyc.jca.Connection)m_connectionFactory.getConnection(
m_connectionRequestInfo);

            java.util.Vector params = new java.util.Vector();
            params.add(this.m_arcServerHostName);

            if(m_connection != null){

                m_serverConnectionObject =
m_connection.getObject("com.esri.arcgis.server.ServerConnection",
                params, this.m_arcServerHostDomain,
this.m_arcServerHostUserName, this.m_arcServerHostPassword);

                if((m_serverConnectionObject != null) &&
(m_serverConnectionObject instanceof
com.esri.arcgis.server.ServerConnection)){
```

```
                    com.esri.arcgis.system.ServerInitializer
serverInitializer = new com.esri.arcgis.system.ServerInitializer();

                serverInitializer.setDefault(this.m_arcServerHostDomain,
this.m_arcServerHostUserName, this.m_arcServerHostPassword);

                serverInitializer.setThreadDefault(
this.m_arcServerHostDomain, this.m_arcServerHostUserName,
this.m_arcServerHostPassword);

                serverInitializer.trackObjectsInCurrentThread();

                m_serverConnection =
(com.esri.arcgis.server.ServerConnection)m_serverConnectionObject;

            }
          }
        }
      catch(Exception exception){
          System.out.println("MapBean initialization failed: " +
exception.getMessage());
        }
    }
```

Accessing a MapServer through the ServerConnection object

Now that you have connected to ArcGIS Server, you need to get some objects that will produce a map image that can be passed to the EJB client code. The client application itself will be addressed later in the scenario.

Having already created an instance of the ServerConnection object, you are now able to obtain an instance of a SOM from an ArcGIS Server. To do so, you must first connect to an ArcGIS Server host. Once connected to the host, you can create an instance of a SOM. Through the SOM, you can, in turn, create an instance of a ServerContext for a particular MapServer by specifying the MapServer name and MapServer type. From the ServerContext, you now have the ability to obtain objects pertaining to the associated ArcGIS Server from the SOM. In the following steps, you will add code to the EJB that performs these tasks.

1. Define a new public method called *getMap*, with a signature that takes all the parameters specific to ArcGIS Server for obtaining a map image. This method exports an image from the MapServer object that is accessible, through a URL, to any client application. The method will return a java.util.Hashtable, which will be used for maintaining the state of the requested map.

```
public java.util.Hashtable getMap(String mapServerHost, String
mapServerName, String mapName, int width, int height, double dpi, int
imageFormat, int imageReturnType) throws Exception{
        try{
            initMap(mapServerHost, mapServerName, mapName, width,
height, dpi, imageFormat, imageReturnType);
```

The code includes a call to another method, *initMap*. This will be defined and discussed shortly.

2. Create a java.util.Hashtable. This collection holds objects representing the parameters essential for maintaining the current map state.

```
java.util.Hashtable hTable = new java.util.Hashtable(3);
```

3. Next, you need to export an image from the MapServer object. This is achieved by calling the object's *exportMapImage()* method. MapDescription and ImageDescription objects that are passed to the exportMapImage method enable the MapServer to generate an image based on the object description parameters held within the Hashtable. The exportMapImage method returns an image, from which you can get a URL.

```
m_mapImage = m_mapServer.exportMapImage(m_mapDescription,
m_imageDescription);
hTable.put("MAPURL", m_mapImage.getURL());


m_mapDescriptionString = m_serverContext.saveObject(m_mapDescription);
m_imageDescriptionString =
m_serverContext.saveObject(m_imageDescription);
hTable.put("MAPDESCRIPTION", m_mapDescriptionString);
hTable.put("IMAGEDESCRIPTION", m_imageDescriptionString);


com.esri.arcgis.geometry.IEnvelope currentExtent =
m_mapDescription.getMapArea().getExtent();
com.esri.arcgis.samples.ejb.value.Envelope envelope = new
com.esri.arcgis.samples.ejb.value.Envelope();
envelope.setMinX(currentExtent.getXMin());
envelope.setMaxX(currentExtent.getXMax());
envelope.setMinY(currentExtent.getYMin());
envelope.setMaxY(currentExtent.getYMax());


hTable.put("EXTENT", envelope);
hTable.put("MAPUNITS", String.valueOf(m_mapServerInfo.getMapUnits()));
return hTable;
}
        catch(Exception exception){
        throw new Exception("getMap failed: " + exception.getMessage());
        }
}
```

4. Define the private method *connectServer*. The *connectServer* method obtains the SOM reference. Once this reference is made, a ServerContext is retrieved and, within this context, server objects can be obtained for direct use within the EJB.

```
private void connectServer(String mapServerHost, String mapServerName,
                  String mapDescription, String imageDescription)
throws MapServerConnectionException{
```

5. Next, connect to the ArcGIS Server host and get the server object manager.

```
try{
        m_serverConnection.connect(mapServerHost);
    com.esri.arcgis.server.IServerObjectManager serverObjectManager =
m_serverConnection.getServerObjectManager();
```

6. Create a *ServerContext* within which your objects will work.

```
m_serverContext =
serverObjectManager.createServerContext(mapServerName, "MapServer");
```

7. Instantiate the *MapDescription* and *ImageDescription* objects. Again, these are used to maintain the state of the map and will eventually be stored in a Hashtable.

```
if(mapDescription != null && imageDescription != null){
m_mapDescription = new com.esri.arcgis.carto.IMapDescriptionProxy(
m_serverContext.loadObject(mapDescription));
m_imageDescription = new
com.esri.arcgis.carto.IImageDescriptionProxy(
m_serverContext.loadObject(imageDescription));
}


m_mapServer = new com.esri.arcgis.carto.IMapServerProxy(
m_serverContext.getServerObject());


}
catch(Exception exception){
throw new MapServerConnectionException(exception.getMessage());
}
}
```

8. Define the first *initMap* private method. This signature will be called when *getMap* is called for the first time. In this step, you will call *connectServer*, and initialize the map properties. Later in the exercise, this method will be overloaded for use during zoom in and zoom out actions.

```
private void initMap(String mapServerHost, String mapServerName, String
mapName, int width, int height, double dpi, int imageFormat, int
imageReturnType) throws Exception{
try{
if(this.m_arcServerHostName.length() > 0 &&
this.m_mapServerName.length() > 0){
connectServer(m_arcServerHostName, m_mapServerName, null, null);
}
else{
connectServer(mapServerHost, mapServerName, null, null);
}
```

9. Now that the *MapDescription* and *ImageDescription* references have been made through the server object manager in the *connectServer* method, get the actual objects from the server using the appropriate method calls on the *IMapServerInfo* object.

```
com.esri.arcgis.carto.IMapServerObjects mapServerObjects = new
com.esri.arcgis.carto.IMapServerObjectsProxy(m_mapServer);

if(mapName == null){
m_mapServerInfo = new
com.esri.arcgis.carto.IMapServerInfoProxy(m_mapServer.getServerInfo(
m_mapServer.getDefaultMapName()));
}
else{
```

```
m_mapServerInfo = new
com.esri.arcgis.carto.IMapServerInfoProxy(m_mapServer.getServerInfo(mapName));
        }


m_mapDescription = new
com.esri.arcgis.carto.IMapDescriptionProxy(m_mapServerInfo.
getDefaultMapDescription());


m_graphicElements = new
com.esri.arcgis.carto.IGraphicElementsProxy(m_serverContext.createObject(
"esricarto.GraphicElements"));


if(m_graphicElements.getCount() > 0){
                    m_graphicElements.removeAll();
        }
```

10. Set properties to the *MapDescription* and *ImageDescription* objects.

```
m_imageType = new
com.esri.arcgis.carto.IImageTypeProxy(m_serverContext.createObject(
"esricarto.ImageType"));
m_imageType.setFormat(imageFormat);
m_imageType.setReturnType(imageReturnType);


m_imageDisplay = new
com.esri.arcgis.carto.IImageDisplayProxy(m_serverContext.createObject(
"esricarto.ImageDisplay"));
m_imageDisplay.setHeight(height);
m_imageDisplay.setWidth(width);
m_imageDisplay.setDeviceResolution(dpi);


m_imageDescription = new
com.esri.arcgis.carto.IImageDescriptionProxy(m_serverContext.createObject(
"esricarto.ImageDescription"));
m_imageDescription.setType(m_imageType);
m_imageDescription.setDisplay(m_imageDisplay);
```

11. Since the bean in this example is a stateless session bean, you need to pass
serialized versions of the *MapDescription* and *ImageDescription* objects back to
the requesting client. The code below serializes the current state of the
MapDescription and *ImageDescription* objects. Once serialized, this information
can be stored within the Hashtable in this format.

```
m_mapDescriptionString = m_serverContext.saveObject(m_mapDescription);


m_imageDescriptionString =
m_serverContext.saveObject(m_imageDescription);
        }
                catch(Exception exception){
        throw new Exception("Error in initMap: " + exception.getMessage());
            }
        }
```

12. As mentioned earlier, you can now overload the *initMap* method by defining a
different signature; this time, you pass in the *MapDescription* and

ImageDescription that were created and serialized for state management above. This feature will be called when the zoom in and zoom out functions are invoked. The body of this method should be familiar to you.

```
private void initMap(String mapServerHost, String mapServerName, String
mapDescription, String imageDescription) throws Exception{
    try{

    if(this.m_arcServerHostName.length() > 0 &&
this.m_mapServerName.length() > 0){
    connectServer(m_arcServerHostName, m_mapServerName, mapDescription,
imageDescription);
    }
    else{
    connectServer(mapServerHost, mapServerName, mapDescription,
imageDescription);
    }

    com.esri.arcgis.carto.IMapServerObjects mapServerObjects = new
com.esri.arcgis.carto.IMapServerObjectsProxy(m_mapServer);

    if(m_mapDescription.getCustomGraphics() != null){
            m_graphicElements = m_mapDescription.getCustomGraphics();
    }
    else{
            m_graphicElements = new
com.esri.arcgis.carto.IGraphicElementsProxy(m_serverContext.createObject(
"esricarto.GraphicElements"));
    }

    m_mapServerInfo = new
com.esri.arcgis.carto.IMapServerInfoProxy(m_mapServer.getServerInfo(
m_mapServer.getDefaultMapName()));

    m_imageType = new
com.esri.arcgis.carto.IImageTypeProxy(m_imageDescription.getType());
    m_imageDisplay = new
com.esri.arcgis.carto.IImageDisplayProxy(m_imageDescription.getDisplay());

    m_mapDescriptionString =
m_serverContext.saveObject(m_mapDescription);

    m_imageDescriptionString =
m_serverContext.saveObject(m_imageDescription);
    }
    catch(Exception msce){
        throw new Exception("Error in initMap: " +
msce.getMessage());
    }
}
```

Zooming and recentering the map image extent

Now that you have taken care of all the connection plumbing for both the JCA and the server object manager, you are ready to program the EJB for interacting with the map itself.

Most mapping functions are accomplished through the *IMapServer* interface. You will add the *zoomMap* method in order to zoom in and out on the map. Pixel coordinate values and a zoom factor are passed to the method, resulting in a change to the extent of the map. The *toMapPoints* method on the MapServer object is used to convert the pixel coordinates to database coordinates. This method returns a collection of *IPoint* objects, which are converted database coordinates drawn from the x and y pixel coordinates contained in the current *MapDescription* and *ImageDescription* objects. Current map envelopes are recentered by passing one of the newly created *IPoint* objects to the *Envelope* object's *centerAt* method. Once the map extent has been recentered, the map envelope can now be expanded to the zoom factor parameter. To zoom out from the current extent, the parameter will be a double of 1.5 or greater. Conversely, to zoom in on the map, the parameter will be a double value less than 1.0.

1. Define and write a public *zoomMap* method. This will return the Hashtable of stateful parameters that was discussed in the previous section.

```
public java.util.Hashtable zoomMap(String mapServerHost, String
mapServerName, String mapDescription, String imageDescription, int
pixelX, int pixelY, double factor) throws Exception{
        try{
        this.initMap(mapServerHost, mapServerName, mapDescription,
imageDescription);
```

2. Add the x and y pixel integers to the respective long arrays.

```
com.esri.arcgis.system.ILongArray longArrayX = new
com.esri.arcgis.system.ILongArrayProxy(
m_serverContext.createObject("esrisystem.LongArray"));
    longArrayX.add(pixelX);


com.esri.arcgis.system.ILongArray longArrayY = new
com.esri.arcgis.system.ILongArrayProxy(m_serverContext.createObject(
"esrisystem.LongArray"));
    longArrayY.add(pixelY);
```

3. The *toMapPoints* method returns an *IPointsCollection*. You only need to get the first IPoint in the collection. This coordinate provides the center point of the new map to be retrieved from the server.

```
com.esri.arcgis.geometry.IPointCollection points =
m_mapServer.toMapPoints(m_mapDescription, m_imageDisplay, longArrayX,
longArrayY);

IPoint point = points.getPoint(0);
```

4. Get the current map extent as an envelope and expand or shrink it by the specified factor.

```
IEnvelope env = m_mapDescription.getMapArea().getExtent();
env.centerAt(point);
env.expand(factor,factor,true);
```

```
com.esri.arcgis.carto.IMapExtent mapExtent = new
com.esri.arcgis.carto.IMapExtentProxy(m_serverContext.createObject(
"esricarto.MapExtent"));
        mapExtent.setExtent(env);


com.esri.arcgis.carto.IMapArea mapArea = new
com.esri.arcgis.carto.IMapAreaProxy(mapExtent);
        m_mapDescription.setMapArea(mapArea);


java.util.Hashtable hTable = new java.util.Hashtable(3);
m_mapImage = m_mapServer.exportMapImage(m_mapDescription,
m_imageDescription);
hTable.put("MAPURL", m_mapImage.getURL());


m_mapDescriptionString = m_serverContext.saveObject(m_mapDescription);
m_imageDescriptionString =
m_serverContext.saveObject(m_imageDescription);


hTable.put("MAPDESCRIPTION", m_mapDescriptionString);
hTable.put("IMAGEDESCRIPTION", m_imageDescriptionString);


com.esri.arcgis.geometry.IEnvelope currentExtent =
m_mapDescription.getMapArea().getExtent();


com.esri.arcgis.samples.ejb.value.Envelope envelope = new
com.esri.arcgis.samples.ejb.value.Envelope();
        envelope.setMinX(currentExtent.getXMin());
        envelope.setMaxX(currentExtent.getXMax());
        envelope.setMinY(currentExtent.getYMin());
        envelope.setMaxY(currentExtent.getYMax());

hTable.put("EXTENT", envelope);
            return hTable;
        }
        catch(Exception exception){
            throw new Exception("Error in zoomMap: " +
exception.getMessage());
        }
    }
```

Releasing JCA and ArcGIS Server resources

At this point, it is evident that some critical and essential connection object references to the ArcGIS Server and the JCA resource adapter have been made. These references will continue, even after the application server takes the EJB out of scope, unless the references are explicitly removed from your code.

Both resource connections are required to be closed when the EJB is removed from the container. Add the following code to the *ejbRemove* method to close the connections as required.

```
try{
      m_serverContext.removeAll();
    m_serverContext.releaseContext();
```

```
        m_connection.closeConnection();
      m_connection.releaseObject(m_serverConnectionObject);
      m_connection.close();
        }
      catch(Exception exception){
        System.out.println(exception.getMessage());
        }
```

Creating home and remote interfaces for StatelessMapImageEJB

The home and remote interfaces are responsible for creating remote instances of
EJBs and exposing the beans' methods to clients remotely. The home interface for
your bean needs to contain one method for creating an instance of it.

1. Create the home interface. This will accompany the *StatelessMapImageBean*
   class in the same package. The code you need for this interface is given below.

   ```
   package com.esri.arcgis.samples.ejb;

   import javax.ejb.EJBHome;
   import javax.ejb.CreateException;
   import javax.ejb.RemoveException;
   import java.rmi.RemoteException;
   import javax.ejb.Handle;

   public interface StatelessMapImageHome extends javax.ejb.EJBHome {

       public com.esri.arcgis.samples.ejb.StatelessMapImage create()throws
   CreateException, RemoteException;

   }
   ```

2. Create the remote interface. This, too, will accompany the
   StatelessMapImageBean class in the same package.

   ```
   package com.esri.arcgis.samples.ejb;

   import com.esri.arcgis.samples.ejb.value.Envelope;
   import java.rmi.RemoteException;

   public interface StatelessMapImage extends javax.ejb.EJBObject {

       public java.util.Hashtable getMap(String mapServerHost, String
   mapServerName, String mapName, int width, int height, double dpi, int
   imageFormat, int imageReturnType) throws RemoteException, Exception;

       public java.util.Hashtable zoomMap(String mapServerHost, String
   mapServerName, String mapDescription, String imageDescription, int
   pixelX, int pixelY, double factor) throws RemoteException, Exception;

   }
   ```

There are many more functions that can be used by the EJB to manipulate the
map you are working with. However, this scenario just examines the essentials;
other important tasks are needed to complete this application.

Creating the ejb-jar.xml

EJBs are typically deployed as EJB modules to J2EE application servers in the form of JAR files. The EJB module contains the compiled classes for the bean and an ejb-jar.xml file. Later in the exercise, you will create a META-INF directory that will contain this file. J2EE application servers read the ejb-jar.xml file for specific instructions on how to deploy the included beans.

Some J2EE application servers may require an additional, vendor-specific XML deployment descriptor, along with the ejb-jar.xml file.

1. Create the ejb-jar.xml file and begin it by adding the following:

Use an XML editor if you need to validate the format of the ejb-jar.xml file.

```
<?xml version="1.0" encoding="UTF-8"?>
<!DOCTYPE ejb-jar PUBLIC "-//Sun Microsystems, Inc.//DTD Enterprise
JavaBeans 2.0//EN" "http://java.sun.com/dtd/ejb-jar_2_0.dtd">
<ejb-jar>
  <enterprise-beans>
```

2. The next part of the file describes the bean and its expected behavior in the EJB container. In this case, you need to tell the application server that the EJB is a session bean and is stateless and that transaction management will be delegated to the application server's own mechanism.

```
<session>
        <display-name>StatelessMapImageEJB</display-name>
        <ejb-name>StatelessMapImageEJB</ejb-name>
        <home>com.esri.arcgis.samples.ejb.StatelessMapImageHome</home>
        <remote>com.esri.arcgis.samples.ejb.StatelessMapImage</remote>
        <ejb-class>com.esri.arcgis.samples.ejb.StatelessMapImageBean</ejb-
class>
        <session-type>Stateless</session-type>
        <transaction-type>Container</transaction-type>
```

3. Next, the descriptor defines some runtime arguments for the EJB to reference. Each entry requires an <env-entry-value>. These include the appropriate string values for the host, domain, username, password, and MapServer name for the machine serving as the ArcGIS Server. Many application servers are sensitive to unpopulated values. In this case, all of the values below are essential for this EJB to work as expected. Some example values are provided below. Edit them for your specific site.

The <env-entry-value> parameter text must be edited to match the appropriate names, domain, username, and password for your ArcGIS Server and the MapServer object you are using.

```
<env-entry>
        <description>Name of host where ArcGIS Server is running</
description>
        <env-entry-name>ArcServerHostName</env-entry-name>
        <env-entry-type>java.lang.String</env-entry-type>
        <env-entry-value>YourArcGISServerHostName</env-entry-value>
</env-entry>
<env-entry>
        <description>Domain of host where ArcGIS Server is running</
description>
        <env-entry-name>ArcServerDomainName</env-entry-name>
        <env-entry-type>java.lang.String</env-entry-type>
        <env-entry-value>YourArcGISServerDomainName</env-entry-value>
</env-entry>
<env-entry>
        <description>ArcGIS Server host machines available system logon
username</description>
```

```
        <env-entry-name>ArcServerHostUserName</env-entry-name>
        <env-entry-type>java.lang.String</env-entry-type>
        <env-entry-value>YourArcGISServerSystemUserName </env-entry-value>
</env-entry>
<env-entry>
        <description>ArcGIS Server host machines available system logon
password</description>
        <env-entry-name>ArcServerHostPassword</env-entry-name>
        <env-entry-type>java.lang.String</env-entry-type>
        <env-entry-value>YourArcGISServerSystemPassword</env-entry-value>
</env-entry>
<env-entry>
        <description>Name of MapServer to access</description>
        <env-entry-name>MapServerName</env-entry-name>
        <env-entry-type>java.lang.String</env-entry-type>
        <env-entry-value>YourMapServerName</env-entry-value>
</env-entry>
</session>
</enterprise-beans>
```

4. Finish editing the descriptor with this section of text.

```
<assembly-descriptor>
        <container-transaction>
                <method>
                    <ejb-name>StatelessMapImageEJB</ejb-name>
                    <method-name>*</method-name>
                </method>
                <trans-attribute>Required</trans-attribute>
        </container-transaction>
</assembly-descriptor>
</ejb-jar>
```

Creating the client application

Since the goal of this exercise is to demonstrate how ArcGIS Server objects are created and used inside an EJB container through the connection pooling and connection objects of the JCA, the client application you create to be used with the EJB is a simple one. The client JSP code simply obtains and displays an image from the MapServer, through the EJB container, and allows you to zoom in and out on that map image.

There are many tools and IDEs available that generate Web applications containing default JSP pages. Any of these can be used for this exercise.

1. Write a simple JSP page called "default.jsp". Begin by setting up the JSP page and HTML essentials.

```
<%@page contentType="text/html"%>
<html>
<head><title>JSP Page</title></head>
<body>
<%

%>
<a href="default.jsp?mapaction=zoomin">Zoom In</a><br>
<a href="default.jsp?mapaction=zoomout">Zoom Out</a><br>
```

```
</body>
</html>
```

The HTML links for Zoom In and Zoom Out will call the application again, telling it what zoom action to take on the map. You will see how to handle this shortly.

2. Inside the JSP directives (<% %>), write the scriptlet code that looks up the EJB home interface and obtains a reference to the EJB itself. This code is used the first time the JSP page is accessed. Once the reference is obtained, call the *getMap* method to return the Hashtable of map parameters. Remember that the actual map URL, which is what will be displayed in this page, is stored inside the returned Hashtable, just as you programmed the EJB to do. The StatelessMapImage reference is stored in the JSP session with the keyword "MAP". "MyServerHost" and "MyMapServer" need to be changed to reflect a valid host name and MapServer.

```
<%
    com.esri.arcgis.samples.ejb.StatelessMapImage m_map;
    java.lang.String mapDescription;
    java.lang.String imageDescription;
    java.lang.String mapurl;

    if(session.getAttribute("MAP") == null){

        try{
        javax.naming.InitialContext initContext = new
javax.naming.InitialContext();
        Object mapObj = initContext.lookup("StatelessMapImageEJB");

        com.esri.arcgis.samples.ejb.StatelessMapImageHome mapHome =
                (com.esri.arcgis.samples.ejb.StatelessMapImageHome)mapObj;

        m_map = mapHome.create();

        java.util.Hashtable hTable =
m_map.getMap("MyServerHost","MyMapServer", null, 400, 400,96.0,1,0);
        mapurl = hTable.get("MAPURL").toString();

        out.println("<img src='" + mapurl + "' />");

                session.setAttribute("MAP",m_map);
          session.setAttribute("MD",hTable.get("MAPDESCRIPTION").toString());
          session.setAttribute("ID",hTable.get("IMAGEDESCRIPTION").toString());


        }catch(Exception exception){
          out.println(exception.getMessage());
        }
```

Edit "MyServerHost" and "MyMapServer" to reflect a valid host name and MapServer.

3. If the session contains the attribute "MAP", then you have already initialized the EJB. Write the following code to zoom in on the map. On the *mapObj* object, call the *zoomMap* method, using the *MapDescription* and *ImageDescription* information to help you with the current state of the map extent.

```
}else{
```

```
        java.util.Hashtable ht =
(java.util.Hashtable)session.getAttribute("HT");
        m_map = (com.esri.arcgis.samples.ejb.StatelessMapImage)
session.getAttribute("MAP");
        mapDescription = session.getAttribute("MD").toString();
        imageDescription = session.getAttribute("ID").toString();

        String action = request.getParameter("mapaction");

        if(action.equals("zoomin")){
                java.util.Hashtable hTable = m_map.zoomMap(
"MyServerHost","MyMapServer", mapDescription, imageDescription,
200,200,0.5);
                mapurl = hTable.get("MAPURL").toString();
        session.setAttribute("MD",hTable.get("MAPDESCRIPTION").toString());
        session.setAttribute("ID",hTable.get("IMAGEDESCRIPTION").toString());
                }
```

Edit "MyServerHost" and "MyMapServer" to reflect a valid host name and MapServer.

4. Next, write similar code to zoom out from the map.

```
        else if(action.equals("zoomout")){
                java.util.Hashtable hTable = m_map.zoomMap(
"MyServerHost","MyMapServer", mapDescription, imageDescription,
200,200,1.5);
                mapurl = hTable.get("MAPURL").toString();
        session.setAttribute("MD",hTable.get("MAPDESCRIPTION").toString());
        session.setAttribute("ID",hTable.get("IMAGEDESCRIPTION").toString());
                }
```

Edit "MyServerHost" and "MyMapServer" to reflect a valid host name and MapServer.

5. You may have a map reference already in the JSP session, without a specified map action. As such, the following retrieves the initial map with the *getMap* method.

```
        else{
                java.util.Hashtable hTable = m_map.getMap("MyServerHost",
"MyMapServer", null, 400, 400,96.0,1,0);
                mapurl = hTable.get("MAPURL").toString();
        session.setAttribute("MD",hTable.get("MAPDESCRIPTION").toString());
        session.setAttribute("ID",hTable.get("IMAGEDESCRIPTION").toString());
                }
```

Edit "MyServerHost" and "MyMapServer" to reflect a valid host name and MapServer.

6. Now that you have the map URL, write code to print the map image to HTML format.

```
out.println("<img src='" + mapurl + "' />");

%>
```

The client code, although not an elegant or scalable approach, is now complete and will sufficiently demonstrate the process. The final step in the client creation process is to build a folder structure for the application.

7. Create the following directory structure, \mapapp\WEB-INF\classes. Copy your default.jsp file to the mapapp folder. Create a Web application descriptor file named "web.xml" and add it into the WEB-INF folder. For your application, the web.xml is defined as follows.

```xml
<?xml version="1.0" encoding="UTF-8"?>
<!DOCTYPE web-app PUBLIC "-//Sun Microsystems, Inc.//DTD Web Application
2.3//EN" "http://java.sun.com/dtd/web-app_2_3.dtd">
<web-app>
        <display-name>MapTester</display-name>
        <session-config>
                <session-timeout>30</session-timeout>
        </session-config>
        <welcome-file-list>
                <welcome-file>default.jsp</welcome-file>
        </welcome-file-list>
        <ejb-ref>
                <ejb-ref-name>StatelessMapImageEJB</ejb-ref-name>
                <ejb-ref-type>Session</ejb-ref-type>
                <home>com.esri.arcgis.samples.ejb.StatelessMapImageHome</
home>
                <remote>com.esri.arcgis.samples.ejb.StatelessMapImage</
remote>
        </ejb-ref>
        <distributable>false</distributable>
</web-app>
```

Now that the structure is in place, the entire application folder structure can be packaged into a Web archive (WAR) file (.war). This occurs during the build process, shown later in this discussion.

Building the application with ANT

ANT is a Java-based tool that can be used to compile and execute any kind of program on any target OS. It is similar to the standard UNIX tool called "make", which is often used by experienced C programmers. ANT utilizes a build file, typically named "build.xml", that tells it what to create and build. The build file, in XML format, uses XML tags, with attributes and subtags defining the details of work to be accomplished, to trigger different actions, commonly referred to as "targets". A target is a set of tasks you want to execute. When starting ANT, you can select which targets you want to execute. When no target is given, the project's default is used. For example, your source code needs to be compiled; the compiled classes need to be archived into JAR files; and finally, the JAR files need to be copied to a specific file system location. There are three targets here that need to be executed in a particular order. ANT will resolve these target dependencies for you, so that if target JAR is called first, it will execute target BUILD, then it will proceed to execute JAR.

Your build project should be arranged in the directory structure that the targets

There are no stringent rules for creating a build structure in this manner; however, this structure will be easiest for this exercise.

can utilize. The EJB source code, deployment descriptors, and client Web application need to be arranged in the directory structure outlined below.

1. Create a workspace that contains the following items:

Copy the ejb-jar.xml descriptor file into both the ejb-jar and META-INF folders. These will be archived into a deployable EJB JAR file by the ANT build.

Copy the source code for the *StatelessMapImageBean* that you created earlier into the src folder.

Copy the entire mapapp folder and its files (default.jsp) and subfolders (WEB-INF) into the webapps folder. The build process will bundle this entire directory into a WAR file.

The build.xml file is the ANT project you will write in the next step.

See the ArcGIS Server Installation Guide for details on setting these variables correctly.

2. Verify that the AGSDEVKITHOME and J2EE_HOME system environment variables have been set and are pointing to valid locations.

3. Write the ANT build.xml script. The first section of the script sets some variables to be used for directory locations and necessary JAR files.

```xml
<?xml version="1.0" encoding="UTF-8"?>
<project basedir="." default="all" name="EJBSample">

    <property environment="env"/>
    <property name="engine.home" value="${env.AGSDEVKITHOME}" />
    <property name="root.dir" location="${basedir}"/>
    <property name="src.dir" location="src"/>
    <property name="build.dir" location="build"/>
    <property name="dist.dir" location="${build.dir}/dist"/>
    <property name="class.dir" location="${build.dir}/classes"/>
    <property name="war.mapapp.name" value="MapApp-war"/>
    <property name="ejb.name" value="MapEJB"/>
    <property name="webapps.dir" location="webapps" />
    <property name="webapp.subdir.mapapp" value="mapapp" />
    <property name="webxml" value="WEB-INF/web.xml" />
    <property name="meta.inf" location="META-INF" />
    <property name="arcgis.java.dir"  location="${engine.home}/../java"/>
    <property name="arcgis.java.subdir" value="opt"/>
    <property name="jintegra.jar" location="${arcgis.java.dir}/
jintegra.jar"/>
    <property name="intrinsycjca.jar" location="${arcgis.java.dir}/
intrinsycjca.jar"/>
    <property name="j2ee.jar" location="${env.J2EE_HOME}/lib/j2ee.jar"/>
    <property name="arcgis_engine.jar" location="${arcgis.java.dir}/
${arcgis.java.subdir}/arcobjects.jar"/>

    <path id="compile.classpath">
      <pathelement location="${j2ee.jar}"/>
```

```
        <pathelement location="${intrinsycjca.jar}"/>
        <pathelement location="${jintegra.jar}"/>
        <pathelement location="${arcgis_engine.jar}"/>
    </path>
```

4. Next, define the following target names and dependencies.

- all
- validate-engine
- compile
- ejbjar

- clean
- init
- make-war

The target named "all" will process all targets in order.

```
    <target name="all" depends="ejbjar" description="build everything">
    </target>


    <target name="clean" description="clean all build products">
        <delete dir="${build.dir}" />
    </target>


    <target name="validate-engine">
        <condition property="engine.available">
            <and>
              <isset property="env.AGSDEVKITHOME" />
            </and>
        </condition>
        <fail message="Missing Dependencies: AGSDEVKITHOME environment
variable not correctly set" unless="engine.available"/>
    </target>



    <target name="init" depends="validate-engine">
        <tstamp/>
        <mkdir dir="${build.dir}"/>
        <mkdir dir="${dist.dir}"/>
        <mkdir dir="${class.dir}"/>
    </target>

    <target name="compile" depends="clean, init">
        <javac srcdir="${src.dir}" destdir="${class.dir}">
            <classpath refid="compile.classpath" />
        </javac>
    </target>

    <target name="make-war" depends="compile" if="webapps.dir">
        <war destfile="${dist.dir}/${war.mapapp.name}.war"
webxml="${webapps.dir}/${webapp.subdir.mapapp}/${webxml}">
            <fileset dir="${webapps.dir}/${webapp.subdir.mapapp}" />
                <classes dir="${class.dir}" includes="**/
StatelessMapImage.class, **/StatelessMapImageHome.class" />
        </war>
    </target>
```

```
        <target name="ejbjar" depends="make-war" description="Builds the ejb-
jar file">
            <copy todir="${class.dir}/META-INF">
                    <fileset dir="${meta.inf}" />
            </copy>
            <jar destfile="${dist.dir}/${ejb.name}.jar"
basedir="${class.dir}" />
        </target>
    </project>
```

Now you are ready to build the EJB and client applications.

5. Open a command prompt and navigate to the location of the build.xml file you just created. If you have properly configured your AGSDEVKITHOME system environment variable and included $AGSDEVKITHOME/tools/ant/bin to your system path, you can type "arcgisant all" to begin the build. ANT will identify the build.xml and begin executing the "all" target. You will see the following output in your console.

```
C:\arcdev\ArcServerEJBDocSample>arcgisant all
Buildfile: build.xml

clean:

validate-engine:

init:
    [mkdir] Created dir: C:\arcdev\ArcServerEJBDocSample\build
    [mkdir] Created dir: C:\arcdev\ArcServerEJBDocSample\build\dist
    [mkdir] Created dir: C:\arcdev\ArcServerEJBDocSample\build\classes

compile:
    [javac] Compiling 7 source files to
C:\arcdev\ArcServerEJBDocSample\build\classes

make-war:
      [war] Building war:
C:\arcdev\ArcServerEJBDocSample\build\dist\MapApp-war.war
      [war] Warning: selected war files include a WEB-INF/web.xml which
will be ignored (please use webxml attribute to war task)

ejbjar:
     [copy] Copying 1 file to
C:\arcdev\ArcServerEJBDocSample\build\classes\META-INF
      [jar] Building jar:
C:\arcdev\ArcServerEJBDocSample\build\dist\MapEJB.jar

all:

BUILD SUCCESSFUL
Total time: 4 seconds
C:\arcdev\ArcServerEJBDocSample>
```

The console information highlights the various tasks performed by ANT. First, the ./build/dist and ./build/classes folders were created. Next, the EJB source code was compiled and the classes copied to ./build/classes. Then, the webapp was packaged into a WAR file and copied to ./build/dist. Finally, the EJB archive was created and copied to the ./build/dist folder.

As a result, you should now have two archive files: the EJB jar, named MapEJB.jar, and the MapApp-war.war, which is the Web archive containing the client application. These archives are now ready for deployment into any J2EE-compliant application server. Remember, you will also need to deploy the intrinsycjca.rar archive to the same application server to which you deploy this sample application. This will be demonstrated in the next section using the WebSphere Application Server.

Deploying the EJB, the JCA resource adapter, and the client

This scenario demonstrates the deployment process using the WebSphere 5.0 Application Server as the EJB container on a Windows platform. Every J2EE-compliant application server will require similar steps for deploying EJB modules, Web applications, and resource adapters.

1. Open the WebSphere 5.0 Application Assembly Tool. From the Start menu click Programs, then IBM WebSphere Application Server. When the Welcome dialog box appears, choose Application on the New tab and click OK.

2. Change the name of the application to "MapApplication.ear" and click Apply.

Enterprise Application Archives, or .ear files, contain all of the files and resources comprising an entire EJB application; these include Web applications, deployment descriptors, and the EJBs themselves. The .ear file provides a single deployable entity for a complete enterprise application.

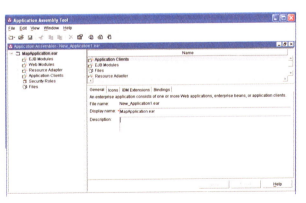

The next step is to add the necessary pieces to the enterprise archive.

3. Right-click the EJB Modules node and click Import. Browse to the location of your newly created MapEJB.jar file and click Open.

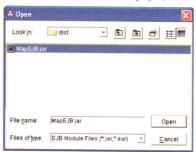

4. When the Confirm values dialog box appears, click OK and accept the default values.

5. Expand the EJB Modules node and highlight the Environment Entries node of the StatelessMapImageEJB. Verify the default values for each of the environment entries listed below. These are the same values as those you entered in the ejb-jar.xml file.

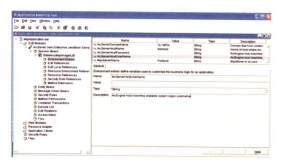

ArcServerDomainName: Domain name of the host where ArcGIS Server is running.

ArcServerHostName: Host name of the ArcGIS Server machine.

ArcServerHostUserName: Authorized username for the host ArcGIS Server machine.

ArcServerHostPassword: Password for the username.

MapServerName: Name of the MapServer object to access, running on the ArcServerHostName.

6. Next, deploy the JCA resource adapter to the application server. Right-click the Resource References node and choose New.

Edit the settings as indicated here.

Name: Name of the Resource Adapter.

Type: Type of resource.

Authentication: Type of authentication to use.

Sharing Scope: Set to Sharable.

7. Click the Bindings tab and type "intrinsycjca" as the JNDI name value. Click Apply to set the assigned values and Cancel to close the dialog box.

8. Right-click the Web Modules node and click Import. Browse to the location of the deployable Web application WAR file that you have just created and click Open. If the Select Module dialog box appears, select MapTester, then click OK.

9. Set the Context root value to "/mapapp" and click OK.

10. Save the enterprise application by clicking the Save button. In the Save dialog box, enter a unique filename with the .ear file extension. Next, click the Generate code for deployment button on the toolbar. This creates an enterprise application archive file.

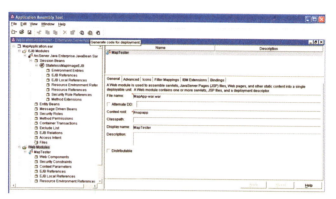

11. In the Deployed module location text box, type the path to the location of your newly created .ear file. This is the enterprise application that will be deployed. Click Generate Now.

12. Once the code generation has completed successfully, click Close and close the Application Assembly tool.

Now that you have created the enterprise archive file, you are ready to perform the deployment to the WebSphere server.

13. Open the WebSphere Administrative Console.

14. Deploy the intrinsycjca resource adapter to the application server. Open the Resources node and click the Resource Adapters link. Click Install RAR. Browse to the location of the intrinsycjca.rar file, <install_location>/java, click Open, and click Next. Click OK to accept the defaults. The intrinsycjca RAR is now listed as an available resource adapter.

15. After installing the intrinsycjca resource adapter in the WebSphere Administrative Console, click the intrinsycjca resource adapter link to view its properties. Scroll to the bottom of the properties page and choose the J2C Connection Factories link to create a new connection factory.

16. Click the New button to create a new connection factory. Name the connection factory "intrinsycjca", assign "intrinsycjca" as the JNDI name, and click Apply. Verify that the new configurations have been saved to the master configuration.

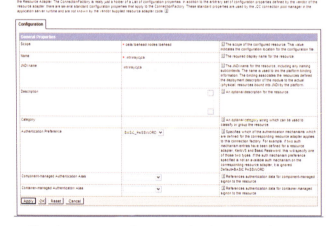

There is now an entry for a connection factory named "intrinsycjca".

17. Click the Environment node, then click the Shared Libraries link. At this point, you need to add the ArcGIS JARs to the application server's classpath. Click the New button to create a new library category. Type "ArcServer" as the name and add the full path names to jintegra.jar, arcobjects.jar, and intrinsycjca.jar in the Classpath section of the form. Click Apply and click OK. Note that these libraries will be available to all applications deployed on the server.

18. Click the Application node, then click the Install New Application link. Browse to the location of the Deployed_ArcServerApplication.ear file and click OK. Click Next.

19. Accept the defaults of the next step.

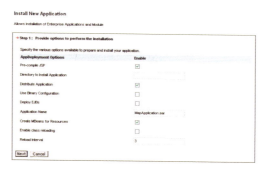

20. Check the Pre-compile JSP check box and click Next.

21. Enter "StatelessMapImageEJB" for the JNDI name value and click Next.

22. Enter "StatelessMapImageEJB" for the Web application EJB reference. Click Next for all other steps through summary, then click Finish. The enterprise application installation summary will be shown at end of installation. Click the Save to Master Configuration link to save the installation settings, then click Save.

Test the client application

Open the Enterprise Applications view, click the MapApplication.ear entry, and click the Start button. The EJB has now been deployed and is ready to be accessed by the client Web application. To start the Web application, browse to http://localhost:9080/mapapp.

ADDITIONAL RESOURCES

This scenario includes functionality and programming techniques covering a few aspects of ArcObjects and the ArcGIS Server API.

You are encouraged to read Chapter 4 of this book, 'Developing ArcGIS Server applications,' to get a better understanding of core ArcGIS Server programming concepts and programming guidelines for working with server contexts and ArcObjects running within those contexts.

ArcGIS Server applications exploit the rich GIS functionality of ArcObjects. This application is no exception. To learn more about these aspects of ArcObjects, refer to ArcGIS Developer Help.

Developer environments

ArcObjects is based on Microsoft's Component Object Model. End users of ArcGIS applications don't necessarily have to understand COM, but if you're a developer intent on developing applications based on ArcObjects or extending the existing ArcGIS applications using ArcObjects, an understanding of COM is a requirement even if you plan to use the .NET or Java API and not COM specifically. The level of understanding required depends on the depth of customization or development you wish to undertake.

This appendix does not cover the entire COM environment; however, it provides both Visual Basic and Visual C++® developers with sufficient knowledge to be effective in using ArcObjects. Later sections of the appendix detail how .NET and Java developers work with ArcObjects. There are many coding tips and guidelines that should make your work with ArcObjects more effective.

Before discussing COM specifically, it is worth considering the wider use of software components in general. There are a number of factors driving the motivation behind software components, but the principal one is the fact that software development is a costly and time-consuming venture.

In an ideal world, it should be possible to write a piece of code once and reuse it again and again using a variety of development tools, even in circumstances that the original developer did not foresee. Ideally, changes to the code's functionality made by the original developer could be deployed without requiring existing users to change or recompile their code.

Early attempts at producing reusable chunks of code revolved around the creation of class libraries, usually developed in C++. These early attempts suffered from several limitations, notably difficulty of sharing parts of the system (it is difficult to share binary C++ components—most attempts have only shared source code), problems of persistence and updating C++ components without recompiling, lack of good modeling languages and tools, and proprietary interfaces and customization tools.

To counteract these and other problems, many software engineers have adopted component-based approaches to system development. A software component is a binary unit of reusable code.

Several different but overlapping standards have emerged for developing and sharing components. For building interactive desktop applications, Microsoft's COM is the de facto standard. On the Internet, JavaBeans is viable technology. At a coarser grain appropriate for application-level interoperability, the Object Management Group (OMG) has specified the common object request broker architecture (CORBA).

ESRI chose COM as the component technology for ArcGIS because it is a mature technology that offers good performance, many of today's development tools support it, and there are a multitude of third-party components that can be used to extend the functionality of ArcObjects.

To understand COM—and, therefore, all COM-based technologies—it's important to realize that it isn't an object-oriented language but a protocol, or standard. COM is more than just a technology; it is a methodology of software development. COM defines a protocol that connects one software component, or module, with another. By making use of this protocol, it's possible to build reusable software components that can be dynamically interchanged in a distributed system.

COM also defines a programming model known as interface-based programming. Objects encapsulate the manipulation methods and the data that characterize each instantiated object behind a well-defined interface. This promotes structured and safe system development, since the client of an object is protected from knowing any details of how a particular method is implemented. COM doesn't specify how an application should be structured. As an application programmer working with COM, language, structure, and implementation details are left up to you.

The key to the success of components is that they implement, in a practical way, many of the object-oriented principles now commonly accepted in software engineering. Components facilitate software reuse because they are self-contained building blocks that can easily be assembled into larger systems.

COM does specify an object model and programming requirements that enable COM objects to interact with other COM objects. These objects can be within a single process, in other processes, or even on remote machines. They can be written in other languages and may have been developed in very different ways. That is why COM is referred to as a binary specification or standard—it is a standard that applies after a program has been translated to binary machine code.

COM allows these objects to be reused at a binary level, meaning that third-party developers do not require access to source code, header files, or object libraries to extend the system, even at the lowest level.

COMPONENTS, OBJECTS, CLIENTS, AND SERVERS

Different texts use the terms components, objects, clients, and servers to mean different things. (To add to the confusion, various texts refer to the same thing using all these terms.) Therefore, it is worthwhile to define some terminology.

COM is a client/server architecture. The server (or object) provides some functionality, and the client uses that functionality. COM facilitates the communication between the client and the object. An object can, at the same time, be a server to a client and a client of some other object's services.

Objects are instances of COM classes that make services available for use by a client. Hence, it is normal to talk of clients and objects instead of clients and servers. These objects are often referred to as COM objects and component objects. This book will refer to them simply as objects.

The client and its servers can exist in the same process or in a different process space. In-process servers are packaged in Dynamic Link Library form, and these DLLs are loaded into the client's address space when the client first accesses the server. Out-of-process servers are packaged in executables (EXE) and run in their own address space. COM makes the differences transparent to the client.

When creating COM objects, the developer must be aware of the type of server that the objects will reside in, but if the creator of the object has implemented them correctly, the packaging does not affect the use of the objects by the client.

There are pros and cons to each method of packaging that are symmetrically opposite. DLLs are faster to load into memory, and calling a DLL function is faster. EXEs, on the other hand, provide a more robust solution (if the server fails, the client will not crash), and security is better handled since the server has its own security context.

In a distributed system, EXEs are more flexible, and it does not matter if the server has a different byte ordering from the client. The majority of ArcObjects servers are packaged as in-process servers (DLLs). Later, you will see the performance benefits associated with in-process servers.

In a COM system, the client, or user of functionality, is completely isolated from the provider of that functionality, the object. All the client needs to know is that the functionality is available; with this knowledge, the client can make method calls to the object and expect the object to honor them. In this way, COM is said to act as a contract between client and object. If the object breaks that contract, the behavior of the system will be unspecified. In this way, COM development is based on trust between the implementer and the user of functionality.

In the ArcGIS applications, there are many objects that provide, via their interfaces, thousands of properties and methods. When you use the ESRI object libraries, you can assume that all these properties and interfaces have been fully implemented, and if they are present on the object diagrams, they are there to use.

Client and server

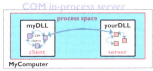

Objects inside an in-process server are accessed directly by their clients.

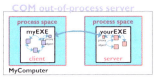

Objects inside an out-of-process server are accessed by COM-supplied proxy objects which make access transparent to the client

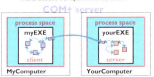

Objects inside an out-of-process server are accessed by COM-supplied proxy objects which make access transparent to the client. The COM run-time handles the remoting layer

CLASS FACTORY

Within each server, there is an object called a class factory that the COM runtime interacts with to instantiate objects of a particular class. For every corresponding COM class, there is a class factory. Normally, when a client requests an object from a server, the appropriate class factory creates a new object and passes out that object to the client.

SINGLETON OBJECTS

Although this is the normal implementation, it is not the only implementation possible. The class factory can also create an instance of the object the first time and, with subsequent calls, pass the same object to clients. This type of implementation creates what is known as a singleton object since there is only one instance of the object per process.

GLOBALLY UNIQUE IDENTIFIERS

A distributed system potentially has many thousands of interfaces, classes, and servers, all of which must be referenced when locating and binding clients and objects together at runtime. Clearly, using human-readable names would lead to the potential for clashes; hence, COM uses Globally Unique Identifiers, 128-bit numbers that are virtually guaranteed to be unique in the world. It is possible to generate 10 million GUIDs per second until the year 5770 A.D., and each one would be unique.

The COM API defines a function that can be used to generate GUIDs; in addition, all COM-compliant development tools automatically assign GUIDs when appropriate. GUIDs are the same as Universally Unique Identifiers (UUIDs), defined by the Open Group's Distributed Computing Environment (DCE) specification. Below is a sample GUID in registry format.

{E6BDAA76-4D35-11D0-98BE-00805F7CED21}

COM CLASSES AND INTERFACES

Developing with COM means developing using interfaces, the so-called interface-based programming model. All communication between objects is made via their interfaces. COM interfaces are abstract, meaning there is no implementation associated with an interface; the code associated with an interface comes from a class implementation. The interface sets out what requests can be made of an object that chooses to implement the interface.

How an interface is implemented differs among objects. Thus the objects inherit the type of interface, not its implementation, which is called type inheritance. Functionality is modeled abstractly with the interfaces and implemented within a class implementation. Classes and interfaces are often referred to as the "what" and "how" of COM. The interface defines what an object can do, and the class defines how it is done.

COM classes provide the code associated with one or more interfaces, thus encapsulating the functionality entirely within the class. Two classes can both have the same interface, but they may implement them quite differently. By implementing these interfaces in this way, COM displays classic object-oriented polymorphic behavior. COM does not support the concept of multiple inheritance; however,

A server is a binary file that contains all the code required by one or more COM classes. This includes both the code that works with COM to instantiate objects into memory and the code to perform the methods supported by the objects contained within the server.

GUIDGEN.EXE is a utility that ships with Microsoft's Visual Studio and provides an easy-to-use user interface for generating GUIDs. It can be found in the directory <VS Install Dir>\Common\Tools.

The acronym GUID is commonly pronounced "gwid".

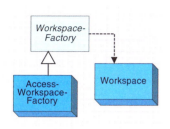

This is a simplified portion of the geodatabase object model showing type inheritance among abstract classes and coclasses and instantiation of classes.

this is not a shortcoming since individual classes can implement multiple interfaces. See the diagram to the lower left on polymorphic behavior.

Within ArcObjects are three types of classes that the developer must be aware of: abstract classes, coclasses, and classes. An abstract class cannot be created; it is solely a specification for instances of subclasses (through type inheritance). ArcObjects Dataset or Geometry classes are examples of abstract classes. An object of type Geometry cannot be created, but an object of type Polyline can. This Polyline object, in turn, implements the interfaces defined within the Geometry base class, hence any interfaces defined within object-based classes are accessible from the coclass.

A coclass is a publicly creatable class. In other words, it is possible for COM to create an instance of that class and give the resultant object to the client to use the services defined by the interfaces of that class. A class cannot be publicly created, but objects of this class can be created by other objects within ArcObjects and given to clients to use.

To the left is a diagram that illustrates the polymorphic behavior exhibited in COM classes when implementing interfaces. Notice that both the *Human* and *Parrot* classes implement the *ITalk* interface. The *ITalk* interface defines the methods and properties, such as *StartTalking*, *StopTalking*, or *Language*, but clearly the two classes implement these differently.

INSIDE INTERFACES

COM interfaces are how COM objects communicate with each other. When working with COM objects, the developer never works with the COM object directly but gains access to the object via one of its interfaces. COM interfaces are designed to be a grouping of logically related functions. The virtual functions are called by the client and implemented by the server; in this way, an object's interfaces are the contract between the client and object. The client of an object is holding an interface pointer to that object. This interface pointer is referred to as an opaque pointer since the client cannot gain any knowledge of the implementation details within an object or direct access to an object's state data. The client must communicate through the member functions of the interface. This allows COM to provide a binary standard through which all objects can effectively communicate.

This diagram shows how common behavior, expressed as interfaces, can be shared among multiple objects, animals in this example, to support polymorphism.

Interfaces allow developers to model functionality abstractly. Visual C++ developers see interfaces as collections of pure virtual functions, while Visual Basic developers see interfaces as collections of properties, functions, and subroutines.

The concept of the interface is fundamental in COM. The COM Specification (Microsoft, 1995) emphasizes these four points when discussing COM interfaces:

1. An interface is not a class. An interface cannot be instantiated by itself since it carries no implementation.

2. An interface is not an object. An interface is a related group of functions and is the binary standard through which clients and objects communicate.

3. Interfaces are strongly typed. Every interface has its own interface identifier, thereby eliminating the possibility of a collision between interfaces of the same human-readable name.

4. Interfaces are immutable. Interfaces are never versioned. Once defined and published, an interface cannot be changed.

Once an interface has been published, it is not possible to change the external signature of that interface. It is possible at any time to change the implementation details of an object that exposes an interface. This change may be a minor bug fix or a complete reworking of the underlying algorithm; the clients of the interface do not care since the interface appears the same to them. This means that when upgrades to the servers are deployed in the form of new DLLs and EXEs, existing clients need not be recompiled to make use of the new functionality. If the external signature of the interface is no longer sufficient, a new interface is created to expose the new functions. Old or deprecated interfaces are not removed from a class to ensure all existing client applications can continue to communicate with the newly upgraded server. Newer clients will have the choice of using the old or new interfaces.

An interface's permanence is not restricted to simply its method signatures, but extends to its semantic behavior as well. For example, an interface defines two methods, A and B, with no restrictions placed on their use. It breaks the COM contract if at a subsequent release Method A requires that Method B be executed first. A change like this would force possible recompilations of clients.

THE IUNKNOWN INTERFACE

All COM interfaces derive from the *IUnknown* interface, and all COM objects must implement this interface. The *IUnknown* interface performs two tasks: it controls object lifetime and provides runtime type support. It is through the *IUnknown* interface that clients maintain a reference on an object while it is in use—leaving the actual lifetime management to the object itself.

Object lifetime is controlled with two methods, *AddRef* and *Release*, and an internal reference counter. Every object must have an implementation of *IUnknown* to control its own lifetime. Anytime an interface pointer is created or duplicated, the *AddRef* method is called, and when the client no longer requires this pointer, the corresponding *Release* method is called. When the reference count reaches zero, the object destroys itself.

The name IUnknown came from a 1988 internal Microsoft paper called Object Architecture: Dealing with the Unknown – or – Type Safety in a Dynamically Extensible Class Library.

Clients also use *IUnknown* to acquire other interfaces on an object. *QueryInterface* is the method that a client calls when another interface on the object is required. When a client calls *QueryInterface*, the object provides an interface and calls *AddRef*. In fact, it is the responsibility of any COM method that returns an interface to increment the reference count for the object on behalf of the caller. The client must call the *Release* method when the interface is no longer needed. The client calls *AddRef* explicitly only when an interface is duplicated.

When developing a COM object, the developer must obey the rules of *QueryInterface*. These rules dictate that interfaces for an object are symmetrical, transitive, and reflexive and are always available for the lifetime of an object. For the client this means that, given a valid interface to an object, it is always valid to ask the object, via a call to *QueryInterface*, for any other interface on that object including itself. It is not possible to support an interface and later deny access to that interface, perhaps because of time or security constraints. Other mechanisms

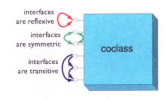

The rules of QueryInterface dictate that interfaces of an object are reflexive, symmetrical, and transitive. It is always possible, holding a valid interface pointer on an object, to get any other interface on that object.

The method QueryInterface is often referred to by the abbreviation QI.

Since IUnknown is fundamental to all COM objects, in general there are no references to IUnknown in any of the ArcObjects documentation and class diagrams.

Smart pointers are a class-based smart type and are covered in detail later in this appendix.

must be used to provide this level of functionality. Some classes support the concept of optional interfaces. Depending on the coclass, they may optionally implement an interface; this does not break this rule since the interface is either always available or always not available on the class.

When requested for a particular interface, the *QueryInterface* method can return an already assigned piece of memory for that requested interface, or it can allocate a new piece of memory and return that. The only case when the same piece of memory must be returned is when the *IUnknown* interface is requested. When comparing two interface pointers to see if they point to the same object, it is important that a simple comparison not be performed. To correctly compare two interface pointers to see if they are for the same object, they both must be queried for their *IUnknown* interface, and the comparison must be performed on the *IUnknown* pointers. In this way, the *IUnknown* interface is said to define a COM object's identity.

It's good practice in Visual Basic to call *Release* explicitly by assigning an interface equal to *Nothing* to release any resources it's holding. Even if you don't call *Release*, Visual Basic will automatically call it when you no longer need the object—that is, when it goes out of scope. With global variables, you must explicitly call *Release*. In Visual Basic, the system performs all these reference-counting operations for you, making the use of COM objects relatively straightforward.

In C++, however, you must increment and decrement the reference count to allow an object to correctly control its own lifetime. Likewise, the *QueryInterface* method must be called when asking for another interface. In C++ the use of smart pointers simplifies much of this. These smart pointers are class based and, hence, have appropriate constructors, destructors, and overloaded operators to automate much of the reference counting and query interface operations.

INTERFACE DEFINITION LANGUAGE

MIDL is commonly referred to simply as IDL.

The IDL defines the public interface that developers use when working with ArcObjects. When compiled, the IDL creates a type library.

Microsoft Interface Definition Language (MIDL) is used to describe COM objects including their interfaces. This MIDL is an extension of the Interface Definition Language (IDL) defined by the Distributed Computing Environment (DCE), where it used to define remote procedure calls between clients and servers. The MIDL extensions include most of the Object Definition Language (ODL) statements and attributes. ODL was used in the early days of OLE automation for the creation of type libraries.

TYPE LIBRARY

A type library is best thought of as a binary version of an IDL file. It contains a binary description of all coclasses, interfaces, methods, and types contained within a server or servers.

There are several COM interfaces provided by Microsoft that work with type libraries. Two of these interfaces are *ITypeInfo* and *ITypeLib*. By utilizing these standard COM interfaces, various development tools and compilers can gain information about the coclasses and interfaces supported by a particular library.

To support the concept of a language-independent development set of components, all relevant data concerning the ArcObjects libraries is shipped inside type libraries. There are no header files, source files, or object files supplied or needed by external developers.

INBOUND AND OUTBOUND INTERFACES

Interfaces can be either inbound or outbound. An inbound interface is the most common kind—the client makes calls to functions within the interface contained on an object. An outbound interface is one in which the object makes calls to the client—a technique analogous to the traditional callback mechanism.

There are differences in the ways these interfaces are implemented. The implementer of an inbound interface must implement all functions of the interface; failure to do so breaks the contract of COM. This is also true for outbound interfaces. If you use Visual Basic, you don't have to implement all functions present on the interface since it provides stub methods for the methods you don't implement. On the other hand, if you use C++, you must implement all the pure virtual functions to compile the class.

Connection points is a specific methodology for working with outbound COM interfaces. The connection point architecture defines how the communication between objects is set up and taken down. Connection points are not the most efficient way of initializing bidirectional object communication, but they are in common use because many development tools and environments support them.

Dispatch event interfaces

There are some objects with ArcObjects that support two outbound event interfaces that look similar to the methods they support. Examples of two such interfaces are the *IDocumentEvents* and the *IDocumentEventsDisp*. The "Disp" suffix denotes a pure Dispatch interface. These dispatch interfaces are used by VBA when dealing with certain application events, such as loading documents. A VBA programmer works with the dispatch interfaces, while a developer using another development language uses the nonpure dispatch interface. Since these dispatch event interfaces are application specific, consult the ArcGIS Developer Help for more details on using the interface.

Default interfaces

Every COM object has a default interface that is returned when the object is created if no other interface is specified. All the objects within the ESRI object libraries have *IUnknown* as their default interface, with a few exceptions.

The default interface of the *Application* object for both ArcCatalog and ArcMap is the *IApplication* interface. These uses of non*IUnknown* default interfaces are a requirement of Visual Basic for Applications and are found on the ArcMap and ArcCatalog application-level objects.

This means that variables that hold interface pointers must be declared in a certain way. For more details, see the coding sections later in this appendix. When COM objects are created, any of the supported interfaces can be requested at creation time.

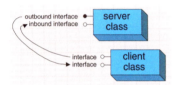

In the diagrams in this book and the ArcObjects object model diagrams, outbound interfaces are depicted with a solid circle on the interface jack.

The reason for making IUnknown the default interface is because the VB object browser hides information for the default interface. The fact that it hides IUnknown is not important for VB developers.

IDispatch interface

COM supports three types of binding:

1. Late. This is where type discovery is left until runtime. Method calls made by the client but not implemented by the object will fail at execution time.

2. ID. Method IDs are stored at compile time, but execution of the method is still performed through a higher-level function.

3. Custom vTable (early). Binding is performed at compile time. The client can then make method calls directly into the object.

Binding is the term given to the process of matching the location of a function given a pointer to an object.

The *IDispatch* interface supports late- and ID-binding languages. The *IDispatch* interface has methods that allow clients to ask the object what methods it supports.

Assuming the required method is supported, the client executes the method by calling the *IDispatch::Invoke* method. This method, in turn, calls the required method and returns the status and any parameters back to the client on completion of the method call.

Binding type	In process DLL	Out of process DLL
Late binding	22,250	5,000
Custom vTable binding	825,000	20,000

This table shows the number of function calls that can be made per second on a typical Pentium® III machine.

Clearly, this is not the most efficient way to make calls on a COM object. Late binding requires a call to the object to retrieve the list of method IDs; the client must then construct the call to the *Invoke* method and call it. The *Invoke* method must then unpack the method parameters and call the function.

All these steps add significant overhead to the time it takes to execute a method. In addition, every object must have an implementation for *IDispatch*, which makes all objects larger and adds to their development time.

ID binding offers a slight improvement over late binding in that the method IDs are cached at compile time, which means the initial call to retrieve the IDs is not required. However, there is still significant call overhead because the *IDispatch::Invoke* method is still called to execute the required method on the object.

Early binding, often referred to as custom vTable binding, does not use the *IDispatch* interface. Instead, a type library provides the required information at compile time to allow the client to know the layout of the server object. At runtime, the client makes method calls directly into the object. This is the fastest method of calling object methods and also has the benefit of compile-time type checking.

These diagrams summarize the custom and IDispatch interfaces for two classes in ArcObjects. The layout of the vTable displays the differences. It also illustrates the importance of implementing all methods—if one method is missing, the vTable will have the wrong layout, and hence, the wrong function pointer would be returned to the client, resulting in a system crash.

Objects that support both *IDispatch* and custom vTable are referred to as dual interface objects. The object classes within the ESRI object libraries do not implement the *IDispatch* interface; this means that these object libraries cannot be used with late-binding scripting languages, such as JavaScript or VBScript, since these languages require that all COM servers accessed support the *IDispatch* interface.

Careful examination of the ArcGIS class diagrams indicates that the *Application* objects support *IDispatch* because there is a requirement in VBA for the *IDispatch* interface.

All ActiveX controls support *IDispatch*. This means it is possible to use the various ActiveX controls shipped with ArcObjects to access functionality from within scripting environments.

INTERFACE INHERITANCE

An interface consists of a group of methods and properties. If one interface inherits from another, then all the methods and properties in the parent are directly available in the inheriting object.

The underlying principle here is interface inheritance, rather than the implementation inheritance you may have seen in languages such as SmallTalk and C++. In implementation inheritance, an object inherits actual code from its parent; in interface inheritance, it's the definitions of the methods of the object that are passed on. The coclass that implements the interfaces must provide the implementation for all inherited interfaces.

Interfaces that directly inherit from an interface other than IUnknown cannot be implemented in VB.

Implementation inheritance is not supported in a heterogeneous development environment because of the need to access source and header files. For reuse of code, COM uses the principles of aggregation and containment. Both of these are binary-reuse techniques.

AGGREGATION AND CONTAINMENT

For a third-party developer to make use of existing objects, using either containment or aggregation, the only requirement is that the server housing the contained or aggregated object is installed on both the developer and target release machines. Not all development languages support aggregation.

The simplest form of binary reuse is containment. Containment allows modification of the original object's method behavior but not the method's signature. With containment, the contained object (inner) has no knowledge that it is contained within another object (outer). The outer object must implement all the interfaces supported by the inner. When requests are made on these interfaces, the outer object simply delegates them to the inner. To support new functionality, the outer object can either implement one of the interfaces without passing the calls on or implement an entirely new interface in addition to those interfaces from the inner object.

COM aggregation involves an outer object that controls which interfaces it chooses to expose from an inner object. Aggregation does not allow modification of the original object's method behavior. The inner object is aware that it is being aggregated into another object and forwards any *QueryInterface* calls to the outer (controlling) object so the object as a whole obeys the laws of COM.

To the clients of an object using aggregation, there is no way to distinguish which interfaces the outer object implements and which interfaces the inner object implements.

Custom features make use of both containment and aggregation. The developer aggregates the interfaces where no customizations are required and contains those that are to be customized. The individual methods on the contained interfaces can then either be implemented in the customized class, thus providing custom functionality, or the method call can be passed to the appropriate method on the contained interface.

Aggregation is important in this case since there are some hidden interfaces defined on a feature that cannot be contained.

Visual Basic 6 does not support aggregation, so it can't be used to create custom features.

THREADS, APARTMENTS, AND MARSHALLING

A thread is a process flow through an application. There are potentially many threads within Windows applications. An apartment is a group of threads that work with contexts within a process. With COM+, a context belongs to one apartment. There are potentially many types of contexts; security is an example of a type of context. Before successfully communicating with each other, objects must have compatible contexts.

COM supports two types of apartments: single-threaded apartment and multithreaded apartment (MTA). COM+ supports the additional thread-neutral apartment (TNA). A process can have any number of STAs; each process creates one STA called the main apartment. Threads that are created as apartments are placed in an STA. All user interface code is placed in an STA to prevent deadlock situations. A process can only have one MTA. A thread that is started as multithreaded is placed in the MTA. The TNA has no threads permanently associated with it; rather, threads enter and leave the apartment when appropriate.

In-process objects have an entry in the registry, the ThreadingModel, that informs the COM service control manager (SCM) into which apartment to place the object. If the object's requested apartment is compatible with the creator's apartment, the object is placed in that apartment; otherwise, the SCM will find or create the appropriate apartment. If no threading model is defined, the object will be placed in the main apartment of the process. The ThreadingModel registry entry can have the following values:

1. Apartment. Object must be executed within the STA. Normally used by UI objects.

2. Free. Object must be executed within the MTA. Objects creating threads are normally placed in the MTA.

3. Both. Object is compatible with all apartment types. The object will be created in the same apartment as the creator.

4. Neutral. Objects must execute in the TNA. Used by objects to ensure there is no thread switch when called from other apartments. This is only available under COM+.

Marshalling enables a client to make interface function calls to objects in other apartments transparently. Marshalling can occur between COM apartments on different machines, between COM apartments in different process spaces, and between COM apartments in the same process space (STA to MTA, for example). COM provides a standard marshaller that handles function calls that use automation-compliant data types (see table below). Nonautomation data types can be handled by the standard marshaller as long as proxy stub code is generated; otherwise, custom marshalling code is required.

Although an understanding of apartments and threading is not essential in the use of ArcObjects, basic knowledge will help you understand some of the implications with certain development environments highlighted later in this appendix.

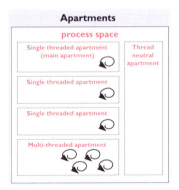

Think of the SCM (pronounced scum) as the COM runtime environment. The SCM interacts with objects, servers, and the operating system and provides the transparency between clients and the objects they work with.

Type	Description
Boolean	Data item that can have the value True or False
unsigned char	8-bit unsigned data item
double	64-bit IEEE floating-point number
float	32-bit IEEE floating-point number
int	Signed integer, whose size is system dependent
long	32-bit signed integer
short	16-bit signed integer
BSTR	Length-prefixed string
CURRENCY	8-byte, fixed-point number
DATE	64-bit, floating-point fractional number of days since Dec 30, 1899
SCODE	For 16-bit systems - Built-in error that corresponds to VT_ERROR
Typedef enum myenum	Signed integer, whose size is system dependent
Interface IDispatch *	Pointer to the IDispatch interface
Interface IUnknown *	Pointer to an interface that does not derive from IDispatch
dispinterface Typename *	Pointer to an interface derived from IDispatch
Coclass Typename *	Pointer to a coclass name (VT_UNKNOWN)
[oleautomation] interface Typename *	Pointer to an interface that derives from IDispatch
SAFEARRAY(TypeName)	TypeName is any of the above types. Array of these types
TypeName*	TypeName is any of the above types. Pointer to a type
Decimal	96-bit unsigned binary integer scaled by a variable power of 10. A decimal data type that provides a size and scale for a number (as in coordinates)

COMPONENT CATEGORY

Component categories are used by client applications to find all COM classes of a particular type that are installed on the system efficiently. For example, a client application may support a data export function in which you can specify the output format—a component category could be used to find all the data export classes for the various formats. If component categories are not used, the application has to instantiate each object and interrogate it to see if it supports the required functionality, which is not a practical approach. Component categories support the extensibility of COM by allowing the developer of the client application to create and work with classes that belong to a particular category. If at a later date a new class is added to the category, the client application need not be changed to take advantage of the new class; it will automatically pick up the new class the next time the category is read.

COM AND THE REGISTRY

COM makes use of the Windows system registry to store information about the various parts that compose a COM system. The classes, interfaces, DLLs, EXEs, type libraries, and so forth, are all given unique identifiers (GUIDs) that the SCM uses when referencing these components. To see an example of this, run regedit, then open HKEY_CLASSES_ROOT. This opens a list of all the classes registered on the system.

ESRI keys in the Windows system registry

COM makes use of the registry for a number of housekeeping tasks, but the most important and most easily understood is the use of the registry when instantiating COM objects into memory. In the simplest case, that of an in-process server, the steps are as follows:

1. Client requests the services of a COM object.

2. SCM looks for the requested objects registry entry by searching on the class ID (a GUID).

3. DLL is located and loaded into memory. The SCM calls a function within the DLL called *DllGetClassObject*, passing the desired class as the first argument.

4. The class object normally implements the interface *IClassFactory*. The SCM calls the method *CreateInstance* on this interface to instantiate the appropriate object into memory.

5. Finally, the SCM asks the newly created object for the interface that the client requested and passes that interface back to the client. At this stage, the SCM drops out of the equation, and the client and object communicate directly.

The function DllGetClassObject is the function that makes a DLL a COM DLL. Other functions, such as DllRegisterServer and DllUnregisterServer, are nice to have but not essential for a DLL to function as a COM DLL.

From the above sequence of steps, it is easy to imagine how changes in the object's packaging (DLL versus EXE) make little difference to the client of the object. COM handles these differences.

AUTOMATION

Automation is the technology used by individual objects or entire applications to provide access to their encapsulated functionality via a late-bound language. Commonly, automation is thought of as writing macros, where these macros can access many applications for a task to be done. ArcObjects, as already stated, does not support the IDispatch interface; hence, it cannot be used alone by an automation controller.

It is possible to instantiate an instance of ArcMap by cocreating the document object and making calls into ArcMap via the document object or one of its connected objects. There are, however, problems with this approach since the automation controller instance and the ArcMap instance are running in separate processes. Many of the objects contained within ArcObjects are process dependent, and therefore, simple automation will not work.

ArcGIS applications are built using ArcObjects and can be developed via several APIs. These include COM (VB, VC++, Delphi™, MainWin), .NET (VB.NET and C#), Java, and C++. Some APIs are more suitable than others for developing certain applications. This is briefly discussed later, but you should also read the appropriate developer guide for the product you are working with for more information and recommendations on which API to use.

The subsequent sections of this appendix cover some general guidelines and considerations when developing with ArcObjects regardless of the API. Some of the more common API languages each have a section describing the development environment, programming techniques, resources, and other issues you must consider when developing with ArcObjects.

CODING STANDARDS

Each of the language-specific sections begins with a section on coding standards for that language. These standards are used internally at ESRI and are followed by the samples that ship with the software.

For simplicity, some samples will not follow the coding standards. As an example, it is recommended that when coding in Visual Basic, all types defined within an ESRI object library are prefixed with the library name, for example, esriGeometry.IPolyline. This is only done in samples in which a name clash will occur. Omitting this text makes the code easier to understand for developers new to ArcObjects.

To understand why standards and guidelines are important, consider that in any large software development project, there are many backgrounds represented by the team members. Each programmer has personal opinions concerning how code should look and be built. If each programmer engineers code differently, it becomes increasingly difficult to share work and ideas. On a successful team, the developers adapt their coding styles to the tone set by the group. Often, this means adapting one's code to match the style of existing code in the system.

Initially, this may seem burdensome, but adopting a uniform programming style and set of techniques invariably increases software quality. When all the code in a project conforms to a standard set of styles and conventions, less time is wasted learning the particular syntactic quirks of individual programmers, and more time can be spent reviewing, debugging, and extending the code. Even at a social level, uniform style encourages team-oriented, rather than individualist, outlooks—leading to greater team unity, productivity, and ultimately, better software.

GENERAL CODING TIPS AND RESOURCES

This section on general coding tips will benefit all developers working with ArcObjects no matter what language they are using. Code examples are shown in VBA, however.

Class diagrams

Getting help with the object model is fundamental to successfully working with ArcObjects. Appendix B, 'Reading the object model diagrams', provides an introduction to the class diagrams and shows many of the common routes through objects. The class diagrams are most useful if viewed in the early learning process in printed form. This allows developers to appreciate the overall structure of the object model implemented by ArcObjects. When you are comfortable with the overall structure, the PDF files included with the software distribution can be more effective to work with. The PDF files are searchable; you can use the Search dialog box in Acrobat Reader to find classes and interfaces quickly.

Object browsers

In addition to the class diagram PDF files, the type library information can be viewed using a number of object browsers, depending on your development platform.

Visual Basic and .NET have built-in object browsers; OLEView (a free utility from Microsoft) also displays type library information. The best object viewer to use in this environment is the ESRI object viewer. This object viewer can be used to view type information for any type library that you reference within it. Information on the classes and interfaces can be displayed in Visual Basic, Visual C++, or object diagram format. The object browsers can view coclasses and classes but cannot be used to view abstract classes. Abstract classes are only viewable on the object diagrams, where their use is solely to simplify the models.

Java and C++ developers should refer to the ArcObjects—JavaDoc or ArcGIS Developer Help.

Component help

All interfaces and coclasses are documented in the component help file. Ultimately, this will be the help most commonly accessed when you get to know the object models better.

For Visual Basic and .NET developers, this is a compiled HTML file that can be viewed by itself or when using an IDE. If the cursor is over an ESRI type when the F1 key is pressed, the appropriate page in the ArcObjects Class Help in the ArcGIS Developer Help system is displayed in the compiled HTML viewer.

For Java and C++ developers, refer to ArcObjects—JavaDoc or ArcGIS Developer Help.

Code wizards

There are a number of code generation wizards available to help with the creation of boilerplate code in Visual Basic, Visual C++, and .NET. Although these wizards are useful in removing the tediousness in common tasks, they do not excuse you as the developer from understanding the underlying principles of the generated code. The main objective should be to read the accompanying documentation and understand the limitations of these tools.

Indexing of collections

All collection-like objects in ArcObjects are zero-based for their indexing. This is not the case with all development environments; Visual Basic has both zero- and one-based collections. As a general rule, if the collection base is not known, assume that the collection base is zero. This ensures that a runtime error will be raised when the collection is first accessed (assuming the access of the collection does not start at zero). Assuming a base of one means the first element of a zero-based collection would be missed and an error would only be raised if the end of the collection were reached when the code is executed.

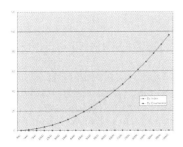

This graph shows the performance benefits of accessing a collection using an enumerator opposed to the elements index. As expected, the graph shows a classic power trend line ($y=cx^b$).

Accessing collection elements

When accessing elements of a collection sequentially, it is best to use an enumerator interface. This provides the fastest method of walking through the collection. The reason for this is that each time an element is requested by index, internally an enumerator is used to locate the element. Hence, if the collection is looped over getting each element in turn, the time taken increases exponentially ($y=cx^b$).

Enumerator use

When requesting an enumerator interface from an object, the client has no idea how the object has implemented this interface. The object may create a new enumerator, or it may decide for efficiency to return a previously created enumerator. If a previous enumerator is passed to the client, the position of the element pointer will be at the last accessed element. To ensure that the enumerator is at the start of the collection, the client should reset the enumerator before use.

Error handling

Exception handling is language specific, and since COM is language neutral, exceptions are not supported.

All methods of interfaces, in other words, methods callable from other objects, should handle internal errors and signify success or failure via an appropriate HRESULT. COM does not support passing exceptions out of interface method calls. COM supports the notion of a COM exception. A COM exception utilizes the COM error object by populating it with relevant information and returning an appropriate HRESULT to signify failure. Clients, on receiving the HRESULT, can then interrogate the COM *Error* object for contextual information about the error. Languages, such as Visual Basic, implement their own form of exception handling. For more information, see the specific section for the language with which you are developing available later in this appendix.

Notification interfaces

There are a number of interfaces in ArcObjects that have no methods. These are known as notification interfaces. Their purpose is to inform the application framework that the class that implements them supports a particular set of functionality. For instance, the application framework uses these interfaces to determine if a menu object is a root-level menu (*IRootLevelMenu*) or a context menu (*IShortcutMenu*).

Client-side storage

Some ArcObjects methods expect interface pointers to point to valid objects prior to making the method call. This is known as client storage since the client allocates the memory needed for the object before the method call. Suppose you have a polygon, and you want to obtain its bounding box. To do this, use the *QueryEnvelope* method on *IPolygon*. If you write the following code:

```
Dim pEnv As IEnvelope
pPolygon.QueryEnvelope pEnv
```

you'll get an error because the *QueryEnvelope* method expects you (the client) to create the *Envelope*. The method will modify the envelope you pass in and return the changed one back to you. The correct code follows

```
Dim pEnv As IEnvelope
Set pEnv = New Envelope
pPolygon.QueryEnvelope pEnv
```

How do you know when to create and when not to create? In general, all methods that begin with "Query", such as *QueryEnvelope*, expect you to create the object. If the method name is *GetEnvelope*, then an object will be created for you. The reason for this client-side storage is performance. When it is anticipated that the method on an object will be called in a tight loop, the parameters need only be created once and simply populated. This is faster than creating new objects inside the method each time.

Property by value and by reference

Occasionally, you will see a property that can be set by value or by reference, meaning that it has both a *put_XXX* and a *putref_XXX* method. On first appearance, this may seem odd—Why does a property need to support both? A Visual C++ developer sees this as simply giving the client the opportunity to pass ownership of a resource over to the server (using the *putref_XXX* method). A Visual Basic developer will see this as quite different; indeed, it is likely because of the Visual Basic developer that both *By Reference* and *By Value* are supported on the property.

To illustrate this, assume there are two text boxes on a form, Text1 and Text2. With a *propput*, it is possible to do the following in Visual Basic:

```
Text1.text = Text2.text
```

It is also possible to write this:

```
Text1.text = Text2
```

or this:

```
Text1 = Text2
```

DISPIDs *are unique IDs given to properties and methods for the IDispatch interface to efficiently call the appropriate method using the Invoke method.*

All these cases make use of the *propput* method to assign the text string of text box Text2 to the text string of text box Text1. The second and third cases work because no specific property is stated, so Visual Basic looks for the property with a *DISPID* of 0.

This all makes sense assuming that it is the text string property of the text box that is manipulated. What happens if the actual object referenced by the variable Text2 is to be assigned to the variable Text1? If there were only a *propput* method, it would not be possible; hence, the need for a *propputref* method. With the *propputref* method, the following code will achieve the setting of the object reference:

Notice the use of the "Set".

```
Set Text1 = Text2
```

Initializing Outbound interfaces

When initializing an Outbound interface, it is important to only initialize the variable if the variable does not already listen to events from the server object. Failure to follow this rule will result in an infinite loop.

As an example, assume there is a variable *ViewEvents* that has been dimensioned as:

```
Private WithEvents ViewEvents As Map
```

To correctly sink this event handler, you can write code within the *OnClick* event of a UI button control, like this:

```
Private Sub UIButtonControl1_Click()
  Dim pMxDoc As IMxDocument
  Set pMxDoc = ThisDocument

  ' Check to see that the map is different than what is currently connected.
  If (Not ViewEvents Is pMxDoc.FocusMap) Then
    ' Sink the event.
    Set ViewEvents = pMxDoc.FocusMap
  End If
End Sub
```

Notice in the above code the use of the *Is* keyword to check for object identity.

DATABASE CONSIDERATIONS

When programming against the database, there are a number of rules that must be followed to ensure that the code will be optimal. These rules are detailed below.

If you are going to edit data programmatically, that is, not use the editing tools in ArcMap, you need to follow these rules to ensure that custom object behavior, such as network topology maintenance or triggering of custom feature-defined methods, is correctly invoked in response to the changes your application makes to the database. You must also follow these rules to ensure that your changes are made within the multiuser editing (long transaction) framework.

Edit sessions

Make all changes to the geodatabase within an edit session, which is bracketed between *StartEditing* and *StopEditing* method calls on the *IWorkspaceEdit* interface found on the *Workspace* object.

This behavior is required for any multiuser update of the database. Starting an edit session gives the application a state of the database that is guaranteed not to change, except for changes made by the editing application.

In addition, starting an edit session turns on behavior in the geodatabase such that a query against the database is guaranteed to return a reference to an existing object in memory if the object was previously retrieved and is still in use.

This behavior is required for correct application behavior when navigating between a cluster of related objects while making modifications to objects. In other words, when you are not within an edit session, the database can create a new instance of a COM object each time the application requests a particular object from the database.

Edit operations

Group your changes into edit operations, which are bracketed between the *StartEditOperation* and *StopEditOperation* method calls on the *IWorkspaceEdit* interface.

You may make all your changes within a single edit operation if so required. Edit operations can be undone and redone. If you are working with data stored in ArcSDE, creating at least one edit operation is a requirement. There is no additional overhead to creating an edit operation.

Recycling and nonrecycling cursors

Use nonrecycling search cursors to select or fetch objects that are to be updated. Recycling cursors should only be used for read-only operations, such as drawing and querying features.

Nonrecycling cursors within an edit session create new objects only if the object to be returned does not already exist in memory.

Fetching properties using query filters

Always fetch all properties of the object; query filters should always use "*". For efficient database access, the number of properties of an object retrieved from the database can be specified. As an example, drawing a feature requires only the *OID* and the *Shape* of the feature; hence, the simpler renderers only retrieve these two columns from the database. This optimization speeds up drawing but is not suitable when editing features.

If all properties are not fetched, then object-specific code that is triggered may not find the properties that the method requires. For example, a custom feature developer might write code to update attributes A and B whenever the geometry of a feature changes. If only the geometry was retrieved, then attributes A and B would be found to be missing within the *OnChanged* method. This would cause the *OnChanged* method to return an error, which would cause the *Store* to return an error and the edit operation to fail.

Marking changed objects

After changing an object, mark the object as changed (and ensure that it is updated in the database) by calling *Store* on the object. Delete an object by calling the *Delete* method on the object. Set versions of these calls also exist and should be used if the operation is being performed on a set of objects to ensure optimal performance.

Calling these methods guarantees that all necessary polymorphic object behavior built into the geodatabase is executed (for example, updating of network topology or updating of specific columns in response to changes in other columns in ESRI-supplied objects). It also guarantees that developer-supplied behavior is correctly triggered.

Update and insert cursors

Never use update cursors or insert cursors to update or insert objects into object and feature classes in an already-loaded geodatabase that has active behavior.

Update and insert cursors are bulk cursor APIs for use during initial database loading. If used on an object or feature class with active behavior, they will bypass all object-specific behavior associated with object creation, such as topology creation, and with attribute or geometry updating, such as automatic recalculation of other dependent columns.

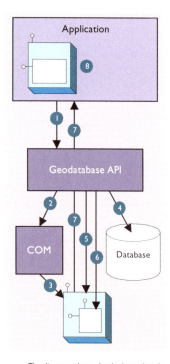

Shape and ShapeCopy geometry property

Make use of a *Feature* object's *Shape* and *ShapeCopy* properties to optimally retrieve the geometry of a feature. To better understand how these properties relate to a feature's geometry, refer to the diagram to the left to see how features coming from a data source are instantiated into memory for use within an application.

Features are instantiated from the data source using the following sequence:

1. The application requests a *Feature* object from a data source by calling the appropriate geodatabase API method calls.

2. The geodatabase makes a request to COM to create a vanilla COM object of the desired COM class (normally this class is *esriGeoDatabase.Feature*).

3. COM creates the *Feature* COM object.

4. The geodatabase gets attribute and geometry data from a data source.

5. The vanilla *Feature* object is populated with appropriate attributes.

6. The *Geometry* COM object is created, and a reference is set in the *Feature* object.

7. The *Feature* object is passed to the application.

8. The *Feature* object exists in the application until it is no longer required.

USING A TYPE LIBRARY

Since objects from ArcObjects do not implement *IDispatch*, it is essential to make use of a type library for the compiler to early-bind to the correct data types. This applies to all development environments; although, for Visual Basic, Visual C++, and .NET, there are wizards that help you set this reference.

The type libraries required by ArcObjects are located within the ArcGIS install folder. For example, the COM type libraries can be found in the COM folder while the .NET Interop assemblies are within the DotNet folder. Many different files can contain type library information, including EXEs, DLLs, OLE custom controls (OCXs), and object libraries (OLBs).

COM DATA TYPES

COM objects talk via their interfaces, and hence, all data types used must be supported by IDL. IDL supports a large number of data types; however, not all languages that support COM support these data types. Because of this, ArcObjects does not make use of all the data types available in IDL but limits the majority of interfaces to the data type supported by Visual Basic. The following table shows the data types supported by IDL and their corresponding types in a variety of languages.

The diagram above clearly shows that the Feature, which is a COM object, has another COM object for its geometry. The Shape property of the feature simply passes the IGeometry interface pointer to this geometry object to the caller that requested the shape. This means that if more than one client requested the shape, all clients point to the same geometry object. Hence, this geometry object must be treated as read-only. No changes should be performed on the geometry returned from this property, even if the changes are temporary. Anytime a change is to be made to a feature's shape, the change must be made on the geometry returned by the ShapeCopy property, and the updated geometry should subsequently be assigned to the Shape property.

Language	IDL	Microsoft C++	Visual Basic	Java
Base types	boolean	unsigned char	unsupported	char
	byte	unsigned char	unsupported	char
	small	char	unsupported	char
	short	short	Integer	short
	long	long	Long	int
	hyper	__int64	unsupported	long
	float	float	Single	float
	double	double	Double	double
	char	unsigned char	unsupported	char
	wchar_t	wchar_t	Integer	short
	enum	enum	Enum	int
	Interface Pointer	Interface Pointer	Interface Ref.	Interface Ref.
Extended types	VARIANT	VARIANT	Variant	ms.com.Variant
	BSTR	BSTR	String	java.lang.String
	VARIANT_BOOL	short (-1/0)	Boolean	[true/false]

Note the extended data types at the bottom of the table: *VARIANT, BSTR,* and *VARIANT_BOOL*. Although it is possible to pass strings using data types such as *char* and *wchar_t*, these are not supported in languages such as Visual Basic. Visual Basic uses *BSTRs* as its text data type. A *BSTR* is a length-prefixed wide character array in which the pointer to the array points to the text contained within it and not the length prefix. Visual C++ maps *VARIANT_BOOL* values onto 0 and –1 for the *False* and *True* values, respectively. This is different from the normal mapping of 0 and 1. Hence, when writing C++ code, be sure to use the correct macros—*VARIANT_FALSE* and *VARIANT_TRUE*—not *False* and *True*.

USING COMPONENT CATEGORIES

Component categories are used extensively in ArcObjects so developers can extend the system without requiring any changes to the ArcObjects code that will work with the new functionality.

ArcObjects uses component categories in two ways. The first requires classes to be registered in the respective component category at all times—for example, ESRI Mx Extensions. Classes, if present in that component category, have an object that implements the *IExtension* interface and is instantiated when the ArcMap application is started. If the class is removed from the component category, the extension will not load, even if the map document (.mxd file) is referencing that extension.

The second use is when the application framework uses the component category to locate classes and display them to a user to allow some user customization to occur. Unlike the first method, the application remembers (inside its map document) the objects being used and will subsequently load them from the map document. An example of this is the commands used within ArcMap. ArcMap reads the ESRI Mx Commands category when the Customization dialog box is displayed to the user. This is the only time the category is read. Once the user selects a command and adds it to a toolbar, the map document is used to determine what commands should be instantiated. Later, when debugging in Visual Basic is covered in 'The Visual Basic 6 development environment' section of this appendix, you'll see the importance of this.

Now that you've seen two uses of component categories, you will see how to get

The Customize dialog box in ArcMap and ArcCatalog

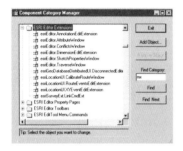

The Component Category Manager

your classes registered into the correct component category. Development environments have various levels of support for component categories; ESRI provides two ways of adding classes to a component category. The first can only be used for commands and command bars that are added to either ArcMap or ArcCatalog. Using the Add From File button on the Customize dialog box (shown on the left), it is possible to choose a server. All classes in that server are then added to either the ESRI Gx Commands or the ESRI Mx Commands, depending on the application being customized. Although this utility is useful, it is limited since it adds all the classes found in the server. It is not possible to remove classes, and it only supports two of the many component categories implemented within ArcObjects.

Distributed with ArcGIS applications is a utility application called the Component Category Manager, shown on the left. This small application allows you to add and remove classes from any of the component categories on your system, not just ArcObjects categories. Expanding a category displays a list of classes in the category. You can then use the Add Object button to display a checklist of all the classes found in the server. You check the required classes, and these checked classes are then added to the category.

Using these ESRI tools is not the only method of interacting with component categories. During the installation of the server on the target user's machine, it is possible to add the relevant information to the registry using a registry script. Below is one such script. The first line tells Windows for which version of regedit this script is intended. The last line, starting with "[HKEY_LOCAL_", executes the registry command; all the other lines are comments in the file.

```
REGEDIT4
; This Registry Script enters coclasses into their appropriate Component
Category
; Use this script during installation of the components


; Coclass: Exporter.ExportingExtension
; CLSID: {E233797D-020B-4AD4-935C-F659EB237065}
; Component Category: ESRI Mx Extensions
[HKEY_LOCAL_MACHINE\SOFTWARE\Classes\CLSID\{E233797D-020B-4AD4-935C-
F659EB237065}\Implemented Categories\{B56A7C45-83D4-11D2-A2E9-080009B6F22B}]
```

The last line in the code above is one continuous line in the script.

The last method is for the self-registration code of the server to add the relevant classes within the server to the appropriate categories. Not all development environments allow this to be set up. Visual Basic has no support for component categories, although there is an add-in that allows this functionality. See the sections on Visual Basic developer add-ins and Active Template Library (ATL) later in this appendix.

The tables below summarize suggested naming standards for the various elements of your Visual Basic projects.

Module Type	Prefix
Form	frm
Class	cls
Standard	bas
Project	prj

Name your modules according to the overall function they provide; do not leave any with default names (such as "Form1", "Class1", or "Module1"). In addition, prefix the names of forms, classes, and standard modules with three letters that denote the type of module, as shown in the table above.

Control Type	Prefix
Check box	chk
Combo box	cbo
Command button	cmd
Common dialog	cdl
Form	frm
Frame	fra
Graph	gph
Grid	grd
Image	img
Image list	iml
Label	lbl
List box	lst
List view	lvw
Map control	map
Masked edit	msk
Menu	mnu
OLE client	ole
Option button	opt
Picture box	pic
Progress bar	pbr
Rich text box	rtf
Scroll bar	srl
Slider	sld
Status bar	sbr
Tab strip	tab
Text box	txt
Timer	tmr
Tool bar	tbr
Tree view	tvw

As with modules, name your controls according to the function they provide; do not leave them with default names since this leads to decreased maintainability. Use the three-letter prefixes above to identify the type of control.

This section is intended for both VB6 and VBA developers. Differences in the development environments are clearly marked throughout the text.

USER INTERFACE STANDARDS

Consider preloading forms to increase the responsiveness of your application. Be careful not to preload too many (preloading three or four forms is fine).

Use resource files (.res) instead of external files when working with bitmap files, icons, and related files.

Make use of constructors and destructors to set variable references that are only set when the class is loaded. These are the VB functions: *Class_Initialize()* and *Class_Terminate()* or *Form_Load()* and *Form_Unload()*. Set all variables to *Nothing* when the object is destroyed.

Make sure the tab order is set correctly for the form. Do not add scroll bars to the tabbing sequence; it is too confusing.

Add access keys to those labels that identify controls of special importance on the form (use the *TabIndex* property).

Use system colors where possible instead of hard-coded colors.

Variable declaration

- Always use *Option Explicit* (or turn on Require Variable Declaration in the VB Options dialog box). This forces all variables to be declared before use and thereby prevents careless mistakes.

- Use *Public* and *Private* to declare variables at module scope and *Dim* in local scope. (*Dim* and *Private* mean the same at *Module* scope; however, using *Private* is more informative.) Do not use *Global* anymore; it is available only for backward compatibility with VB 3.0 and earlier.

- Always provide an explicit type for variables, arguments, and functions. Otherwise, they default to *Variant*, which is less efficient.

- Only declare one variable per line unless the type is specified for each variable.

This line causes *count* to be declared as a *Variant*, which is likely to be unintended.

```
Dim count, max As Long
```

This line declares both *count* and *max* as *Long*, the intended type.

```
Dim count As Long, max As Long
```

These lines also declare *count* and *max* as *Long* and are more readable.

```
Dim count As Long
Dim max As Long
```

Parentheses

Use parentheses to make operator precedence and logic comparison statements easier to read.

```
Result = ((x * 24) / (y / 12)) + 42
If ((Not pFoo Is Nothing) And (Counter > 200)) Then
```

Use the following notation for naming variables and constants:

[<libraryName.>][<scope_>]<type><name>

<name> describes how the variable is used or what it contains. The <scope> and <type> portions should always be lowercase, and the <name> should use mixed case.

Library Name	Library
esriGeometry	ESRI Object Library
stdole	Standard OLE COM Library
<empty>	Simple variable data type

<libraryName>

Prefix	Variable scope
c	constant within a form or class
g	public variable defined in a class form or standard module
m	private variable defined in a class or form
<empty>	local variable

<scope>

Prefix	Data Type
b	Boolean
by	byte or unsigned char
d	double
fn	function
h	handle
i	int (integer)
l	long
p	a pointer
s	string

<type>

Order of conditional determination

Visual Basic, unlike languages such as C and C++, performs conditional tests on all parts of the condition, even if the first part of the condition is False. This means you must not perform conditional tests on objects and interfaces that had their validity tested in an earlier part of the conditional statement.

```
' The following line will raise a runtime error if pFoo is NULL.
If ((Not pFoo Is Nothing) And (TypeOf pFoo.Thing Is IBar)) then
End If


' The correct way to test this code is
If (Not pFoo Is Nothing) Then
  If (TypeOf pFoo.Thing Is IBar) Then
    ' Perform action on IBar thing of Foo.
  End If
End If
```

Indentation

Use two spaces or a tab width of two for indentation. Since there is always only one editor for VB code, formatting is not as critical an issue as it is for C++ code.

Default properties

Avoid using default properties except for the most common cases. They lead to decreased legibility.

Intermodule referencing

When accessing intermodule data or functions, always qualify the reference with the module name. This makes the code more readable and results in more efficient runtime binding.

Multiple property operations

When performing multiple operations against different properties of the same object, use a With … End With statement. It is more efficient than specifying the object each time.

```
With frmHello
  .Caption = "Hello world"
  .Font = "Playbill"
  .Left = (Screen.Width - .Width) / 2
  .Top  = (Screen.Height - .Height) / 2
End With
```

Arrays

For arrays, never change Option Base to anything other than zero (which is the default). Use LBound and UBound to iterate over all items in an array.

```
myArray = GetSomeArray
For i = LBound(myArray) To UBound(myArray)
  MsgBox cstr(myArray(i))
Next I
```

Bitwise operators

Since *And*, *Or*, and *Not* are bitwise operators, ensure that all conditions using them test only for Boolean values (unless, of course, bitwise semantics are what is intended).

```
If (Not pFoo Is Nothing) Then
  ' Valid Foo do something with it
End If
```

Type suffixes

Refrain from using type suffixes on variables or function names (such as *myString$* or *Right$(myString)*), unless they are needed to distinguish 16-bit from 32-bit numbers.

Ambiguous type matching

For ambiguous type matching, use explicit conversion operators, such as *CSng*, *CDbl*, and *CStr* instead of relying on VB to pick which one will be used.

Simple image display

Use an *ImageControl* rather than a *PictureBox* for simple image display. It is much more efficient.

Error handling

Recovery Statement	Frequency	Meaning
Exit Sub	usually	Function failed, pass control back to caller
Raise	often	Raise a new error code in the caller's scope
Resume	rarely	Error condition removed, reattempt offending statement
Resume Next	very rarely	Ignore error and continue with next statement

Always use *On Error* to ensure fault-tolerant code. For each function that does error checking, use *On Error* to jump to a single error handler for the routine that deals with all exceptional conditions that are likely to be encountered. After the error handler processes the error—usually by displaying a message—it should proceed by issuing one of the recovery statements shown on the table to the left.

Error handling in Visual Basic is not the same as general error handling in COM (see the section 'Working with HRESULTs' in this appendix).

Event functions

Refrain from placing more than a few lines of code in event functions to prevent highly fractured and unorganized code. Event functions should simply dispatch to reusable functions elsewhere.

Memory management

To ensure efficient use of memory resources, the following points should be considered:

- Unload forms regularly. Do not keep many forms loaded but invisible since this consumes system resources.

- Be aware that referencing a form-scoped variable causes the form to be loaded.

- Set unused objects to *Nothing* to free up their memory.

- Make use of *Class_Initialize()* and *Class_Terminate()* to allocate and destroy resources.

While Wend constructs

Avoid *While … Wend* constructs. Use the *Do While … Loop* or *Do Until … Loop* instead because you can conditionally branch out of this construct.

```
pFoos.Reset
Set pFoo = pFoos.Next
Do While (Not pFoo Is Nothing)
  If (pFoo.Answer = "Done") Then Exit Loop
  Set pFoo = pFoos.Next
Loop
```

The Visual Basic Virtual Machine

The VBVM was called the VB Runtime in earlier versions of the software.

The Visual Basic Virtual Machine (VBVM) contains the intrinsic Visual Basic controls and services, such as starting and ending a Visual Basic application, required to successfully execute all Visual Basic developed code.

The VBVM is packaged as a DLL that must be installed on any machine wanting to execute code written with Visual Basic, even if the code has been compiled to native code. If the dependencies of any Visual Basic compiled file are viewed, the file msvbvm60.dll is listed; this is the DLL housing the Virtual Machine.

For more information on the services provided by the VBVM, see the sections 'Interacting with the IUnknown interface' and 'Working with HRESULTs' in this appendix.

Interacting with the IUnknown interface

The section 'The Microsoft Component Object Model' in this appendix contains a lengthy section on the *IUnknown* interface and how it forms the basis on which all of COM is built. Visual Basic hides this interface from developers and performs the required interactions (*QueryInterface*, *AddRef*, and *Release* function calls) on the developer's behalf. It achieves this because of functionality contained within the VBVM. This simplifies development with COM for many developers, but to work successfully with ArcObjects, you must understand what the VBVM is doing.

Visual Basic developers are accustomed to dimensioning variables as follows:

```
Dim pColn as New Collection  'Create a new collection object.
PColn.Add "Foo", "Bar"       'Add element to collection.
```

It is worth considering what is happening at this point. From a quick inspection of the code, it appears that the first line creates a collection object and gives the developer a handle on that object in the form of *pColn*. The developer then calls a method on the object *Add*. Earlier in the appendix you learned that objects talk via their interfaces, never through a direct handle on the object itself. Remember, objects expose their services via their interfaces. If this is true, something isn't adding up.

What is actually happening is some "VB magic" performed by the VBVM and some trickery by the Visual Basic Editor in the way that it presents objects and interfaces. The first line of code instantiates an instance of the collection class, then assigns the default interface for that object, *_Collection*, to the variable *pColn*. It is this interface, *_Collection*, that has the methods defined on it. Visual Basic has hidden the interface-based programming to simplify the developer experience. This is not an issue if all the functionality implemented by the object can be accessed via one interface, but it is an issue when there are multiple interfaces on an object that provides services.

The Visual Basic Editor backs this up by hiding default interfaces from the IntelliSense completion list and the object browser. By default, any interfaces that begin with an underscore, "_", are not displayed in the object browser (to display these interfaces, turn Show Hidden Member on, although this will still not display default interfaces).

You have already learned that the majority of ArcObjects have *IUnknown* as their default interface and that Visual Basic does not expose any of *IUnknown*'s methods, namely, *QueryInterface*, *AddRef*, and *Release*. Assume you have a class *Foo* that supports three interfaces, *IUnknown* (the default interface), *IFoo*, and *IBar*. This means that if you were to dimension the variable *pFoo* as below, the variable *pFoo* would point to the *IUnknown* interfaces.

```
Dim pFoo As New Foo     ' Create a new Foo object.
pFoo.??????
```

Since Visual Basic does not allow direct access to the methods of *IUnknown*, you would immediately have to *QI* for an interface with methods on it that you can call. Because of this, the correct way to dimension a variable that will hold pointers to interfaces is as follows:

```
Dim pFoo As IFoo  ' Variable will hold pointer to IFoo interface.
Set pFoo = New Foo ' Create Instance of Foo object and QI for IFoo.
```

Now that you have a pointer to one of the object's interfaces, it is an easy matter to request from the object any of its other interfaces.

```
Dim pBar as IBar  'Dim variable to hold pointer to interface.
Set pBar = pFoo   'QI for IBar interface.
```

By convention, most classes have an interface with the same name as the class with an "I" prefix; this tends to be the interface most commonly used when working with the object. You are not restricted to which interface you request when instantiating an object. Any supported interface can be requested; hence, the code below is valid.

```
Dim pBar as IBar
Set pBar = New Foo  'CoCreate Object.
Set pFoo = pBar     'QI for interface.
```

Objects control their own lifetime, which requires clients to call *AddRef* anytime an interface pointer is duplicated by assigning it to another variable and to call *Release* anytime the interface pointer is no longer required. Ensuring that there are a matching number of *AddRefs* and *Releases* is important, and fortunately, Visual Basic performs these calls automatically. This ensures that objects do not "leak". Even when interface pointers are reused, Visual Basic will correctly call release on the old interface before assigning the new interface to the variable. The following code illustrates these concepts; note the reference count on the object at the various stages of code execution.

See the Visual Basic Magic sample in the server guide samples on disk for this code. You are encouraged to run the sample and use the code. This object also uses an ATL C++ project to define the SimpleObject and its interfaces; you are encouraged to look at this code to learn a simple implementation of a C++ ATL object.

```
Private Sub VBMagic()
  ' Dim a variable to the IUnknown interface on the simple object.
  Dim pUnk As IUnknown

  ' Co Create simpleobject asking for the IUnknown interface.
  Set pUnk = New SimpleObject 'refCount = 1

  ' QI for a useful interface.
  ' Define the interface.
  Dim pMagic As ISimpleObject

  ' Perform the QI operation.
  Set pMagic = punk 'refCount = 2

  ' Dim another variable to hold another interface on the object.
  Dim pMagic2 As IAnotherInterface

  ' QI for that interface.
  Set pMagic2 = pMagic 'refCount = 3

  ' Release the interface pointer.
  Set pMagic2 = Nothing 'refCount = 2

  ' Release the interface
  Set pMagic = Nothing 'refCount = 1

  ' Now reuse the pUnk variable - what will VB do for this?
  Set pUnk = New SimpleObject 'refCount = 1, then 0, then 1

  ' Let the interface variable go out of scope and let VB tidy up.
End Sub 'refCount = 0
```

Often interfaces have properties that are actually pointers to other interfaces. Visual Basic allows you to access these properties in a shorthand fashion by chaining interfaces together. For instance, assume that you have a pointer to the *IFoo* interface, and that interface has a property called *Gak* that is an *IGak* interface with the method *DoSomething()*. You have a choice on how to access the *DoSomething* method. The first method is the long-handed way.

```
Dim pGak as IGak
Set pGak = pFoo      'Assign IGak interface to local variable.
pGak.DoSomething     'Call method on IGak interface.
```

Alternatively, you can chain the interfaces and accomplish the same thing on one line of code.

```
pFoo.Gak.DoSomething  'Call method on IGak interface.
```

When looking at the sample code, you will see both methods. Normally, the former method is used on the simpler samples, as it explicitly tells you what interfaces are being worked with. More complex samples use the shorthand method.

This technique of chaining interfaces together can always be used to get the value of a property, but it cannot always be used to set the value of a property. Interface chaining can only be used to set a property if all the interfaces in the chain are set by reference. For instance, the code below would execute successfully.

```
Dim pMxDoc As ImxDocument
Set pMxDoc = ThisDocument
pMxDoc.FocusMap.Layers(0).Name = "Foo"
```

The above example works because both the *Layer* of the *Map* and the *Map* of the document are returned by reference. The lines of code below would not work since the *Extent* envelope is set by value on the active view.

```
pMxDoc.ActiveView.Extent.Width = 32
```

The reason that this does not work is that the VBVM expands the interface chain to get the end property. Because an interface in the chain is dealt with by value, the VBVM has its own copy of the variable, not the one chained. To set the *Width* property of the extent envelope in the above example, the VBVM must write code similar to this:

```
Dim pActiveView as IActiveView
Set pActiveView = pMxDoc.ActiveView

Dim pEnv as IEnvelope
Set pEnv = pActiveView.Extent  ' This is a get by value.

PEnv.Width = 32   ' The VBVM has set its copy of the Extent and not
                  ' the copy inside the ActiveView.
```

For this to work, the VBVM requires the extra line below.

```
pActiveView.Extent = pEnv  ' This is a set by value.
```

Accessing ArcObjects

You will now see some specific uses of the create instance and query interface operations that involve ArcObjects. To use an ArcGIS object in Visual Basic or VBA, you must first reference the ESRI library that contains that object. If you are using VBA inside ArcMap or ArcCatalog, most of the common ESRI object libraries are already referenced for you. In standalone Visual Basic applications or components you will have to manually reference the required libraries.

To find out what library an ArcObjects component is in, review the object model diagrams in the developer help or use the LibraryLocator tool in your developer kit tools directory.

You will start by identifying a simple object and an interface that it supports. In this case, you will use a *Point* object and the *IPoint* interface. One way to set the coordinates of the point is to invoke the *PutCoords* method on the *IPoint* interface and pass in the coordinate values.

```
Dim pPt As IPoint
Set pPt = New Point
pPt.PutCoords 100, 100
```

IID is short for Interface Identifier, a GUID.

The first line of this simple code fragment illustrates the use of a variable to hold a reference to the interface that the object supports. The line reads the *IID* for the *IPoint* interface from the ESRI object library. You may find it less ambiguous (as per the coding guidelines), particularly if you reference other object libraries in the same project, to precede the interface name with the library name, for example:

```
Dim pPt As esriGeometry.IPoint
```

Coclass is an abbreviation of component object class.

A QI is required since the default interface of the object is IUnknown. Since the pPt variable was declared as type IPoint, the default IUnknown interface was QI'd for the IPoint interface.

This is the compilation error message shown when a method or property is not found on an interface.

That way, if there happens to be another *IPoint* referenced in your project, there won't be any ambiguity as to which one you are referring to.

The second line of the fragment creates an instance of the object or coclass, then performs a *QI* operation for the *IPoint* interface that it assigns to *pPt*.

With a name for the coclass as common as *Point*, you may want to precede the coclass name with the library name, for example:

```
Set pPt = New esriGeometry.Point
```

The last line of the code fragment invokes the *PutCoords* method. If a method can't be located on the interface, an error will be shown at compile time.

Working with HRESULTs

So far, you have seen that all COM methods signify success or failure via an HRESULT that is returned from the method; no exceptions are raised outside the interface. You have also learned that Visual Basic raises exceptions when errors are encountered. In Visual Basic, HRESULTs are never returned from method calls, and to confuse you further when errors do occur, Visual Basic throws an exception. How can this be? The answer lies with the Visual Basic Virtual Machine. It is the VBVM that receives the HRESULT; if this is anything other than *S_OK*, the VBVM throws the exception. If it was able to retrieve any worthwhile error information from the COM error object, it populates the Visual Basic *Err* object with that information. In this way, the VBVM handles all HRESULTs returned from the client.

When implementing interfaces in Visual Basic, it is good coding practice to raise an HRESULT error to inform the caller that an error has occurred. Normally, this is done when a method has not been implemented.

```
' Defined in module
Const E_NOTIMPL = &H80004001 'Constant that represents HRESULT
'Added to any method not implemented
On Error GoTo 0
Err.Raise E_NOTIMPL
```

You must also write code to handle the possibility that an HRESULT other than *S_OK* is returned. When this happens, an error handler should be called and the error dealt with. This may mean simply notifying the user, or it may mean automatically dealing with the error and continuing with the function. The choice depends on the circumstances. Below is a simple error handler that will catch any error that occurs within the function and report it to the user. Note the use of the *Err* object to provide the user with some description of the error.

```
Private Sub Test()
  On Error GoTo ErrorHandler
  ' Do something here.
  Exit Sub    ' Must exit sub here before error handler
ErrorHandler:
  Msgbox "Error In Application – Description " & Err.Description
End Sub
```

Working with properties

Some properties refer to specific interfaces in the ESRI object library, and other properties have values that are standard data types, such as strings, numeric expressions, and Boolean values. For interface references, declare an interface variable and use the *Set* statement to assign the interface reference to the property. For other values, declare a variable with an explicit data type or use Visual Basic's *Variant* data type. Then, use a simple assignment statement to assign the value to the variable.

Properties that are interfaces can be set either by reference or by value. Properties that are set by value do not require the *Set* statement.

```
Dim pEnv As IEnvelope
Set pEnv = pActiveView.Extent    'Get extent property of view.
pEnv.Expand 0.5, 0.5, True       'Shrink envelope.
pActiveView.Extent = pEnv        'Set By Value extent back on IActiveView.


Dim pFeatureLayer as IfeatureLayer
Set pFeatureLayer = New FeatureLayer    'Create New Layer.
Set pFeatureLayer.FeatureClass = pClass 'Set ByRef a class into layer.
```

As you might expect, some properties are read-only, others are write-only, and still others are read/write. All the object browsers and the ArcObjects Class Help (found in the ArcGIS Developer Help system) provide this information. If you attempt to use a property and either forget or misuse the *Set* keyword, Visual Basic will fail the compilation of the source code with a "method or data member not found error message". This error may seem strange since it may be given for trying to assign a value to a read-only property. The reason for the message is that Visual Basic is attempting to find a method in the type library that maps to the property name. In the above examples, the underlying method calls in the type library are *put_Extent* and *putref_FeatureClass*.

Working with methods

Methods perform some action and may or may not return a value. In some instances, a method returns a value that's an interface; for example, in the code fragment below, *EditSelection* returns an enumerated feature interface.

```
Dim pApp As IApplication
Dim pEditor As IEditor
Dim pEnumFeat As IEnumFeature 'Holds the selection
Dim pID As New UID
'Get a pointer to the Editor extension.
pID = "esriEditor.Editor"
Set pApp = Application
Set pEditor = pApp.FindExtensionByCLSID(pID)
'Get the selection
Set pEnumFeat = pEditor.EditSelection
```

In other instances, a method returns a Boolean value that reflects the success of an operation or writes data to a parameter; for example, the *DoModalOpen* method of *GxDialog* returns a value of True if a selection occurs and writes the selection to an *IEnumGxObject* parameter.

Be careful not to confuse the idea of a Visual Basic return value from a method call with the idea that all COM methods must return an HRESULT. The VBVM is able to read type library information and set up the return value of the VB method call to be the appropriate parameter of the COM method.

Working with events

Events let you know when something has occurred. You can add code to respond to an event. For example, a command button has a *Click* event. You add code to perform some action when the user clicks the control. You can also add events that certain objects generate. VBA and Visual Basic let you declare a variable with the keyword *WithEvents*. *WithEvents* tells the development environment that the object variable will be used to respond to the object's events. This is sometimes referred to as an "event sink". The declaration must be made in a class module or a form. Here's how you declare a variable and expose the events of an object in the *Declarations* section:

```
Private WithEvents m_pViewEvents as Map
```

Visual Basic only supports one outbound interface (marked as the default outbound interface in the IDL) per coclass. To get around this limitation, the coclasses that implement more than one outbound interface have an associated dummy coclass that allows access to the secondary outbound interface. These coclasses have the same name as the outbound interface they contain, minus the I.

```
Private WithEvents m_pMapEvents as MapEvents
```

Once you've declared the variable, search for its name in the Object combo box at the top left of the Code window. Then, inspect the list of events to which you can attach code in the Procedure/Events combo box at the top right of the Code window.

Not all procedures of the outbound event interface need to be stubbed out, as Visual Basic will stub out any unimplemented methods. This is different from inbound interfaces, in which all methods must be stubbed out for compilation to occur.

Before the methods are called, the hookup between the event source and sink must be made. This is done by setting the variable that represents the sink to the event source.

```
Set m_pMapEvents = pMxDoc.FocusMap
```

Pointers to valid objects as parameters

Some ArcGIS methods expect interfaces for some of their parameters. The interface pointers passed can point to an instanced object before the method call or after the method call is completed.

For example, if you have a polygon (*pPolygon*) whose center point you want to find, you can write code as follows:

```
Dim pArea As IArea
Dim pPt As IPoint
Set pArea = pPolygon        ' QI for IArea on pPolygon.
Set pPt = pArea.Center
```

You don't need to create *pPt* because the *Center* method creates a *Point* object for you and passes back a reference to the object via its *IPoint* interface. Only methods that use client-side storage require you to create the object prior to the method call.

Passing data between modules

When passing data between modules it is best to use accessor and mutator functions that manipulate some private member variable. This provides data encapsulation, which is a fundamental technique in object-oriented programming. Public variables should never be used.

For instance, you might have decided that a variable has a valid range of 1–100. If you were to allow other developers direct access to that variable, they could set the value to an illegal value. The only way of coping with these illegal values is to check them before they get used. This is both error prone and tiresome to program. The technique of declaring all variables as private member variables of the class and providing accessor and mutator functions for manipulating these variables will solve this problem.

In the example below, these properties are added to the default interface of the class. Notice the technique used to raise an error to the client.

```
Private m_lPercentage As Long


Public Property Get Percentage() As Long
  Percentage = m_lPercentage
End Property


Public Property Let Percentage(ByVal lNewValue As Long)
  If (lNewValue >= 0) And (lNewValue <= 100) Then
    m_lPercentage = lNewValue
  Else
    Err.Raise vbObjectError + 29566, "MyProj.MyObject", _
    "Invalid Percentage Value. Valid values (0 -> 100)"
  End If
End Property
```

When you write code to pass an object reference from one form, class, or module to another, for example:

```
Private Property Set PointCoord(ByRef pPt As IPoint)
  Set m_pPoint = pPt
End Property
```

your code passes a pointer to an instance of the *IPoint* interface. This means that you are only passing the reference to the interface, not the interface itself; if you add the *ByVal* keyword (as follows), the interface is passed by value.

```
Private Property Let PointCoord(ByVal pPt As IPoint)
  Set m_pPoint = pPt
End Property
```

In both of these cases the object pointed to by the interfaces is always passed by reference. To pass the object by value, a clone of the object must be made, and that clone is passed.

Using the TypeOf keyword

To check whether an object supports an interface, you can use Visual Basic's *TypeOf* keyword. For example, given an item selected in the ArcMap table of contents, you can test whether it is a *FeatureLayer* using the following code:

```
Dim pDoc As IMxDocument
Dim pUnk As IUnknown
Dim pFeatLyr As IGeoFeatureLayer
Set pDoc = ThisDocument
Set pUnk = pDoc.SelectedItem
If TypeOf pUnk Is IGeoFeatureLayer Then  ' can we QI for IGeoFeatureLayer?
  Set pFeatLyr = pUnk                    ' actually QI happens here
  ' Do something with pFeatLyr.
End If
```

Using the Is operator

If your code requires you to compare two interface reference variables, you can use the *Is* operator. Typically, you can use the *Is* operator in the following circumstances:

- To check if you have a valid interface—For example, see the following code:
```
Dim pPt As IPoint
Set pPt = New Point
If (Not pPt Is Nothing) Then 'a valid pointer?
  ...' do something with pPt
End If
```

- To check if two interface variables refer to the same actual object—imagine that you have two interface variables of type *IPoint*, *pPt1*, and *pPt2*. Are they pointing to the same object? If they are, then *pPt1* Is *pPt2*.

The *Is* keyword works with the COM identity of an object. Below is an example that illustrates the use of the *Is* keyword when finding out if a certain method on an interface returns a copy of or a reference to the same real object.

In the following example, the *Extent* property on a map (*IMap*) returns a copy, while the *ActiveView* property on a document (*IMxDocument*) always returns a reference to the real object.

```
Dim pDoc As IMxDocument
Dim pEnv1 As IEnvelope, pEnv2 as IEnvelope
Dim pActView1 As IActiveView
Dim pActView2 as IActiveView
Set pDoc = ThisDocument
Set pEnv1 = pDoc.ActiveView.Extent
Set pEnv2 = pDoc.ActiveView.Extent
Set pActView1 = pDoc.ActiveView
Set pActView2 = pDoc.ActiveView
' Extent returns a copy,
' so pEnv1 Is pEnv2 returns False
Debug.Print pEnv1 Is pEnv2
' ActiveView returns a reference,
```

```
' so pActView1 Is pActView2
Debug.Print pActView1 Is pActView2
```

Iterating through a collection

Enumerators can support other methods, but these two methods are common among all enumerators.

In your work with ArcMap and ArcCatalog, you'll discover that, in many cases, you'll be working with collections. You can iterate through these collections with an enumerator. An enumerator is an interface that provides methods for traversing a list of elements. Enumerator interfaces typically begin with *IEnum* and have two methods: *Next* and *Reset*. *Next* returns the next element in the set and advances the internal pointer, and *Reset* resets the internal pointer to the beginning.

Here is some VBA code that loops through the selected features (*IEnumFeature*) in a map. To try the code, add the States sample layer to the map and use the Select tool to select multiple features (drag a rectangle to do this). Add the code to a VBA macro, then execute the macro. The name of each selected state will be printed in the debug window.

```
Dim pDoc As IMxDocument
Dim pEnumFeat As IEnumFeature
Dim pFeat As IFeature
Set pDoc = ThisDocument
Set pEnumFeat = pDoc.FocusMap.FeatureSelection
Set pFeat = pEnumFeat.Next
Do While (Not pFeat Is Nothing)
  Debug.Print pFeat.Value(pFeat.Fields.FindField("state_name"))
  Set pFeat = pEnumFeat.Next
Loop
```

Some collection objects, the Visual Basic Collection being one, implement a special interface called *_NewEnum*. This interface, because of the _ prefix, is hidden, but Visual Basic developers can still use it to simplify iterating through a collection. The Visual Basic *For Each* construct works with this interface to perform the *Reset* and *Next* steps through a collection.

```
Dim pColn as Collection
Set pColn = GetCollection()' Collection returned from some function

Dim thing as Variant     ' VB uses methods on _NewEnum to step through
For Each thing in pColn  ' an enumerator.
  MsgBox Cstr(thing)
Next
```

In the previous section of this appendix, the focus was primarily on how to write code in the VBA development environment embedded within the ArcGIS Desktop applications. This section focuses on particular issues related to creating ActiveX DLLs that can be added to the applications and writing external standalone applications using the Visual Basic development environment.

CREATING COM COMPONENTS

Most developers use Visual Basic to create a COM component that works with ArcMap or ArcCatalog. Earlier in this appendix you learned that since the ESRI applications are COM clients—their architecture supports the use of software components that adhere to the COM specification—you can build components with different languages, including Visual Basic. These components can then be added to the applications easily. For information about packaging and deploying COM components that you've built with Visual Basic, see the last section of this appendix.

This section is not intended as a Visual Basic tutorial; rather, it highlights aspects of Visual Basic that you should know to be effective when working with ArcObjects.

In Visual Basic you can build a COM component that will work with ArcMap or ArcCatalog by creating an ActiveX DLL. This section will review the rudimentary steps involved. Note that these steps are not all-inclusive. Your project may involve other requirements.

1. Start Visual Basic. In the New Project dialog box, create an ActiveX DLL Project.

2. In the Properties window, make sure that the Instancing property for the initial class module and any other class modules you add to the Project is set to 5—MultiUse.

3. Reference the ESRI Object Libraries that you will require.

4. Implement the required interfaces. When you implement an interface in a class module, the class provides its own versions of all the public procedures specified in the type library of the interface. In addition to providing mapping between the interface prototypes and your procedures, the *Implements* statement causes the class to accept COM *QueryInterface* calls for the specified interface ID. You must include all the public procedures involved. A missing member in an implementation of an interface or class causes an error. If you don't put code in one of the procedures in a class you are implementing, you can raise the appropriate error (*Const E_NOTIMPL* = &H80004001). That way, if someone else uses the class, they'll understand that a member is not implemented.

The ESRI VB Add-In interface implementer can be used to automate Steps 3 and 4.

5. Add any additional code that's needed.

6. Establish the Project Name and other properties to identify the component. In the Project Properties dialog box, the project name you specify will be used as the name of the component's type library. It can be combined with the name of each class the component provides to produce unique class names (these names are also called ProgIDs). These names appear in the Component Category Manager. Save the project.

7. Compile the DLL.

8. Set the component's Version Compatibility to binary. As your code evolves, it's good practice to set the components to Binary Compatibility so, if you make changes to a component, you'll be warned that you're breaking compatibility. For additional information, see the 'Binary compatibility mode' help topic in the Visual Basic online help.

9. Save the project.

10. Make the component available to the application. You can add a component to a document or template by clicking the Add from file button in the Customize dialog box's Commands tab. In addition, you can register a component in the Component Category Manager.

Visual Basic automatically generates the necessary GUIDs for the classes, interfaces, and libraries. Setting binary compatibility forces VB to reuse the GUIDs from a previous compilation of the DLL. This is essential since ArcMap stores the GUIDs of commands in the document for subsequent loading.

IMPLEMENTING INTERFACES

You implement interfaces differently in Visual Basic depending on whether they are inbound or outbound interfaces. An outbound interface is seen by Visual Basic as an event source and is supported through the *WithEvents* keyword. To handle the outbound interface, *IActiveViewEvents*, in Visual Basic (the default outbound interface of the *Map* class), use the *WithEvents* keyword and provide appropriate functions to handle the events.

```
Private WithEvents ViewEvents As Map


Private Sub ViewEvents_SelectionChanged()
  ' User changed feature selection update my feature list form
  UpdateMyFeatureForm
End Sub
```

Inbound interfaces are supported with the *Implements* keyword. However, unlike the outbound interface, all the methods defined on the interface must be stubbed out. This ensures that the vTable is correctly formed when the object is instantiated. Not all of the methods have to be fully coded, but the stub functions must be there. If the implementation is blank, an appropriate return code should be given to any client to inform them that the method is not implemented (see the section 'Working with HRESULTs'). To implement the *IExtension* interface, code similar to below is required. Note that all the methods are implemented.

```
Private m_pApp As IApplication
Implements IExtension
 Private Property Get IExtension_Name() As String
  IExtension_Name = "Sample Extension"
End Property


Private Sub IExtension_Startup(ByRef initializationData As Variant)
  Set m_pApp = initializationData
End Sub


Private Sub IExtension_Shutdown()
  Set m_pApp = Nothing
End Sub
```

SETTING REFERENCES TO THE ESRI OBJECT LIBRARIES

The principal difference between working with the VBA development environment embedded in the applications and working with Visual Basic is that the latter environment requires that you load the appropriate object libraries so that any object variables that you declare can be found. If you don't add the reference, you'll get the error message to the left. In addition, the global variables *ThisDocument* and *Application* are not available to you.

Adding a reference to an object library

Depending on what you want your code to do, you may need to add several ESRI object and extension libraries. You can determine what library an object belongs to by reviewing the object model diagrams in the developer help or by using the LibraryLocator tool located in the tools directory of your developer kit.

To display the References dialog box in which you can set the references you need, select References in the Visual Basic Project menu.

After you set a reference to an object library by selecting the check box next to its name, you can find a specific object and its methods and properties in the object browser.

If you are not using any objects in a referenced library, you should clear the check box for that reference to minimize the number of object references Visual Basic must resolve, thus reducing the time it takes your project to compile. You should not remove a reference for an item that is used in your project.

You can't remove the "Visual Basic for Applications" and "Visual Basic objects and procedures" references because they are necessary for running Visual Basic.

REFERRING TO A DOCUMENT

Each VBA project (Normal, Project, TemplateProject) has a class called *ThisDocument*, which represents the document object. Anywhere you write code in VBA you can reference the document as *ThisDocument*. Further, if you are writing your code in the *ThisDocument* Code window, you have direct access to all the methods and properties on *IDocument*. This is not available in Visual Basic. You must first get a reference to the *Application*, then the document. When adding both extensions and commands to ArcGIS applications, a pointer to the *IApplication* interface is provided.

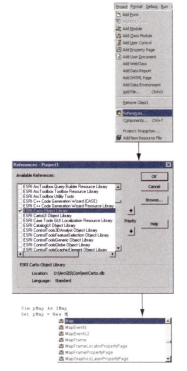

After the applicable ESRI object libraries are referenced, all the types contained within them are available to Visual Basic. IntelliSense will also work with the contents of the object libraries.

```
Implements IExtension
Private m_pApp As IApplication

Private Sub IExtension_Startup(ByRef initializationData As Variant)
  Set m_pApp = initializationData    ' Assign IApplication.
End Sub


Implements ICommand
Private m_pApp As IApplication

Private Sub ICommand_OnCreate(ByVal hook As Object)
  Set m_pApp = hook                  ' QI for IApplication.
End Sub
```

Now that a reference to the application is in an *IApplication* pointer member variable, the document and, hence, all other objects can be accessed from any method within the class.

```
Dim pDoc as IDocument
Set pDoc = m_pApp.Document
MsgBox pDoc.Name
```

GETTING TO AN OBJECT

In the previous example, navigating around the objects within ArcMap was a straightforward process since a pointer to the *Application* object, the root object of most of the ArcGIS application's objects, was passed to the object via one of its interfaces. This, however, is not the case with all interfaces that are implemented within the ArcObjects application framework. There are cases when you may implement an object that exists within the framework and there is no possibility to traverse the object hierarchy from that object. This is because very few objects support a reference to their parent object (the *IDocument* interface has a property named *Parent* that references the *IApplication* interface). To give developers access to the application object, there is a singleton object that provides a pointer to the running application object. The code below illustrates its use.

Singletons are objects that only support one instance of the object. These objects have a class factory that ensures that anytime an object is requested, a pointer to an already existing object is returned.

```
Dim pAppRef As New AppRef
Dim pApp as IApplication
Set pApp = pAppRef
```

You must be careful to ensure that this object is only used where the implementation will always only run within ArcMap and ArcCatalog. For instance, it would not be a good idea to make use of this function from within a custom feature since that would restrict what applications could be used to view the feature class.

RUNNING ArcMap WITH A COMMAND LINE ARGUMENT

You can start ArcMap from the command line and pass it an argument that is either the pathname of a document (.mxd) or the pathname of a template (.mxt). In the former case, ArcMap will open the document; in the latter case, ArcMap will create a new document based on the template specified.

You can also pass an argument and create an instance of ArcMap by supplying arguments to the Win32 API's *ShellExecute* function or Visual Basic's *Shell* function as follows:

```
Dim ret As Variant
ret = Shell("d:\arcgis\bin\arcmap.exe _
  d:\arcgis\bin\templates\LetterPortrait.mxt", vbNormalFocus)
```

In Visual Basic, it is not possible to determine the command line used to start the application. There is a sample on disk that provides this functionality. It can be found at <ArcGIS Developer Kit install>\samples\COM Techniques\Command Line.

By default, *Shell* runs other programs asynchronously. This means that ArcMap might not finish executing before the statements following the *Shell* function are executed.

To execute a program and wait until it is terminated, you must call three Win32 API functions. First, call the *CreateProcessA* function to load and execute ArcMap. Next, call the *WaitForSingleObject* function, which forces the operating system to wait until ArcMap has been terminated. Finally, when the user has terminated the application, call the *CloseHandle* function to release the application's 32-bit identifier to the system pool.

DEBUGGING VISUAL BASIC CODE

Visual Basic has a debugger integrated into its development environment. This is in many cases a valuable tool when debugging Visual Basic code; however, in some cases it is not possible to use the VB debugger. The use of the debugger and these special cases are discussed below.

Running the code within an application

It is possible to use the Visual Basic debugger to debug your ArcObjects software-based source code even when ActiveX DLLs are the target server. The application that will host your DLL must be set as the Debug application. To do this, select the appropriate application, ArcMap.exe, for instance, and set it as the Start Program in the Debugging Options of the Project Properties.

Using commands on the Debug toolbar, ArcMap can be started and the DLL loaded and debugged. Break points can be set, lines stepped over, functions stepped into, and variables checked. Moving the line pointer in the left margin can also set the current execution line.

Visual Basic debugger issues

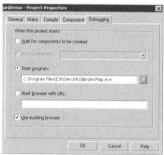

In many cases, the Visual Basic debugger will work without any problems; however, there are two problems when using the debugger that is supplied with Visual Basic 6. Both of these problems exist because of the way that Visual Basic implements its debugger.

Normally when running a tool within ArcMap, the DLL is loaded into ArcMap address space, and calls are made directly into the DLL. When debugging, this is not the case. Visual Basic makes changes to the registry so that the class identifier (CLSID) for your DLL does not point to your DLL but, instead, points to the Visual Basic Debug DLL (VB6debug.dll). The Debug DLL must then support all the interfaces implemented by your class on the fly. With the VB Debug DLL loaded into ArcMap, any method calls that come into the DLL are forwarded to Visual Basic, where the code to be debugged is executed. The two problems with this are caused by the changes made to the registry and the cross-process space method calling. When these restrictions are first encountered, it can be confusing since the object works outside the debugger until it hits a line of problem code.

Since the method calls made from ArcMap to the custom tool are across apartments, there is a requirement for the interfaces to be marshalled. This marshalling causes problems in certain circumstances. Most data types can be automatically marshalled by the system, but there are a few that require custom code because the standard marshaller does not support the data types. If one of these data types is used by an interface within the custom tool and there is no custom marshalling code, the debugger will fail with an "Interface not supported" error.

The registry manipulation also breaks the support for component categories. Any time there is a request on a component category, the category manager within COM will be unable to find your component because, rather than asking whether your DLL belongs to the component category, COM is asking whether the VB debugger DLL belongs to the component category, and it doesn't. What this means is that anytime a component category is used to automate the loading of a DLL, the DLL cannot be debugged using the Visual Basic debugger.

This causes problems for many of the ways to extend the framework. The most common way to extend the framework is to add a command or tool. Previously, it was discussed how component categories were used in this instance. Remember the component category was only used to build the list of commands in the dialog box. This means that if the command to be debugged is already present on a toolbar, the Visual Basic debugger can be used. Hence, the procedure for debugging Visual Basic objects that implement the *ICommand* interface is to ensure that the command is added to a toolbar when ArcMap is executed standalone and, after saving the document, load ArcMap through the debugger.

In some cases, such as extensions and property pages, it is not possible to use the Visual Basic debugger. If you have access to the Visual C++ debugger, you can use one of the options outlined below. Fortunately, there are a number of ESRI Visual Basic Add-ins that make it possible to track down the problem quickly and effectively. The add-ins, described in ArcGIS Developer Help in the section 'Visual Basic Developer Add-ins', provide error log information including line and module details. A sample output from an error log is given below; note the call stack information along with line numbers.

Error Log saved on : 8/28/2000 – 10:39:04 AM
Record Call Stack Sequence – Bottom line is error line.

    chkVisible_MouseUp C:\Source\MapControl\Commands\frmLayer.frm Line : 196
    RefreshMap C:\Source\MapControl\Commands\frmLayer.frm Line : 20

Description
    Object variable or With block variable not set

Alternatives to the Visual Basic debugger

If the Visual Basic debugger and add-ins do not provide enough information, the Visual C++ debugger can be used, either on its own or with C++ ATL wrapper classes. The Visual C++ debugger does not run the object to be debugged out of process from ArcMap, which means that none of the above issues apply. Common debug commands are given in the Visual C++ section 'Debugging tips in Developer Studio'. Both of the techniques below require the Visual Basic project to be compiled with debug symbol information.

The Visual C++ Debugger can work with this symbolic debug information and the source files.

Visual C++ debugger

It is possible to use the Visual C++ debugger directly by attaching to a running process that has the Visual Basic object to be debugged loaded and setting a break point in the Visual Basic file. When the line of code is reached, the debugger will halt execution and step you into the source file at the correct line. The required steps are as follows:

1. Start an appropriate application, such as ArcMap.exe.

2. Start Microsoft Visual C++.

Create debug symbol information using the Create Symbolic Debug info option on the Compile tab of the Project Properties dialog box.

3. Attach to the ArcMap process using Menu option Build > Start Debug > Attach to process.

4. Load the appropriate Visual Basic Source file into the Visual C++ debugger and set the break point.

5. Call the method within ArcMap.

No changes can be made to the source code within the debugger, and variables cannot be inspected, but code execution can be viewed and altered. This is often sufficient to determine what is wrong, especially with logic-related problems.

ATL wrapper classes

Using the ATL, you can create a class that implements the same interfaces as the Visual Basic class. When you create the ATL object, you create the Visual Basic object. All method calls are then passed to the Visual Basic object for execution. You debug the contained object by setting a break point in the appropriate C++ wrapper method, and when the code reaches the break point, the debugger is stepped into the Visual Basic code. For more information on this technique, look at the ATL debugger sample in the developer guide scenarios\development environment samples in the ArcGIS Developer Help system.

Developing in Visual C++ is a large and complex subject, as it provides a much lower level of interaction with the underlying Windows APIs and COM APIs when compared to other development environments.

While this can be a hindrance for rapid application development, it is the most flexible approach. A number of design patterns, such as COM aggregation and singletons that are possible in Visual C++ are not possible in Visual Basic 6. By using standard class libraries, such as Active Template Library, the complex COM plumbing code can be hidden. However, it is still important to have a thorough understanding of the underlying ATL COM implementation.

The documentation in this section is based on Microsoft Visual C++ version 6 and provides some guidance for ArcGIS development in this environment. With the release of Visual Studio C++ .NET, (also referred to as VC7), many new enhancements are available to the C++ developer. While VC7 can work with the managed .NET environment, and it is possible to work with the ArcGIS .NET API, this will only add overhead to access the underlying ArcGIS COM objects. So for the purposes of ArcGIS development in VC7, it is recommended to work the "traditional" way—that is, directly with the ArcGIS COM interfaces and objects.

There are many enhancements to ATL in VC7. Some of the relevant changes are covered in the section 'ATL in Visual C++ .NET', later in this appendix.

With the addition of the Visual C# .NET language, it is worth considering porting Visual C++ code to this environment and using the ArcGIS .NET API. The syntax of C# is not unlike C++, but the resulting code is generally simpler and more consistent.

This section is intended to serve two main purposes:

1. To familiarize you with general Visual C++ coding style and debugging, beginning with a discussion on ATL

2. To detail specific usage requirements and recommendations for working with the ArcObjects programming platform in Visual C++

WORKING WITH ATL

This section cannot cover all the topics that a developer working with ATL should know to be effective, but it will serve as an introduction to ATL. ATL helps you implement COM objects and it saves typing, but it does not excuse you from knowing C++ and how to develop COM objects.

ATL is the recommended framework for implementing COM objects. The ATL code can be combined with Microsoft Foundation Class Library (MFC) code, which provides more support for writing applications. An alternative to MFC is the Windows Template Library (WTL), which is based on the ATL template methodology and provides many wrappers for window classes and other application support for ATL. WTL is available for download from Microsoft; at the time of writing, version 7.1 is the latest and can be used with Visual C++ version 6 and Visual C++ .NET.

ATL in brief

ATL is a set of C++ template classes designed to be small, fast, and extensible, based loosely on the Standard Template Library (STL). STL provides generic template classes for C++ objects, such as vectors, stacks, and queues. ATL also provides a set of wizards that extend the Visual Studio development environment. These wizards automate some of the tedious plumbing code that all ATL projects must have. The wizards include, but are not limited to, the following:

- Application—Used to initialize an ATL C++ project.

- Object—Used to create COM objects. Both C++ and IDL code is generated, along with the appropriate code to support the creation of the objects at runtime.

- Property—Used to add properties to interfaces.

- Method—Used to add methods to interfaces; both the Property and Method Wizards require you to know some IDL syntax.

- Interface Implementation—Used to implement stub functions for existing interfaces.

- Connection Point Implement—Used to implement outbound events' interfaces.

Typically these are accessed by a right-click on a project, class, or interface in Visual Studio Workspace/Class view.

ATL provides base classes for implementing COM objects as well as implementations for some of the common COM interfaces, including *IUnknown*, *IDispatch*, and *IClassFactory*. There are also classes that provide support for ActiveX controls and their containers.

ATL provides the required services for exposing ATL-based COM objects including registration, server lifetime, and class objects.

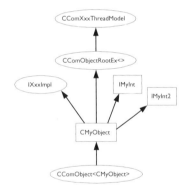

The hierarchical layers of ATL

These template classes build a hierarchy that sandwiches your class. These inheritances are shown to the left. The *CComxxxThreadModel* class supports thread-safe access to global, instance, and static data. The *CComObjectRootEx* class provides the behavior for the *IUnknown* methods. The interfaces at the second level represent the interfaces that the class will implement; these come in two varieties. The *IXxxImpl* interface contains ATL-supplied interfaces that also include an implementation; the other interfaces have pure virtual functions that must be fully implemented within your class. The *CComObject* class inherits your class; this class provides the implementation of the *IUnknown* methods along with the object instantiation and lifetime control.

ATL and DTC

A more detailed discussion on Direct-To-COM follows in the section 'Direct-To-COM smart types'.

Along with smart types, covered later in this appendix, Direct-To-COM (DTC) provides some useful compiler extensions you can use when creating ATL-based objects. The functions *__declspec* and *__uuidof* are two such functions, but the most useful is the *#import* command.

COM interfaces are defined in IDL, then compiled by the Microsoft IDL compiler (MIDL.exe). This results in the creation of a type library and header files. The project uses these files automatically when compiling software that references these interfaces. This approach is limited in that when working with interfaces, you must have access to the IDL files. As a developer of ArcGIS, you only have access to the ArcGIS type library information contained in .olb and .ocx files. While it is possible to engineer a header file from a type library, it is a tedious process. The *#import* command automates the creation of the necessary files required by the compiler. Since the command was developed to support DTC, when using it to import ArcGIS type libraries, there are a number of parameters that must be passed so that the correct import takes place. For further information on this process, see the later section 'Importing ArcGIS type libraries'.

Handling errors in ATL

It is possible to just return an E_FAIL HRESULT code to indicate the failure within a method; however, this does not give the caller any indication of the nature of the failure. There are a number of standard Windows HRESULTs available, for example, E_INVALIDARG (one or more arguments are invalid) and E_POINTER (invalid pointer). These error codes are listed in the window header file *winerror.h*. Not all development environments have comprehensive support for HRESULT; Visual Basic clients often see error results as "Automation Error – Unspecified Error". ATL provides a simple mechanism for working with the COM error information object that can provide an error string description, as well as an error code.

When creating an ATL object, the Object Wizard has an option to support *ISupportErrorInfo*. If you toggle the option on, when the wizard completes, your object will implement the interface *ISupportErrorInfo*, and a method will be added that looks something like this:

```
STDMETHODIMP MyClass::InterfaceSupportsErrorInfo(REFIID riid)
{
    static const IID* arr[] =
    {
        &IID_IMyClass,
    };

    for (int i = 0; i < sizeof(arr) / sizeof(arr[0]); i++)
    {
        if (InlineIsEqualGUID(*arr[i], riid))
            return S_OK;
    }

    return S_FALSE;
}
```

It is now possible to return rich error messages by calling one of the ATL error functions. These functions even work with resource files to ensure easy internationalization of the message strings.

```
// Return a simple string
AtlReportError(CLSID_MyClass, _T("No connection to Database."),
IID_IMyClass, E_FAIL);
// Get the Error Text from a resource string
AtlReportError(CLSID_MyClass, IDS_DBERROR, IID_IMyClass, E_FAIL,
_Module.m_hInstResource);
```

To extract an error string from a failed method, use the Windows function *GetErrorInfo*. This is used to retrieve the last IErrorInfo object on the current thread and clears the current error state.

Although Visual C++ does support an exception mechanism (try ... catch), it is not recommended to mix this with COM code. If an exception unwinds out of a COM interface, there is no guarantee the client will be able to catch this, and the most likely result is a crash.

Linking ATL code

One of the primary purposes of ATL is to support the creation of small fast objects. To support this, ATL gives the developer a number of choices when compiling and linking the source code. Choices must be made about how to link or dynamically access the C runtime (CRT) libraries, the registration code, and the various ATL utility functions. If no CRT calls are made in the code, this can be removed from the link. If CRT calls are made and the linker switch _ATL_MIN_CRT is not removed from the link line, the following error will be generated during the build:

```
LIBCMT.lib(crt0.obj) : error LNK2001: unresolved external symbol _main
ReleaseMinSize/History.dll : fatal error LNK1120: 1 unresolved externals
Error executing link.exe.
```

When compiling a debug build, there will probably not be a problem; however, depending on the code written, there may be problems when compiling a release build. If you receive this error either remove the CRT calls or change the linker switches.

If the utilities code is dynamically loaded at runtime, you must ensure that the appropriate DLL (ATL.DLL) is installed and registered on the user's system. The ArcGIS 9 runtime installation will install ATL.dll. The table below shows the various choices and the related linker switches.

	Symbols	CRT	Utilities	Registrar
Debug		yes	static	dynamic
RelMinDepend	_ATL_MIN_CRT _ATL_STATIC_REGISTRY	no	static	static
RelMinSize	_ATL_MIN_CRT _ATL_DLL	no	dynamic	dynamic

By default, there are build configurations for ANSI and Unicode builds. A component that is built with ANSI compilation will run on Windows 9x; however, considering that ArcGIS is only supported on Unicode operating systems (Windows NT®, Windows 2000, and Windows XP), these configurations are redundant. To delete a configuration in Visual Studio, click Build / Configurations. Then delete *Win32 Debug*, *Win32 Release MinSize*, and *Win32 Release MinDependency*.

Registration of a COM component

The ATL project wizard generates the standard Windows entry points for registration. This code will register the DLLs type library and execute a registry script file (.rgs) for each COM object within the DLL. Additional C++ code to perform other registration tasks can be inserted into these functions.

```
STDAPI DllRegisterServer(void)
{
  // registers object in .rgs, typelib and all interfaces in typelib
  // TRUE instructs the type library to be registered
  return _Module.RegisterServer(TRUE);
}

STDAPI DllUnregisterServer(void)
{
```

```
        return _Module.UnregisterServer(TRUE);
}
```

ATL provides a text file format, .rgs, that is parsed by the ATL's registrar component when a DLL is registered and unregistered. The .rgs file is built into a DLL as a custom resource. The file can be edited to add additional registry entries and contains ProgID, ClassID, and component category entries to place in the registry. The syntax describes keys, values, names, and subkeys to be added or removed from the registry. The format can be summarized as follows:

```
[NoRemove | ForceRemove | val] Name | [ = s 'Value' | d ' Value' | b 'Value' ]
{
   .. optional subkeys for the registry
}
```

NoRemove signifies that the registry key should not be removed on unregistration. *ForceRemove* will ensure the key and subkeys are removed before registering the new keys. The *s*, *d*, and *b* values indicate string (enclosed with apostrophes), double word (32-bit integer value), and binary registry values. A typical registration script is shown below.

```
HKCR
{
   SimpleObject.SimpleCOMObject.1 = s 'SimpleCOMObject Class'
   {
      CLSID = s '{2AFFC10E-ECFB-4697-8B3D-0405650B7CFB}'
   }
   SimpleObject.SimpleCOMObject = s 'SimpleCOMObject Class'
   {
      CLSID = s '{2AFFC10E-ECFB-4697-8B3D-0405650B7CFB}'
      CurVer = s 'SimpleObject.SimpleCOMObject.1'
   }
   NoRemove CLSID
   {
      ForceRemove {2AFFC10E-ECFB-4697-8B3D-0405650B7CFB} = s 'SimpleCOMObject
Class'
      {
         ProgID = s 'SimpleObject.SimpleCOMObject.1'
         VersionIndependentProgID = s 'SimpleObject.SimpleCOMObject'
         InprocServer32 = s '%MODULE%'
         {
            val ThreadingModel = s 'Apartment'
         }
         'TypeLib' = s '{855DD226-5938-489D-986E-149600FEDD63}'
         'Implemented Categories'
         {
            {7DD95801-9882-11CF-9FA9-00AA006C42C4}
         }
      }
   }
}
```

NoRemove CLSID ensures the registry key *CLSID* is never removed. COM objects register their ProgIDs and GUIDs below this subkey, so its removal would result in a serious corruption of the registry. *InprocServer32* is the standard COM mechanism that relates a component GUID to a DLL; ATL will insert the correct module name using the %MODULE% variable. Other entries under the GUID specify the ProgID, threading model, and type library to use with this component.

To register a COM coclass into a component category, there are two approaches. The recommended approach is illustrated above: place GUIDs for component categories beneath an Implemented Categories key, which in turn is under the GUID of the coclass. The second approach is to use ATL macros in an objects header file: BEGIN_CATEGORY_MAP, IMPLEMENTED_CATEGORY, and END_CATEGORY_MAP. However, these macros do not correctly remove registry entries as explained in MSDN article *Q279459 BUG: Component Category Registry Entries Not Removed in ATL Component*. A header file is supplied with the GUIDs of all the component categories used by ArcGIS; this is available in \Program Files\ArcGIS\include\CatIDs\ArcCATIDs.h.

If the GUID of a component is changed during development or the type library name is changed, then it is important to keep the .rgs content consistent with these changes. Otherwise, the registry will be incorrect and object creation can fail.

Debugging ATL code

In addition to the standard Visual Studio facilities, ATL provides a number of debugging options with specific support for debugging COM objects. The output of these debugging options is displayed in the Visual C++ Output window. The *QueryInterface* call can be debugged by setting the symbol *_ATL_DEBUG_QI*, *AddRef* and *Release* calls with the symbol *_ATL_DEBUG_INTERFACES*, and leaked objects can be traced by monitoring the list of leaked interfaces at termination time when the *_ATL_DEBUG_INTERFACES* symbol is defined. The leaked interfaces list has entries like the following:

```
INTERFACE LEAK: RefCount = 1, MaxRefCount = 3, {Allocation = 10}
```

On its own, this does not tell you much apart from the fact that one of your objects is leaking because an interface pointer has not been released. However, the *Allocation* number allows you to automatically break when that interface is obtained by setting the *m_nIndexBreakAt* member of the *CComModule* at server startup. This in turn calls the function *DebugBreak()* to force the execution of the code to stop at the relevant place in the debugger. For this to work the program flow must be the same.

```
extern "C"
BOOL WINAPI DllMain(HINSTANCE hInstance, DWORD dwReason, LPVOID /
*lpReserved*/)
{
  if (dwReason == DLL_PROCESS_ATTACH)
  {
    _Module.Init(ObjectMap, hInstance, &LIBID_HISTORYLib);
    DisableThreadLibraryCalls(hInstance);
    _Module.m_nIndexBreakAt = 10;
  }
  else if (dwReason == DLL_PROCESS_DETACH)
  {
    _Module.Term();
  }
  return TRUE;
}
```

Boolean types

Historically, ANSI C did not have a Boolean data type and used int value instead, where 0 represents false and nonzero represents true. However, the bool data-type has now become part of ANSI C++. COM APIs are language independent and define a different Boolean type, VARIANT_BOOL. In addition, Win32 API uses a different bool type. It is important to use the correct type at the appropriate time. The following table summarizes their usage:

Type	True Value	False Value	Where Defined	When to Use
bool	true (1)	false (0)	Defined by compiler	This is an intrinsic compiler type so there is more potential for the compiler to optimise its use. This type can also be promoted to an int value. Expressions (e.g. if=0) returns a type of bool. Typically used for class member variables and local variables.
BOOL (int)	TRUE (1)	FALSE (0)	Windows Data Type (defined in windef.h)	Used with windows API functions, often as a return value to indicate success or failure.
VARIANT_BOOL (16bit short)	VARIANT_ TRUE (-1)	VARIANT_ FALSE (0)	COM boolean values (wtypes.h)	Used in COM APIs for boolean values. Also used within VARIANT types, if the VARIANT type is VT_BOOL, then the VARIANT value (boolVal) is populated with a VARIANT_BOOL. Take care to convert a bool class member variable to the correct VARIANT_BOOL value. Often the conditional test "hook - colon" operator is used. For example where bRes is defined as a bool, then to set an result type: *pVal = bRes ? VARIANT_TRUE : VARIANT_FALSE;

String types

Considering that strings (sequences of text characters) are a simple concept, they have unfortunately become a complex and confusing topic in C++. The two main reasons for this confusion are the lack of C++ support for variable length strings combined with the requirement to support ANSI and Unicode character sets within the same code. As ArcGIS is only available on Unicode platforms, it may simplify development to remove the ANSI requirements.

The C++ convention for strings is an array of characters terminated with a 0. This is not always good for performance when calculating lengths of large strings. To support variable length strings, the character arrays can be dynamically allocated and released on the heap, typically using *malloc* and *free* or *new* and *delete*. Consequently, a number of wrapper classes provide this support; CString defined in MFC and WTL is the most widely used. In addition, for COM usage the BSTR type is defined and the ATL wrapper class CComBSTR is available.

To allow for international character sets, Microsoft Windows migrated from an 8-bit ANSI character string (8-bit character) representation (found on Windows 95, Windows 98, and Windows Me platforms) to a 16-bit Unicode character string (16-bit unsigned short). Unicode is synonymous with wide characters (wchar_t). In COM APIs, OLECHAR is the type used and is defined to be wchar_t on Windows. Windows operating systems, such as Windows NT, Windows 2000, and Windows XP, natively support Unicode characters. To allow the same C++ code to be compiled for ANSI and Unicode platforms, compiler switches are used to change Windows API functions (for example, SetWindowText) to resolve to an ANSI version (SetWindowTextA) or a Unicode version (SetWindowTextW). In addition, character-independent types (TCHAR defined in tchar.h) were introduced to represent a character; on an ANSI build this is defined to be a *char* and on a Unicode build this is a *wchar_t*, a typedef defined as unsigned short. To perform standard C string manipulation, there are typically three different definitions of the same function; for example, for a case-insensitive comparison, *strcmp* provides the ANSI version, *wcscmp* provides the Unicode version, and *_tcscmp* provides the TCHAR version. There is also a fourth version, *_mbscmp*, which is a variation of the 8-bit ANSI version that will interpret multibyte character sequences (MBCS) within the 8-bit string.

```
// Initialize some fixed length strings
char*   pNameANSI = "Bill";        // 5 bytes (4 characters plus a terminator)
wchar_t* pNameUNICODE = L"Bill";  // 10 bytes (4 16-bit characters plus a
                                  //          16-bit terminator)
TCHAR*  pNameTCHAR = _T("Bill"); // either 5 or 10 depending on compiler
                                  //          settings
```

COM APIs represent variable length strings with a BSTR type; this is a pointer to a sequence of OLECHAR characters, which is defined as Unicode characters and is the same as a wchar_t. A BSTR must be allocated and released with the SysAllocString and SysFreeString windows functions. Unlike C strings, they can contain embedded zero characters, although this is unusual. The BSTR also has a count value, which is stored four bytes before the BSTR pointer address. The CComBSTR wrappers are often used to manage the lifetime of a string.

Do not pass a pointer to a C style array of Unicode characters (OLECHAR or wchar_t) to a function expecting a BSTR. The compiler will not raise an error as the types are identical. However, the function receiving the BSTR can behave incorrectly or crash when accessing the string length, which will be random memory values.

```
ipFoo->put_WindowTitle(L"Hello");          // This is bad!
ipFoo->put_WindowTitle(CComBSTR(L"Hello")); // This correctly initializes
                                            //        and passes a BSTR
```

ATL provides conversion macros to switch strings between ANSI (A), TCHAR (T), Unicode (W), and OLECHAR (OLE). In addition, the types can have a const modifier (C). These macros use the abbreviations shown in brackets with a "2" between them. For example, to convert between OLECHAR (for example, an input BSTR) to const TCHAR (for use in a Windows function), use the OLE2CT conversion macro. To convert ANSI to Unicode, use A2W. These macros require the USES_CONVERSION macro to be placed at the top of a method; this will create some local variables that are used by the conversion macros. When the source and destination character sets are different and the destination type is not a BSTR, the macro allocates the destination string on the call stack (using the _alloca runtime function). It's important to realize this especially when using these macros within a loop; otherwise the stack may grow large and run out of stack space.

```
STDMETHODIMP CFoo::put_WindowTitle(BSTR bstrTitle)
{
  USES_CONVERSION;
  if (::SysStringLen(bstrTitle) == 0)
    return E_INVALIDARG;

  ::SetWindowText(m_hWnd, OLE2CT(bstrTitle));

  return S_OK;
}
```

To check if two CComBSTR strings are different, do not use the not equal ("!=") operator. The "==" operator performs a case-sensitive comparison of the string contents; however, "!=" will compare pointer values and not the string contents, typically returning false.

Implementing noncreatable classes

Noncreatable classes are COM objects that cannot be created by *CoCreateInstance*. Instead, the object is created within a method call of a different object and an interface pointer to the noncreatable class is returned. This type of object is found in abundance in the geodatabase model. For example, *FeatureClass* is noncreatable and can only be obtained by calling one of a number of methods; one example is the *IFeatureWorkspace::OpenFeatureClass* method.

One advantage of a noncreatable class is that it can be initialized with private data using method calls that are not exposed in a COM API. Below is a simplified example of returning a noncreatable object:

```
// Foo is a cocreatable object.
IFooPtr ipFoo;
HRESULT hr = ipFoo.CreateInstance(CLSID_Foo);


// Bar is a noncreatable object, cannot use
ipBar.CreateInstance(CLSID_Bar).
IBarPtr ipBar;
// Use a method on Foo to create a new Bar object.
hr = ipFoo->CreateBar(&ipBar);
ipBar->DoSomething();
```

The steps required to change a cocreatable ATL class into a noncreatable class are shown below:

1. Add "noncreatable" to the IDL file's coclass attributes.

```
[
    uuid(DCB87952-0716-4873-852B-F56AE8F9BC42),
    noncreatable
]
coclass Bar
{
    [default] interface IUnknown;
    interface IBar;
};
```

2. Change the class factory implementation to fail any cocreate instance of the noncreatable class. This happens via ATL's object map in the main DLL module.

```
BEGIN_OBJECT_MAP(ObjectMap)
    OBJECT_ENTRY(CLSID_Foo, CFoo)                  // Creatable object
    OBJECT_ENTRY_NON_CREATEABLE(CLSID_Bar, CBar)  // Noncreatable object
END_OBJECT_MAP()
```

3. Optionally, the registry entries can be removed. First, remove the registry script for the object from the resources (Bar.rgs in this example). Then change the class definition DECLARE_REGISTRY_RESOURCEID(IDR_BAR) to DECLARE_NO_REGISTRY().

4. To create the noncreatable object inside a method, use the CComObject template to supply the implementation of CreateInstance.

```
// Get NonCreatable object Bar (implementing IBar) from COM object Foo.
STDMETHODIMP CFoo::CreateBar(IBar **pVal)
{
```

```
        if (pVal==0) return E_POINTER;

    // Smart pointer to noncreatable object Bar
    IBarPtr ipBar = 0;

    // C++ Pointer to Bar, with ATL template to supply CreateInstance
implementation
    CComObject<CBar>* pBar = 0;

    HRESULT hr = CComObject<CBar>::CreateInstance(&pBar);
    if (SUCCEEDED(hr))
    {
        // Increment the ref count from 0 to 1 to protect the object
        // from being released in any initialization code.
        pBar->AddRef();

        // Call C++ methods (not exposed to COM) to initialize the Bar object.
        pBar->InitialiseBar(10);

        // QI to IBar and hold a smart pointer reference to the object Bar.
        hr = pBar->QueryInterface(IID_IBar, (void**)&ipBar);

        pBar->Release();
    }

    // Return IBar pointer to the caller.
    *pVal = ipBar.Detach();

    return S_OK;
}
```

ATL in Visual C++ .NET

Visual C++ version 6 is used for the majority of this help. However, with the release of Visual C++ .NET, there are enhancements and changes that are relevant to the ArcGIS ATL developer. Some of these are summarized below:

Attribute-based programming—This is a major change introduced in VC7. Attributes are inserted in the source code enclosed in square brackets—for example, [coclass]. Attributes are designed to simplify COM programming and .NET framework common language runtime development. When you include attributes in your source files, the compiler works with provider DLLs to insert code or modify the code in the generated object files. There are attributes that aid in the creation of .idl files, interfaces, type libraries, and other COM elements. In the IDE, attributes are supported by the wizards and by the Properties window. The ATL wizards make extensive use of attributes to inject the ATL boilerplate code into the class. Consequently, typical COM coclass header files in VC7 contain much less ATL code than at VC6. As IDL is generated from attributes, there is typically no .idl file present in COM projects as before, and the IDL file is generated at compile time.

Build configurations—There are only two default build configurations in VC7; these are ANSI Debug- and Release-based builds. As ArcGIS is only available on

Unicode platforms, it is recommended to change these by modifying the project properties. The general project properties page has an option for "Character Set". Change this from "Use Multi-Byte Character Set" to "Use Unicode Character Set".

Character conversion macros—The character conversion macros (USES_CONVERSION, W2A, W2CT, and so forth) have improved alternative versions. These no longer allocate space on the stack, so they can be used in loops without running out of stack space. The USES_CONVERSION macro is also no longer required. These macros are now implemented as classes and begin with a "C"—for example, CW2A, CW2CT.

Safe array support—This is available with CComSafeArray and CComSafeArrayBound classes.

Module level global—The module level global CComModule _module has been split into a number of related classes, for example, CAtlComModule and CAtlWinModule. To retrieve the resource module instance, use the following code: `_AtlBaseModule.GetResourceInstance();`

String support—General variable length string support is now available through CString in ATL. This is defined in the header file atlstr.h and cstringt.h. If ATL is combined with MFC, this defaults to MFC's CString implementation.

Filepath handling—A collection of related functions for processing the components of filepaths is available through the CPath class defined in atlpath.h.

ATLServer—This is a new selection of ATL classes designed for writing Web applications, XML Web services, and other server applications.

#import issues—When using *#import*, a few modifications are required. For example, the *#import* of esriSystem requires an exclude or rename of *GetObject*, and the *#import* of esriGeometry requires an exclude or rename of *ISegment*.

ATL REFERENCES

The Microsoft Developer Network (MSDN) provides a wealth of documentation, articles, and samples that are installed with Visual Studio products. ATL reference documentation for Visual Studio version 6 is under:

MSDN Library - October 2001 / Visual Tools and Languages / Visual Studio 6.0 Documentation / Visual C++ Documentation / Reference / Active Template Library

Additional documentation is also available on the MSDN Web site at http://www.msdn.microsoft.com.

You may also find the following books to be useful:

Grimes, Richard. *ATL COM Programmer's Reference*. Chicago: Wrox Press Inc., 1988.

Grimes, Richard. *Professional ATL COM Programming*. Chicago: Wrox Press Inc., 1988.

Grimes, Richard, Reilly Stockton, Alex Stockton, and Julian Templeman. *Beginning ATL 3 COM Programming*. Chicago: Wrox Press Inc. 1999.

King, Brad and George Shepherd. *Inside ATL*. Redmond, WA: Microsoft Press, 1999.

Rector, Brent, Chris Sells, and Jim Springfield. *ATL Internals*. Reading, MA: Addison–Wesley, 1999.

SMART TYPES

Smart types are objects that behave as types. They are C++ class implementations that encapsulate a data type, wrapping it with operators and functions that make working with the underlying type easier and less error prone. When these smart types encapsulate an interface pointer, they are referred to as *smart pointers*. Smart pointers work with the *IUnknown* interface to ensure that resource allocation and deallocation is correctly managed. They accomplish this by various functions, construct and destruct methods, and overloaded operators. There are numerous smart types available to the C++ programmer. The two main smart types covered here are Direct-To-COM and Active Template Library.

Smart types can make the task of working with COM interfaces and data types easier, since many of the API calls are moved into a class implementation; however, they must be used with caution and never without a clear understanding of how they are interacting with the encapsulated data type.

Direct-To-COM smart types

The smart type classes supplied with DTC are known as the Compiler COM Support Classes and consist of:

- *_com_error*—This class represents an exception condition in one of the COM support classes. This object encapsulates the HRESULT and the *IErrorInfo* COM exception objects.

- *_com_ptr_t*—This class encapsulates a COM interface pointer. See below for common uses.

- *_bstr_t*—This class encapsulates the *BSTR* data type. The functions and operators on this class are not as rich as the ATL *CComBSTR* smart type; hence, this is not normally used.

- *_variant_t*—This class encapsulates the *VARIANT* data type. The functions and operators on this class are not as rich as the ATL *CComVariant* smart type; hence, this is not normally used.

To define a smart pointer for an interface, you can use the macro *_COM_SMARTPTR_TYPEDEF* like this:

```
_COM_SMARTPTR_TYPEDEF(IFoo, __uuidof(IFoo));
```

The compiler expands this as follows:

```
typedef _com_ptr_t< _com_IIID<IFoo, __uuidof(IFoo)> > IFooPtr;
```

Once declared, it is simply a matter of declaring a variable as the type of the interface and appending *Ptr* to the end of the interface. Below are some common uses of this smart pointer that you will see in the numerous C++ samples.

```
// Get a CLSID GUID constant.
extern "C" const GUID __declspec(selectany) CLSID_Foo = \
    {0x2f3b470c,0xb01f,0x11d3,{0x83,0x8e,0x00,0x00,0x00,0x00,0x00,0x00}};
```

```
// Declare Smart Pointers for IFoo, IBar, and IGak interfaces.
_COM_SMARTPTR_TYPEDEF(IFoo, __uuidof(IFoo));
_COM_SMARTPTR_TYPEDEF(IBar, __uuidof(IBar));
_COM_SMARTPTR_TYPEDEF(IGak, __uuidof(IGak));

STDMETHODIMP SomeClass::Do()
{
  // Create Instance of Foo class and QueryInterface (QI) for IFoo interface.
  IFooPtr ipFoo;
  HRESULT hr = ipFoo.CreateInstance(CLSID_Foo);
  if (FAILED(hr)) return hr;

  // Call method on IFoo to get IBar.
  IBarPtr ipBar;
  hr = ipFoo->get_Bar(&ipBar);
  if (FAILED(hr)) return hr;

  // QI IBar interface for IGak interface.
  IGakPtr ipGak(ipBar);

  // Call method on IGak.
  hr = ipGak->DoSomething();
  if (FAILED(hr)) return hr;

  // Explicitly call Release().
  ipGak = 0;
  ipBar = 0;

  // Let destructor call IFoo's Release.
  return S_OK;
}
```

One of the main advantages of using the DTC smart pointers is that they are automatically generated from the *#import* compiler statement for all interface and coclass definitions in a type library. For more details on this functionality, see the later section 'Importing ArcGIS type libraries'.

It is possible to create an object implicitly in a DTC smart pointer's constructor, for example:

```
IFooPtr ipFoo(CLSID_Foo)
```

However, this will raise a C++ exception if there is an error during object creation—for example, if the DLL containing the object implementation was accidentally deleted. This exception will typically be unhandled and cause a crash. A more robust approach is to avoid exceptions in COM, call CreateInstance explicitly, and handle the failure code, for example:

```
IFooPtr ipFoo;
HRESULT hr = ipFoo.CreateInstance(CLSID_Foo);
if (FAILED(hr))
  return hr; // Return object creation failure code to caller.
```

Active Template Library smart types

ATL defines various smart types, as seen in the list below. You are free to combine both the ATL and DTC smart types in your code. However, it is typical to use the DTC for smart pointers, as they are easily generated by importing type libraries. For BSTR and VARIANT types, the ATL versions for CComBSTR, CComVariant are typically used.

ATL smart types include:

- *CComPtr*—encapsulates a COM interface pointer by wrapping the *AddRef* and *Release* methods of the *IUnknown* interface

- *CComQIPtr*—encapsulates a COM interface and supports all three methods of the *IUnknown* interface: *QueryInterface*, *AddRef*, and *Release*

- *CComBSTR*—encapsulates the *BSTR* data type

- *CComVariant*—encapsulates the *VARIANT* data type

- *CRegKey*—provides methods for manipulating Windows registry entries

- *CComDispatchDriver*—provides methods for getting and setting properties, and calling methods through an object's *IDispatch* interface

- *CSecurityDescriptor*—provides methods for setting up and working with the Discretionary Access Control List (DACL)

This section examines the first four smart types and their uses. The example code below, written with ATL smart pointers, looks like the following:

```
// Get a CLSID GUID constant.
extern "C" const GUID __declspec(selectany) CLSID_Foo = \
    {0x2f3b470c,0xb01f,0x11d3,{0x83,0x8e,0x00,0x00,0x00,0x00,0x00,0x00}};

STDMETHODIMP SomeClass::Do ()
{
  // Create Instance of Foo class and QI for IFoo interface.
  CComPtr<IFoo> ipFoo;
  HRESULT hr = CoCreateInstance(CLSID_Foo, NULL, CLSCTX_INPROC_SERVER,
IID_IFoo, (void **)&ipFoo);
   if (FAILED(hr)) return hr;

  // Call method on IFoo to get IBar.
  CComPtr<IBar> ipBar;
  HRESULT hr = ipFoo->get_Bar(&ipBar);
  if (FAILED(hr)) return hr;

  // IBar interface for IGak interface
  CComQIPtr<IGak> ipGak(ipBar);

  // Call method on IGak.
  hr = ipGak->DoSomething();
  if (FAILED(hr)) return hr;

  // Explicitly call Release().
  ipGak = 0;
```

The equality operator ("==") may have different implementations when used during smart pointer comparisons. The COM specification states object identification is performed by comparing the pointer values of IUnknown. The DTC smart pointers will perform necessary QI and comparison when using the "==" operator. However, the ATL smart pointers will not do this, so you must use the ATL IsEqualObject() method.

```
  ipBar = 0;

  // Let destructor call Foo's Release.
  return S_OK;
}
```

The most common smart pointer seen in the Visual C++ samples is the DTC type. In the examples below, which illustrate the *BSTR* and *VARIANT* data types, the DTC pointers are used. When working with *CComBSTR*, use the text mapping L to declare constant *OLECHAR* strings. *CComVariant* derives directly from the VARIANT data type, meaning that there is no overloading with its implementation, which in turn simplifies its use. It has a rich set of constructors and functions that make working with *VARIANT*s straightforward; there are even methods for reading and writing from streams. Be sure to call the *Clear* method before reusing the variable.

```
  ipFoo->put_Name(CComBSTR(L"NewName"));
  if FAILED(hr)) return hr;

  // Create a VT_I4 variant (signed long).
  CComVariant vValue(12);

  // Change its data type to a string.
  hr = vValue.ChangeType(VT_BSTR);
  if (FAILED(hr)) return hr;
```

Some method calls in IDL are marked as being optional and take a variant parameter. However in VC++, these parameters still have to be supplied. To signify that a parameter value is not supplied, a variant is passed specifying an error code or type DISP_E_PARAMNOTFOUND:

```
  CComBSTR documentFilename(L"World.mxd");

  CComVariant noPassword;
  noPassword.vt = VT_ERROR;
  noPassword.scode = DISP_E_PARAMNOTFOUND;
  HRESULT hr = ipMapControl->LoadMxFile(documentFilename, noPassword);
```

When working with *CComBSTR* and *CComVariant*, the *Detach()* function releases the underlying data type from the smart type so it can be used when passing a result as an [out] parameter of a method. The use of the Detach method with *CComBSTR* is shown below:

```
STDMETHODIMP CFoo::get_Name(BSTR* name)
{
  if (name==0) return E_POINTER;
  CComBSTR bsName(L"FooBar");
  *name = bsName.Detach();
}
```

CComVariant(VARIANT_TRUE) will create a short integer variant (type VT_I2) and not a Boolean variant (type VT_BOOL) as expected. You can use CComVariant(true) to create a Boolean variant.

CComVariant myVar(ipSmartPointer) will result in a variant type of Boolean (VT_BOOL) and not a variant with an object reference (VT_UNKNOWN) as expected. It is better to pass unambiguous types to constructors, that is, types that are not themselves smart types with overloaded cast operators.

```
// Perform QI if IUnknown.
IUnknownPtr ipUnk = ipSmartPointer;
// Ensure IUnknown* constructor of CComVariant is used.
CComVariant myVar2(ipUnk.GetInterfacePtr());
```

A common practice with smart pointers is to use *Detach()* to return an object from a method call. When returning an interface pointer, the COM standard is to increment reference count of the [out] parameter inside the method implementation. It is the caller's responsibility to call Release when the pointer is no longer required. Consequently, care must be taken to avoid calling *Detach()* directly on a member variable. A typical pattern is shown below:

```
STDMETHODIMP CFoo::get_Bar(IBar **pVal)
{
  if (pVal==0) return E_POINTER;

  // Constructing a local smart pointer using another smart pointer
  // results in an AddRef (if pointer is not 0).
  IBarPtr ipBar(m_ipBar);

  // Detach will clear the local smart pointer, and the
  // interface is written into the output parameter.
  *pVal = ipBar.Detach();

  // This can be combined into one line
  // *pVal = IBarPtr(m_ipBar).Detach();

  return S_OK;
}
```

The above pattern has the same result as the following code; note that a conditional test for a zero pointer is required before AddRef can be called. Calling AddRef (or any method) on a zero pointer will result in an access violation exception and typically crash the application:

```
STDMETHODIMP CFoo::get_Bar(IBar **pVal)
{
  if (pVal==0) return E_POINTER;

  // Copy the interface pointer (no AddRef) into the output parameter.
  *pVal = m_ipBar;

  // Make sure interface pointer is nonzero before calling AddRef.
  if (*pVal)
    *pVal->AddRef();

  return S_OK;
}
```

When using a smart pointer to receive an object from an [out] parameter on a method, use the smart pointer "&" dereference operator. This will cause the previous interface pointer in the smart pointer to be released. The smart pointer is then populated with the new [out] value. The implementation of the method will have already incremented the object reference count. This will be released when the smart pointer goes out of scope:

```
{
  IFooPtr ipFoo1, ipFoo2;
  ipFoo1.CreateInstance(CLSID_Foo);
  ipFoo2.CreateInstance(CLSID_Foo);

  // Initialize ipBar Smart pointer from Foo1.
  IBarPtr ipBar;
  ipFoo1->get_Bar(&ipBar);

  // The "&" dereference will call Release on ipBar.
  // ipBar is then repopulated with a new instance of IBar.
  ipFoo2->get_Bar(&ipBar);
}
// ipBar goes out of scope, and the smart pointer destructor calls Release.
```

Naming conventions

Type names

All type names (*class*, *struct*, *enum*, and *typedef*) begin with an uppercase letter and use mixed case for the rest of the name:

```
class Foo : public CObject { . . .};
struct Bar { . . .};
enum ShapeType { . . . };
typedef int* FooInt;
```

Typedefs for function pointers (callbacks) append Proc to the end of their names.

```
typedef void (*FooProgressProc)(int step);
```

Enumeration values all begin with a lowercase string that identifies the project; in the case of ArcObjects this is esri, and each string occurs on a separate line:

```
typedef enum esriQuuxness
{
  esriQLow,
  esriQMedium,
  esriQHigh
} esriQuuxness;
```

Function names

Name functions using the following conventions:

- For simple accessor and mutator functions, use Get<Property> and Set<Property>:

```
int GetSize();
void SetSize(int size);
```

- If the client is providing storage for the result, use Query<Property>:

```
void QuerySize(int& size);
```

[<scope>_]<type><name>

Prefix	Variable scope
m	Instance class members
c	Static class member (including constants)
g	Globally static variable
<empty>	local variable or struct or public class member

<type>

Prefix	Data Type
b	Boolean
by	byte or unsigned char
cx / cy	short used as size
d	double
dw	DWORD, double word or unsigned long
f	float
fn	function
h	handle
i	int (integer)
ip	smart pointer
l	long
p	a pointer
s	string
sz	ASCIIZ null-terminated string
w	WORD unsigned int
x, y	short used as coordinates

<name> describes how the variable is used or what it contains. The <scope> and <type> portions should always be lowercase, and the <name> should use mixed case:

Variable Name	Description
m_hWnd	a handle to a HWND
ipEnvelope	a smart pointer to a COM interface
m_pUnkOuter	a pointer to an object
c_isLoaded	a static class member
g_pWindowList	a global pointer to an object

- For state functions, use Set<State> and Is<State> or Can<State>:

```
bool IsFileDirty();
void SetFileDirty(bool dirty);
bool CanConnect();
```

- Where the semantics of an operation are obvious from the types of arguments, leave type names out of the function names.

Instead of:

```
AddDatabase(Database& db);
```

consider using:

```
Add(Database& db);
```

Instead of:

```
ConvertFoo2Bar(Foo* foo, Bar* bar);
```

consider using:

```
Convert(Foo* foo, Bar* bar)
```

- If a client relinquishes ownership of some data to an object, use Give<Property>. If an object relinquishes ownership of some data to a client, use Take<Property>:

```
void GiveGraphic(Graphic* graphic);
Graphic* TakeGraphic(int itemNum);
```

- Use function overloading when a particular operation works with different argument types:

```
void Append(const CString& text);
void Append(int number);
```

Argument names

Use descriptive argument names in function declarations. The argument name should clearly indicate what purpose the argument serves:

```
bool Send(int messageID, const char* address, const char* message);
```

DEBUGGING TIPS IN DEVELOPER STUDIO

Visual C++ comes with a feature-rich debugger. These tips will help you get the most from your debugging session.

Backing up after failure

When a function call has failed and you'd like to know why (by stepping into it), you don't have to restart the application. Use the Set Next Statement command to reposition the program cursor back to the statement that failed (right-click on the statement to bring up the debugging context menu). Then step into the function.

Edit and Continue

Visual Studio 6 allows changes to source code to be made during a debugging session. The changes can be recompiled and incorporated into the executing code without stopping the debugger. There are some limitations to the type of changes that can be made; in this case, the debug session must be restarted. This feature is enabled by default; the settings are available in the Settings command of the project menu. Click the C/C++ tab, then choose General from the Category dropdown list. In the Debug info dropdown list, click Program Database for Edit and Continue.

Unicode string display

To set your debugger options to display Unicode strings, click the Tools menu, click Options, click Debug, then check the Display Unicode Strings check box.

Variable value display

Pause the cursor over a variable name in the source code to see its current value. If it is a structure, click the Eyeglasses icon or press Shift+F9 to bring up the QuickWatch dialog box or drag and drop it into the Watch window.

Undocking windows

If the Output window (or any docked window, for that matter) seems too small to you, try undocking it to make it a real window by right-clicking it and toggling the Docking View item.

Conditional break points

Use conditional break points when you need to stop at a break point only once some condition is reached—for instance, when a for loop reaches a particular counter value. To do so, set the break point normally, then bring up the Breakpoints window (Ctrl+B or Alt+F9). Select the specific break point you just set and click the Condition button to display a dialog box in which you specify the break point condition.

Preloading DLLs

You can preload DLLs that you want to debug before executing the program. This allows you to set break points up front rather than wait until the DLL has been loaded during program execution. To do this, click Project, click Settings, click Debug, click Category, then click Additional DLLs. Then, click in the list area to add any DLLs you want to preload.

Changing display formats

You can change the display format of variables in the QuickWatch dialog box or in the Watch window using the formatting symbols in the following table.

Symbol	Format	Value	Displays
d, i	signed decimal integer	0xF000F065	-268373915
u	unsigned decimal integer	0x0065	101
o	unsigned octal integer	0xF065	0170145
x, X	hexadecimal integer	61541	0x0000F065
l, h	long or short prefix for d, i, u, o, x, X	00406042, hx	0x0C22
f	signed floating-point	3./2.	1.500000
e	signed scientific notation	3./2.	1.500000e+00
g	e or f, whichever is shorter	3./2.	1.5
c	single character	0x0065	'e'
s	string	0x0012FDE8	"Hello"
su	Unicode string		"Hello"
hr	string	0	S_OK

To use a formatting symbol, type the variable name followed by a comma and the appropriate symbol. For example, if var has a value of 0x0065, and you want to see the value in character form, type "var,c" in the Name column on the tab of the Watch window. When you press Enter, the character format value appears: var,c = 'e'. Likewise, assuming that *hr* is a variable holding HRESULTs, view a human-readable form of the HRESULT by typing "hr,hr" in the Name column.

You can use the formatting symbols shown in the following table to format the contents of memory locations.

Symbol	Format	Value
ma	64 ASCII characters	0x0012ffac .4...0...".0W&.. ......IW&.0..W..I ....".I.JO&.I.2 ..".I...0y....I
m	16 bytes in hex, followed by 16 ASCII characters	0x0012ffac B3 34 CB 00 84 30 94 80 FF 22 8A 30 57 26 00 00 .4...0....".0W&..
mb	16 bytes in hex, followed by 16 ASCII characters	0x0012ffac B3 34 CB 00 84 30 94 80 FF 22 8A 30 57 26 00 00 .4...0....".0W&..
mw	8 words	0x0012ffac 34B3 00CB 3084 8094 22FF 308A 2657 0000
md	4 double-words	0x0012ffac 00CB34B3 80943084 308A22FF 00002657
mu	2-byte characters (Unicode)	0x0012fc60 8478 77f4 ffff ffff 0000 0000 0000 0000

With the memory location formatting symbols, you can type any value or expression that evaluates a location. To display the value of a character array as a string, precede the array name with an ampersand, &*yourname*. A formatting character can also follow an expression:

- *rep+1,x*
- *alps[0],mb*
- *xloc,g*
- *count,d*

To watch the value at an address or the value to which a register points, use the *BY*, *WO*, or *DW* operators:

- *BY* returns the contents of the byte pointed at.
- *WO* returns the contents of the word pointed at.
- *DW* returns the contents of the doubleword pointed at.

Follow the operator with a variable, register, or constant. If the *BY*, *WO*, or *DW* operator is followed by a variable, then the environment watches the byte, word, or doubleword at the address contained in the variable.

You can also use the context operator { } to display the contents of any location.

To display a Unicode string in the Watch window or the QuickWatch dialog box, use the su format specifier. To display data bytes with Unicode characters in the Watch window or the QuickWatch dialog box, use the mu format specifier.

Keyboard shortcuts

There are numerous keyboard shortcuts that make working with the Visual Studio Editor faster. Some of the more useful keyboard shortcuts follow.

The text editor uses many of the standard shortcut keys used by Windows applications, such as Word. Some specific source code editing shortcuts are listed below.

Shortcut	Action
Alt+F8	Correctly indent selected code based on surrounding lines.
Ctrl+]	Find the matching brace.
Ctrl+J	Display list of members.
Ctrl+Spacebar	Complete the word, once the number of letters entered allows the editor to recognize it. Use full when completing function and variable names.
Tab	Indents selection one tab stop to the right.
Shift+Tab	Indents selection one tab to the left.

Below is a table of common keyboard shortcuts used in the debugger.

Shortcut	Action
F9	Add or remove breakpoint from current line.
Ctrl+Shift+F9	Remove all breakpoints.
Ctrl+F9	Disable breakpoints.
Ctrl+Alt+A	Display auto window and move cursor into it.
Ctrl+Alt+C	Display call stack window and move cursor into it.
Ctrl+Alt+L	Display locals window and move cursor into it.
Ctrl+Alt+A	Display auto window and move cursor into it.
Shift+F5	End debugging session.
F11	Execute code one statement at a time, stepping into functions.
F10	Execute code one statement at a time, stepping over functions.
Ctrl+Shift+F5	Restart a debugging session.
Ctrl+F10	Resume execution from current statement to selected statement.
F5	Run the application.
Ctrl+F5	Run the application without the debugger.
Ctrl+Shift+F10	Set the next statement.
Ctrl+Break	Stop execution.

Loading the following shortcuts can greatly increase your productivity with the Visual Studio development environment.

Shortcut	Action
ESC	Close a menu or dialog box, cancel an operation in progress, or place focus in the current document window.
CTRL+SHIFT+N	Create a new file.
CTRL+N	Create a new project.
CTRL+F6 or CTRL+TAB	Cycle through the MDI child windows one window at a time.
CTRL+ALT+A	Display the auto window and move the cursor into it.
CTRL+ALT+C	Display the call stack window and move the cursor into it.
CTRL+ALT+T	Display the document outline window and move the cursor into it.
CTRL+H	Display the find window.
CTRL+F	Display the find window. If there is no current Find criteria, put the word under your cursor in the find box.
CTRL+ALT+I	Display the immediate window and move the cursor into it. Not available if you are in the text editor window.
CTRL+ALT+L	Display the locals window and move the cursor into it.
CTRL+ALT+O	Display the output window and move the cursor into it
CTRL+ALT+J	Display the project explorer window and move the cursor into it.
CTRL+ALT+P	Display the properties window and move the cursor into it.
CTRL+SHIFT+O	Open a file.
CTRL+O	Open a project.
CTRL+P	Print all or part of the document.
CTRL+SHIFT+S	Save all of the files, projects, or documents.
CTRL+S	Select all.
CTRL+A	Save the current document or selected item or items.

Navigating through online help topics

Right-click a blank area of a toolbar to display a list of all the available toolbars. The Infoviewer toolbar contains up and down arrows that allow you to cycle through help topics in the order in which they appear in the table of contents. The left and right arrows cycle through help topics in the order that you visited them.

IMPORTING ArcGIS TYPE LIBRARIES

To reference ArcGIS interfaces, types, and objects, you will need to import the definitions into Visual C++ types. The *#import* command automates the creation of the necessary files required by the compiler. The *#import* was developed to support Direct-To-Com. When importing ArcGIS library types, there are a number of parameters that must be passed.

```
#pragma warning(push)
#pragma warning(disable : 4192) /* Ignore warnings for types that are
                                   duplicated in win32 header files. */
#pragma warning(disable : 4146) /* Ignore warnings for use of minus on
                                   unsigned types. */

#import "\Program Files\ArcGIS\com\esriSystem.olb"
                            /* Type library to generate C++ wrappers. */ \
    raw_interfaces_only,    /* Don't add raw_ to method names. */ \
    raw_native_types,       /* Don't map to DTC smart types. */ \
    no_namespace,           /* Don't wrap with C++ name space. */ \
    named_guids,            /* Named guids and declspecs. */ \
    exclude("OLE_COLOR", "OLE_HANDLE", "VARTYPE")
                            /* Exclude conflicting types. */
#pragma warning(pop)
```

The main use of *#import* is to create C++ code for interface definitions and GUID constants (LIBID, CLSID, and IID) and to define smart pointers. The exclude ("OLE_COLOR", "OLE_HANDLE", "VARTYPE") is required because Windows defines these to be unsigned longs, which conflicts with the ArcGIS definition of long—this was required to support Visual Basic as a client of ArcObjects, since Visual Basic has no support for unsigned types. There are no issues with excluding these.

You can view the code generated by *#import* in the type library header (.tlh) files, which are similar in format to a .h file. You may also find a type library implementation (.tli) file, which corresponds to a .cpp file. These files can be large but are only regenerated when the type libraries change.

There are many type libraries at ArcGIS 9 for different functional areas, you can start by importing those that contain the definitions that you require. However *#import* does not automatically include all other definitions that the imported type library requires. For example, when importing the type library esriGeometry, it will contain references to types that are defined in esriSystem, so esriSystem must be imported before esriGeometry.

A complete list of library dependencies can be found in the Overview topic for each library.

Choosing the minimum set of type libraries helps reduce compilation time, although this is not always significant. Here are some steps to help determine the minimum number of type libraries required:

1. Do a compilation and look at the "missing type definition" errors generated from code, for example, ICommand not found.

2. Place a *#import* statement for the library you need a reference for into your stdafx.h file. Use the LibraryLocator utility or component help to assist in this task.

3. Compile the project a second time.

4. The compiler will issue errors for types it cannot resolve in the imported type libraries; these are typically type definitions, such as WKSPoint or interfaces that are inherited into other interfaces. For example, if working with geometry objects such as points, start by importing esriGeometry. The compiler will issue various error messages, such as:
   ```
   c:\temp\sample\debug\esrigeometry.tlh(869) : error C2061: syntax error :
   identifier 'WKSPoint'
   ```
   Looking up the definition of WKSPoint, you see it is defined in esriSystem. Therefore, importing esriSystem before esriGeometry will resolve all these issues.

Below is a typical list of imports for working with the ActiveX controls.

```
#pragma warning(push)
#pragma warning(disable : 4192) /* Ignore warnings for types that are
                                   duplicated in win32 header files. */
#pragma warning(disable : 4146) /* Ignore warnings for use of minus on
                                   unsigned types. */
```

```
     #import "\Program Files\ArcGIS\com\esriSystem.olb" raw_interfaces_only,
raw_native_types, no_namespace, named_guids, exclude("OLE_COLOR",
"OLE_HANDLE", "VARTYPE")
     #import "\Program Files\ArcGIS\com\esriSystemUI.olb" raw_interfaces_only,
raw_native_types, no_namespace, named_guids
     #import "\Program Files\ArcGIS\com\esriGeometry.olb" raw_interfaces_only,
raw_native_types, no_namespace, named_guids
     #import "\Program Files\ArcGIS\com\esriDisplay.olb" raw_interfaces_only,
raw_native_types, no_namespace, named_guids
     #import "\Program Files\ArcGIS\com\esriOutput.olb" raw_interfaces_only,
raw_native_types, no_namespace, named_guids
     #import "\Program Files\ArcGIS\com\esriGeoDatabase.olb"
raw_interfaces_only, raw_native_types, no_namespace, named_guids
     #import "\Program Files\ArcGIS\com\esriCarto.olb" raw_interfaces_only,
raw_native_types, no_namespace, named_guids

     // Some of the Engine controls
     #import "\Program Files\ArcGIS\bin\TOCControl.ocx" raw_interfaces_only,
raw_native_types, no_namespace, named_guids
     #import "\Program Files\ArcGIS\bin\ToolbarControl.ocx"
raw_interfaces_only, raw_native_types, no_namespace, named_guids
     #import "\Program Files\ArcGIS\bin\MapControl.ocx" raw_interfaces_only,
raw_native_types, no_namespace, named_guids
     #import "\Program Files\ArcGIS\bin\PageLayoutControl.ocx"
raw_interfaces_only, raw_native_types, no_namespace, named_guids

     // Additionally for 3D controls
#import "\Program Files\ArcGIS\com\esri3DAnalyst.olb" raw_interfaces_only,
raw_native_types, no_namespace, named_guids
#import "\Program Files\ArcGIS\com\esriGlobeCore.olb" raw_interfaces_only,
raw_native_types, no_namespace, named_guids
#import "\Program Files\ArcGIS\bin\SceneControl.ocx" raw_interfaces_only,
raw_native_types, no_namespace, named_guids
#import "\Program Files\ArcGIS\bin\GlobeControl.ocx" raw_interfaces_only,
raw_native_types, no_namespace, named_guids
```

A similar issue arises when writing IDL that contains definitions from other type libraries. In this situation, use importlib just after the library definition. For example, writing an external command for ArcMap would require you to create a COM object implementing *ICommand*. This definition is in *esriSystemUI* and is imported into the IDL as follows:

```
library WALKTHROUGH1CPPLib
{
  importlib("stdole32.tlb");
  importlib("stdole2.tlb");
  importlib("\Program Files\ArcGIS\com\esriSystemUI.olb");

  coclass ZoomIn
   {
    [default] interface IUnknown;
    interface ICommand;
   }
};
```

ATL AND THE ActiveX CONTROLS

For a general discussion of ATL, see the earlier section 'ATL in Brief'.

This section covers how to use ATL to add controls to a dialog box. Although ATL is focused on providing COM support, it also supplies some useful Windows programming wrapper classes. One of the most useful is CWindow, a wrapper around a window handle (HWND). The method names on CWindow correspond to the Win32 API functions. For example:

```
HWND buttonHWnd = GetDlgItem( IDC_BUTTON1 );        // Get window handle of
                                                       button.
CWindow myButtonWindow( buttonHWnd );               // Attach window handle
                                                       to CWindow class.
myButtonWindow.SetWindowText(_T("Button Title"));   // Win32 function to
                                                       change button caption
```

CWindow is a generic wrapper for all window handles, so for specific Windows messages to window common controls, such as buttons, tree views, or edit boxes, one approach is to send window messages direct to the window, for example:

```
// Set button to be checked (pushed in or checkmarked, depending on button style)
myButtonWindow.SendMessage(BM_SETCHECK, BST_CHECKED);
```

However, there are some wrapper classes for these standard window common controls in a header file *atlcontrols.h*. This is available as part of an ATL sample ATLCON supplied in MSDN. See the article "*HOWTO: Using Class Wrappers to Access Windows Common Controls in ATL*". This header file is an early version of Windows Template Libraries (WTL), available for download from Microsoft.

The Visual Studio Resource Editor can be used to design and position windows common controls and ActiveX® controls on a dialog box. To create and manipulate the dialog box, a C++ class is typically created that inherits from *CAxDialogImpl*. This class provides the plumbing to create and manage the ActiveX control on a window. The ATL wizard can be used to supply the majority of the boilerplate code. The steps to create a dialog box and add an ActiveX control in an ATL project are discussed below.

1. Click the menu comman Insert/New ATL Object.

2. Click the Miscellaneous category, then click the Dialog object.

3. A dialog box resource and a class inheriting from CAxDialogImpl will be added to your project.

4. Right-click the dialog box in resource view and click Insert ActiveX Control. This will display a list of available ActiveX controls.

5. Double-click a control in the list to add that control to the dialog box.

6. Right-click the control and click Properties to set the control's design-time properties.

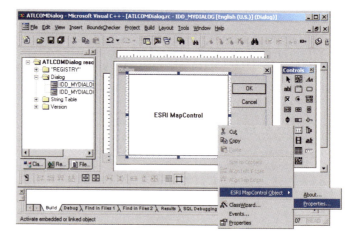

Make sure dialog boxes that host ActiveX controls inherit from CAxDialogImpl and not CDialogImpl. If this mistake is made, the DoModal method of the dialog box simply exits with no obvious cause.

Make sure applications that use windows common controls, such as treeview, correctly call InitCommonControlsEx to load the window class. Otherwise, the class will not function correctly.

Accessing a control on a dialog box through a COM interface

To retrieve a handle to the control that is hosted on a form, use the *GetDlgControl* ATL method that is inherited from *CAxDialogImpl* to take a resource ID and return the underlying control pointer:

Make sure applications using COM objects call CoInitialize. This initializes COM in the application. Without this call, any CoCreate calls will fail.

```
ITOCControlPtr ipTOCControl;
GetDlgControl(IDC_TOCCONTROL1, IID_ITOCControl, (void**) &ipTOCControl);
ipTOCControl->AboutBox();
```

Listening to events from a control

The simplest way to add events is to use the class wizard. Simply right-click the control and choose Events. Next, click the resource ID of the control, then click the event (for example, *OnMouseDown*). Next click Add Handler. Finally, ensure the dialog box begins listening to events by adding AtlAdviseSinkMap(this,TRUE) to the *OnInitDialog*. To finish listening to events, add a message handler for *OnDestroy* and add a call to AtlAdviseSinkMap(this, FALSE).

For a detailed discussion on handling events in ATL, see the later section 'Handling COM events in ATL'.

Creating a control at run time

The CAxWindow class provides a mechanism to create and host ActiveX controls in a similar manner to any other window class. This may be desirable if the parent window of the control is also created at runtime.

```
AtlAxWinInit();
CAxWindow wnd;
//m_hWnd is the parent window handle.
//rect is the size of ActiveX control in client coordinates.
//IDC_MYCTL is a unique ID to identify the controls window.
RECT rect = {10,10,400,300};
wnd.Create(m_hWnd, rect, _T("esriReaderControl.ReaderControl"),
WS_CHILD|WS_VISIBLE, 0, IDC_MYCTL);
```

Setting the buddy control property

The *ToolbarControl* and *TOCControl* need to be associated with a "buddy" control on the dialog box. This is typically performed in the *OnInitDialog* windows message handler of a dialog box.

```
LRESULT CEngineControlsDlg::OnInitDialog(UINT uMsg, WPARAM wParam, LPARAM
lParam, BOOL& bHandled)
{
  // Get the Control's interfaces into class member variables.
  GetDlgControl(IDC_TOOLBARCONTROL, IID_IToolbarControl, (void **)
&m_ipToolbarControl);
  GetDlgControl(IDC_TOCCONTROL, IID_ITOCControl, (void **) &m_ipTOCControl);
  GetDlgControl(IDC_PAGELAYOUTCONTROL, IID_IPageLayoutControl, (void **)
&m_ipPageLayoutControl);

  // Connect to the controls.
  AtlAdviseSinkMap(this, TRUE);

  // Set buddy controls.
  m_ipTOCControl->SetBuddyControl(m_ipPageLayoutControl);
  m_ipToolbarControl->SetBuddyControl(m_ipPageLayoutControl);

  return TRUE;
}
```

Known limitations of Visual Studio C++ Resource Editor and ArcGIS ActiveX controls

Disabled buddy property on property page

In Visual Studio C++ you cannot set the Buddy property of the *TOCControl* and the *ToolbarControl* through the General property page. Visual C++ does not support controls finding other controls at design time. However, this step can be performed in code in the *OnInitDialog* method.

ToolbarControl not resized to the height of one button

In other environments (Visual Basic 6, .NET) the ToolbarControl will automatically resize to be one button high. However, in Visual Studio C++ 6 it can be any size. In MFC and ATL the ActiveX host classes do not allow controls to determine their own size.

Design-time property pages disappearing when displaying context-sensitive help

When viewing the controls property page at design time, right-clicking and clicking What's This? will cause the help tip to display; however, the property pages will then close. This is a limitation of the Visual Studio floating windows combined with the floating tip window from HTML help. Clicking the Help button provides the same text for the whole property page.

MFC AND THE ActiveX CONTROLS

There are many choices for how to work with ArcGIS ActiveX Controls in Visual C++, the first of which is what framework to use to host the controls (for example, ATL or MFC). A second decision is where the control will be hosted

(Dialog, MDI, and so forth). This section discusses MFC and hosting the control on a dialog box.

Creating an MFC dialog box-based application

If you do not have a dialog box in your application or component, here are the steps to create an MFC dialog box application.

1. Launch Visual Studio C++ 6 and click New.

2. Click the Projects tab and choose MFC AppWizard (exe). Enter the Project name and location and click OK

3. For Step 1 of the wizard: From the buttons change the application type to Dialog Based. Click Next.

4. For Step 2 of the wizard: The default project features are fine, although you can uncheck AboutBox to simplify the application. Ensure that the option to support ActiveX Controls is checked. Click Next.

5. For Step 3 of the wizard: The default settings on this page are fine. The MFC DLL is shared. Click Next.

6. For Step 4 of the wizard: This shows you what the wizard will generate. Click Finish.

You should now have a simple dialog box-based application. In the resource view, you will see "TODO: Place Dialog Controls Here". You can place buttons, list boxes, and so forth, in this dialog box. The dialog box can also host ActiveX controls; there are two approaches to doing this, as discussed below. You can also compile and run this application.

Hosting controls on an MFC dialog box and accessing them using IDispatch

Inserting ActiveX controls on a dialog box in Visual Studio C++ design time. The TOCControl and MapControl have been added to the dialog box. The ToolbarControl is next.

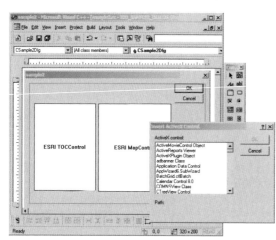

1. Right-click the MFC dialog box and click Insert ActiveX control.

2. Double-click a control from the list box. The control appears on the dialog box with a default size.

3. Size and position the control as required.

4. Repeat Steps 1 through 3 for each control.

5. You can right-click the control and choose Properties to set the control's design-time properties.

6. To access the control in code, you will need ArcGIS interface definitions for IMapControl, for example. To do this, use the *#import* command in your stdafx.h file. See the section 'Importing ArcGIS type libraries' on how to do this.

7. MFC provides control hosting on a dialog box; this will translate Windows messages, such as WM_SIZE into appropriate control method calls. However, to be able to make calls on a control, there are a few steps you must perform to go from a resource ID to a controls interface. The following code illustrates setting the TOCControl's Buddy to be the MapControl:

```
// Code to set the Buddy property of the TOCControl to be the MapControl.

// Get a pointer to the PageLayoutControl and TOCControl
IPageLayoutControlPtr ipPageLayoutControl;
GetDlgControl(IDC_PAGELAYOUTCONTROL1, IID_IPageLayoutControl, (void**)
          &ipPageLayoutControl);
ITOCControlPtr ipTOCControl;
GetDlgControl(IDC_TOCCONTROL1, IID_ITOCControl, (void**) &ipTOCControl);

// Get the IDispatch of the PageLayoutControl.
IDispatchPtr ipBuddyDisp = ipPageLayoutControl;

// Set the TOCControls Buddy to the map control.
ipTOCControl->putref_Buddy(ipBuddyDisp);
```

8. To catch events from the controls, double click the control on the form and supply the name of a method to be called. By default, the wizard will add an extra word "On" to the beginning of the event handler. Remove this to avoid the event handler's name from becoming "OnOnMouseDownMapcontrol1". The wizard will then automatically generate the necessary MFC sink map macros to listen to events.

Adding controls to an MFC dialog box using IDispatch wrappers

As all ActiveX controls support *IDispatch*, this is the typical approach to add an ActiveX control to an MFC project:

1. Click Project, click Add, then click Components and Controls.

2. Click Registered ActiveX Controls.

3. Double-click to select a control (for example, ESRI TOCControl), then click OK to insert a component. Click OK to generate wrappers. This will add an icon for the control to the Controls toolbar in Visual Studio.

4. Additional source files are added to your project (for example, toccontrol.cpp and toccontol.h). These files contain a wrapper class (for example, CTOCControl) to provide methods and properties to access the control. This class will invoke the control through the *IDispatch* calling mechanism. Note that *IDispatch* does incur some performance overhead to package parameters when making method and property calls. The wrapper class inherits from a MFC *CWnd* class that hosts an ActiveX control.

5. Repeat Steps 1 through 4 to add each control to the project's Controls toolbar.

6. Select a control from the Controls toolbar and drag it onto the dialog box.

7. Right-click the control and click Properties. This will allow design-time properties to be set on the control. NOTE: In Visual Studio C++, you cannot set the Buddy property of the *TOCControl* and the *ToolbarControl*.

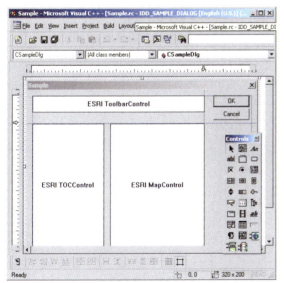

The design environment showing the TOCControl, MapControl, and ToolbarControl has been added to the Controls toolbar and to the dialog box.

This environment does not support controls finding other controls at design time. However, this step can be performed in code using the *OnInitDialog* method.

```
// Note no addref performed with GetControlUnknown, so no need to release
// this pointer.
LPUNKNOWN pUnk = m_mapcontrol.GetControlUnknown();
LPDISPATCH pDisp =
0;pUnk->QueryInterface(IID_IDispatch, (void **) &pDisp);

// Set TOCControls buddy to be MapControl.
m_toccontrol.SetRefBuddy(pDisp);
pDisp->Release();
```

8. Right-click the control and choose Class Wizard to launch the class wizard. Click the Member Variables tab and click the resource ID corresponding to the control to give the control member variable name. The dialog box class member variable can now be used to invoke methods and properties on the control.

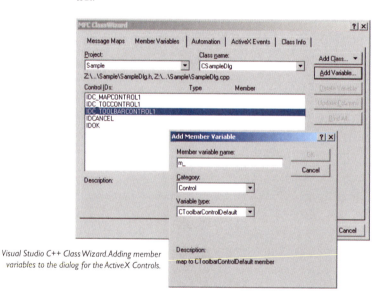

Visual Studio C++ Class Wizard. Adding member variables to the dialog for the ActiveX Controls.

Do not use the method GetIDispatch (inherited from MFC's CCmdTarget) on the wrapper classes; it is intended for objects implementing IDispatch and not the wrapper classes that are calling IDispatch. Instead, to get a control's IDispatch, use m_mapcontrol.GetControlUnknown() and QueryInterface to IDispatch. See the above example of setting the Buddy property.

9. To catch control events, click the Message Maps tab of the class wizard and choose the resource ID of the control. In the list of messages, click the event to catch—for example, *OnBeginLabelEdit*. Double-click this event and a handler for it will be added to your dialog box class. By default, the wizard will add an extra word "On" to the beginning of the event handler. Remove this to avoid the event handler name becoming *OnOnBeginLabelEditToccontrol1*.

HANDLING COM EVENTS IN ATL

Here is a summary of terminology used here when discussing COM events in Visual C++ and ATL.

Inbound interface—This is the normal case where a COM object implements a predefined interface.

Outbound interface—This is an interface of methods that a COM object will fire at various times. For example, the *Map* coclass will fire an event on the *IActiveViewEvents* in response to changes in the map.

Event source—The source COM object will fire events to an outbound interface when certain actions occur. For example the Map coclass is a source of *IActiveViewEvents* and will fire the *IActiveViewEvents::ItemAdded* event when a new layer is added to the map. The source object can have any number of clients, or event sink objects, listening to events. Also, a source object may have more than one outbound interface; for example, the *MapCoClass* also fires events on an *IMapEvents* interface. An event source will typically declare its outbound interfaces in IDL with the *[source]* tag.

Event sink—A COM object that listens to events is said to be a "sink" for events. The sink object implements the outbound interface; this is not always advertised in the type libraries because the sink may listen to events internally. An event sink typically uses the *connection point* mechanism to register its interest in the events of a source object.

Connection point—COM objects that are the source of events typically use the connection point mechanism to allow sinks to hook up to a source. The connection point interfaces are the standard COM interfaces *IConnectionPointContainer* and *IConnectionPoint*.

Fire event—When a source object needs to inform all the sinks of a particular action, the source is said to "fire" an event. This results in the source iterating all the sinks and making the same method call on each. For example, when a layer is added to a map, The *Map* coclass is said to fire the *ItemAdded* event. So all the objects listening to the *Map*'s outbound *IActiveViewEvents* interface will be called on their implementation of the *ItemAdded* method.

Advise and unadvise events—To begin receiving events a sink object is said to "advise" a source object that it needs to receive events. When events are no longer required, the sink will "unadvise" the source.

The ConnectionPoint mechanism

The source object implements the *IConnectionPointContainer* interface to allow sinks to query a source for a specific outbound interface. The following steps are performed to begin listening to an event. ATL implements this with the *AtlAdvise* method.

1. The sink will QI the source object's *IConnectionPointContainer* and call *FindConnectionPoint* to supply an interface ID for outbound interfaces. To be able to receive events, the sink object must implement this interface.

2. The source may implement many outbound interfaces and will return a pointer to a specific connection point object implementing *IConnectionPoint* to represent one outbound interface.

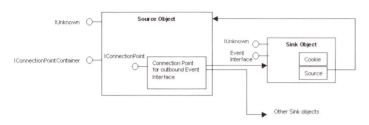

3. The sink calls *IConnectionPoint::Advise*, passing a pointer to its own *IUnknown* implementation. The source will store this with any other sinks that may be listening to events. If the call to *Advise* was successful, the sink will be given an identifier—a simple unsigned long value, called a cookie—to give back to the source at a later point when it no longer needs to listen to events.

Connection point mechanism for hooking source to sink objects

The connection is now complete; methods will be called on any listening sinks by the source. The sink will typically hold onto an interface pointer to the source, so when a sink has finished listening it can be released from the source object by calling *IConnectionPoint::Unadvise*. This is implemented with *AtlUnadvise*.

IDispatch events versus pure COM events

An outbound interface can be a pure dispatch interface. This means instead of the source calling directly onto a method in a sink, the call is made via the *IDispatch::Invoke* mechanism. The *IDispatch* mechanism has a performance overhead to package parameters compared to a pure vtable COM call. However there are some situations where this must be used. ActiveX controls must implement their default outbound interface as a pure *IDispatch* interface; for example, *IMapControlEvents2* is a pure dispatch interface. Second, Microsoft Visual Basic 6 can only be a source of pure *IDispatch* events. The connection point mechanism is the same as for pure COM mechanisms, the main difference being in how the events are fired.

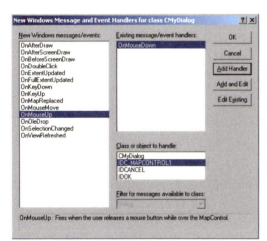

ATL provides some macros to assist with listening to *IDispatch* events; this is discussed on MSDN under 'Event Handling and ATL'. There are two templates available, *IDispEventImpl* and *IDispEventSimpleImpl*, that are discussed in the following sections.

Visual Studio C++ Class Wizard. Adding event handler to an ActiveX control on a dialog box.

Using IDispEventImpl to listen to events

The ATL template *IDispEventImpl* will use a type library to "crack" the *IDispatch* calls and process the arguments into C++ method calls. The Visual Studio Class wizard can provide this mechanism automatically when adding an ActiveX control to a dialog box. Right-click the control and click Events. In the Class wizard, choose the resource ID of the control, then choose the event, then click Add Handler.

There is a bug in the wizard: it does not add the advise and unadvise code to the dialog box. To fix this issue, add a message handler for OnDestroy. Then in the OnInitDialog handler, call AtlAdviseSinkMap with a TRUE second parameter to begin listening to events. Place a corresponding call to AtlAdviseSinkMap (with FALSE as the second parameter) in the OnDestroy handler. This is discussed further in the MSDN article "BUG:ActiveX Control Events Are Not Fired in ATL Dialog (Q190530)".

The following code illustrates the event handling code added by the wizard, with some modifications to ensure advise and unadvise are performed.

```
#pragma once

#include "resource.h"        // Main symbols
#include <atlhost.h>

//////////////////////////////////////////////////////////////////////
// CMyDialog
class CMyDialog :
 public CAxDialogImpl<CMyDialog>,
 public IDispEventImpl<IDC_MAPCONTROL1, CMyDialog>
{

public
  enum { IDD = IDD_MYDIALOG };

BEGIN_MSG_MAP(CMyDialog)
 MESSAGE_HANDLER(WM_INITDIALOG, OnInitDialog)

 // Add a handler to ensure event unadvise occurs.
 MESSAGE_HANDLER(WM_DESTROY, OnDestroy)

 COMMAND_ID_HANDLER(IDOK, OnOK)
 COMMAND_ID_HANDLER(IDCANCEL, OnCancel)
END_MSG_MAP()

 LRESULT OnInitDialog(UINT uMsg, WPARAM wParam, LPARAM lParam, BOOL&
bHandled)
  {
   // Calls IConnectionPoint::Advise() for each control on the dialog box
    // with sink map entry
   AtlAdviseSinkMap(this, TRUE);
   return 1;  // Let the system set the focus.
  }

 LRESULT OnDestroy(UINT uMsg, WPARAM wParam, LPARAM lParam, BOOL& bHandled)
  {
   // Calls IConnectionPoint::Unadvise() for each control on the dialog box
    // with sink map entry
   AtlAdviseSinkMap(this, FALSE);
   return 0;
  }

 LRESULT OnOK(WORD wNotifyCode, WORD wID, HWND hWndCtl, BOOL& bHandled)
  {
   EndDialog(wID);
   return 0;
  }

 LRESULT OnCancel(WORD wNotifyCode, WORD wID, HWND hWndCtl, BOOL& bHandled)
  {
   EndDialog(wID);
```

The following issues with events are documented on the MSDN Knowledge Base when using IDispEventImpl. Fixes to ATL code are shown in MSDN for these issues; however, it is not always desirable to modify or copy ATL header files. In this case, the IDispEventSimpleImpl can be used instead.
BUG: Events Fail in ATL Containers when Enum Used as Event Parameter (Q237771)
BUG: IDispEventImpl Event Handlers May Give Strange Values for Parameters (Q241810)

```
  return 0;
}

// ATL callback from SinkMap entry
VOID __stdcall OnMouseDownMapcontrol1(LONG button, LONG shift, LONG x,
LONG y, DOUBLE mapX, DOUBLE mapY)
{
  MessageBox(_T("MouseDown!"));
}

BEGIN_SINK_MAP(CMyDialog)
  // Make sure the Event Handlers have __stdcall calling convention.
  // The 0x1 is the Dispatch ID of the OnMouseDown method.
  SINK_ENTRY(IDC_MAPCONTROL1, 0x1, OnMouseDownMapcontrol1)
END_SINK_MAP()
};
```

Using IDispEventSimpleImpl to listen to events

As the name of this template suggests, it is a simpler version of *IDispEventImpl*. The type library is no longer used to turn the *IDispatch* arguments into a C++ method call. While this may be a simpler implementation, it now requires the developer to supply a pointer to a structure describing the format of the event parameters. This structure is typically placed in the .cpp file. For example, here is the structure describing the parameters of an *OnMouseDown* event for the *MapControl*:

```
_ATL_FUNC_INFO g_ParamInfo_MapControl_OnMouseDown =
{
  CC_STDCALL,                            // Calling convention
  VT_EMPTY,                              // Return type
  6,                                     // Number of arguments
  {VT_I4, VT_I4, VT_I4, VT_I4, VT_R8, VT_R8}  // VariantArgument types
};
```

The header file now inherits from *IDispEventSimpleImpl* and uses a different macro, SINK_ENTRY_INFO, in the SINK_MAP. Also, the events interface ID is required; *#import* can be used to define this symbol. Note that a dispatch interface is normally prefixed with DIID instead of IID.

See the 'Importing ArcGIS type libraries' section earlier in this appendix for an explanation of #import.

```
#pragma once

#include "resource.h"      // Main symbols
#include <atlhost.h>

// reference to structure defining event parameters
extern _ATL_FUNC_INFO g_ParamInfo_MapControl_OnMouseDown;

/////////////////////////////////////////////////////////////////////////
// CMyDialog2
class CMyDialog2 :
  public CAxDialogImpl<CMyDialog2>,
  public IDispEventSimpleImpl<IDC_MAPCONTROL1, CMyDialog2,
&DIID_IMapControlEvents2>
```

```
{
public:

// Message handler code removed, it is the same as CMyDialog using
IDispEventSimple

BEGIN_SINK_MAP(CMyDialog2)
  // Make sure the Event Handlers have __stdcall calling convention.
  // The 0x1 is the Dispatch ID of the OnMouseDown method.
  SINK_ENTRY_INFO(IDC_MAPCONTROL1,      // ID of event source
       DIID_IMapControlEvents2,         // interface to listen to
       0x1,                             // dispatch ID of MouseDown
       OnMapControlMouseDown,           // method to call when event arrives
       &g_ParamInfo_MapControl_OnMouseDown) // parameter info for method call

END_SINK_MAP()
};
```

Listening to more than one IDispatch event interface on a COM object

If a single COM object needs to receive events from more than one *IDispatch* source, then this can cause compiler issues with ambiguous definitions of the *DispEventAdvise* method. This is not normally a problem in a dialog box, as *AtlAdviseSinkMap* will handle all the connections. The ambiguity can be avoided by introducing different typedefs each time *IDispEventSimpleImpl* is inherited. The following example illustrates a COM object called *CListen*, which is a sink for dispatch events from a *MapControl* and a *PageLayoutControl*.

```
#pragma once

#include "resource.h"      // Main symbols

// This is the parameter information
extern _ATL_FUNC_INFO g_ParamInfo_MapControl_OnMouseDown;
extern _ATL_FUNC_INFO g_ParamInfo_PageLayoutControl_OnMouseDown;

//
// Define some typedefs of the dispatch template.
//
class CListen; // Forward definition

typedef IDispEventSimpleImpl<0, CListen, &DIID_IMapControlEvents2>
     IDispEventSimpleImpl_MapControl;

typedef IDispEventSimpleImpl<1, CListen, &DIID_IPageLayoutControlEvents>
     IDispEventSimpleImpl_PageLayoutControl;

/////////////////////////////////////////////////////////////////////////
// CListen

class ATL_NO_VTABLE CListen :
```

```
            public CComObjectRootEx<CComSingleThreadModel>,
            public CComCoClass<CListen,&CLSID_Listen>,
            public IDispEventSimpleImpl_MapControl,
            public IDispEventSimpleImpl_PageLayoutControl,
            public IListen
{
public:
   CListen()
   {
   }

DECLARE_REGISTRY_RESOURCEID(IDR_LISTEN)

DECLARE_PROTECT_FINAL_CONSTRUCT()

BEGIN_COM_MAP(CListen)
  COM_INTERFACE_ENTRY(IListen)
END_COM_MAP()

// Associated source and dispatchID to a method call
BEGIN_SINK_MAP(CListen)
   SINK_ENTRY_INFO(0,                  // ID of event source
            DIID_IMapControlEvents2,   // Interface to listen to
            0x1,                       // Dispatch ID to receive
            OnMapControlMouseDown,     // Method to call when event arrives
            &g_ParamInfo_MapControl_OnMouseDown) // parameter info for
                                                 // method call

   SINK_ENTRY_INFO(1,
            DIID_IPageLayoutControlEvents,
               0x1,
            OnPageLayoutControlMouseDown,
            &g_ParamInfo_PageLayoutControl_OnMouseDown)
END_SINK_MAP()

// IListen
public:
   STDMETHOD(SetControls)(IUnknown* pMapControl, IUnknown*
pPageLayoutControl);
   STDMETHOD(Clear)();

private:
   void __stdcall OnMapControlMouseDown(long button, long shift, long x, long
y, double mapX, double mapY);
   void __stdcall OnPageLayoutControlMouseDown(long button, long shift, long
x, long y, double pageX, double pageY);

   IUnknownPtr m_ipUnkMapControl;
   IUnknownPtr m_ipUnkPageLayoutControl;
};
```

The implementation of CListen contains the following code to start listening to the controls; the typedef avoids the ambiguity of the *DispEventAdvise* implementation.

```
// Start listening to the MapControl.
IUnknownPtr ipUnk = pMapControl;
HRESULT hr = IDispEventSimpleImpl_MapControl::DispEventAdvise(ipUnk);
if (SUCCEEDED(hr))
  m_ipUnkMapControl = ipUnk;   // Store pointer to MapControl for Unadvise.

// Start listening to the PageLayoutControl.
ipUnk = pPageLayoutControl;
hr = IDispEventSimpleImpl_PageLayoutControl::DispEventAdvise(ipUnk);
if (SUCCEEDED(hr))
  m_ipUnkPageLayoutControl = ipUnk; // Store pointer to PageLayoutControl
                                    //  for Unadvise.
```

The implementation of CListen also contains the following code to UnAdvise and stop listening to the controls.

```
// Stop listening to the MapControl.
if (m_ipUnkMapControl!=0)
  IDispEventSimpleImpl_MapControl::DispEventUnadvise(m_ipUnkMapControl);
m_ipUnkMapControl = 0;

if (m_ipUnkPageLayoutControl!=0)
  IDispEventSimpleImpl_PageLayoutControl::DispEventUnadvise(m_ipUnkPageLayoutControl);
m_ipUnkPageLayoutControl= 0;
```

Creating a COM events source

For an object to be a source of events, it will need to provide an implementation of *IConnectionPointContainer* and a mechanism to track which sinks are listening to which *IConnectionPoint* interfaces. ATL provides this through the *IConnectionPointContainerImpl* template. In addition, ATL provides a wizard to generate code to fire *IDispatch* events for all members of a given dispatch events interface. Below are the steps to modify an ATL COM coclass to support a connection point:

1. First ensure that your ATL coclass has been compiled at least once. This will allow the wizard to find an initial type library.

2. In Class view, right-click the COM object and click Implement Connection Point.

3. Either use a definition of events from the IDL in the project, or click Add Type Lib for browse to another definition.

4. Check the outbound interface to be implemented in the coclass.

5. Clicking OK will modify your ATL class and generate the proxy classes in a header file, with a name ending in CP, for firing events.

If the wizard fails to run, use the following example, which illustrates a coclass that is a source of *ITOCControlEvents*, a pure dispatch interface.

```cpp
#pragma once

#include "resource.h"      // Main symbols
#include "TOCControlCP.h"  // Include generated connection point class
                           // for firing events.


//////////////////////////////////////////////////////////////////////////
// CMyEventSource
class ATL_NO_VTABLE CMyEventSource :
  public CComObjectRootEx<CComSingleThreadModel>,
  public CComCoClass<CMyEventSource,&CLSID_MyEventSource>,
  public IMyEventSource,
  public CProxyITOCControlEvents< CMyEventSource >,    // Generated
                                        // ConnectionPoint class
  public IConnectionPointContainerImpl< CMyEventSource > // Implementation
                                        // of Connection point Container
{
public:
  CMyEventSource()
  {
  }

DECLARE_REGISTRY_RESOURCEID(IDR_MYEVENTSOURCE)

DECLARE_PROTECT_FINAL_CONSTRUCT()

BEGIN_COM_MAP(CMyEventSource)
  COM_INTERFACE_ENTRY(IMyEventSource)
  COM_INTERFACE_ENTRY(IConnectionPointContainer) // Allow QI to this
                                        // interface.
END_COM_MAP()
```

```
// List of available connection points
BEGIN_CONNECTION_POINT_MAP(CMyEventSource)
  CONNECTION_POINT_ENTRY(DIID_ITOCControlEvents)
END_CONNECTION_POINT_MAP()
};
```

The connection point class (*TOCControlEventsCP.h* in the above example) contains code to fire an event to all sink objects on a connection point.

There is one method in the class for each event beginning "Fire". Each method will build a parameter list of variants to pass as an argument to the dispatch Invoke method. Each sink is iterated, and a pointer to the sink is stored in a vector m_vec member variable inherited from *IConnectionPointContainerImpl*. Note that m_vec can contain pointers to zero; and this must be checked before firing the event.

```
template <class T>
class CProxyITOCControlEvents : public IConnectionPointImpl<T,
&DIID_ITOCControlEvents, CComDynamicUnkArray>
{
public:
  VOID Fire_OnMouseDown(LONG button, LONG shift, LONG x, LONG y)
  {
    // Package each of the parameters into an IDispatch argument list.
    T* pT = static_cast<T*>(this);
    int nConnectionIndex;
    CComVariant* pvars = new CComVariant[4];
    int nConnections = m_vec.GetSize();

    // Iterate each sink object.
    for (nConnectionIndex = 0; nConnectionIndex < nConnections;
nConnectionIndex++)
      {
        pT->Lock();
        CComPtr<IUnknown> sp = m_vec.GetAt(nConnectionIndex);
        pT->Unlock();
        IDispatch* pDispatch = reinterpret_cast<IDispatch*>(sp.p);

        // Note m_vec can contain 0 entries so it is important to check for
        // this.
        if (pDispatch != NULL)
         {
          // Build up the argument list.
          pvars[3] = button;
          pvars[2] = shift;
          pvars[1] = x;
          pvars[0] = y;
          DISPPARAMS disp = { pvars, NULL, 4, 0 };

          // Fire the dispatch method, 0x1 is the DispatchId for MouseDown.
          pDispatch->Invoke(0x1, IID_NULL, LOCALE_USER_DEFAULT,
DISPATCH_METHOD, &disp, NULL, NULL, NULL);
```

```
     }
    }
    delete[] pvars; // Clean up the parameter list.

  }
  VOID Fire_OnMouseUp(LONG button, LONG shift, LONG x, LONG y)
  {
    // ... Other events
```

To fire an event from the source, simply call the Fire_OnMouseDown when required.

A similar approach can be used for firing events to a pure COM (non IDispatch) interface. The wizard will not generate the connection point class, so this must be written by hand; the following example illustrates a class that will fire an *ITOCBuddyEvents::ActiveViewReplaced* event; *ITOCBuddyEvents* is a pure COM, non-IDispatch interface. The key difference is that there is no need to package the parameters. A direct method call can be made.

```
template < class T >
class CProxyTOCBuddyEvents : public IConnectionPointImpl< T,
&IID_ITOCBuddyEvents, CComDynamicUnkArray >
{
  // This class based on the ATL-generated connection point class.
public:
  void Fire_ActiveViewReplaced(IActiveView* pNewActiveView)
  {
    T* pT = static_cast< T* >(this);
    int nConnectionIndex;
    int nConnections = this->m_vec.GetSize();
    for (nConnectionIndex = 0; nConnectionIndex < nConnections;
nConnectionIndex++)
    {
      pT->Lock();
      CComPtr< IUnknown > sp=this->m_vec.GetAt(nConnectionIndex);
      pT->Unlock();
      ITOCBuddyEvents* pTOCBuddyEvents = reinterpret_cast< ITOCBuddyEvents*
>(sp.p);
      if (pTOCBuddyEvents)
        pTOCBuddyEvents->ActiveViewReplaced(pNewActiveView);
    }
  }
};
```

IDL declarations for an object that supports events

When an object is exported to a type library, the event interfaces are declared by using the *[source]* tag against the interface name. For example, an object that fires *ITOCBuddyEvents* declares

```
[source] interface ITOCBuddyEvents;
```

If the outbound interface is a dispatch events interface, *dispinterface* is used instead of *interface*. Additionally, a coclass can have a default outbound interface;

this is specified with the *[default]* tag. Default interfaces are identified by some design environments (for example, Visual Basic 6). Following is the declaration for the default outbound events interface:

```
[default, source] dispinterface IMyEvents2;
```

Event circular reference issues

After a sink has performed an advise on the source, there is typically a COM circular reference. This occurs because the source has an interface pointer to a sink to fire events, and this keeps the sink alive. Similarly, a sink object has a pointer back to the source so it can perform the unadvise at a later point. This keeps the source alive. Therefore, these two objects will never be released and may cause substantial memory leaks. There are a number of ways to tackle this issue:

1. Ensure the advise and unadvise are made on a method or windows message that is guaranteed to happen in pairs and is independent of an object's life cycle. For example, in a coclass that is also receiving windows messages, use the windows messages *OnCreate* (WM_CREATE) and *OnDestroy* (WM_DESTROY) to advise and unadvise.

2. If an ATL dialog box class needs to listen to events, one approach is to make the dialog box a private COM class and implement the events interface directly on the dialog box. ATL allows this without much extra coding. This approach is illustrated below. The dialog box class creates a *CustomizeDialog* coclass and listens to *ICustomizeDialogEvents*. The *OnInitDialog* and *OnDestroy* methods (corresponding to Windows messages) are used to advise and unadvise on the *CustomizeDialog*.

```
class CEngineControlsDlg :
  public CAxDialogImpl<CEngineControlsDlg>,
  public CComObjectRoot, // Make Dialog Class a COM Object as well.
  public ICustomizeDialogEvents // Implement this interface directly on
                                // this object.

CEngineControlsDlg() : m_dwCustDlgCookie(0) {} // Initialize cookie for
                                               // event listening.

// ... Event handlers and other standard dialog code has been removed ...

BEGIN_COM_MAP(CEngineControlsDlg)
  COM_INTERFACE_ENTRY(ICustomizeDialogEvents) // Make sure QI works for
                                              // this event interface.
END_COM_MAP()

  // ICustomizeDialogEvents implementation to receive events on this
  // dialog box.
  STDMETHOD(OnStartDialog)();
  STDMETHOD(OnCloseDialog)();

  ICustomizeDialogPtr    m_ipCustomizeDialog; // The source of events
  DWORD                  m_dwCustDlgCookie;   // Cookie for
                                              // CustomizeDialogEvents
}
```

The dialog box needs to be created like a noncreatable COM object, rather than on the stack as a local variable. This allocates the object on the heap and allows it to be released through the COM reference counting mechanism.

```
// Create dialog class on the heap using ATL CComObject template.
CComObject<CEngineControlsDlg> *myDlg;
CComObject<CEngineControlsDlg>::CreateInstance(&myDlg);

myDlg->AddRef();    // Keep dialog box alive until you're done with it.
myDlg->DoModal();   // Launch the dialog box; when method returns, dialog
                    // box has exited.
myDlg->Release();   // Typically, the refcount now goes to 0 and frees the
                       dialog object.
```

3. Implement an intermediate COM object for use by the sink; this is sometimes called a listener or event helper object. This object typically contains no implementation but simply uses C++ method calls to forward events to the sink object. The listener has its reference count incremented by the source, but the sink's reference count is unaffected. This breaks the cycle, allowing the sink's reference count to reach 0 when all other references are released. As the sink executes its destructor code, it instructs the listener to unadvise and release the source.

An alternative to using C++ pointers to communicate between listener and sink is to use an interface pointer that is a weak reference. That is the listener contains a COM pointer to the sink but does not increment the sink's reference count. It is the responsibility of the sink to ensure that this pointer is not accessed after the sink object has been released.

WHAT IS THE .NET FRAMEWORK?

This section, 'What is the .NET Framework?' summarizes the Microsoft overview of the .NET Framework available online as part of MSDN Library. The complete text is available at http://www.msdn.microsoft.com/library/default.asp?url=/library/en-us/cpguide/html/cpovrintroductiontonetframeworksdk.asp

The .NET Framework is an integral Windows component that supports building and running the next generation of applications and XML Web services. The .NET Framework is designed to fulfill the following objectives:

- Provide a consistent object-oriented programming environment whether object code is stored and executed locally, executed locally but Internet-distributed, or executed remotely.

- Provide a code execution environment that minimizes software deployment and versioning conflicts.

- Provide a code execution environment that guarantees safe execution of code, including code created by an unknown or semitrusted third party.

- Provide a code execution environment that eliminates the performance problems of scripted or interpreted environments.

- Make the developer experience consistent across widely varying types of applications, such as Windows-based applications and Web-based applications.

- Build all communication on industry standards to ensure that code based on the .NET Framework can integrate with any other code.

The .NET Framework has two main components: the common language runtime and the .NET Framework class library. The common language runtime is the foundation of the .NET Framework. You can think of the runtime as an agent that manages code at execution time, providing core services such as memory management, thread management, and remoting, while also enforcing strict type safety and other forms of code accuracy that ensure security and robustness. In fact, the concept of code management is a fundamental principle of the runtime. Code that targets the runtime is known as managed code, while code that does not target the runtime is known as unmanaged code. The class library, the other main component of the .NET Framework, is a comprehensive, object-oriented collection of reusable types that you can use to develop applications ranging from traditional command-line or graphical user interface applications to applications based on the latest innovations provided by ASP.NET, such as Web Forms and XML Web services.

The .NET Framework can be hosted by unmanaged components that load the common language runtime into their processes and initiate the execution of managed code, thereby creating a software environment that can exploit both managed and unmanaged features. The .NET Framework not only provides several runtime hosts but also supports the development of third-party runtime hosts.

For example, ASP.NET hosts the runtime to provide a scalable, server-side environment for managed code. ASP.NET works directly with the runtime to enable ASP.NET applications and XML Web services, both of which are discussed later in this topic.

Internet Explorer is an example of an unmanaged application that hosts the runtime (in the form of a MIME type extension). Using Internet Explorer to host the runtime enables you to embed managed components or Windows Forms controls in HTML documents. Hosting the runtime in this way makes managed mobile code (similar to Microsoft ActiveX controls) possible, but with significant improvements that only managed code can offer, such as semitrusted execution and secure isolated file storage.

The following sections describe the main components and features of the .NET Framework in greater detail.

Features of the common language runtime

The common language runtime manages memory, thread execution, code execution, code safety verification, compilation, and other system services. These features are intrinsic to the managed code that runs on the common language runtime.

Regarding security, managed components are awarded varying degrees of trust, depending on a number of factors that includes their origin, such as the Internet, enterprise network, or local computer. This means that a managed component might or might not be able to perform file access operations, registry access operations, or other sensitive functions, even if it is being used in the same active application.

The runtime enforces code access security. For example, users can trust that an executable embedded in a Web page can play an animation onscreen or sing a song but cannot access their personal data, file system, or network. The security features of the runtime thus enable legitimate Internet-deployed software to be exceptionally feature rich.

The runtime also enforces code robustness by implementing a strict type-and-code-verification infrastructure called the common type system (CTS). The CTS ensures that all managed code is self-describing. The various Microsoft and third-party language compilers generate managed code that conforms to the CTS. This means that managed code can consume other managed types and instances, while strictly enforcing type fidelity and type safety.

In addition, the managed environment of the runtime eliminates many common software issues. For example, the runtime automatically handles object layout and manages references to objects, releasing them when they are no longer being used. This automatic memory management resolves the two most common application errors: memory leaks and invalid memory references.

The runtime also accelerates developer productivity. For example, programmers can write applications in their development language of choice, yet take full advantage of the runtime, the class library, and components written in other languages by other developers. Any compiler vendor who chooses to target the runtime can do so. Language compilers that target the .NET Framework make the features of the .NET Framework available to existing code written in that language, greatly easing the migration process for existing applications.

While the runtime is designed for the software of the future, it also supports software of today and yesterday. Interoperability between managed and unmanaged code enables developers to continue to use necessary COM components and DLLs.

The runtime is designed to enhance performance. Although the common language runtime provides many standard runtime services, managed code is never interpreted. A feature called just-in-time (JIT) compiling enables all managed code to run in the native machine language of the system on which it is executing. Meanwhile, the memory manager removes the possibilities of fragmented memory and increases memory locality-of-reference to further increase performance.

Finally, the runtime can be hosted by high-performance, server-side applications, such as Microsoft SQL Server™ and Internet Information Services (IIS). This infrastructure enables you to use managed code to write your business logic, while still enjoying the superior performance of the industry's best enterprise servers that support runtime hosting.

.NET Framework class library

The .NET Framework class library is a collection of reusable types that tightly integrate with the common language runtime. The class library is object oriented, providing types from which your own managed code can derive functionality. This not only makes the .NET Framework types easy to use, but also reduces the time associated with learning new features of the .NET Framework. In addition, third-party components can integrate seamlessly with classes in the .NET Framework.

For example, the .NET Framework collection classes implement a set of interfaces that you can use to develop your own collection classes. Your collection classes will blend seamlessly with the classes in the .NET Framework.

As you would expect from an object-oriented class library, the .NET Framework types enable you to accomplish a range of common programming tasks, including string management, data collection, database connectivity, and file access. In addition to these common tasks, the class library includes types that support a variety of specialized development scenarios. For example, you can use the .NET Framework to develop the following types of applications and services:

- Console applications
- Windows GUI applications (Windows Forms)
- ASP.NET applications
- XML Web services
- Windows services

For example, the Windows Forms classes are a comprehensive set of reusable types that vastly simplify Windows GUI development. If you write an ASP.NET Web Form application, you can use the Windows Forms classes.

Client application development

Client applications are the closest to a traditional style of application in Windows-based programming. These are the types of applications that display windows or forms on the desktop, enabling a user to perform a task. Client applications include applications such as word processors and spreadsheets as well as custom business applications such as data entry and reporting tools. Client applications usually employ windows, menus, buttons, and other GUI elements, and they likely access local resources, such as the file system, and peripherals such as printers.

Another kind of client application is the traditional ActiveX control (now replaced by the managed Windows Forms control) deployed over the Internet as a Web page. This application is much like other client applications: it is executed natively, has access to local resources, and includes graphical elements.

In the past, developers created such applications using C or C++ in conjunction with the Microsoft Foundation Classes or with a rapid application development (RAD) environment such as Microsoft Visual Basic. The .NET Framework incorporates aspects of these existing products into a single, consistent development environment that drastically simplifies the development of client applications.

The Windows Forms classes contained in the .NET Framework are designed to be used for GUI development. You can easily create command windows, buttons, menus, toolbars, and other screen elements with the flexibility necessary to accommodate shifting business needs.

For example, the .NET Framework provides simple properties to adjust visual attributes associated with forms. In some cases, the underlying operating system does not support changing these attributes directly, and in these cases the .NET Framework automatically re-creates the forms. This is one of many ways in which the .NET Framework integrates the developer interface, making coding simpler and more consistent.

Unlike ActiveX controls, Windows Forms controls have semitrusted access to a user's computer. This means that binary or natively executing code can access some of the resources on the user's system, such as GUI elements and limited file access, without being able to access or compromise other resources. Because of code access security, many applications that once needed to be installed on a user's system can now be safely deployed through the Web. Your applications can implement the features of a local application while being deployed like a Web page.

Server application development

Server-side applications in the managed world are implemented through runtime hosts. Unmanaged applications host the common language runtime, which allows your custom managed code to control the behavior of the server. This model provides you with all the features of the common language runtime and class library while gaining the performance and scalability of the host server.

Server-side managed code

ASP.NET is the hosting environment that enables developers to use the .NET Framework to target Web-based applications. However, ASP.NET is more than a runtime host; it is a complete architecture for developing Web sites and Internet-distributed objects using managed code. Both Web Forms and XML Web services use IIS and ASP.NET as the publishing mechanism for applications, and both have a collection of supporting classes in the .NET Framework.

XML Web services, an important evolution in Web-based technology, are distributed, server-side application components similar to common Web sites. However, unlike Web-based applications, XML Web services components have no UI and are not targeted for browsers, such as Internet Explorer and Netscape Navigator.

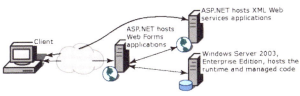

Instead, XML Web services consist of reusable software components designed to be consumed by other applications, such as traditional client applications, Web-based applications, or even other XML Web services. As a result, XML Web services technology is rapidly moving application development and deployment into the highly distributed environment of the Internet.

This diagram illustrates a basic network schema with managed code running in different server environments. Servers, such as IIS and SQL Server, can perform standard operations while your application logic executes the managed code.

If you have used earlier versions of ASP technology, you will immediately notice the improvements that ASP.NET and Web Forms offer. For example, you can develop Web Forms pages in any language that supports the .NET Framework. In addition, your code no longer needs to share the same file with your HTTP text (although it can continue to do so if you prefer). Web Forms pages execute in native machine language because, like any other managed application, they take full advantage of the runtime. In contrast, unmanaged ASP pages are always scripted and interpreted. ASP.NET pages are faster, more functional, and easier to develop than unmanaged ASP pages because they interact with the runtime like any managed application.

The .NET Framework also provides a collection of classes and tools to aid in development and consumption of XML Web services applications. XML Web services are built on standards such as SOAP, a remote procedure-call protocol; XML, an extensible data format; and WSDL, the Web Services Description Language. The .NET Framework is built on these standards to promote interoperability with non-Microsoft solutions.

For example, the Web Services Description Language tool included with the .NET Framework SDK can query an XML Web service published on the Web, parse its WSDL description, and produce C# or Visual Basic source code that your application can use to become a client of the XML Web service. The source code can create classes derived from classes in the class library that handle all the underlying communication using SOAP and XML parsing. Although you can use the class library to consume XML Web services directly, the Web Services Description Language tool and the other tools contained in the SDK facilitate your development efforts with the .NET Framework.

If you develop and publish your own XML Web service, the .NET Framework provides a set of classes that conform to all the underlying communication standards, such as SOAP, WSDL, and XML. Using those classes enables you to focus on the logic of your service, without concerning yourself with the communications infrastructure required by distributed software development.

Finally, like Web Forms pages in the managed environment, your XML Web service will run with the speed of native machine language using the scalable communication of IIS.

INTEROPERATING WITH COM

Code running under the .NET Framework's control is called managed code; conversely, code executing outside the .NET Framework is termed unmanaged code. COM is one example of unmanaged code. The .NET Framework interacts with COM via a technology known as COM Interop.

For COM Interop to work, the CLR requires metadata for all the COM types. This means that the COM type definitions normally stored in the type libraries need to be converted to .NET metadata. This is easily accomplished with the Type Library Importer utility (tlbimp.exe), which ships with the .NET Framework SDK. This utility generates interop assemblies containing the metadata for all the COM definitions in a type library. Once metadata is available, .NET clients can seamlessly create instances of COM types and call its methods as though they were native .NET instances.

Primary interop assemblies

Primary interop assemblies (PIAs) are the official, vendor-supplied, .NET type definitions for interoperating with underlying COM types. Primary interop assemblies are strongly named by the COM library publisher to guarantee uniqueness.

ESRI provides primary interop assemblies for all the ArcObjects type libraries that are implemented with COM. ArcGIS .NET developers should only use these primary interop assemblies that are installed in the Global Assembly Cache (GAC) during install if version 1.1 of the .NET Framework is detected. ESRI only supports the interop assemblies that ship with ArcGIS. You can identify a valid ESRI assembly by its public key (8FC3CC631E44AD86).

COM wrappers

The .NET runtime provides wrapper classes to make both managed and unmanaged clients believe they are communicating with objects within their respective environment. When managed clients call a method on a COM object, the runtime creates a runtime callable wrapper (RCW) that handles the marshalling between the two environments. Similarly, the .NET runtime creates COM callable wrappers for the reverse case, COM clients communicating with .NET components.

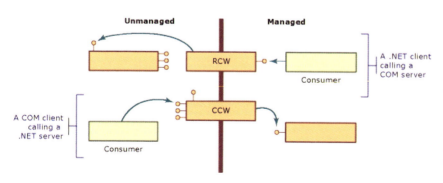

Exposing .NET components to COM

When creating .NET components that COM clients will make use of, follow the guidelines listed below to ensure interoperability.

- Avoid using parameterized constructors.

- Avoid using static methods.

- Define event source interfaces in managed code.

- Include HRESULTs in user-defined exceptions.

- Supply GUIDs for types that require them.

- Expect inheritance differences.

For more information, review 'Interoperating with Unmanaged Code' in the MSDN help collection.

Performance considerations

COM Interop clearly adds a new layer of overhead to applications but the overall cost of interoperating between COM and .NET is small and often unnoticeable. However, the cost creating wrappers and having them marshall between environments does add up; if you suspect COM Interop is the bottleneck in your application's performance, try creating a COM worker class that wraps all the chatty COM calls into one function which managed code can invoke. This improves performance by limiting the marshalling between the two environments.

COM to .NET type conversion

Generally speaking, the type library importer imports types with the same name they originally had in COM. All imported types are additionally added to a namespace that has the following naming convention: ESRI.ArcGIS plus the name of the library. For example, the namespace for the Geometry library is ESRI.ArcGIS.Geometry. All types are identified by their complete namespace and type name.

Classes, Interfaces, and Members

All COM coclasses are converted to managed classes; the managed classes have the same name as the original with 'Class' appended. For example, the Point coclass is PointClass.

All classes additionally have an interface with the same name as the coclass that corresponds to the default interface for the coclass. For example, the PointClass has a Point interface. The type library importer adds this interface so clients can register event sinks.

The .NET classes additionally have class members that .NET supports but COM does not. Each member of each interface the class implements is added as a class member. Any property or method a class implants can be accessed directly from the class rather than having to cast to a specific interface. Since interface member names are not unique, name conflicts are resolved by prefixing the interface name and an underscore to the name of each conflicting member. When member names conflict, the first interface listed with the coclass remains unchanged.

Properties in C# that have by-reference or multiple parameters are not supported with the regular property syntax. In these cases, it is necessary to use the accessor methods instead. The following code excerpt shows an example.

```
        ILayer layer = mapControl.get_Layer(0);
    MessageBox.Show(layer.Name);
```

Events

The type library importer creates several types that enable managed applications to sink to events fired by COM classes. The first type is a delegate that is named after the event interface plus an underscore followed by the event name, then the word EventHandler. For example, the *SelectionChanged* event defined on the *IActiveViewEvents* interface has the following delegate defined: *IActiveViewEvents_SelectionChangedEventHandler*. The importer additionally creates an event interface with a '_Event' suffix added to the end of the original interface name. For example, *IActiveViewEvents* generates *IActiveViewEvents_Event*. Use the event interfaces to set up event sinks.

Non-OLE Automation Compliant Types

COM types that are not OLE automation compliant generally do not work in .NET. ArcGIS contains a few noncompliant methods and these cannot be used in .NET. However, in most cases, supplemental interfaces have been added that have the offending members rewritten compliantly. For example, when defining an envelope via a point array, you can't use *IEnvelope::DefineFromPoints*; instead, you must use *IEnvelopeGEN::DefineFromPoints*.

[VB.NET]
```
    Dim pointArray(1) As IPoint
    pointArray(0) = New PointClass
    pointArray(1) = New PointClass
    pointArray(0).PutCoords(0, 0)
    pointArray(1).PutCoords(100, 100)

    Dim env As IEnvelope
    Dim envGEN As IEnvelopeGEN
    env = New EnvelopeClass
    envGEN = New EnvelopeClass

    'Won't compile
    'env.DefineFromPoints(2, pointArray)

    'Doesn't work
    env.DefineFromPoints(2, pointArray(0))

    'Works
    envGEN.DefineFromPoints(pointArray)
```

[C#]
```
    IPoint[] pointArray = new IPoint[2];
    pointArray[0] = new PointClass();
    pointArray[1] = new PointClass();
    pointArray[0].PutCoords(0,0);
    pointArray[1].PutCoords(100,100);
```

```
IEnvelope env = new EnvelopeClass();
IEnvelopeGEN envGEN = new EnvelopeClass();

// Won't compile
env.DefineFromPoints(3, ref pointArray);

// Doesn't work
env.DefineFromPoints(3, ref pointArray[0]);

// Works
envGEN.DefineFromPoints(ref pointArray);
```

.NET PROGRAMMING TECHNIQUES AND CONSIDERATIONS

This section contains several programming tips and techniques to help developers who are moving to .NET.

Casting between interfaces (QueryInterface)

.NET uses casting to jump from one interface to another interface on the same class. In COM this is called QueryInterface. VB.NET and C# cast differently.

VB.NET

There are two types of casts, implicit and explicit. Implicit casts require no additional syntax, whereas explicit casts require cast operators.

```
geometry = point                      'Implicit cast
geometry = CType(point, IGeometry)    'Explicit cast
```

When casting between interfaces, it perfectly acceptable to use implicit casts because there is no chance of data loss as there is when casting between numeric types. However, when casts fail, an exception (System.InvalidCastException) is thrown; to avoid handling unnecessary exceptions, it's best to test if the object implements both interfaces beforehand. The recommended technique is to use the *TypeOf* keyword, which is a comparison clause that tests whether an object is derived from or implements a particular type, such as an interface. The example below performs an implicit conversion from an *IPoint* to an *IGeometry* only if at runtime it is determined that the Point class implements *IGeometry*.

```
Dim point As New PointClass
Dim geometry As IGeometry
If (TypeOf point Is IGeometry) Then
  geometry = point
End If
```

If you prefer using the Option Strict On statement to restrict implicit conversions, use the CType function to make the cast explicit. The example below adds an explicit cast to the code sample above.

```
Dim point As New PointClass
Dim geometry As IGeometry
```

```
If (TypeOf point Is IGeometry) Then
   geometry = CType(point, IGeometry)
End If
```

C#

In C#, the best method for casting between interfaces is to use the as operator. Using the as operator is a better coding strategy than a straight cast because it yields a null on a conversion failure rather than raising an exception.

The first line of code below is a straight cast. This is acceptable practice if you are absolutely certain the object in question implements both interfaces; if the object does not implement the interface you are attempting to get a handle to, .NET will throw an exception. A safer model is to use the as operator that returns a null if the object cannot return a reference to the desired interface.

```
IGeometry geometry = point;                 // Straight cast
IGeometry geometry = point as IGeometry;  // As operator
```

The example below shows how to handle the possibility of a returned null interface handle.

```
IPoint point = new PointClass();
IGeometry geometry = point;
IGeometry geometry = point as IGeometry;
if (geometry != null)
{
   Console.WriteLine(geometry.GeometryType.ToString());
}
```

Binary compatibility

Most existing ArcGIS Visual Basic 6 developers are familiar with the notion of binary compatibility. This compiler flag in Visual Basic ensures that components maintain the same GUID each time they are compiled. When this flag is not set, a new GUID is generated for each class every time the project is compiled. This has the adverse side effect of having to then re-register the components in their appropriate component categories.

To keep from having the same problem in .NET, you can use the *GUIDAttribute* class to manually specify a GUID for a class. Explicitly specifying a GUID guarantees that it will never change. If you do not specify a GUID, the type library exporter will automatically generate one when you first export your components to COM and, although the exporter is meant to keep using the same GUIDs on subsequent exports, it's not guaranteed to do so.

The example below shows a GUID attribute being applied to a class.

```
[VB.NET]
<GuidAttribute("9ED54F84-A89D-4fcd-A854-44251E925F09")> _
Public Class SampleClass
   '
End Class
```

```
[C#]
[GuidAttribute("9ED54F84-A89D-4fcd-A854-44251E925F09")]
Public class SampleClass
```

```
{
//
}
```

Events

An event is a message sent by an object to signal the occurrence of an action. The action could be caused by user interaction, such as a mouse click, or it could be triggered by some other program logic. The object that raises (triggers) the event is called the event sender. The object that captures the event and responds to it is called the event receiver.

In event communication, the event sender class does not know which object or method will receive (handle) the events it raises. What is needed is an intermediary (or pointer-like mechanism) between the source and the receiver. The .NET Framework defines a special type (<Delegate>) that provides the functionality of a function pointer.

A delegate is a class that can hold a reference to a method. Unlike other classes, a delegate class has a signature, and it can hold references only to methods that match its signature. A delegate is thus equivalent to a type-safe function pointer or a callback.

To consume an event in an application, you must provide an event handler (an event-handling method) that executes program logic in response to the event and register the event handler with the event source the event handler must have the same signature as the event delegate. This process is referred to as event wiring.

The ArcObjects code excerpt below shows a custom ArcMap command wiring up to the *Map* object's selection changed event. For simplicity, the event is wired up in the *OnClick* event.

[VB.NET]
```
'Can't use WithEvents because the outbound interface is not the
'default interface.

'IActiveViewEvents is the sink event interface.
'SelectionChanged is the name of the event.
'IActiveViewEvents_SelectionChangedEventHandler is the delegate name.

'Declare the delegate.
Private SelectionChanged As IActiveViewEvents_SelectionChangedEventHandler

Private m_mxDoc As IMxDocument

Public Overloads Overrides Sub OnCreate(ByVal hook As Object)
   Dim app As IApplication
   app = hook
   m_mxDoc = app.Document
End Sub

Public Overrides Sub OnClick()
   Dim map As Map
   map = m_mxDoc.FocusMap
```

```vb
    'Create an instance of the delegate and add it to SelectionChanged
event.
    SelectionChanged = New
IActiveViewEvents_SelectionChangedEventHandler(AddressOf OnSelectionChanged)
    AddHandler map.SelectionChanged, SelectionChanged

  End Sub

  'Event handler
  Private Sub OnSelectionChanged()
    MessageBox.Show("Selection Changed")
  End Sub
```

[C#]
```csharp
  // IActiveViewEvents is the sink event interface.
  // SelectionChanged is the name of the event.
  // IActiveViewEvents_SelectionChangedEventHandler is the delegate name.
  IActiveViewEvents_SelectionChangedEventHandler m_selectionChanged;
  private ESRI.ArcGIS.ArcMapUI.IMxDocument m_mxDoc;

  public override void OnCreate(object hook)
    {
      IApplication app = hook as IApplication;
      m_mxDoc = app.Document as IMxDocument;

    }

    public override void OnClick()
    {
      IMap map = m_mxDoc.FocusMap;

      // Create a delegate instance and add it to SelectionChanged event.
      m_selectionChanged = new
IActiveViewEvents_SelectionChangedEventHandler(SelectionChanged);
      ((IActiveViewEvents_Event)map).SelectionChanged += m_selectionChanged;
    }
  // Event handler
  private void SelectionChanged()
  {
    MessageBox.Show("Selection changed");
  }
```

Error handling

The error handling construct in Visual Studio .NET is known as structured exception handling. The constructs used may be new to Visual Basic users but should be familiar to users of C++ or Java.

Structured exception handling is straightforward to implement, and the same concepts are applicable to either VB.NET or C#. VB.NET allows backward compatibility by also providing unstructured exception handling, via the familiar

OnError GoTo statement and *Err* object, although this model is not discussed in this section.

Exceptions

Exceptions are used to handle error conditions in Visual Studio .NET. They provide information about the error condition.

An exception is an instance of a class that inherits from the System.Exception base class. Many different types of exception classes are provided by the .NET Framework, and it is also possible to create your own exception classes. Each type extends the basic functionality of the System.Exception class by allowing further access to information about the specific type of error that has occurred.

An instance of an Exception class is created and thrown when the .NET Framework encounters an error condition. You can deal with exceptions by using the Try, Catch, Finally construct.

Try, Catch, Finally

This construct allows you to catch errors that are thrown within your code. An example of this construct is shown below. An attempt is made to rotate an envelope, which throws an error.

```
[VB.NET]
  Dim env As IEnvelope = New EnvelopeClass()
  env.PutCoords(0D, 0D, 10D, 10D)
  Dim trans As ITransform2D = env
  trans.Rotate(env.LowerLeft, 1D)
Catch ex As System.Exception
  MessageBox.Show("Error: " + ex.Message)

  ' Perform any tidy up code.
End Try
```

```
[C#]
{
  IEnvelope env = new EnvelopeClass();
  env.PutCoords(0D, 0D, 10D, 10D);
  ITransform2D trans = (ITransform2D) env;
  trans.Rotate(env.LowerLeft, 1D);
}
catch (System.Exception ex)
{
  MessageBox.Show("Error: " + ex.Message);
}

{
  // Perform any tidy up code.
}
```

You place a try block around code that may fail. If the application throws an error within the Try block, the point of execution will switch to the first Catch block.

The Catch block handles a thrown error. The application executes the Catch block when the Type of a thrown error matches the Type of error specified by the Catch block. You can have more than one Catch block to handle different kinds of errors. The code shown below checks first if the exception thrown is a *DivideByZeroException*.

[VB.NET]

```
...
Catch divEx As DivideByZeroException
  ' Perform divide by zero error handling.
Catch ex As System.Exception
  ' Perform general error handling.
...
```

[C#]

```
...
catch (DivideByZeroException divEx)
{
  // Perform divide by zero error handling.
}
catch (System.Exception ex)
{
  // Perform general error handling.
}
...
```

If you do have more than one Catch block, note that the more specific exception, Types, should precede the general System.Exception, which will always succeed the type check.

The application always executes the Finally block, either after the Try block completes, or after a Catch block, if an error was thrown. The Finally block should, therefore, contain code that must always be executed, for example, to clean up resources such as file handles or database connections.

If you do not have any cleanup code, you do not need to include a Finally block.

Code without exception handling

If a line of code not contained in a Try block throws an error, the .NET runtime searches for a Catch block in the calling function, continuing up the call stack until a Catch block is found.

If no Catch block is specified in the call stack at all, the exact outcome may depend on the location of the executed code and the configuration of the .NET runtime. Therefore, it is advisable to include at least a Try, Catch, Finally construct for all entry points to a program.

Errors from COM components

The structured exception handling model differs from the HRESULT model used by COM. C++ developers can easily ignore an error condition in an HRESULT if they want; in Visual Basic 6, however, an error condition in an HRESULT populates the *Err* object and raises an error.

The .NET runtime's handling of errors from COM components is somewhat similar to the way COM errors were handled at VB6. If a .NET program calls a function in a COM component (through the COM interop services) and returns an error condition as the HRESULT, the HRESULT is used to populate an instance of the *COMException* class. This is then thrown by the .NET runtime, where you can handle it in the usual way, by using a Try, Catch, Finally block.

Therefore, it is advisable to enclose all code that may raise an error in a COM component within a Try block with a corresponding Catch block to catch a *COMException*. Below is the first example rewritten to check for an error from a COM component.

```
[VB.NET]
  Dim env As IEnvelope = New EnvelopeClass()
  env.PutCoords(0D, 0D, 10D, 10D)
  Dim trans As ITransform2D = env
  trans.Rotate(env.LowerLeft, 1D)
Catch COMex As COMException
  If (COMex.ErrorCode = -2147220984) Then
    MessageBox.Show("You cannot rotate an Envelope")

    MessageBox.Show _
      ("Error " + COMex.ErrorCode.ToString() + ": " + COMex.Message)
  End If
Catch ex As System.Exception
  MessageBox.Show("Error: " + ex.Message)
...
```

```
[C#]
{
  IEnvelope env = new EnvelopeClass();
  env.PutCoords(0D, 0D, 10D, 10D);
  ITransform2D trans = (ITransform2D) env;
  trans.Rotate(env.LowerLeft, 1D);
}
catch (COMException COMex)
{
  if (COMex.ErrorCode == -2147220984)
    MessageBox.Show("You cannot rotate an Envelope");

    MessageBox.Show ("Error " + COMex.ErrorCode.ToString() + ": " +
COMex.Message);
}
catch (System.Exception ex)
{
  MessageBox.Show("Error: " + ex.Message);
}
...
```

The *COMException* class belongs to the System.Runtime.InteropServices namespace. It provides access to the value of the original HRESULT via the

ErrorCode property, which you can test to find out which error condition occurred.

Throwing errors and the exception hierarchy

If you are coding a user interface, you may want to attempt to correct the error condition in code and try the call again. Alternatively, you may want to report the error to the user to let them decide which course of action to take; here you can make use of the Message property of the Exception class to identify the problem.

However, if you are writing a function that is only called from other code, you may want to deal with an error by creating a specific error condition and propagating this error to the caller. You can do this using the Throw keyword.

To throw the existing error to the caller function, write your error handler using the Throw keyword, as shown below.

```
[VB.NET]
Catch ex As System.Exception
...
```

```
[C#]
catch (System.Exception ex)
{
  throw;
}
...
```

If you wish to propagate a different or more specific error back to the caller, you should create a new instance of an *Exception* class, populate it appropriately, and throw this exception back to the caller. The example shown below uses the *ApplicationException* constructor to set the *Message* property.

```
[VB.NET]
Catch ex As System.Exception
  Throw New ApplicationException _
    ("You had an error in your application")
...
```

```
[C#]
catch (System.Exception ex)
{
  throw new ApplicationException("You had an error in your application");
}
...
```

If you do this, however, the original exception is lost. To allow complete error information to be propagated, the *Exception* class includes the *InnerException* property. This property should be set to equal the caught exception, before the new exception is thrown. This creates an error hierarchy. Again, the example shown below uses the *ApplicationException* constructor to set the *InnerException* and *Message* properties.

```
[VB.NET]
Catch ex As System.Exception
  Dim appEx As System.ApplicationException = _
    New ApplicationException("You had an error in your application", ex)
```

```
    Throw appEx
...
```

```
[C#]
catch (System.Exception ex)
{
   System.ApplicationException appEx =
      new ApplicationException("You had an error in your application", ex);
   throw appEx;
}
...
```

In this way, the function that eventually deals with the error condition can access all the information about the cause of the condition and its context.

If you throw an error, the application will execute the current function's Finally clause before control is returned to the calling function.

Working with resources

Using strings and embedded images directly (no localization)

If your customization does not support localization now, and you do not intend for it to support localization later, you can use strings and images directly without the need for resource files. For example, strings can be specified and used directly in your code:

```
[VB.NET]
Me.TextBox1.Text = "My String"
```

```
[C#]
this.textBox1.Text = "My String";
```

Image files (BMPs, JPEGs, PNGs, and so forth) can be embedded in your assembly as follows:

1. Right-click the Project in the Solution Explorer, click Add, then click Add Existing Item.

2. In the Add Existing Item dialog box, browse to your image file and click Open.

3. In the Solution Explorer, select the image file you just added, then press F4 to display its properties.

4. Set the Build Action property to Embedded Resource.

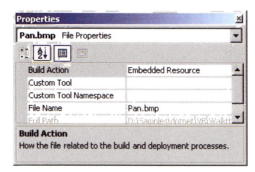

Now you can reference the image in your code. For example, the following code creates a bitmap object from the first embedded resource in the assembly:

```
[VB.NET]
Dim res() As String = GetType(Form1).Assembly.GetManifestResourceNames()
If (res.GetLength(0) > 0)
  Dim bmp As System.Drawing.Bitmap = New System.Drawing.Bitmap( _
    GetType(Form1).Assembly.GetManifestResourceStream(res(0)))
  ...
```

```
[C#]
string[] res = GetType().Assembly.GetManifestResourceNames();
if (res.GetLength(0) > 0)
{
  System.Drawing.Bitmap bmp = new System.Drawing.Bitmap(
    GetType().Assembly.GetManifestResourceStream(res[0]));
  ...
```

Creating resource files

Before attempting to provide localized resources, you should ensure you are familiar with the process of creating resource files for your .NET projects. Even if you do not intend to localize your resources, you can still use resources files, instead of using images and strings directly as described above.

Visual Studio .NET projects use an XML-based file format to contain managed resources. These XML files have the extension .resx and can contain any kind of data (images, cursors, and so forth) so long as the data is converted to ASCII format. RESx files are compiled to .resources files, which are binary representations of the resource data. Binary .resources files can be embedded by the compiler into either the main project assembly or a separate satellite assembly that contains only resources.

The following options are available to create your resource files. Each is discussed below.

- Creating a .resx file for string resources

- Creating resource files for image resources

- Compiling a .resx file into a .resources file

Creating a .resx file for string resources

If all you need to localize is strings—not images or cursors—you can use Visual Studio.NET to create a new .resx file that will be compiled automatically into a .resources module embedded in the main assembly.

1. Right-click the Project name in the Solution Explorer, click Add, then click Add New Item.

2. In the Add New Item dialog box, click Assembly Resource File.

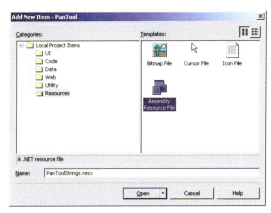

3. Open the new .resx file in Visual Studio, and add name–value pairs for the culture-specific strings in your application.

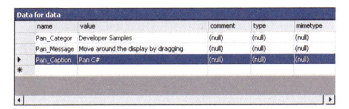

	name	value	comment	type	mimetype
	Pan_Categor	Developer Samples	(null)	(null)	(null)
	Pan_Message	Move around the display by dragging	(null)	(null)	(null)
▶	Pan_Caption	Pan C#	(null)	(null)	(null)
*					

4. When you compile your project, the .resx file will be compiled into a .resources module inside your main assembly.

Creating resource files for image resources

The process of adding images, icons, or cursors to a resources file in .NET is more complex than creating a file containing only string values, because the tools currently available in the Visual Studio .NET IDE can only be used to add string resources.

However, a number of sample projects are available with the Visual Studio .NET Framework SDK that can help you work with resource files. One such sample is the Resource Editor (ResEditor).

The ResEditor sample can be used to add images, icons, imagelists, and strings to a resource file. The tool cannot be used to add cursor resources. Files can be saved as either .resx or .resource files.

A list of tools useful for working with resources can be found in the Microsoft .NET Framework documentation.

Additional information on the ResEditor sample can be found in the Microsoft .NET Framework documentation.

The ResEditor sample is provided by Microsoft as source code. You must build the sample first if you want to create resource files using this tool. You can find information on building the SDK samples under the SDK subdirectory of your Visual Studio .NET installation.

Creating resource files programmatically

You can create XML .resx files containing resources programmatically by using the *ResXResourceWriter* class (part of the .NET framework). You can create binary .resources files programmatically by using the *ResourceWriter* class (also part of the .NET framework). These classes will allow more flexibility to add the kind of resources you require.

These classes may be particularly useful if you want to add resources that cannot be handled by the .NET Framework SDK samples and tools, for example, cursors. The basic usage of the two classes is similar: first, create a new resource writer class specifying the filename, then add resources individually by using the *AddResource* method.

The code below demonstrates how you could create a new .resx file using the *ResXResourceWriter* class and add a bitmap and cursor to the file.

```
[VB.NET]
Dim img As System.Drawing.Image = CType(New
System.Drawing.Bitmap("ABitmap.bmp"), System.Drawing.Image)
Dim cur As New System.Windows.Forms.Cursor("Pencil.cur")

Dim rsxw As New System.Resources.ResXResourceWriter("en-AU.resx")
rsxw.AddResource("MyBmp_jpg", img)
rsxw.AddResource("Mycursor_cur", cur)
rsxw.Close()
[C#]
System.Drawing.Image img = (System.Drawing.Bitmap) new
System.Drawing.Bitmap("ABitmap.bmp");
System.Windows.Forms.Cursor cur = new
System.Windows.Forms.Cursor("Pencil.cur");
```

```
System.Resources.ResXResourceWriter rsxw = new
System.Resources.ResXResourceWriter("en-GB.resx");
rsxw.AddResource("MyBmp_jpg", img);
rsxw.AddResource("Mycursor_cur", cur);
rsxw.Close();
```

The PanTool developer sample (Samples\Map Analysis\Tools) includes a script—MakeResources—that shows you how to use the *ResXResourceWriter* class to write bitmap, cursor files, and strings into a .resx file. It also shows you how to read from a .resx file using the *ResXResourceReader* class. The sample includes a .resx file that holds a bitmap, two cursors, and three strings.

Compiling a .resx file into a .resources file

XML-based .resx files can be compiled to binary .resources files either by using the Visual Studio IDE or the ResX Generator (ResXGen) sample in the tutorial.

More information on the ResXGen can be found in the Microsoft .NET Framework documentation.

- Any .resx file included in a Visual Studio project will be compiled to a .resources module when the project is built. See the 'Using resources with localization' section below for more information on how multiple resource files are used for localization.

- You can convert a .resx file into a .resources file independently of the build process using the .NET Framework SDK command resgen, for example:

```
resgen PanToolCS.resx PanToolCS.resources
```

Using resources with localization

This section explains how you can localize resources for your customizations.

How to use resources with localization

In .NET, a combination of a specific Language and Country/Region is called a *culture*. For example, the American dialect of English is indicated by the string "en-US", and the Swiss dialect of French is indicated by "fr-CH".

If you want your project to support various cultures (languages and dialects), you should construct a separate .resources file containing culture-specific strings and images for each culture.

When you build a .NET project that uses resources, .NET embeds the default .resources file in the main assembly. Culture-specific .resources files are compiled into satellite assemblies (using the naming convention <Main Assembly Name>.resources.dll) and placed in subdirectories of the main build directory. The subdirectories are named after the culture of the satellite assembly they contain. For example, Swiss–French resources would be contained in an fr-CH subdirectory.

When an application runs, it automatically uses the resources contained in the satellite assembly with the appropriate culture. The appropriate culture is determined from the Windows settings. If a satellite assembly for the appropriate culture cannot be found, the default resources (those embedded in the main assembly) will be used instead.

The following sections give more information on creating your own .resx and .resources files.

The Visual Basic .NET and C# flavors of the Pan Tool developer sample illustrate how to localize resources for German language environments. The sample can be found in the Developer Samples\ArcMap\Commands and Tools\Pan Tool folder. Strictly speaking, the sample only requires localized strings, but the images have been changed for the "de" culture as well, to serve as illustration.

A batch file named buildResources.bat has been provided in the Pan Tool sample to create the default .resources files and the culture-specific satellite assemblies.

Embedding a default .Resources file in your project

1. Right-click the Project name in the Solution Explorer, click Add, then click Add Existing Item to navigate to your .resx or .resources file.

2. In the Solution Explorer, choose the file you just added and click F4 to display its Properties.

3. Set the Build Action property to Embedded Resource.

   This will ensure that your application always has a set of resources to fall back on if there isn't a resource DLL for the culture your application runs in.

Creating .Resources.dll files for cultures supported by your project

1. First, ensure you have a default .resx or .resources file in your project.

2. Take the default .resx or .resources file and create a separate localized file for each culture you want to support.

   • Each file should contain resources with the same Names; the Value of each resource in the file should contain the localized value.

   • Localized resource files should be named according to their culture, for example, <BaseName>.<Culture>.resx or <BaseName>.<Culture>.resources.

3. Add the new resource files to the project, ensuring each one has its Build Action set to Embedded Resource.

4. Build the project.

   The compiler and linker will create a separate satellite assembly for each culture. The satellite assemblies will be placed in subdirectories under the directory holding your main assembly. The subdirectories will be named by culture, allowing the .NET runtime to locate the resources appropriate to the culture in which the application runs.

   The main (default) resources file will be embedded in the main assembly.

Assembly versioning and redirection

Applications that are built using a specific version of a strongly named assembly require the same assembly at run time. For example, if you create an application that uses ESRI.ArcGIS.System version 9.0.452, you will not be able to run this application on a system that has a newer version of ESRI.ArcGIS.System (for example, 9.0.0.692) installed. This may be the case if someone has installed a newer version of ArcGIS; however, using configuration files you can redirect an application to use a newer version of an assembly.

You have two choices for redirecting assemblies:

• Application configuration files

• Machine configuration files

Application configuration files

Application configuration files contain settings specific to an application. This file contains configuration settings that the common language runtime reads, such as assembly binding policy or remoting objects, and settings that the application can read.

The name and location of the application configuration file depend on the application's host, which can be one of the following:

- Executable-hosted application—The configuration file for an application hosted by the executable host is in the same directory as the application. The name of the configuration file is the name of the application with a .config extension. For example, an application called myApp.exe can be associated with a configuration file called myApp.exe.config.

- ASP.NET-hosted application—ASP.NET configuration files are called Web.config. Configuration files in ASP.NET applications inherit the settings of configuration files in the URL path. For example, given the URL www.esri.com/aaa/bbb, where www.esri.com/aaa is the Web application, the configuration file associated with the application is located at www.esri.com/aaa. ASP.NET pages that are in the subdirectory /bbb use both the settings that are in the configuration file at the application level and the settings in the configuration file that are in /bbb.

- Internet Explorer-hosted application—If an application hosted in Internet Explorer has a configuration file, the location of this file is specified in a <link> tag with the following syntax:

  ```
  <link rel="ConfigurationFileName" href="location">
  ```

  In this tag, *location* is a URL to the configuration file. This sets the application base. The configuration file must be located on the same Web site as the application.

Machine configuration files

The machine configuration file, Machine.config, contains settings that apply to an entire computer. This file is located in the %runtime install path%\Config directory. Machine.config contains configuration settings for machinewide assembly binding, built-in remoting channels, and ASP.NET.

The configuration system first looks in the machine configuration file for the <appSettings> element and other configuration sections that a developer might define. It then looks in the application configuration file. To keep the machine configuration file manageable, it is best to put these settings in the application configuration file. However, putting the settings in the machine configuration file can make your system more maintainable. For example, if you have a third-party component that both your client and server application use, it is easier to put the settings for that component in one place. In this case, the machine configuration file is the appropriate place for the settings, so you don't have the same settings in two different files.

Deploying an application using XCOPY will not copy the settings in the machine configuration file.

The configuration file below shows how to bind to an assembly and redirect it to a newer version.

```
<configuration>
    <runtime>
```

```
        <assemblyBinding xmlns="urn:schemas-microsoft-com:asm.v1">
            <dependentAssembly>
                <assemblyIdentity name="ESRI.ArcGIS.System"
                        publicKeyToken="8fc3cc631e44ad86"
                        culture="neutral" />
                <!– Assembly versions can be redirected in application, publisher
    policy, or machine configuration files. –>
                <bindingRedirect oldVersion="9.0.0.452" newVersion="9.0.0.692"/>
            </dependentAssembly>
        </assemblyBinding>
    </runtime>
</configuration>
```

ARCGIS DEVELOPMENT USING .NET

Using .NET, you can customize the ArcGIS applications, create standalone applications that use ESRI's types, and extend ESRI's types. For example, you can create a custom tool for ArcMap, create a standalone application that uses the MapControl, or create a custom layer. This section discusses several key issues related to developing with ArcGIS and .NET.

Registering .NET components with COM

Extending ArcGIS applications with custom .NET components requires registering the components in the COM registry and exporting the .NET assemblies to a type library (TLB). When developing a component, there are two ways to perform this task: you can use the RegAsm utility that ships with the .NET Framework SDK or Visual Studio.NET, which has a Register for COM Interop compiler flag.

The example below shows an EditTools assembly being registered with COM. The /tlb parameter specifies that a type library should additionally be generated and the /codebase option indicates that the path to the assembly should be included in the registry settings. Both of these parameters are required when extending the ArcGIS applications with .NET components.

`regasm EditTools.dll /tlb:EditTools.tlb /codebase`

Visual Studio.NET performs this same operation automatically if you set the Register for COM Interop compiler flag; this is the simplest way to perform the registration on a development machine. To check a project's settings, click Project Properties from the Project menu, then look at the Build property under Configuration Properties. The last item, Register for COM Interop, should be set to True.

Registering .NET classes in COM component categories

Much of the extensibility of ArcGIS relies on COM component categories. In fact, most custom ArcGIS components must be registered in component categories appropriate to their intended context and function for the host application to make use of their functionality. For example, all ArcMap commands and tools must be registered in the *ESRI Mx Commands* component category. There are a few different ways you can register a .NET component in a particular category but before doing so, the .NET components must be registered with COM. See the 'Registering .NET Components with COM' section above for details.

Customize dialog box

Custom .NET ArcGIS commands and tools can quickly be added to toolbars via the Add From File button on the Customize dialog box. In this case, you simply have to browse for the TLB and open it. The ArcGIS framework will automatically add the classes you select in the type library to the appropriate component category.

Categories utility

Another option is to use the Component Categories Manager (Categories.exe). In this case you select the desired component category in the utility, browse for your type library, and choose the appropriate class.

COMRegisterFunction

The final and recommended solution is to add code to your .NET classes that will automatically register them in a particular component category whenever the component is registered with COM. The .NET Framework contains two attribute classes (*ComRegisterFunctionAttribute* and *ComUnregisterFunctionAttribute*) that allow you to specify methods that will be called whenever your component is being registered or unregistered. Both methods are passed the CLSID of the class currently being registered and with this information you can write code inside the methods to make the appropriate registry entries or deletions. Registering a component in a component category requires that you also know the component category's unique ID (CATID).

The code excerpt below shows a custom ArcMap command that automatically registers itself in the *MxCommands* component category whenever the .NET assembly in which it resides is registered with COM.

```
public sealed class AngleAngleTool: BaseTool
{

    [ComRegisterFunction()]
    static void Reg(String regKey)
    {
        Microsoft.Win32.Registry.ClassesRoot.CreateSubKey(regKey.
Substring(18)+ "\\Implemented Categories\\" + "{B56A7C42-83D4-11D2-A2E9-
080009B6F22B}");
    }
    [ComUnregisterFunction()]
    static void Unreg(String regKey)
    {
    Microsoft.Win32.Registry.ClassesRoot.DeleteSubKey(regKey.Substring(18)+
"\\Implemented Categories\\" + "{B56A7C42-83D4-11D2-A2E9-080009B6F22B}");
    }
```

To simplify this process, ESRI provides classes for each component category ArcGIS exposes with static functions to register and unregister components. Each class knows the GUID of the component category it represents, so registering custom components is greatly simplified. For more details on using these classes, see the 'Working with the ESRI .NET component category classes' section below.

Simplifying your code using the ESRI.ArcGIS.Utility assembly

Part of the ArcGIS Developer Kit includes a number of .NET utility classes that facilitate .NET development by taking advantage of a few .NET capabilities including object inheritance and static functions.

Working with the ESRI .NET Base Classes

ESRI provides two abstract base classes (BaseCommand and BaseTool) to help you create new custom commands and tools for ArcGIS. The classes are abstract classes (marked as MustInherit in Visual Basic .NET), which means that although the class may contain some implementation code, it cannot itself be instantiated directly and can only be used by being inherited by another class. Both base classes are defined in the ESRI.ArcGIS.Utility assembly and belong to the ESRI.ArcGIS.Utility.BaseClasses namespace.

These base classes simplify the creation of custom commands and tools by providing a default implementation for each of the members of *ICommand* and *ITool*. Instead of stubbing out each member and providing implementation code, you only have to override the members that your custom command or tool requires. The exception is *ICommand::OnCreate*; this member must be overridden in your derived class.

Using these base classes is the recommended way to create commands and tools for ArcGIS applications in .NET languages. You can create similar COM classes from first principles; however, you should find the base class technique to be a quicker, simpler, less error-prone method of creating commands and tools.

Syntax

Both base classes have an overloaded constructor, allowing you to quickly set many of the properties of a command or tool, such as Name and Category, via constructor parameters.

The overloaded *BaseCommand* constructor has the following signature:

```
[VB.NET]
Public Sub New( _
  ByVal bitmap As System.Drawing.Bitmap _
  ByVal caption As String _
  ByVal category As String _
  ByVal helpContextId As Integer _
  ByVal helpFile As String _
  ByVal message As String _
  ByVal name As String _
  ByVal tooltip As String)


[C#]
  public BaseCommand(
    System.Drawing.Bitmap bitmap,
    string caption,
    string category,
    int helpContextId,
    string helpFile,
```

```
      string message,
      string name,
      string toolTip,
  );
```

The overloaded BaseTool constructor has the following signature:

```
[VB.NET]
Public Sub New( _
  ByVal bitmap As System.Drawing.Bitmap _
  ByVal caption As String _
  ByVal category As String _
  ByVal cursor As System.Windows.Forms.Cursor _
  ByVal helpContextId As Integer _
  ByVal helpFile As String _
  ByVal message As String _
  ByVal name As String _
  ByVal tooltip As String _
)
```

```
[C#]
  public BaseTool(
    System.Drawing.Bitmap bitmap,
    string caption,
    string category,
    System.Windows.Forms.Cursor cursor,
    int helpContextId,
    string helpFile,
    string message,
    string name,
    string toolTip,
  );
```

Inheriting the base classes

You can use these parameterized constructors when you write your new classes, for example, as shown below for a new class called PanTool that inherits the BaseTool class.

```
[VB.NET]
Public Sub New()
  MyBase.New( Nothing, "Pan", "My Custom Tools", _
    System.Windows.Forms.Cursors.Cross,    0, "", "Pans the map.",
"PanTool", "Pan")
End Sub
```

```
[C#]
public PanTool() : base ( null,"Pan", "My Custom Tools",
  System.Windows.Forms.Cursors.Cross, 0, "","Pans the map.", "PanTool",
"Pan")
{
  ...
}
```

Setting base class members directly

As an alternative to using the parameterized constructors, you can set the members of the base class directly.

The base classes expose their internal member variables to the inheritor class, one per property, so you can directly access them in your derived class. For example, instead of using the constructor to set the Caption or overriding the Caption function, you can set the m_caption class member variable declared in the base class.

```
[VB.NET]
Public Sub New()
  MyBase.New()
  MyBase..m_bitmap = New
System.Drawing.Bitmap([GetType]().Assembly.GetManifestResourceStream("Namespace.Pan.bmp"))
  MyBase..m_cursor = System.Windows.Forms.Cursors.Cross
  MyBase..m_category = "My Custom Tools"
  MyBase..m_caption = "Pan"
  MyBase..m_message = "Pans the map."
  MyBase..m_name = "PanTool"
  MyBase..m_toolTip = "Pan"
End Sub
```

```
[C#]
public PanTool()
{
  base.m_bitmap = new
System.Drawing.Bitmap(GetType().Assembly.GetManifestResourceStream("Namespace.Pan.bmp"));
  base.m_cursor = System.Windows.Forms.Cursors.Cross;
  base.m_category = "My Custom Tools";
  base.m_caption = "Pan";
  base.m_message = "Pans the map.";
  base.m_name = "PanTool";
  base.m_toolTip = "Pan";
}
```

Overriding members

When you create custom commands and tools that inherit a base class, you will more than likely need to override a few members. When you override a member in your class, the implementation code that you provide for that member will be executed instead of the default member implementation inherited from the base class. For example, the *OnClick* method in the *BaseCommand* has no implementation code at all, as *OnClick* will not do anything by default. This may be suitable for a tool, but is probably not for a command.

To override any member, you can right-click the member of the base class in the Solution Explorer Window, click Add, then click Override to stub out the member as overridden. Note that if you right-click the member of the underlying interface (*ICommand* or *ITool*) instead of the base class member, the overridden member will not include the overrides keyword, and the method will instead be shadowed.

```
[VB.NET]
Public Overrides Sub OnClick()
   ' Your OnClick
End Sub
```

```
[C#]
public override void OnClick()
{
   // Your OnClick
}
```

Alternatively, to override a member of the base class, click Overrides from the dropdown list on the right on the Code Window Wizard bar, then choose the member you want to override from the left dropdown list. This will stub out the member as overridden.

What do the base classes do by default?

The table below shows the base class members that have a significant base class implementation, along with a description of that implementation. Override these members when the base class behavior is not consistent with your customization. For example, Enabled is set to True by default; if you want your custom command enabled only when a specific set of criteria has been met, you must override this property in your derived class.

Member	Description
ICommand::Bitmap	The given bitmap is made transparent based on the pixel value at position 1,1. The bitmap is null until set by the derived class.
ICommand::Category	If null, sets the category "Misc."
ICommand::Checked	Set to False.
ICommand::Enabled	Set to True.
ITool::OnContextMenu	Set to False.
ITool::Deactivate	Set to True.

Working with the ESRI .NET component category classes

To help register .NET components in COM component categories, ESRI provides the ESRI.ArcGIS.Utility.CATIDs namespace, which has classes that represent each of the ArcGIS component categories. Each class knows its CATID and exposes static methods (Register and Unregister) for adding and removing components. Registering your component becomes as easy as adding COM registration methods with the appropriate attributes and passing the received CLSID to the appropriate static method.

The example below shows a custom Pan tool that registers itself in the ESRI Mx Commands component category. Notice in this example MxCommands.Register and MxCommands.Unregister are used instead of Microsoft.Win32.Registry.ClassesRoot.CreateSubKey and Microsoft.Win32.Registry.ClassesRoot.DeleteSubKey.

```
[VB.NET]
Public NotInheritable Class PanTool
  Inherits BaseTool

  <ComRegisterFunction()> _
  Public Shared Sub Reg(ByVal regKey As [String])
    MxCommands.Register(regKey)
  End

  <ComUnregisterFunction()> _
  Public Shared Sub Unreg(ByVal regKey As [String])
    MxCommands.Unregister(regKey)
  End Sub

[C#]
public sealed class PanTool : BaseTool
{
  [ComRegisterFunction()]
  static void Reg(string regKey)
  {
    MxCommands.Register(regKey);
  }

  [ComUnregisterFunction()]
  static void Unreg(string regKey)
  {
    MxCommands.Unregister(regKey);
  }
}
```

Extending the server

When using .NET to create a COM object for use in the GIS server, there are some specific guidelines you need to follow to ensure that you can use your object in a server context and that it will perform well in that environment. The guidelines below apply specifically to COM objects you create to run within the server.

- You must explicitly create an interface that your COM class implements. Unlike Visual Basic 6, .NET will not create an implicit interface for your COM class that you can use when creating the object in a server context.

- Your COM class should be marshalled using the Automation marshaller. You specify this by adding *AutomationProxyAttribute* to your class with a value of true.

- Your COM class should generate a dual class interface. You specify this by adding *ClassInterfaceAttribute* to your class with a value of ClassInterfaceType.AutoDual.

- To ensure that your COM object performs well in the server, it must inherit from *ServicedComponent*, which is in the System.EnterpriseServices assembly. This is necessary due to the current COM interop implementation of the .NET Framework.

For more details and an example of a custom Server COM object written in .NET, see Chapter 4, 'Developing ArcGIS Server applications'.

Releasing COM references

ArcGIS Engine and ArcGIS Desktop applications

An unexpected crash may occur when a standalone application attempts to shut down. For example, an application hosting a MapControl with a loaded map document will crash on exit. The crashes result from COM objects hanging around longer than expected. To avoid crashes, all COM references must be unloaded prior to shutdown. To help unload COM references, a static Shutdown function has been added to the ESRI.ArcGIS.Utility assembly. The following code excerpt shows the function in use.

```
[VB.NET]
Private Sub Form1_Closing(ByVal sender As Object, ByVal e As
System.ComponentModel.CancelEventArgs) Handles MyBase.Closing
    ESRI.ArcGIS.Utility.COMSupport.AOUninitialize.Shutdown()
End Sub
```

```
[C#]
private void Form1_Closing(object sender, CancelEventArgs e)
{
    ESRI.ArcGIS.Utility.COMSupport.AOUninitialize.Shutdown();
}
```

The *AOUninitialize.Shutdown* function handles most of the shutdown problems in standalone applications but you may still experience problems as there are COM objects that require explicit releasing; in these cases, call *System.Runtime.InteropServices.Marshal.ReleaseComObject()* to decrement the reference count, allowing the application to terminate cleanly. The StyleGallery is one such object, and the following example documents how to handle references to this class.

```
[VB.NET]
  Dim styleGallery As IStyleGallery
  styleGallery = New StyleGalleryClass
  MessageBox.Show(styleGallery.ClassCount)
  Marshal.ReleaseComObject(styleGallery)
[C#]
  IStyleGallery sg = new StyleGalleryClass() as IStyleGallery;
  MessageBox.Show(sg.ClassCount.ToString());
  Marshal.ReleaseComObject(sg);
```

Working with geodatabase cursors in ArcGIS Server

Some objects that you can create in a server context may lock or use resources that the object frees only in its destructor. For example, a geodatabase cursor may acquire a shared schema lock on a file-based feature class or table on which it is based or may hold onto an SDE stream.

While the shared schema lock is in place, other applications can continue to query or update the rows in the table, but they cannot delete the feature class or modify its schema. In the case of file-based data sources, such as shapefiles, update cursors acquire an exclusive write lock on the file, which will prevent other applications from accessing the file for read or write. The effect of these locks is that the data may be unavailable to other applications until all of the references on the cursor object are released.

In the case of SDE data sources, the cursor holds onto an SDE stream, and if the application has multiple clients, each may obtain and hold onto an SDE stream, eventually exhausting the maximum allowable streams. The effect of the number of SDE streams exceeding the maximum is that other clients will fail to open their own cursors to query the database.

Because of the above reasons, it's important to ensure that your reference to any cursor your application opens is released in a timely manner. In .NET, your reference on the cursor (or any other COM object) will not be released until garbage collection kicks in. In a Web application or Web service servicing multiple concurrent sessions and requests, relying on garbage collection to release references on objects will result in cursors and their resources not being released in a timely manner.

To ensure a COM object is released when it goes out of scope, the *WebControls* assembly contains a helper object called *WebObject*. Use the *ManageLifetime* method to add your COM object to the set of objects that will be explicitly released when the *WebObject* is disposed. You must scope the use of *WebObject* within a *Using* block. When you scope the use of *WebObject* within a using block, any object (including your cursor) that you have added to the *WebObject* using the *ManageLifetime* method will be explicitly released at the end of the using block.

The following example demonstrates this coding pattern:

[VB.NET]

```
  Private Sub doSomething_Click(ByVal sender As System.Object, ByVal e As
System.EventArgs) Handles doSomething.Click
     Dim webobj As WebObject = New WebObject
     Dim ctx As IServerContext = Nothing
     Try
        Dim serverConn As ServerConnection = New ServerConnection("doug", True)
        Dim som As IServerObjectManager = serverConn.ServerObjectManager

        ctx = som.CreateServerContext("Yellowstone", "MapServer")
        Dim mapsrv As IMapServer = ctx.ServerObject
        Dim mapo As IMapServerObjects = mapsrv
        Dim map As IMap = mapo.Map(mapsrv.DefaultMapName)

        Dim flayer As IFeatureLayer = map.Layer(0)
        Dim fClass As IFeatureClass = flayer.FeatureClass

        Dim fcursor As IFeatureCursor = fClass.Search(Nothing, True)
        webobj.ManageLifetime(fcursor)

        Dim f As IFeature = fcursor.NextFeature()
        Do Until f Is Nothing
           ' Do something with the feature.
           f = fcursor.NextFeature()
        Loop

     Finally
        ctx.ReleaseContext()
        webobj.Dispose()
```

```
      End Try
   End Sub

[C#]
   private void doSomthing_Click(object sender, System.EventArgs e)
   {
      using (WebObject webobj = new WebObject())
      {
         ServerConnection serverConn = new ServerConnection("doug",true);
         IServerObjectManager som = serverConn.ServerObjectManager;

         IServerContext ctx =
som.CreateServerContext("Yellowstone","MapServer");
         IMapServer mapsrv = ctx.ServerObject as IMapServer;
         IMapServerObjects mapo = mapsrv as IMapServerObjects;
         IMap map = mapo.get_Map(mapsrv.DefaultMapName);

         IFeatureLayer flayer = map.get_Layer(0) as IFeatureLayer;
         IFeatureClass fclass = flayer.FeatureClass;

         IFeatureCursor fcursor = fclass.Search(null, true);
         webobj.ManageLifetime(fcursor);

         IFeature f = null;
         while ((f = fcursor.NextFeature()) != null)
         {
            // Do something with the feature.
         }

         ctx.ReleaseContext();
      }
   }
```

The *WebMap*, *WebGeocode*, and *WebPageLayout* objects also have a *ManageLifetime* method. If you are using, for example, a *WebMap* and scope your code in a using block, you can rely on these objects to explicitly release objects you add with *ManageLifetime* at the end of the using block.

Deploying .NET ArcGIS customizations

All ArcGIS Engine and Desktop customizations require an ArcGIS installation on all client machines. The ArcGIS installation must include the ESRI primary interop assemblies, which the setup program installs in the global assembly cache. For example, deploying a standalone GIS application that only requires an ArcGIS Engine license requires an ArcGIS Engine installation on all target machines.

Standalone applications

Deploying standalone applications to either ArcGIS Engine or Desktop clients involves copying over the executable to the client machine. Copying over the executable can be as simple as using xcopy or more involved, such as creating a custom install or setup program. Note that aside from the ArcGIS primary interop assemblies and the .NET Framework assemblies, all dependencies must additionally be packaged and deployed.

ArcGIS components

Components that extend the ArcGIS applications are trickier to deploy than standalone applications because they must be registered with COM and in specific component categories. As discussed earlier, implementing *COMRegisterFunction* and *COMUnregisterFunctions* facilitates deployment by providing self category registration, but this only occurs when the components are registered.

There are two techniques for registering components with COM. One option is to run the register assembly utility (RegAsm.exe) that ships with the .NET Framework SDK. This is typically not a viable solution as client machines may or may not have this utility and it's difficult to automate. The second and recommended approach is to add an automatic registration step to a custom setup or install program.

The key to creating a custom install program that both deploys and registers components is the *System.Runtime.InteropServices.RegistrationServices* class. This class has the members *RegisterAssembly* and *UnregisterAssembly*, which register and unregister managed classes with COM. These are the same functions the RegAsm utility uses. Using these functions inside a custom installer class along with a setup program is the complete solution.

The basic steps below outline the creation of a deployable solution. NOTE: the steps assume you are starting with a solution that already contains a project with at least one COM-enabled class.

1. In Visual Studio.NET, add a new Installer Class and name it accordingly.

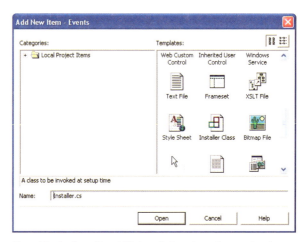

Override the Install and Uninstall functions that are implemented in the Installer base class and use the RegistrationServices class's *RegisterAssembly* and *UnregisterAssembly* methods to register the components. Make sure you use the SetCodeBase flag; this indicates that the code base key for the assembly should be set in the registry.

```
[VB.NET]
Public Overrides Sub Install(ByVal stateSaver As
System.Collections.IDictionary)
    MyBase.Install(stateSaver)
    Dim regsrv As New RegistrationServices
    regsrv.RegisterAssembly(MyBase.GetType().Assembly,
AssemblyRegistrationFlags.SetCodeBase)
End Sub

Public Overrides Sub Uninstall(ByVal savedState As
System.Collections.IDictionary)
    MyBase.Uninstall(savedState)
    Dim regsrv As New RegistrationServices
    regsrv.UnregisterAssembly(MyBase.GetType().Assembly)
  End Sub
End Class

[C#]
public override void Install(IDictionary stateSaver)
{
  base.Install (stateSaver);
  RegistrationServices regSrv = new RegistrationServices();
  regSrv.RegisterAssembly(base.GetType().Assembly,
AssemblyRegistrationFlags.SetCodeBase);
}

public override void Uninstall(IDictionary savedState)
{
  base.Uninstall (savedState);
  RegistrationServices regSrv = new RegistrationServices();
  regSrv.UnregisterAssembly(base.GetType().Assembly);
}
```

2. Add a setup program to your solution.

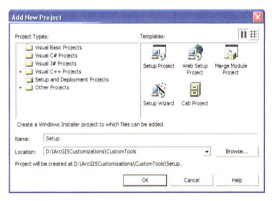

a. In the Solution Explorer, right-click the new project and click Add >
    Project Output. Choose the project you want to deploy and choose Primary
    output.

b. From the list of detected dependencies that is regenerated, remove all references to ESRI primary interop assemblies (for example, ESRI.ArcGIS.System) and stdole.dll. The only items typically left in the list are your TLB and Primary output from <AssemblyName><Version>M which represent the DLL or EXE you are compiling.

c. The final steps involve associating the custom installation steps configured in the new installer class with the setup project. To do this, right-click the setup project in the Solution Explorer and click View Custom Actions.

d. In the resulting view, right-click the Install folder and click Add Custom Action. Double-click the Application folder, then double-click the Primary output from the <AssemblyName><Version> item. This step associates the custom install function created earlier with the setup's custom install action.

e. Repeat the last step for the setup's uninstall.

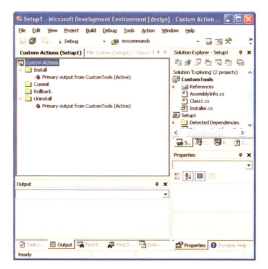

3. Finally, rebuild the entire solution to generate the setup executable file. Running the executable on a target machine installs the components and registers them with COM. The *COMRegisterFunction* routines then register the components in the appropriate component categories.

ArcGIS Server deployments

To deploy Web applications developed on a development server to product production servers, use the built-in Visual Studio.NET tools.

1. In the Solution Explorer, click your project.
2. Click the Project menu, then click Copy Project.
3. In the Copy Project dialog box, specify the deployment location.
4. Click OK.

In addition to copying the project, you must copy and register any related DLLs containing custom COM objects onto your Web server and all the GIS server's server object container machines.

The ArcGIS version 9.0 Java API is a programming interface which interoperates with ArcObjects and is specifically designed to target Java developers. Java technology is both a platform and an object-oriented programming language developed by Sun Microsystems which comes in three versions and consists of two components:

Versions:

- Java 2 platform, Standard Edition (J2SE)

- Java 2 platform, Enterprise Edition (J2EE)

- Java 2 platform, Micro Edition (J2ME)

Components:

- Java Virtual Machine (JVM) - Java runtime and client/server compilers.

- Java Application Programming Interface (API) - Suite of core, integration, and user interface toolkits.

The Java language is important because it is an open standard. All implementations of the programming language must meet the standards provided by the JVM. This enables applications to run on any hardware platforms that host the JVM.

PLATFORM CONFIGURATION

This section will describe all the necessary configurations needed to be productive with the Java API including classpath and environment settings.

Java Developer Kit

In order to develop with ArcObjects using the Java API, you must have the Java 2 Platform Standard Software Developer Kit (J2SDK) installed. All of your J2SDK tools are located in the install directory. You can either explicitly invoke them from that directory or add it to your PATH environment variable. Adding the directory to your PATH variable involves two steps:

Setting the JAVA_HOME variable is not absolutely necessary; however, some Java IDEs and Java tools require it be set.

1 Create a new environment variable named JAVA_HOME.

    JAVA_HOME=[path to JDK install directory]

    For example:

        JAVA_HOME=c:\j2sdk

Adding the PATH variable allows you to run executables (javac, java, javadoc, etc.) from any directory on your system.

2. Edit the PATH variable to include the bin directory of JAVA_HOME.

    PATH=..;%JAVA_HOME%\bin

In order to compile server-based applications like servlets and EJBs, you will also need to install and include the Java 2 Enterprise Edition toolkit in your classpath. Java application servers generally provide this or you can get the reference implementation provided by Sun Microsystems.

ArcGIS Engine

The ArcGIS Engine developer kit uses standard Java Native Interface (JNI) to access core ArcObjects components. This requires some native libraries to be in the developer's path when compiling and running applications. You must be sure to include the correct paths to invoke interoperability into native ArcObjects.

The native *.dlls are located in the following location:

`..\ArcGIS\bin`

ESRI recommends setting an ARCENGINEHOME environment variable and added the bin directory to your PATH variable. Although this is not a requirement to use the ArcGIS Engine developer kit, the developer samples all use this variable to ensure your classpath setting are accurate.

Setting the ARCENGINEHOME variable:

`ARCENGINEHOME=[path to ArcGIS install directory]`

For example:

`ARCENGINEHOME=c:\ArcGIS`

Editing the PATH variable enables your system to use the native resource libraries that ship with the ArcGIS Engine Runtime. Edit the PATH directory to include the jre\bin directory of ARCENGINEHOME.

`PATH=..%ARCENGINEHOME%\java\jre\bin`

Classpath

The Java API provides Java Archive (JAR) files (*.jar) for ArcObjects and the native runtime libraries. These JAR files are located on disk at %ARCENGINEHOME%\java

All Java applications built with any of the ArcGIS developer kits must have the following JARs referenced in the respective application's classpath:

- arcobjects.jar—contains all of the non-UI classes in one complete archive

- jintegra.jar—contains all classes for the runtime library that handles interop to COM

In addition, individual arcgis_xxx.jar files should be added to your classpath as needed. For example, applications that leverage the Java visual beans included with ArcObjects, require that the arcgis_visualbeans.jar file be added in the classpath.

JRE

The ArcGIS Engine and Server developer kits include a version of the Java Runtime Environment (JRE). This enables you to run any ArcGIS Java application as long as all the necessary settings described above are local to the runtime. You will notice the necessary *.dlls in the bin directory and the necessary *.jars in the library extension directory. All you need to do to get started with this runtime environment is ensure that the bin directory is added to your PATH environment variable:

`PATH=..\ArcGIS\java\jre\bin`

JAVA PROGRAMMING TECHNIQUES

This section provides you with some fundamental concepts of the Java programming language. It assumes you understand general programming concepts but are relatively new to Java.

Features of the Java Virtual Machine

The Java Virtual Machine (JVM) specification provides a platform independent, abstract computer for executing code. The JVM knows nothing about the Java language, instead it understands a particular binary format, the class file that contains instructions in the form of bytecodes. The Java Virtual Machine specification provides an environment that both compiles and interprets programs. The compiler creates a .java file, produces a series of bytecodes, and stores them in a .class file, and the Java interpreter executes the bytecodes stored in the .class file.

Each implementation of the JVM is platform specific as it actually interacts with the specific operating system. The JVM handles things such as memory allocation, garbage collection, and security monitoring.

Java Native Interfaces

Even though Java programs are designed to run on multiple platforms, there may be times where the standard Java class library doesn't support platform-dependent features needed by a particular application or a Java program needs to implement a lower-level program and have the Java program call it. The JNI is a standard cross-platform programming interface provided by the Java language. It enables you to write Java programs that can operate with applications and libraries written in other programming languages, such as C or C++. This is the technology used to bridge native ArcObjects with the Java API in ArcGIS.

In order to initialize your Java environment for native usage of ArcObjects, every ArcGIS Engine Java application must call the static *initializeEngine()* method on the *EngineInitializer* class. This should be the first call you do, even before *AoInitialize*.

```
public static void main(String[] args){
  /* always initialize ArcGIS Engine for native usage */
  EngineInitializer.initializeEngine();

  ...
}
```

To see how the initializeEngine method is used as the first call, refer to the Java developer samples in ArcGIS Developer Help.

ArcGIS DEVELOPMENT USING JAVA

This section is intended for developers using the Java SDK for ArcGIS Engine. The SDK provides interoperability with ArcObjects, allowing a developer to access ArcObjects as though they were Java objects. The API is not limited to any specific Java Virtual Machine or platform and uses standard Java Native Interface to access ArcObjects. The API exposes the complete functionality of ArcObjects via Java classes and interfaces, which allows Java developers to write once, run anywhere, and also benefit from ArcObjects component reuse. The Java API provides "Proxy" classes that are generated from ArcObjects components type libraries (TLBs) which allow interoperability with all of the underlying components. These proxy classes expose ArcObjects properties, methods, and events via their Java equivalents.

Interfaces

Native ArcObjects uses an interface-based programming model. The concept of an interface is fundamental to ArcObjects and emphasizes four points:

1. An interface is not a class.

2. An interface is not an object.

3. Interfaces are strongly typed.

4. Interfaces are immutable.

ArcObjects interfaces are abstract, meaning there is no implementation associated with an interface. Objects use type inheritance; the code associated with an interface comes from the class implementation.

This model shares some features of the Java interface model, which was introduced to enhance Java's single-inheritance model. An interface in the Java language is a specification of methods that an object declares it implements. A Java interface does not include instance variables or implementation code.

The Java API has two objects for every ArcObjects interface: a corresponding interface and an interface proxy class. The interface is named in the ArcObjects style, prefixed with an I. The interface proxy class appends the term proxy to the name. An example of this mapping is provided below:

```
interface IArea : IUnknown    public interface IArea{}
        public class IAreaProxy implements IArea{}
```

The proxy classes are used internally by the Java API to provide implementation to respective interfaces. An application developer should never use the default constructor of these classes as it holds no implementation. ArcObjects requires developers to go through an interface to access objects. The Java language does not use this model; subsequently, the Java API to ArcObjects has two ways of accessing objects—by interface or by class.

```
/* use the class implementing IPoint */
IPoint iPoint = new Point();
```

```
/* access object through class */
Point cPoint = new Point();
```

You cannot access objects through the default interface proxy class:

```
IPointProxy proxyPoint = new IPointProxy(); // incorrect usage
```

This will be discussed in more depth in subsequent sections.

ArcObjects interfaces are immutable and subsequently never versioned. An interface is never changed once it is defined and published. When an interface requires additional methods, the API defines a new interface by the same name with a version number appended to it as described in the following table.

```
interface IGeometry : IUnknown        public interface IGeometry{}
interface IGeometry2 : IGeometry      public interface IGeometry2 extends
                                                        IGeometry{}

interface IGeometry3 : IGeometry2     public interface IGeometry3 extends
                                                        IGeometry2{}

                                      public interface IGeometry4 extends
                                                        IGeometry3{}
```

Classes

In the ArcObjects model, classes provide the implementation of the defined interfaces. ArcObjects provides three types of classes: *abstract classes*, *classes*, and *coclasses*. These classes can be distinguished through the object model diagrams provided in ArcGIS Developer Help. It is important to be familiar with them before you begin to use the three class types.

In ArcObjects, an *abstract class* cannot be used to create new objects and are absent in the Java API. These classes are specifications in ArcObjects for instances of subclasses through type inheritance. An abstract class enumerates what interfaces are to be implemented by the implementing subclass but does not provide an implementation to those interfaces. For each abstract class in ArcObjects there are subclasses that provide the implementation.

A *class* cannot be publicly created in ArcObjects; however, objects of this class can be created as a property of another class or instantiated by objects from another class. In the Java API, the default constructor normally used to create a class is undefined for ArcObjects classes.

```
/* the constructor for FeatureClass() is unsupported*/
FeatureClass fc = new FeatureClass();
```

The following example illustrates this behavior while stepping you through the process of opening a feature class.

```
IWorkspaceFactory wf = new ShapefileWorkspaceFactory();
IFeatureWorkspace fw = new
IFeatureWorkspaceProxy(wf.openFromFile("\path\to\data", 0));
/* Create a Feature Class from FeatureWorkspace. */
IFeatureClass fc = fw.openFeatureClass("featureclass name");
```

In ArcObjects, a *coclass* is a publicly creatable class. This means that you can create your own objects merely by declaring a new object as shown below.

```
/* Create an Envelope from the Envelope coclass. */
Envelope env = new Envelope();
```

Structs

A structure defines a new data type made up of elements called members. Java does not have structures as complex data types. The Java language provides this functionality though classes; you can simply declare a class with the appropriate instance variables. For each structure in ArcObjects, there is a representative Java class with publicly declared instance variables matching the structure members as outlined below.

```
struct WKSPointZ           public class _WKSPointZ ... {
   double X          public double x;
   double Y          public double y;
   double Z          public double z;
       }
```

You can work with these classes like any other class in Java:

```
_WKSPointZ pt = new _WKSPointZ();
pt.x = 2.23;
pt.y = -23.14;
pt.z = 4.85;

System.out.println(pt.x + " " + pt.y + " " + pt.z);
```

Enumerations

Java does not have enum types. To emulate enumerations in Java, a class or interface must be created that holds constants. For each enumeration in native ArcObjects, there is a Java interface with publicly declared static integers representing the enumeration value.

```
enum esri3DAxis              public interface esri3DAxis {
  esriXAxis = 0                public static final int esriXAxis = 0;
  esriYAxis = 1                public static final int esriYAxis = 1;
  esriZAxis = 2                public static final int esriZAxis = 2;
       }
```

You can now refer to the *esriXAxis* constant using the following notation:

```
esri3DAxis.esriXAxis;
```

Variants

The variant data type can contain a wide array of subtypes. With variants all types can be contained within a single type variant. Everything in the Java programming language is an object. Even primitive data types can be encapsulated inside objects if required. Every class in Java extends *java.lang.Object;* consequently, methods in ArcObjects that take variants as parameters can be passed any object type in the Java API.

Calling methods with "variant" objects as parameters

For methods that take variants as parameters, any object types can be passed, as all objects derive from *java.lang.Object*. As this is considered a "widening cast", an explicit cast to *Object* is not needed. If you want to pass primitives as parameters to methods, when variants are required, the corresponding primitive wrapper class can be used.

Using methods that return variants

When using variant objects returned by methods, explicitly "downcast" those objects to the corresponding wrapper object. For example, if expecting a *String*, downcast to *java.lang.String*; if expecting a *short*, downcast to short's wrapper class, that is, *java.lang.Short*, as shown in the code below.

```
ICursor spCursor = spTable.ITable_search(spQueryFilter, false);
/*Iterate over the rows*/
IRow spRow = spCursor.nextRow();
while (spRow != null) {
  Short ID = (Short) (spRow.getValue(1));
  String name = (String) (spRow.getValue(2));
  Short baseID =    (Short) (spRow.getValue(3));

    System.out.println("ID="+ ID +"\t name="+ name +"\tbaseID="+ baseID);
  /* Move to the next row.*/
  spRow = spCursor.nextRow();
}
```

IRowBuffer is a superinterface of *IRow* and defines the *getValue(int)* method as:

```
public Object getValue(int index)
                throws IOException,
                    AutomationException
```

The value of the field with the specified index.

Parameters:

index - The index (in)

Returns:

return value. A Variant

The return value is an *Object*, specified by the javadoc as "variant". Therefore, the value can be downcasted to *String* or *Short*, depending upon their type in the geodatabase being queried.

Casting

ArcObjects follows an interface-based programming style. Many methods use interface types as parameters and have interfaces as return value. When the return value of a method is an interface type, the method returns an object implementing that interface. When a method takes an interface type as parameter, it can take in any object implementing that interface. This style of programming has the advantage that the same method can work with many different object types, provided they all implement the said interface.

For example, *IFeature.getShape()* method returns an object implementing *IGeometry*. The object returned could potentially be any one of the following classes that implement *IGeometry*: BezierCurve, CircularArc, EllipticArc, Envelope, GeometryBag, Line, MultiPatch, Multipoint, Path, Point, Polygon, Polyline, Ray, Ring, Sphere, TriangleFan, Triangles, or TriangleStrip.

Casting is used to convert between types. There are three types of potential casts you, as a developer, may be tempted to use with the Java API:

1. Interface to concrete class casting

2. Interface cross-casting

3. Interface downcasting

It is important to understand that objects returned from methods within ArcObjects can behave differently than objects implicitly defined because the object reference is not held in the JVM.

If you have a method, *doSomeProcessingOnPolygon(Polygon p)*, that operates only on *Polygon* objects, and you want to pass the object obtained as a result of *IFeature.getShape()*, you need a way to convert the "type" of the object from *IGeometry* to *Polygon*. In Java, this is done using a class cast operation:

```
/* incorrect usage: will give ClassCastException */
Polygon poly = (Polygon)geom;
```

However, use the same code with the ArcObjects Java API, and you will get a *ClassCastException*. The reason for the exception is that the "geom" object reference is actually a reference to the native ArcObjects component. As a consequence of the interoperability between Java and the native ArcObjects components, the logic of casting this object reference to the *Polygon* object resides in the constructor of the *Polygon* object, and not in the JVM.

Every class in the Java API has a constructor that takes in a single object as parameter. This constructor can create the corresponding object using the reference to the ArcObjects component. Therefore, to achieve the equivalent of a class casting when using the Java API, use the "object constructor" of the class being casted to.

```
Polygon poly = new Polygon(geom);
```

The following code illustrates the object constructor being used to "cast" the geom object to a *Polygon*:

```
IFeature feature = featureClass.getFeature(i);
IGeometry geom = feature.getShape();
if (geom.getGeometryType() == esriGeometryType.esriGeometryPolygon){
  /*Note: "Polygon p = (Polygon) geom;" will give ClassCastException*/
  Polygon poly = new Polygon(geom);
  doSomeProcessingOnPolygon(poly);
}
```

The *Polygon* object thus constructed will implement all interfaces implemented by the *Polygon* class. Consequently, you can call methods belonging to any of the implemented interfaces on the *poly* object.

You could write all your code using the object constructors alone, but there are times when it might be better to cast an object implementing a particular interface, not to a class type, but to another interface implemented by that object.

Continuing the previous example, suppose you want to use the *doSomeProcessingOnPolygon(Polygon p)* method not only on *Polygon* objects but on other objects implementing *IArea*, such as *Envelope* and *Ring*. You could write a generic *doSomeProcessingOnArea(IArea area)* method that works on all objects implementing *IArea*. As *Polygon*, *Envelope*, and *Ring* objects all implement the *IArea* interface, you could pass in those objects to this generic method, thereby preventing the need to write additional methods for each object type, such as *doSomeProcessingOnEnvelope(Envelope env)* and *doSomeProcessingOnRing(Ring ring)*. To accomplish this, you would need to cast from the *IGeometry* type to the *IArea* type. In Java, this is typically done using interface cross-casting.

```
/* incorrect usage: will give ClassCastException */
IArea area = (IArea) geom ;
```

However, for the same reason noted in the class-cast above, such a cast would fail with a *ClassCastException*. To be able to cast to the ArcObjects interface, you will need to use the interface proxy classes discussed earlier in this section. In the Java API, you achieve the equivalent of an interface cross-casting by using the *InterfaceProxy* of the interface being casted to.

```
IArea area = new IAreaProxy(geom);
```

The following code shows the use of an *InterfaceProxy* class to cross-cast the geom object to *IArea*:

```
IFeature feature = featureClass.getFeature(i);
IGeometry geom = feature.getShape();
/*Note: "IArea area = (IArea) geom;" will give ClassCastException*/
IArea area = new IAreaProxy(geom);
doSomeProcessingOnArea(area);
```

Using the *IAreaProxy* class as shown in the code above allows you to access the object through its *IArea* interface so that it can then be passed to a method that takes an argument of type *IArea*. Thus, in this particular example, one method can deal with three different object types. However, only methods belonging to the *IArea* interface will be valid for the area object. To call other methods of the object, you will need to either class-cast to the appropriate object type using its object constructor or get a reference to the other interfaces using the *InterfaceProxy* classes.

Instanceof

The *instanceof* operator in Java allows a developer to determine if its first operand is an instance of its second.

```
operand1 instanceof operand2
```

You can use instanceof in ArcObjects—Java when the logic behind the type is held in Java. You cannot use instanceof when the type is held in ArcObjects as the logic of determining whether an object is an instance of a specified type resides in the constructors of that object type and not the JVM.

```
Point point = new Point();
point.putCoords(0, 0);
point.putCoords(10, 10);
point.putCoords(20, 20);

if(point instanceof IGeometry){
    System.out.println(" point is a IGeometry");
    geom = point;
}
if(point instanceof IClone){
    System.out.println(" point is a IClone");
}
```

The above code works since the type information is held in Java for *Point*. When you construct a *Point* object, a proxy class for each implementing interface is also constructed. This allows you to use instanceof on any of these types. Developers would have access to any methods on *Point* implementing the *IGeometry* or *IClone* interfaces.

This is backwards compatible as well:

```
if(geom instanceof Polyline){
    System.out.println(" geom is a Polyline");
}
else if(geom instanceof Point){
    System.out.println(" geom is a Point");
    pnt = (IPoint)geom;  // allowable cast as the type is held in JVM
}
```

Since a direct cast of the *geom* object into *Point* was created, the *geom* object is of type *Point* and instanceof can be used to check this information. However, since the type information was known before it was checked above, it is not extremely useful. What would be useful is to apply the above logic on methods that return objects of super interfaces.

Consider the *IWorkspaceFactory.openFromFile()* method which returns an *IWorkspace*. Since the object returned is a Java object which implements *IWorkspace*, you cannot check if the returned object is of any of the known implementing classes that implement *IWorkspace*. In this case, to check for type information, you should call a method on the returned object which is expected. If the method does not throw an exception, it is of that type. This occurs because the logic on this object is declared at runtime and is held inside the underlying ArcObjects component.

```
RasterWorkspaceFactory rasterWkspFactory = new RasterWorkspaceFactory();
IWorkspace wksp = rasterWkspFactory.openFromFile( aPath, 0 );

if(wksp instanceof RasterWorkspace){
   /*code does not execute as logic is in ArcObjects*/
   System.out.println(" wksp is a RasterWorkspace");
   rasWksp = (RasterWorkspace)wksp;
}
else{
    try{
       rasWksp = (RasterWorkspace)wksp;
       rasWksp.openRasterDataset( aRaster );

    }catch(Exception e){
       /*code executes if wksp is not a RasterWorkspace*/
       System.out.println(" wksp is not a RasterWorkspace");
    }
}
```

Methods that take out parameters

ArcObjects provides many methods that return more than one value. The Java API requires sending single element arrays as parameters to such methods. Basically, you pass in single element arrays of the object that you want to be returned, and ArcObjects fills in the first elements of those arrays with the return value. Upon returning from the method call, the first element of the array contains the value that has been set during the method call. One such method, that you will be using in this section is the *toMapPoint* of *IARMap* interface. Take a look at the javadoc of this method:

```
public void toMapPoint(int x,
                int y,
             double[] xCoord,
             double[] yCoord)
    throws IOException,
          AutomationException
```

Converts a point in device coordinates (typically pixels) to coordinates in map units.
Converts the x and y screen coordinates supplied in pixels to x and y map coordinates. The returned map coordinates will be in MapUnits.

Parameters:
x – The x (in)

y - The y (in)

xCoord - The xCoord (in/out: use single element array)

yCoord - The yCoord (in/out: use single element array)

Notice that the parameters *xCoord* and *yCoord* are marked as "in/out: use single element array". To use this method, the first two parameters are the *x* and *y* coordinates in pixel units. The next two parameters are actually used to get return values from the method call. You pass in single dimensional single element double arrays:

```
double [] dXcoord = {0.0};
double [] dYcoord = {0.0};
```

When the method call completes, you can query the values of *dXcoord[0]* and *dYcoord[0]*. These values will be modified by the method and will actually refer to the x and y coordinates in map units. A practical example of this method call is to update the status bar with the current map coordinates as the mouse moves over the control.

```
public void updateStatusBar(IARControlEventsOnMouseMoveEvent params)
    throws IOException {
    /*create single dimensional array of doubles
     *the values of the first element of the arrays will be filled in
     *by the arMap.toMapPoint(...) method
     */
    double [] dXcoord = {0.0};
    double [] dYcoord = {0.0};
    int screenX = params.esri_getX();
    int screenY = params.esri_getY();
    IARMap arMap = arControl.getARPageLayout().getFocusARMap();
    arMap.toMapPoint(screenX, screenY, dXcoord, dYcoord);
    /*set the statusLabel*/
    statusLabel.setText("Map x,y: " + dXcoord[0]+", "+dYcoord[0]);
}
```

The Java API will not allow developers to populate an array with a superclass type, even when it has been cast to a superclass type. Consider the following Java example:

```
Integer[] integers = { new Integer(0), new Integer(1), new Integer(2)};
Object[] integersAsObjects = (Object[])integers;
integersAsObjects[0] = new Object();
```

The above is not allowed and will cause an *ArrayStoreException*. Consider the following ArcObjects example:

```
Polyline[] polyline = {new Polyline()};
tin.interpolateShape( breakline, polyline, null );
Polyline firstPolyLine = polyline[0];
```

The above is not allowed and will cause the same *ArrayStoreException* as the earlier example. Take a look at the *interpolateShape()* method of *ISurface* and analyze what is going on here.

```
public void interpolateShape(IGeometry pShape,
        IGeometry[] ppOutShape,
```

```
          Object pStepSize)
       throws IOException,
               AutomationException
```

Parameters:

    pShape - A reference to a com.esri.arcgis.geometry.IGeometry (in)

    ppOutShape - A reference to a com.esri.arcgis.geometry.IGeometry
                                      (out: use single element array)

    pStepSize - A Variant (in, optional, pass null if not required)

Throws:

    IOException - If there are communications problems.

    AutomationException - If the remote server throws an exception.

IGeometry is a super-interface to *IPolyline*, and the *Polyline* class implements both interfaces. In the first attempt you tried to send a single element *Polyline* array into a method which requires an in/out *IGeometry* parameter. This causes an *ArrayStoreException* as ArcObjects is attempting to populate an *IPolyline* array with an *IGeometry* object, attempting to place a super-class type into a subclass array. The correct way to use this method is outlined below:

```
/*Set up the array and call the method*/
IGeometry[] geoArray = {new Polyline()};
tin.interpolateShape( breakline, geoArray, null );
/* "Cast" the first array element as a Polyline - this is
 * the equivalent of calling QueryInterface on IGeometry
 */
IPolyline firstPolyLine = new IPolylineProxy(geoArray[0]);
```

Non-OLE Automation Compliant Types

ArcObjects types that are not OLE automation compliant types do not work within the Java API. ArcObjects contains a few of these methods and unfortunately such methods are not usable in the Java API. However, in most cases, supplemental interfaces have been added which have the offending methods overwritten as automation compatible. These new interfaces are appended with the letters "GEN", implying that they are generic for all supported APIs.

```
IPoint[] points = new Point[2];
points[0] = new Point();
points[1] = new Point();

points[0].putCoords(0, 0);
points[1].putCoords(10, 10);

IEnvelope env = new Envelope();
IEnvelopeGEN envGEN = new Envelope();
/*not automation compatible - throws exception*/
env.defineFromPoints(2, points[0]);
/*automation compatible*/
envGEN.defineFromPoints(points);
```

Using visual beans

The Java API provides a set of reusable components as prebuilt pieces of software code designed to provide graphical functions. As a visual beans developer, you only need to write code to "buddy" them into your ArcGIS Engine application. The use of beans creates a bridge between Java and the ActiveX controls provided by ArcGIS. These visual components are heavyweight AWT components and conform to the JavaBeans component architecture, allowing them to be used as drag and drop components for designing Java GUIs in JavaBean-compatible IDEs.

Mixing heavyweight and lightweight components

One of the primary goals of the Swing architecture was that it be based on the existing AWT architecture. This allows developers to mix both kinds of components in the same application. When using the ArcObjects JavaBeans with Swing components, care should be taken while mixing the heavyweight and lightweight components. For guidelines, refer to the article 'Mixing heavy and light components' at *http://java.sun.com/products/jfc/tsc/articles/mixing/*.

If using Swing components, disable lightweight popups where the option is available, using code similar to:

```
jComboBox.setLightWeightPopupEnabled(false);
jPopupMenu.setLightWeightPopupEnabled(false);
```

Listening to events

All ArcObjects JavaBean are capable of firing events. For instance, the *ARControl* bean fires the following events:

```
void onAction(IARControlEventsOnActionEvent theEvent)
void onAfterScreenDraw(IARControlEventsOnAfterScreenDrawEvent theEvent)
void onBeforeScreenDraw(IARControlEventsOnBeforeScreenDrawEvent theEvent)
void onCurrentViewChanged(IARControlEventsOnCurrentViewChangedEvent
theEvent)
void onDocumentLoaded(IARControlEventsOnDocumentLoadedEvent theEvent)
void onDocumentUnloaded(IARControlEventsOnDocumentUnloadedEvent
theEvent)
void onDoubleClick(IARControlEventsOnDoubleClickEvent theEvent)
void onFocusARMapChanged(IARControlEventsOnFocusARMapChangedEvent
theEvent)
void onKeyDown(IARControlEventsOnKeyDownEvent theEvent)
void onKeyUp(IARControlEventsOnKeyUpEvent theEvent)
void onMouseDown(IARControlEventsOnMouseDownEvent theEvent)
void onMouseMove(IARControlEventsOnMouseMoveEvent theEvent)
void onMouseUp(IARControlEventsOnMouseUpEvent theEvent)
```

To add and remove listeners for the events, the beans have methods of the form *addXYZEventListener* and *removeXYZEventListener*. Adapter classes are provided as a convenience for creating listener objects.

```
public void addIARControlEventsListener(IARControlEvents theListener)
    throws IOException
```

```
public void removeIARControlEventsListener(IARControlEvents
        theListener)
    throws IOException
```

The following code shows using an anonymous inner class with the *IARControlEventsAdapter* to add event listeners for *onDocumentLoaded* and *onDocumentUnloaded* events to the *arControl* object:

```
arControl = new ARControl();
...

/*wire up the events for arControl*/
arControl.addIARControlEventsListener(new IARControlEventsAdapter(){
   public void onDocumentLoaded(IARControlEventsOnDocumentLoadedEvent evt)
      throws IOException{
         /*set the statusbar text to point to the currently loaded
document*/
         statusLabel.setText(" Document filename: "+
arControl.getDocumentFilename());
         /*Determine whether permission to toggle TOC visibility*/
         if (arControl.hasDocumentPermission(
esriARDocumentPermissions.esriARDocumentPermissionsViewTOC))
            {
             tocVisibilityCheckBox.setEnabled(true);
             tocVisibilityCheckBox.setSelected(arControl.isTOCVisible());
            } else {
             JOptionPane.showMessageDialog((Component)arg0.getSource(),
"You do not have permission toggle TOC visibility");
            }
   }

   public void onDocumentUnloaded(IARControlEventsOnDocumentUnloadedEvent
evt)
      throws IOException{
         /*set the statusbar text to empty string*/
         statusLabel.setText("");
   }
});
```

It is worthwhile to note that the events fired by the beans are custom events for which the listeners are provided as part of the Java API. Adding listeners from the *java.awt.event* package (such *MouseListener*) to the beans will not be helpful as the JavaBeans do not fire those events. Instead, you could use similar events, such as *onMouseDown*, *onMouseUp*, and *onMouseMove*, provided by the corresponding event listener, which in the case of *ARControl* is *IARControlEvents*.

Reading the object model diagrams

The ArcObjects object model diagrams are an important supplement to the information you receive in object browsers. This chapter describes the diagram notation used throughout this book and in the object model diagrams that are accessed through ArcGIS Developer Help.

The diagram notation used in this book and the ArcObjects object model diagrams is based on the Unified Modeling Language (UML) notation, an industry-diagramming standard for object-oriented analysis and design, with some modifications for documenting COM-specific constructs.

The object model diagrams are an important supplement to the information you receive in object browsers. The development environment, Visual Basic or other, lists all of the classes and members but does not show the structure or relationships of those classes. These diagrams complete your understanding of the ArcObjects components.

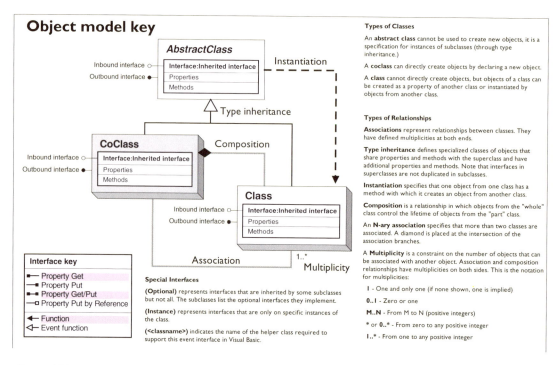

Object model key

Types of Classes

An **abstract class** cannot be used to create new objects, it is a specification for instances of subclasses (through type inheritance.)

A **coclass** can directly create objects by declaring a new object.

A **class** cannot directly create objects, but objects of a class can be created as a property of another class or instantiated by objects from another class.

Types of Relationships

Associations represent relationships between classes. They have defined multiplicities at both ends.

Type inheritance defines specialized classes of objects that share properties and methods with the superclass and have additional properties and methods. Note that interfaces in superclasses are not duplicated in subclasses.

Instantiation specifies that one object from one class has a method with which it creates an object from another class.

Composition is a relationship in which objects from the "whole" class control the lifetime of objects from the "part" class.

An **N-ary association** specifies that more than two classes are associated. A diamond is placed at the intersection of the association branches.

A **Multiplicity** is a constraint on the number of objects that can be associated with another object. Association and composition relationships have multiplicities on both sides. This is the notation for multiplicities:

1 - One and only one (if none shown, one is implied)

0..1 - Zero or one

M..N - From M to N (positive integers)

* or 0..* - From zero to any positive integer

1..* - From one to any positive integer

Special Interfaces

(Optional) represents interfaces that are inherited by some subclasses but not all. The subclasses list the optional interfaces they implement.

(Instance) represents interfaces that are only on specific instances of the class.

(<classname>) indicates the name of the helper class required to support this event interface in Visual Basic.

Object model diagram key showing the types of ArcObjects and the relationships between them

CLASSES AND OBJECTS

There are three types of classes shown in the UML diagrams: abstract classes, coclasses, and classes.

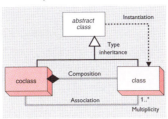

A *coclass* represents objects that you can directly create using the object declaration syntax in your development environment. In Visual Basic, this is written with the *Dim pFoo As New FooObject* syntax.

A *class* cannot directly create new objects, but objects of a class can be created as a property of another class or by functions from another class.

An *abstract class* cannot be used to create new objects; it is a specification for subclasses. An example is that a "line" could be an abstract class for "primary line" and "secondary line" classes. Abstract classes are important for developers who wish to create a subclass of their own since it shows which interfaces are required and which are optional for the type of class they are implementing. Required interfaces must be implemented on any subclass of the abstract class to ensure the new class behaves correctly in the ArcObjects system.

RELATIONSHIPS

Among abstract classes, coclasses, and classes, there are several types of class relationships possible.

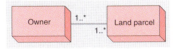

In this diagram, an owner can own one or many land parcels, and a land parcel can be owned by one or many owners.

Associations represent relationships between classes. They have defined multiplicities at both ends.

A *multiplicity* is a constraint on the number of objects that can be associated with another object. This is the notation for multiplicities:

1—One and only one. Showing this multiplicity is optional; if none is shown, "1" is implied.

0..1—Zero or one

M..N—From M to N (positive integers)

* or 0..*—From zero to any positive integer

1..*—From one to any positive integer

TYPE INHERITANCE

Type inheritance defines specialized classes that share properties and methods with the superclass and have additional properties and methods.

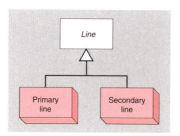

This diagram shows that a primary and secondary lines (creatable classes) are types of a line (abstract class).

INSTANTIATION

Instantiation specifies that one object from one class has a method with which it creates an object from another class.

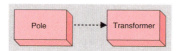

A pole object might have a method to create a transformer object.

COMPOSITION

Composition is a stronger form of aggregation in which objects from the "whole" class control the lifetime of objects from the "part" class.

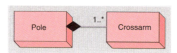

A pole contains one or many crossarms. In this design, a crossarm cannot be recycled when the pole is removed. The pole object controls the lifetime of the crossarm object.

Glossary

The following is a glossary of common terms used throughout this book. While it is not meant to be an all-encompassing list, its should provide you with a quick reference to ArcGIS Server-specific terminology.

abstract class   A specification for subclasses that is often shown on object model diagrams to help give structure to the diagram. An abstract class is not defined in a type library and cannot be instantiated.

Active Server Pages   A Microsoft server-side scripting environment that can be used to create and run dynamic, interactive Web server applications, which are typically coded in JavaScript or VBScript. An ASP file contains not only text and HTML tags, similar to standard Web documents, but also commands written in a scripting language, which can be carried out on the server.

Active Template Library   A set of C++ template classes, designed to be small, fast, and extensible.

add-in   An extension to a development environment that performs a custom task. ESRI provides various developer add-ins as part of the ArcGIS developer kit.

ADF   Application Developer Framework. The set of custom Web controls and templates that can be used to build Web applications that communicate with a GIS server. ArcGIS Server includes an ADF for both .NET and Java.

ADF runtime   The components required to run an application built with the ADF. See also ADF.

apartment   A group of threads working within a process that work within the same context. See also MTA, STA, thread, TNA.

API   See application programming interface.

application programming interface   A set of routines, protocols, and tools that application developers use to build or customize a program or set of programs. APIs make it easier to develop a program by providing building blocks for a preconstructed interface instead of requiring direct programming of a device or piece of software. They also guarantee that all programs using a common API will have similar interfaces. APIs can be built for programming languages such as C, COM, Java, and so on.

application Web service   A Web service that solves a particular problem, for example, a Web service that finds all of the hospitals within a certain distance of an address. An application Web service can be implemented using the native Web service framework of a Web server, for example, an ASP.NET Web service (WebMethod) or Java Web service (Axis).

ArcGIS Server Web service   A Web service processed and executed from within an ArcGIS Server. Each Web service is a distinct HTTP endpoint (URL). Administrators can expose MapServer and GeocodeServer objects as generic ArcGIS Server Web services for access across the Internet. See also Web service catalog.

arcgisant   The command, provided with the Java ADF, that starts the Apache ANT tool that builds and deploys Web applications. See also ADF.

ArcObjects   A library of software components that makes up the foundation of ArcGIS. ArcGIS Desktop, ArcGIS Engine, and ArcGIS Server are all built on top of the ArcObjects libraries.

ASCII   American Standard Code for Information Interchange. The de facto standard for the format of text files in computers and on the Internet. Each alphabetic, numeric, or special character is represented with a 7-bit binary number (a string of seven 1s and 0s). 128 possible characters are defined.

ASP   See Active Server Pages.

ASP.NET A programming framework built on the Common Language Runtime (CLR) that can be used on a server to build Web applications in any programming language supported by .NET. See also Active Server Pages.

assembly A package of software and its associated resources. Typically, an ArcGIS Win32 assembly will include executables and DLLs, object libraries, registry files, and help files for a unit of software. A .NET assembly is a unit of software built with a .NET language that uses the .NET Framework and the CLR to execute.

ATL See Active Template Library.

authentication The process of obtaining identification credentials, such as a name and password, from a user and validating those credentials against some authority. If the credentials are valid, the entity that submitted the credentials is considered an authenticated identity. Authentication can be used to determine whether an entity has access to a given resource.

.bat file Sometimes referred to as a batch file, a file that contains commands that can be run in a command window. It is used to perform repetitive tasks and to run scheduled commands.

big endian A computer hardware architecture in which, within a multibyte numeric representation, the most significant byte has the lowest address and the remaining bytes are encoded in decreasing order of significance. See also little endian.

binary Any file format for digital data encoded as a sequence of bits (1s and 0s) but not consisting of a sequence of printable characters (ASCII format). The term is often used for executable machine code, such as a DLL or EXE file that contains information that can be directly loaded or executed by the computer.

by value A way of passing a parameter to a function such that a temporary copy of the value of the parameter is created. The function makes changes to this temporary copy, which is discarded after the function exits. If the parameter is a reference to an underlying object, any changes made to the underlying object will be preserved after the function exits.

C++ A common object-oriented programming language, with many different implementations designed for different platforms.

Cascading Style Sheets A standard for defining the layout or presentation of an HTML or XML document. Style information includes font size, background color, text alignment, and margins. Multiple stylesheets may be applied to "cascade" over previous style settings, adding to or overriding them. The World Wide Web Consortium (W3C) maintains the CSS standard. See also World Wide Web Consortium.

CASE Computer-aided software engineering. A category of software that provides a development environment for programming teams. CASE systems offer tools to automate, manage, and simplify the development process. Complex tasks that often require many lines of code are simplified with CASE user interfaces and code generators.

class A template for a type of object in an object-oriented programming language. A class may be considered to be a set of objects that share a common structure and behavior.

class identifier	A COM term referring to the globally unique number that is used by the system registry and the COM framework to identify a particular coclass. See also GUID.
client	An application, computer, or device in a client/server model that makes requests to a server.
cloning	The process of creating a new instance of a class with the same state as an existing instance.
CLR	Common Language Runtime. The execution engine for .NET Framework applications, providing services such as code loading and execution and memory management.
CLSID	See class identifier.
coclass	A template for an object that can be instantiated in memory.
COM	See Component Object Model.
COM contract	The COM requirement that interfaces, once published, cannot be altered.
COM interface	A grouping of logically related virtual functions, implemented by a server object, allowing a client to interact with the server object. Interfaces form the basis of COM's communication between objects and the basis of the COM contract.
COM-compliant language	A language that can be used to create COM components.
command	Any class in an ArcGIS system that implements the ICommand interface and can therefore be added to a menu or toolbar in an ArcGIS application.
command bar	A toolbar, menu bar, menu, or context menu in an ArcGIS application.
command line	An onscreen interface in which the user types in commands at a prompt. In geoprocessing, any tool added to the ArcToolbox™ window can be run from the command line.
component	A binary unit of code that can be used to create COM objects.
component category	A section of the registry that can be used to categorize classes by their functionality. Component categories are used extensively in ArcGIS to allow extensibility of the system.
Component Category Manager	An ArcGIS utility program (Categories.exe) that can be used to view and manipulate component category information.
Component Object Model	A binary standard that enables software components to interoperate in a networked environment regardless of the language in which they were developed. Developed by Microsoft, COM technology provides the underlying services of interface negotiation, lifecycle management (determining when an object can be removed from a system), licensing, and event services (putting one object into service as the result of an event that has happened to another object). The ArcGIS system is created using COM objects.
computer-aided software engineering	See CASE.

container account The operating system account that server object container processes run as, which is specified by the GIS server post installation utility. Objects running in a server container process have the same access rights to system resources as the container account.

container process A process in which one or more server objects is running. Container processes run on SOC machines and are started and shut down by the SOM.

Content Standard for Digital Geospatial Metadata A publication authored by the Federal Geographic Data Committee (FGDC) that specifies the information content of metadata for a set of digital geospatial data. The purpose of the standard is to provide a common set of terminology and definitions for concepts related to the metadata. All U.S. government agencies (federal, state, and local) that receive federal funds to create metadata must follow this standard.

control A component with a user interface. In ArcGIS, the term often refers to the MapControl, PageLayoutControl, TOCControl, ToolbarControl, or ArcReaderControl, which are parts of ArcGIS Engine.

control points See control.

creation time The time it takes to initialize an instance of a server object when server objects are created in the GIS server either as a result of the server starting or in response to a request for a server object by a client.

CSDGM See Content Standard for Digital Geospatial Metadata.

CSS See Cascading Style Sheets.

custom Functionality provided or created by a party who is not the original software developer.

data type The attribute of a variable, field, or column in a table that determines the kind of data it can store. Common data types include character, integer, decimal, single, double, and string.

database management system A set of computer programs that organizes the information in a database according to a conceptual schema and provides tools for data input, verification, storage, modification, and retrieval.

database support The proprietary database platforms supported by a program or component.

DBMS See database management system.

DCOM Distributed Component Object Model. Extends COM to support communication among objects on different computers on a network.

debug To test a program or component in order to determine the cause of faults.

deeply stateful application An application that uses the GIS server to maintain application state by changing the state of a server object or its related objects. Deeply stateful applications require non-pooled server objects.

default interface When a COM object is created, the interface that is returned automatically if no other interface is specified. Most ArcObjects classes specify IUnknown as the default interface.

deployment The installation of a component or application to a target machine.

developer sample A sample contained in the ArcGIS Developer Help system.

development environment A software product used to write, compile, and debug components or applications.

device context Represents a surface that can be drawn to, for example, a screen, bitmap, or printer. In ArcGIS, the Display abstract class is used to abstract a device context.

display Often used to refer to subclasses of the Display abstract class. For example, "when drawing to the display" means when drawing to any of the display coclasses; "the display pipeline" refers to the sequence of calls made when drawing occurs.

DLL See dynamic link library.

dockable window A window that can exist in a floating state or be attached to the main application window.

dynamic link library Modules of code containing a set of routines that are called from procedures. A DLL is loaded and linked to an application at runtime by its calling modules (EXE or DLL).

early binding A technique that an application uses to access an object. In early binding, an object's properties and methods are defined from a class, instead of being checked at runtime as in late binding. This difference often gives early binding performance benefits over late binding. See also late binding.

EJB See Enterprise JavaBeans.

EMF Enhanced Metafile. A spool file format used in printing by the Windows operating system.

Enterprise JavaBeans The server-side component architecture for the J2EE platform. EJB enables development of distributed, transactional, secure, and portable Java applications.

EOBrowser An ArcGIS utility application that can be used to investigate the contents of object libraries.

event handling Sinking an event interface raised by another class.

executable file A binary file containing a program that can be implemented or run. Executable files are designated with a .exe extension.

extension In ArcGIS, an optional software module that adds specialized tools and functionality to ArcGIS Desktop. ArcGIS Network Analyst, StreetMap, and ArcGIS Business Analyst are examples of ArcGIS extensions.

Federal Geographic Data Committee An organization established by the United States Federal Office of Management and Budget responsible for coordinating the development, use, sharing, and dissemination of surveying, mapping, and related spatial data. The committee is comprised of representatives from federal and state government agencies, academia, and the private sector. The FGDC defines spatial data metadata standards for the United States in its Content Standard for Digital Geospatial Metadata and manages the development of the National Spatial Data Infrastructure (NSDI).

FGDC   See Federal Geographic Data Committee.

framework   The existing ArcObjects components that comprise the ArcGIS system.

GDB   See geodatabase.

GDI   Graphical Device Interface. A standard for representing graphical objects and transmitting them to output devices, such as a monitor. GDI generally refers to the Windows GDI API.

GeocodeServer   An ArcGIS Server software component that provides programmatic access to an address locator and performs single and batch address matching. It is designed for use in building Web services and Web applications using ArcGIS Server.

geodatabase   An object-oriented data model introduced by ESRI that represents geographic features and attributes as objects and the relationships between objects but is hosted inside a relational database management system. A geodatabase can store objects, such as feature classes, feature datasets, nonspatial tables, and relationship classes.

geometry   The measures and properties of points, lines, and surfaces. In a GIS, geometry is used to represent the spatial component of geographic features. An ArcGIS geometry class is one derived from the Geometry abstract class to represent a shape, such as a polygon or point.

geoprocessing tool   An ArcGIS tool that can create or modify spatial data, including analysis functions (overlay, buffer, slope), data management functions (add field, copy, rename), or data conversion functions.

GIS server   The components of ArcGIS Server that host and run server objects. A GIS server consists of a server object manager and one or more server object containers.

GUID   Globally Unique Identifier. A string used to uniquely identify an interface, class, type library, or component category. See also class identifier.

hexadecimal   A number system using base 16 notation.

HKCR   HKEY_CLASSES_ROOT registry hive. A Windows registry root key that points to the HKEY_LOCAL_MACHINE\Software\Classes registry key. It displays essential information about OLE and association mappings to support drag-and-drop operations, Windows shortcuts, and core aspects of the Windows user interface.

HRESULT   A 32-bit integer returned from any member of a COM interface indicating success or failure, often written in hexadecimal notation. An HRESULT can also give information about the error that occurred when calling a member of a COM interface. Visual Basic translates HRESULTS into errors; Visual C++ developers work directly with HRESULT values.

IDE   See integrated development environment.

IDispatch   A generic COM interface that has methods allowing clients to ask which members are supported. Classes that implement IDispatch can be used for late binding and ID binding.

IDL   See Interface Definition Language.

IID Interface Identifier. A string that provides the unique name of an interface. An IID is a type of Globally Unique Identifier. See also GUID.

impersonation A process by which a Web application assumes the identity of a particular user and thus gains all the privileges to which that user is entitled.

implement Regarding an interface, to provide code for each of the members of an interface (the interface is defined separately).

inbound interface An interface implemented by a class, on which a client can call members. See also outbound interface.

inheritance In object-oriented programming, the means to derive new classes or interfaces from existing classes or interfaces. New classes or interfaces contain all the methods and properties of another class or interface, plus additional methods and properties. Inheritance is one of the defining characteristics of an object-oriented system.

in-process Within the process space of a client application, a class contained in a DLL is in-process, as objects are loaded into the process space of the client EXE. A component contained in a separate EXE is out-of-process.

integrated development environment A software development tool for creating applications, such as desktop and Web applications. IDEs blend user interface design and layout tools with coding and debugging tools, which allows a developer to easily link functionality to user interface components.

Interface Definition Language A language used to define COM interfaces. The Microsoft implementation of IDL may be referred to as MIDL or Microsoft IDL.

IUnknown All COM interfaces inherit from the IUnknown interface, which controls object lifetime and provides runtime type support.

JavaServer Faces A framework for building user interfaces for Java Web applications. JSF is designed to ease the burden of writing and maintaining applications that run on a Java application server and render their user interfaces back to a target client.

JavaServer Pages A Java technology that enables rapid development of platform-independent Web-based applications. JSP separates the user interface from content generation, enabling designers to change the overall page layout without altering the underlying dynamic content.

JavaServer Pages Standard Tag Library A Java technology that encapsulates core functionality common to many Web-based applications as simple tags. JSTL includes tags for structural tasks such as iteration and conditionals, manipulation of XML documents, internationalization and locale-sensitive formatting, and SQL.

JSF See JavaServer Faces.

JSP See JavaServer Pages.

JSTL See JavaServer Pages Standard Tag Library.

late binding A technique that an application uses for determining data type at runtime using the IDispatch interface, rather than when the code is compiled. Late binding is generally used by scripting languages. See also early binding.

LIBID   Library Identifier. A type of GUID consisting of a unique string assigned to a type library. See also GUID.

library   In object-oriented programming, generic, platform independent term indicating a logical grouping of classes. ArcGIS is composed of approximately 50 libraries. Although the term library refers to a conceptual grouping of ArcGIS types, libraries do have multiple representations on disk: one per development environment. In COM, OLBs contain all the type information; in .NET, Assemblies contain the type information; and in Java, JAR files contain the type information.

license   The grant to a party of the right to use a software package or component.

little endian   A computer hardware architecture in which, within a multibyte numeric representation, the least significant byte has the lowest address and the remaining bytes are encoded in increasing order of significance. See also big endian.

macro   A computer program, usually a text file, containing a sequence of commands that are executed as a single command. Macros are used to perform commonly used sequences of commands or complex operations.

map document   In ArcMap, the file that contains one map; its layout; and its associated layers, tables, charts, and reports. Map documents can be printed or embedded in other documents. Map document files have a .mxd extension.

MapServer   An ArcGIS Server software component that provides programmatic access to the contents of a map document on disk and creates images of the map contents based on user requests. It is designed for use in building map-based Web services and Web applications using ArcGIS Server.

marshaling   The process that enables communication between a client object and server object in different apartments of the same process, between different processes, or between different processes on different machines by specifying how function calls and parameters are to be passed over these boundaries.

members   Refers collectively to the properties and methods, or functions, of an interface or class.

memory leak   When an application or component allocates a section of memory and does not free the memory when finished with it, it is said to have a memory leak; the memory cannot then be used by any other application.

MTA   Multiple threaded apartment. An apartment that can have multiple threads running. A process can only have one MTA. See also apartment, STA, thread, TNA.

network   1. A set of edge, junction, and turn elements and the connectivity between them, also known as a logical network. In other words, an interconnected set of lines representing possible paths from one location to another. A city streets layer is an example of a network. 2. In computing, a group of computers that share software, data, and peripheral devices, as in a LAN or WAN.

object   In object-oriented programming, an instance of a class.

Object Definition Language   Similar to Interface Definition Language but used to define the objects contained in an object library. See also Interface Definition Language, object library.

object library
A binary file that stores information about a logical collection of COM objects and their properties and methods in a form that is accessible to other applications at runtime. Using a type library, an application or browser can determine which interfaces an object supports and invoke an object's interface methods.

object model diagram
A graphical representation of the types in a library and their relationships.

object pooling
The process of precreating a collection of instances of classes, such that the instances can be shared between multiple application sessions at the request level. Pooling objects allows the separation of potentially costly initialization and aquisition of resources from the actual work the object does. Pooled objects are used in a stateless manner.

object-oriented programming
A programming model in which developers define the data type of a data structure as well as the functions, or types of operations, that can be applied to the data structure. Developers can also create relationships between objects. For example, objects can inherit characteristics from other objects.

OCX
See OLE custom control.

ODL
See Object Definition Language.

OGIS
Open Geodata Interoperability Specification. A specification, developed by the Open GIS Consortium, to support interoperability of GIS systems in a heterogeneous computing environment.

OLB
See object library.

OLE
Object Linking and Embedding. A distributed object system and protocol from Microsoft that allows applications to exchange information. Applications using OLE can create compound documents that link to data in other applications. The data can be edited from the document without switching between applications. Based on the Component Object Model, OLE allows the development of reusable objects that are interoperable across multiple applications.

OLE Custom Control
Also known as an ActiveX Control, an OLE custom control is contained in a file with the extension .ocx. The ArcGIS controls are ActiveX Controls.

OleView
A utility, available as part of Microsoft Visual Studio, that can be used to view type information stored in a type library or object library or inside a DLL.

out-of-process
Within the process space of a client application, a component contained in an EXE is out-of-process; instantiated classes are loaded into the process space of the EXE in which they are defined rather than into that of the client. See also in-process.

outbound interface
An interface implemented by a class, on which that object can make calls to its clients; analogous to a callback mechanism. See also inbound interface.

PDF
Portable Document Format. A proprietary file format from Adobe that creates lightweight text-based, formatted files for distribution to a variety of operating systems.

performance
A measure of the speed at which a computer system works. Factors affecting performance include availability, throughput, and response time.

persistence The process by which information indicating the current state of an object is written to a storage medium such as a file on disk. In ArcObjects, persistence is achieved via the standard COM interfaces *IPersist* and *IPersistStream* or the ArcObjects interface *IPersistVariant*.

pixel type See data type.

platform A generic term often referring to the operating system of a machine. May also refer to a programming language or development environment, such as COM, .NET, or Java.

plug-in data source An additional read-only data source provided by either ESRI or a third party developer. It may be a data source forming part of the core ArcObjects or an extension.

PMF See Published Map File.

ProgID A string value, stored in the system registry, identifying a class by library and class name, for example, *esriCarto.FeatureLayer*. The ProgID registry key also contains the human-readable name of a class, the current version number of the class, and a unique class identifier. ProgIDs are used in VB object instantiation. See also class identifier, IID.

property page A user interface component that provides access to change the properties of an object or objects.

proxy object A local representation of a remote object, supporting the same interfaces as the remote object. All interaction with the remote object from the local process is forced via the proxy object. A local object makes calls on the members of a proxy object as if it were working directly with the remote object.

Published Map File A file exported by the Publisher extension that can be read by ArcReader. Publisher Map Files end with a .pmf extension.

query interface A client may request a reference to a different interface on an object by calling the QueryInterface method of the IUnknown interface.

raster A spatial data model that defines space as an array of equally sized cells arranged in rows and columns. Each cell contains an attribute value and location coordinates. Unlike a vector structure, which stores coordinates explicitly, raster coordinates are contained in the ordering of the matrix. Groups of cells that share the same value represent geographic features. See also vector.

recycling The process by which objects in an object pool are replaced by new instances of objects. Recycling allows for objects that have become unusable to be destroyed and replaced with fresh server objects and to reclaim resources taken up by stale server objects.

reference A pointer to an object, interface, or other item allocated in memory. COM objects keep a running total of the references to themselves via the IUnknown interface methods AddRef and Release.

Regedit A utility, part of the Windows operating system, that allows you to view and edit the system registry.

register To add information about a component to the system registry, generally performed using RegSvr32.

registry   Stores information about system configuration for a Windows machine. COM uses the registry extensively, storing details of COM components including ProgIDs and ClassIDs, file location of the binary code, marshaling information, and categories in which they participate.

registry file   A file containing information in Windows Registry format. Double clicking a .reg file in Windows will enter the information in the file to the system registry. Often used to register components to component categories.

RegSvr32   A Windows utility that can add information about a component to the system registry. A component must be registered before it can be used.

rehydrate   To reinstantiate an object and its state from persisted storage.

render   To draw to a display. The conversion of the geometry, coloring, texturing, lighting, and other characteristics of an object into a display image.

runtime environment   The host that provides the services required for compiled code to execute. The Service Control Manager is effectively the runtime environment for COM. The Visual Basic Virtual Machine (VBVM) is the runtime environment that runs Visual Basic code.

scalable   A system that does not show negative effects when its size or complexity grows greater.

SCM   Service Control Manager. An administrative tool that enables the creation and modification of system services. It effectively serves as the runtime environment for COM.

script   A set of instructions in plain text, usually stored in a file and interpreted, or compiled, at runtime. In geoprocessing, scripts can be used to automate tasks, such as data conversion, or generate geodatabases and can be run from their scripting application or added to a toolbox. Geoprocessing scripts can be written in any COM-compliant scripting language, such as Python, JScript, or VBScript.

serialization   A form of persistence, in which an object is written out in sequence to a target, usually a stream. See also persistence.

server   1. A computer in a network that is used to provide services, such as access to files or e-mail routing, to other computers in the network. Servers may also be used to host Web sites or applications that can be accessed remotely. 2. An item that provides functionality to a client—for example, a COM component or object to a user application using components or to a database client utility using a database on a server machine.

server account   The operating system account that the server object manager service runs as. The server account is specified by the GIS server post installation utility.

server context   A space on the GIS server where a server object and its associated objects are running. A server context runs within a server container process. A developer gets a reference to a server object through the server object's server context and can create other objects within a server object's context.

server directory   A location on a file system used by a GIS server for temporary files that are cleaned up by the GIS server.

server object
A coarse-grained ArcObjects component that manages and serves a GIS resource, such as a map or a locator. A server object is a high-level object that simplifies the programming model for doing certain operations and hides the fine-grained ArcObjects that do the work. Server objects also have SOAP interfaces, which makes it possible to expose server objects as Web services that can be consumed by clients across the Internet.

server object isolation
Describes whether server objects share processes with other server objects. Server objects with high isolation run dedicated processes, whereas server objects with low isolation share processes with other server objects of the same type.

server object type
Defines what a server object's initialization parameters are and what methods and properties it exposes to developers. At ArcGIS version 9.0, there are two server object types: MapServer and GeocodeServer.

session state
The process by which a Web application maintains information across a sequence of requests by the same client to the same Web application.

shallowly stateful application
An application that uses the session state management capabilities of a Web server to maintain application state and makes stateless use of server objects in the GIS server. Shallowly stateful applications can use pooled server objects.

singleton
A class for which there can only be one instance in any process.

smart pointer
A Visual C++ class implementation that encapsulates an interface pointer, providing operators and functions that can make working with the underlying type easier and less error prone.

SOAP
Simple Object Access Protocol. An XML-based protocol developed by Microsoft/Lotus/IBM for exchanging information between peers in a decentralized, distributed environment. SOAP allows programs on different computers to communicate independently of an operating system or platform by using the World Wide Web's HTTP and XML as the basis of information exchange. SOAP is now a W3C specification. See also XML, World Wide Web Consortium.

SOC
Server object container. A process in which one or more server objects is running. SOC processes are started and shut down by the SOM. The SOC processes run on the GIS server's container machines. Each container machine is capable of hosting multiple SOC processes. See also SOM.

SOM
Server object manager. A Windows service that manages the set of server objects that are distributed across one or more server object container machines. When an application makes a connection to an ArcGIS Server over a LAN, it is making a connection to the SOM. See also SOC.

SQL
See Structured Query Language.

STA
Single threaded apartment. An apartment that only has a single thread. User interface code is usually placed in an STA. See also apartment, MTA, thread, TNA.

standalone application
An application that runs by itself, not within an ArcGIS application.

state
The current data contained by an object.

stateful operation   An operation that makes changes to an object or one of its associated objects—for example, removing a layer from a map. See also stateless operation.

stateless   An object that stores no state data in between member calls.

stateless operation   An operation that does not make changes to an object—for example, drawing a map. See also stateful operation.

stream   A mode of data delivery in which objects provide data storage. Stream objects can contain any type of data in any internal structure. See also persistence.

Structured Query Language   A syntax for defining and manipulating data from a relational database. Developed by IBM in the 1970s, SQL has become an industry standard for query languages in most relational database management systems.

SXD   Scene Document. A document saved by ArcScene™ that has the extension .sxd.

synchronization   The process of automatically updating certain elements of a metadata file.

target computer   A computer to which an application is deployed.

thread   A process flow through an application. An application can have many threads. See also apartment, MTA, STA, TNA.

TNA   Thread neutral apartment. An apartment that has no threads permanently associated with it; threads enter and leave the apartment as required. See also apartment, MTA, STA, thread.

tool   A command that requires interaction with the user interface before an action is performed. For example, with the Zoom In tool, you must click or draw a box over the geographic data or map before it is redrawn at a larger scale. Tools can be added to any toolbar.

type inheritance   A kind of inheritance in which an interface may inherit from a parent interface. A client may call the child interface as if it were the parent, as all the same members are supported.

type library   A collection of information about classes, interfaces, enumerations, and so on, that is provided to a compiler for inclusion in a component. Type libraries are also used to allow features such as IntelliSense to function correctly. Type libraries usually have the extension .tlb.

UI   User interface. The portion of a computer's hardware and software that facilitates human interaction. The UI includes items that can be displayed on screen, and interacted with by using the keyboard, mouse, video, printer, and data capture.

UML   Unified Modeling Language. A graphical language for object modeling. See also CASE.

URL   Uniform Resource Locator. A standard format for the addresses of Web sites. A URL looks like this: www.esri.com. The first part of the address indicates what protocol to use, while the second part specifies the IP address or the domain name where the Web site is located.

usage time   The amount of time between when a client gets a reference to a server object and when the client releases it.

utility COM object   A COM object that encapsulates a large number of fine-grained ArcObjects method calls and exposes a single coarse-grained method call. Utility COM objects are installed on a GIS server and called by server applications to minimize the round-trips between the client application and the GIS server. See also Component Object Model.

variant   A data type that can contain any kind of data.

VB   Visual Basic. A programming language developed by Microsoft based on an object-oriented form of the BASIC language and intended for application development. Visual Basic runs on Microsoft Windows platforms.

VBA   Visual Basic for Applications. The embedded programming environment for automating, customizing, and extending ESRI applications, such as ArcMap and ArcCatalog. It offers the same tools as Visual Basic in the context of an existing application. A VBA program operates on objects that represent the application and can be used to create custom symbols, workspace extensions, commands, tools, dockable windows, and other objects that can be plugged in to the ArcGIS framework.

VBVM   Visual Basic Virtual Machine. The runtime environment used by Visual Basic code when it runs.

vector   1. A coordinate-based data model that represents geographic features as points, lines, and polygons. Each point feature is represented as a single coordinate pair, while line and polygon features are represented as ordered lists of vertices. Attributes are associated with each feature, as opposed to a raster data model, which associates attributes with grid cells. 2. Any quantity that has both magnitude and direction. See also raster.

virtual directory   A directory name, used as a URL, that corresponds to a physical directory on a Web server.

Visual C++   A Microsoft implementation of the C++ language, which is used in the Microsoft application Visual Studio, producing software that can be used on Windows machines.

W3C   See World Wide Web Consortium.

wait time   The amount of time it takes between a client requesting and receiving a server object.

Web application   An application created and designed specifically to run over the Internet.

Web application template   A file that contains a user interface as well as all the code and necessary files to use as a starting point for creating a new customized Web application. ArcGIS Server contains a number of Web application templates.

Web control   The visual component of a Web form that executes its own action on the server. Web controls are designed specifically to work on Web forms and are similar in appearance to HTML elements.

Web form   Based on ASP.NET technology, Web forms allow the creation of dynamic Web pages in a Web application. Web forms present their user interface to a client in a Web browser or other device but generally execute their actions on the server.

Web server   A computer that manages Web documents, Web applications, and Web services and makes them available to the rest of the world.

Web service   A software component accessible over the World Wide Web for use in other applications. Web services are built using industry standards such as XML and SOAP and thus are not dependent on any particular operating system or programming language, allowing access through a wide range of applications.

Web service catalog   A collection of ArcGIS Server Web services. A Web service catalog is itself a Web service with a distinct endpoint (URL) and can be queried to obtain the list of Web services in the catalog and their URLs. See also ArcGIS Server Web service.

World Wide Web Consortium   An organization that develops standards for the World Wide Web and promotes interoperability between Web technologies, such as browsers. Members from around the world contribute to standards for XML, XSL, HTML, and many other Web-based protocols.

WSDL   Web Service Description Language. The standard format for describing the methods and types of a Web service, expressed in XML.

XMI   See XML Metadata Interchange.

XML   Extensible Markup Language. Developed by the World Wide Web Consortium, XML is a standard for designing text formats that facilitates the interchange of data between computer applications. XML is a set of rules for creating standard information formats using customized tags and sharing both the format and the data across applications.

XML Metadata Interchange   A standard produced by the Object Management Group that specifies how to store a UML model in an XML file. ArcGIS can read models in XMI files.

XSL   Extensible Style Language. A set of standards for defining XML document presentation and transformation. An XSL stylesheet may contain information about how to display tagged content in an XML document, such as font size, background color, and text alignment. An XSL stylesheet may also contain XSLT code that describes how to transform the tagged content in an XML document into an output document with another format. The World Wide Web Consortium maintains the XSL standards. See also XML, World Wide Web Consortium.

XSLT   Extensible Style Language Transformations. A language for transforming the tagged content in an XML document into an output document with another format. An XSL stylesheet contains the XSLT code that defines each transformation to be applied. Transforming a document requires the original XML document, an XSL document containing XSLT code, and an XSLT parser to execute the transformations. The World Wide Web Consortium maintains the XSLT standard. See also XML, XSL, World Wide Web Consortium.

Index

Microsoft Component Object Model (COM) (continued)
objects
managing lifetime 95
releasing 95
Microsoft Interface Definition Language. See IDL
Microsoft Visual Studio .NET 135
namespace 173
templates 137
MIME data 66, 176, 186, 220, 222
Model-View-Controller pattern
described 275–278
Multiplicities 709

N

Namespace 173
Natural breaks classification 139, 230
.NET
assembly
defined 713
Common Language Runtime (CLR)
defined 714
NetworkAnalysis library 82
Non-pooled server object
described 41
example 321, 343
North arrow control
character index 218, 305
described 218, 304
font 218, 305
lifecycle 220
MIME data 220
session state 220
NorthArrow class 218
NorthArrowRenderer class 304
NorthArrowTag class 304
Notification interface 588

O

Object browser utility 587
Object Definition Language. See IDL
Object dictionary
described 108, 488
Object library. See Type library
Object model
Web control 172, 262
OLE automation. See Automation
Open Group's Distributed Computing Environment. See DCE
Outbound interface. See Interface: outbound
Output directory. See Server directory
Output library 79

Overview control 295
events
OVClick 297
OVDragUp 297
interacting with 297
Overview map control
area of interest 196
described 194
events 198
example 323
extent 195
lifecycle 197
session state 196
OverviewEventArgs class 295
OverviewRenderer class 295
OverviewTag class 295

P

Page layout control
described 184, 288
events 192
interacting with
tool 291
toolbar 187
ToolItems 188
JavaScript 189
lifecycle 192
listener 293
MIME data 186
PageClientToolAction enumeration 189
PageToolItem class 185
session state 191
WebPageLayout class 184
Page Layout template 138, 229
PageClientToolAction enumeration 189
PageDescription class 167, 271
PageLayout class 185
PageLayoutRenderer class 288
PageLayoutTag class 288
PageToolItem class 185
Point coclass 601, 602, 604
Polymorphism 577
Pooled server object
described 41
Pooling
described 41
scalability 43
stateful vs stateless application 91
Programable identifier. See Prog ID
Programming language
supported by ArcGIS Server 5
Property by reference 589, 601, 603
Property by value 589, 603
Proxy object 106

Proxy server 68
Published map document (.pmf) 41

Q

QI. *See* QueryInterface
Quantile classification 139, 230
Query
    performance 591
QueryInterface 578–579

R

Recycling
    server object 45
Regedit 594
Registry 584, 594
    regedit. *See* Regedit
    script 594
Release method. *See* IUnknown
ReloadDescriptions method 272
Runas 56

S

Scale bar control
    described 221, 307
    divisions 222
    lifecycle 222
    MIME data 222
    session state 222
    subdivisions 222
ScaleBar class 221
ScaleBarRenderer class 307
ScaleBarTag class 307
SCM 583, 584
Search template 138, 229
Security
    GIS server 46, 85, 161, 258
    Web application 46, 161, 258
Serialization 489
Server account 47
Server context
    ArcObjects and 104
    copying object between 107
    creating 484
        empty 105, 486
    dereferencing JCA resources 553
    discussed 104
    example
        deeply stateful application 321, 343
    garbage collection and 88
    getting
        described 104

Server context (continued)
    object
        lifetime 89
        loading 108, 489
        managing 111
        proxy 106
        saving 108, 489
    object dictionary
        described 108, 488
        example 324, 349
    releasing
        described 104
        in Session_End 166
        non-pooled object 165
        pooled object 164
        ServerContext class 88, 487
    session state and 101
    Web application and 163
Server control. *See* Web control
Server directory
    adding 66, 502
    deleting 503
    deleting files in 66, 109, 529
    described 36
    in Server.cfg 528
    ServerDirectory class 517
    ServerDirectoryInfo class 492
    virtual directory 67
Server library 79
Server object
    accessing 87
    adding
        described 17
        example 322, 344
        many 61
        one 58
    configuration file 530
        tags 532
    configuring 40, 509
    creation time 43
        defined 715
    defined 33
    deleting 61
    described 40, 87
    garbage collection and 88
    getting statistics 63
    GIS resource 41
    isolation
        described 44
        high 44
        low 44
    lifetime 89
    limiting
        image size 132
        queries 130

X